DEVELOPMENTS IN

THEORETICAL AND APPLIED MECHANICS

VOLUME XI

Selected Papers of The Eleventh Southeastern

Conference on Theoretical and Applied Mechanics

sponsored by

The University of Alabama in Huntsville

held in Huntsville, Alabama, April 8-9, 1982

Executive Chairman: T. J. Chung

Editorial Committee: Gerald R. Karr, Chairman

George J. Dvorak | Kwan Rim

Kerry S. Havner | Charles E. Taylor

David J. McGill | D. P. Telionis

Published

by

Department of Mechanical Engineering
The University of Alabama in Huntsville
Huntsville, Alabama 35899

ISBN 0-942166-00-0

PREFACE

This volume, a collection of selected papers of the Eleventh Southeastern Conference on Theoretical and Applied Mechanics (SECTAM XI) contains 48 papers carefully reviewed and selected from 150 presented in SECTAM XI, held at the Huntsville Hilton Hotel, hosted by The University of Alabama in Huntsville, April 8-9, 1982.

Professors C. Truesdell and E. H. Lee highlighted the Conference with their keynote addresses.

The Southeastern Conference on Theoretical and Applied Mechanics was formed in 1962. The founding members were W. L. Greenstreet, R. L. Maxwell, C. H. Parr, C. E. Stoneking, W. A. Shaw, and W. F. Swinson. The purpose of this Conference is to stimulate interest in Mechanics and provide the means for formal and informal exchange of ideas along with the shainrg of the latest research results. Held on a biennial basis, each succeeding Conference has grown in stature and has received increased national and international recognition.

The papers in this volume represent the areas of Stability, Elasticity, Plasticity, Flexural Members, Experimental Mechanics, Vibrations and Dynamics, Composites, Special Topics in Fluids, Turbulence, Multiphase Flow, and Numerical Methods.

We wish to express appreciation to the Authors, Session Organizers, Editorial Committee members, and over one hundred reviewers for their outstanding contributions.

The University
of Alabama in
Huntsville
Huntsville, AL

T. J. Chung
Executive Chairman

TABLE OF CONTENTS

TABLE OF CONTENTS

EXPERIMENTAL MECHANICS

VIBRATIONS AND DYNAMICS

TABLE OF CONTENTS

TABLE OF CONTENTS

TURBULENCE

MULTIPHASE FLOW

TABLE OF CONTENTS

STABILITY

ON THE MORPHOLOGY OF STRUCTURAL CONTROL

H. H. E. Leipholz
Professor, Depts. of Civil and Mechanical Engineering
Solid Mechanics Division
University of Waterloo
Waterloo, Ontario, Canada, N2L 3G1

1. Introduction

Recently, active, automatic control of Civil Engineering structures like bridges, tall buildings, towers, antennas, etc., has gained some interest in the literature, and in the context of practical applications. J.Roorda has investigated tendon control in tall structures [1], J.N.Yang and F.Giannopoulos have dealt with the control of cable bridges and slender structures [2], J.N.Yang and H.T.P.Yao have considered the question of how to formulate structural control [3], C.R.Martin and T.T.Soong have investigated the model control of multi-story structures [4], L.Meirovitch has also contributed to the question of modal control by studying feedback control of distributed systems [5], M.Abdel-Rohman and H.H.E.Leipholz have paid attention to active control of flexible structures [6], S.Toussi and H.T.P.Yao have looked at the problem of identification for multi-story buildings in the context of building control [7], N.R.Petersen has reported on the design of large-scale tuned mass dampers [8], and last not least, W.Zuk has given stron impulses to the application of active structural control by popularizing the idea of "the kinetic structure" [9].

The idea of structural control has received further impulses from a number of papers presented at an IUTAM Symposium held June 4-7, 1979, at the University of Waterloo, Waterloo, Ontario, Canada, [10]. Some of the speakers of the symposium have stressed overall the fact that in the case of structural control, one is faced with the control of continuous systems involving infinitely many degrees of freedom. Although most of the attempts made so far in the literature to deal with structural control have consisted in discretizing the structures involved and in applying to these approximate models the already existing and well developed methods of modern control theory for systems with a finite number of degrees of freedom [10], one should be aware that one had in reality to apply methods of control suitable for *continuous* (distributed) systems as for example presented by J.L.Lions [11].

The following deliberations will center around the problem of properly identifying and defining the specifics of control of continuous systems and the question of how to make those methods derived for continuous systems practicable. In this context, an important point will be how to properly substitute a continuous system by a discrete model for the purpose of obtaining a convenient approximate solution in place of a cumbersome "exact" one.

2. On the Nature of Structural Control

In order to outline the mathematical structure of the control of a continuous body, let the control of a beam be considered. This is a simple enough problem so that mathematical difficulties will not unnecessarily overshadow the essentials, but it is also a general enough one as to involve all those features characteristic for the control of continua. The differential equation of the problem reads

$$\mu \frac{\partial^2 w(t,x)}{\partial t^2} + \alpha \frac{\partial^4 w(t,x)}{\partial x^4} + q \frac{\partial}{\partial x} [(\ell - x) \frac{\partial w(t,x)}{\partial x}] = p(t,x). \qquad (1)$$

In (1), t is the time, x the spatial coordinate (measured along the beam's axis), μ the mass per unit length, α the bending stiffness, ℓ the beam's length, and w the transversal deflection of the vibrating beam. Moreover, q is a uniform axial load per unit length, while p is a distributed transversal load. In addition, a set of boundary conditions are described by

$$U[w]_B = 0, \qquad (2)$$

where U is a "vector" of differential operators. The subscript B in (1) shall indicate that the operations on w are to be carried out at the "boundary" of the system, that is in this case, at the two ends of the beam, respectively.

Let two different classes of problems be distinquished:

(i) $p(t,x) = e^{i\omega t} f(x)$, (3)

i.e., the beam is subjected to forced, harmonic vibrations. In this case,

$$w(t,x) = e^{i\omega t} y(x). \qquad (4)$$

Using (4) in (1) and (2) yields

$$-\mu\omega^2 y + \alpha y'^{V} + q[(\ell - x)y']' = f(x), \qquad (5)$$

$$U[y]_B = 0. \qquad (6)$$

Introducing the operator A defined by

$$Ay = -\mu\omega^2 y + \alpha y'^{V} + q[(\ell - x)y']' \quad , \qquad (7)$$

enables one to represent the forced vibrations of the beam by

$$Ay = f, \; U[y]_B = 0. \qquad (8)$$

Obviously, only the boundary conditions of the problem play a role; initial conditions are irrelevant.

Note, that operators A and U in (8) are ordinary differential operators only for the beam problem. If one deals with plates or shells, the operators in (8) become partial differential operators.

In order to turn the problem into a control problem, one must add to (8) the "control law", a step which changes (8) into

$$Ay - f + Cu, \; U y_B = 0. \qquad (9)$$

The control law is the term Cu on the right side of the differential equation in (9), C being an operator and u being the control. For further specification of the problem, the "observation equation"

$$u = Dy \qquad (10)$$

must be introduced, in which D is again an operator. Setting CD = F, and using (10) in (9) yields finally

$$Ay = f + Fy, \; U y_B = 0. \qquad (11)$$

It should be noted, that the operator F in (11) is yet undetermined. It becomes a specific quantity, once the operators C and D have been chosen by the designer of the control system.

Writing (11) in the form

$$A^*y - f, \; u y_B = 0, \; A^* = A - F, \qquad (12)$$

and comparing the (8), makes it evident that controlling a structure means replacing the original structure governed by operator A, by a new structure ruled by operator A*. It is left to the skill of the designer to choose control law and observation equation in such a way, that the thus generated operator A* warrants a desired or prescribed behaviour of the controlled system.

From (12), the very nature of a control problem can be read off: *for a given perturbance* f, *an operator* A* *is to be determined which ensures a prescribed output* y *of the system.* Usually, perturbance f and the system itself (i.e. operator A*) are given, while the task of the engineer consists in calculating the output y. As compared to this usual problem, the control problem then represents the so called "*inverse problem*". One should realize that this inverse problem (control problem) is by its very nature an "improperly posed" problem, and it may be worthwhile to study control problems from this point of view [12]. In order to remove the ill-posedness of the problems, assumptions concerning control law, observation equation may be made, and a "performance requirement" for the controlled structure may be set up, the latter commonly in the form of a "cost functional". One may thus arrive at "optimal control". However, one should be aware of the fact that such optimization of the controlled structure is only a relative value. Since the additional assumptions removing the ill-posedness of the original problem (specifically the definition of the cost functional), are rather arbitrary ones, any optimization can be claimed only with respect to these more or less arbitrary assumptions and may become irrelevant and even detrimental if these assumptions are improper ones.

Let again the example of the simple beam be used to indicate in what sense design of a control is actually design of a hopefully improved structure: Assume for instance $C \equiv E$ in (9), where E is the identity operator. Moerover, let several possibilities for operator D in (10) be considered. If the choice is

$$Dy = Emy, \quad m = \text{const.}, \tag{13}$$

then, *inertia* is added to the original system (7), (8). If

$$Dy = ky'^{V}, \; k = \text{const.}, \tag{14}$$

the *stiffness* of the original system is affected. If

$$Dy = i\omega\beta y, \quad \beta = \text{const.}, \tag{15}$$

damping has been added to (7), (8), etc. All these additions to (7), (8), by means of control will naturally have an influence on the response of the original system and it must be the concern of the designer to ensure that this influence leads to a modified response satisfying the imposed performance requirements.

On additional remark may be made: Changing inertia and stiffness of the original system by means of control means actually changing the *natural frequencies* of (7), (8). Hence, one is faced with a kind of problem closely related to the classical "inverse problem" of the theory of vibrations, which consists in designing a beam (through proper mass and stiffness distribution) so that it possesses a prescribed spectrum of natural frequencies. The prescribed spectrum indirectly warrants a desired beam performance. This example shows clearly, how broadly the notion of control may be interpreted. Control being basically an inverse problem, may therefore include other inverse problems which in the past have not been considered control problems as such. This broader interpretation of control can be found in [10], for example in the paper by T. Garbacki.

Let problem (9), (10), or (11), respectively, be reformulated so that its generality becomes more apparent. Assume, the operator A possesses for the prescribed boundary conditions an inverse. Thinking of the beam as an example, such inverse consists in an integral operator

$$A^{-1} = \int_0^{\ell} G(x,\xi) \; \ldots \; d\xi \tag{16}$$

which involves the kernel G. From the point of view of mechanics, G is the function of the continuous system, which characterizes its inherent, structural properties. By means of (16), the differential equation in (11) can be rewritten to yield

$$y(x) = \int_0^{\ell} G(x,\xi)\ f(\xi)d\xi + \int_0^{\ell} G(x,\xi)\ F[y(\xi)]d\xi. \tag{17}$$

By inverting correspondingly the differential equation in (8), it becomes obvious that

$$y_o(x) = \int_0^{\ell} G(x,\xi)\ f(\xi)d\xi \tag{18}$$

is the response of the original, uncontrolled system. Using (18) in (17) results in

$$y(x) - y_o(x) = \int_0^{\ell} G(x,\xi)\ F[y(\xi)]d\xi. \tag{19}$$

Let

$$Ty = \int_0^{\ell} G(x,\xi)\ F[y(\xi)]d\xi. \tag{20}$$

Then, (19) can be rewritten as an *operator equation*, that is to say,

$$y - y_o = \Delta y = Ty. \tag{21}$$

Quantity y is a performance measure for the controlled system. The control problem can now be formulated as follows: *Let in* (21) *the operator* T *be determined such that the performance measure* Δy *satisfies certain predetermined requirements.* Let it again be emphasized that (21) represents an *inverse problem: Any problem like*(21), for which not the vector y for given vector Δy and given operator T is to be determined, but *for which* on the contrary *the operator* T *is to be determined for given vectors* Δy *and* y, *is to be called an inverse problem.*

(ii) p(t,x) is an aperiodic function, i.e., the beam is subjected to a forced, aperiodic motion under given initial conditions. The, equation (1) is valid, i.e.,

$$\mu \frac{\partial^2 w}{\partial t^2} + \tilde{A}w = p, \quad \tilde{A} = \alpha \frac{\partial^4}{\partial x^4} + q \frac{\partial}{\partial x} [(\ell - x)] \frac{\partial}{\partial x}\ , \tag{22}$$

together with the boundary conditions

$$U[w]_B = 0. \tag{23}$$

and the initial conditions

$$w(o,x) = h(x),\ \frac{w}{t}\ (o,x) = g(x). \tag{24}$$

Let a Laplace transformation be carried out yielding

$$\bar{w}(,x) = \int_0^{\infty} w(t,x)e^{-st}\ dt,\ \bar{P}(s,x) = \int_0^{\infty} p(t,x)e^{-st}\ dt. \tag{25}$$

Using (25), problem (22), (23), (24) can be changed into

$$\mu s^2 \bar{w} + \tilde{A}\bar{w} = \bar{P} + sh + g. \tag{26}$$

Let the series expansions

$$\left.\begin{aligned} \bar{w} &= \sum_n A_n^*(s) W_n \ , \quad \bar{P} = \sum_n P_n^*(s) W_n, \\ h &= \sum_n h_n^* \, W_n \ , \quad g = \sum_n g_n^* \, W_n, \end{aligned}\right\} \tag{27}$$

be introduced, for which

$$\left.\begin{aligned} A_n^* &= \int_o^\ell \bar{w} \, W_n \, dx, \quad P_n^* = \int_o^\ell \bar{P} \, W_n \, dx, \\ h_n^* &= \int_o^\ell h \, W_n \, dx, \quad g_n^* = \int_o^\ell g \, W_n \, dx, \end{aligned}\right\} \tag{28}$$

holds, and where the W_n are the eigenfunctions of the auxiliary problem (free vibrations of the compressed beam)

$$\tilde{A} \, W_n = \lambda_n \, W_n \ , \quad U[W_n]_B = 0. \tag{29}$$

Using the expansions (27) in (26) and comparing coefficients with the W_n, one arrives at the equation

$$A_n^* \, (\mu \, s^2 + \lambda_n) = P_n^* + s \, h_n^* + g_n^*$$

which yields

$$A_n^* = \frac{P_n^* + s \, h_n^* + g_n^*}{\mu \, s^2 + \lambda_n} \tag{30}$$

Setting

$$\lambda_n / \mu = \omega_n^2 \ , \tag{31}$$

and using (30, (31) in the series expansion of $\bar{w}$ results in

$$\bar{w} = \frac{1}{\ } \sum_n \frac{P_n^* + s \, h_n^* + g_n^*}{s^2 + \omega_n^2} \, W_n . \tag{32}$$

Replacing in (32) the coefficients P_n^*, h_n^*, g_n^* by those given in (28) yields

$$\bar{w} = \frac{1}{\ } \sum_n \frac{W_n}{s + {}_n} \int_o^\ell [\bar{P}(s,\xi) + s \, h(\xi) + g(\xi)] W_n(\xi) d\xi . \tag{33}$$

Inverse Laplace transformation, using also the convolution theorem, changes (33) into

$$w(t,x) = \frac{1}{\mu}\left\{\sum_n W_n(x) \int_o^\ell W_n(\xi) \int_o^t p(\xi,\tau) \frac{1}{\omega_n} \sin\omega_n(t-\tau)d\tau \, d\xi\right.$$

$$\left. + \sum_n W_n(x) \int_o^\ell W_n(\xi)[h(\xi) \cos\omega_n t + \frac{1}{\omega_n} g(\xi) \sin\omega_n t]d\xi \right. . \qquad (34)$$

Let

$$\frac{1}{\mu} \sum_n W_n(x) \, W_n(\xi) \frac{1}{\omega_n} \sin\omega_n(t-\tau) = \tilde{G}(x,\xi,t,\tau). \qquad (35)$$

Then

$$w(t,x) = \int_o^t\int_o^\ell G(x,\xi,t,\tau)p(\xi,\tau)d\xi d\tau + \int_o^\ell [\tilde{G}(x,\xi,t,0)g(\xi) + \frac{\partial\tilde{G}}{\partial t}(x,\xi,t,0)h(\xi)]d\xi. \qquad (36)$$

The original problem (22), (23), (24) becomes a control problem if for example, the differential equation in (22) is replaced by

$$\mu \frac{\partial^2 w}{\partial t^2} + \tilde{A}w = p + Cu \qquad (37)$$

and if an "observation equation" for the control u

$$u = Dw \qquad (38)$$

is added. Setting again CD = F yields the controlled system

$$\mu \frac{\partial^2 w}{\partial t^2} + \tilde{A}w = p + Fw. \qquad (39)$$

This relationship can be transformed on the basis of the preceding calculations in an obvious manner into

$$\int_o^t\int_o^\ell G(x,\xi,t,\tau) \; F \; w(\xi,\tau) \; d\xi d\tau + w_o(t,x), \qquad (40)$$

where $w_o(t,x)$ is the uncontrolled response of the beam given by the right side of eq. (36).

Introducing the operator

$$Tw = \int_o^t\int_o^\ell \tilde{G}(x,\xi,t,\tau) \; F[w(\xi,\tau)]d\xi d\tau \qquad (41)$$

and the performance measure

$$w - w_o = \Delta w, \qquad (42)$$

the control problem can be rewritten as

$$w - w_o = \Delta w = Tw. \tag{43}$$

Relationship (32) indicates that also in this case, one is formally faced with the same problem as in the first case: for the sake of control, one has to solve the *inverse problem of determining an operator* T *which warrants a required response* w of the controlled system *for a prescribed performance measure* Δw.

A specific feature of structural control is due to the fact that a continuous system offers many more possibilities for the implementation of observers and controllers. This point has extensively been discussed by J.L. Lions in 11 , and it alone justifies studying structural control as a separate topic beside common control theory of discrete systems; one may have pointwise observation and control, distributed observation and control, boundary observation and control, observation of the final state, control via initial conditions, etc. All these possibilities place additional demands on the designer for achieving an optimal control.

3. Methods of Solution

After having recognized the very nature of structural control, one is prepared to deal with the question of how to solve the control problem. The most straightforward approach to this problem would be to choose arbitrarily, though intelligently, the operator involved in (21) or (43). Let (43) be taken as an example. One would assume an operator $\tilde{T}$. Then, one had to prescribe a performance measure Δw. The final step would be to solve for the response w. This may by no means be a step of trivial simplicity. The partial differential equation (or integral equation, respectively) related with the control problem may pose significant mathematical challenges. However, at least methodically, it is the most obvious approach. If the thus obtained solution for w is not satisfactory, the whole procedure may be repeated. Hence, one is faced with a trial and error method.

It has been shown that for a beam, the relationship

$$w = \tilde{T}w + w_o \tag{44}$$

holds true. Also,

$$w_o = \tilde{S}p \tag{45}$$

can be claimed to be the relationship which involves the perturbation p. This follows from (36), assuming for the sake of simplicity homogeneous initial conditions. Hence,

$$w = \tilde{T}w + \tilde{S}p \tag{46}$$

is the operator equation describing the behaviour of the system.

A more sophisticated approach to the control problem as the one suggested before would consist in assuming in (46) the perturbation p to be given, a desired system response w to be prescribed, and the operators $\tilde{T}$ and $\tilde{S}$ to be determined accordingly. This is a *mapping problem* from the p-space into the w-space, the mapping operators to be calculated. Also, one may view this problem as a *system identification* problem (see for example[14]) for which the output error happens to be zero. Looking at the control problem in this generality, one has of course to expect that solutions may not exist if care has not been taken to ensure that certain consistency conditions have been met by the prescribed quantities p and w. In other words, the problem has to be posed in such a way that *realizability of the system* is quaranteed. A question like this is a matter of general systems theory and requires to be thoroughly studied 15 .

Yet, in practice, one may follow a simpler route by assuming operators $\tilde{T}$ and $\tilde{S}$ known up to only some factors or terms kept at the beginning undetermined. Assume for example the problem given by (43) and operator T specified by (41). Then, (43) had to be solved for a prescribed performance measure w and a given response w. Proceeding along this line, one realizes immediately that one has to deal with an "*underdetermined problem*" when it comes to specifying those terms in operator $\tilde{T}$ left initially undetermined. This underdeterminancy of the problem is of course a reflection of the fact discussed earlier, that control problems are basically "improperly posed problems" and have to be made properly posed ones by adopting appropriate additional assumptions.

Let with respect to this approach of working with only partly undetermined operators some specific strategies be considered:

(i) Functional Analytical Procedure:

As mentioned above, one has for a beam the relationship

$$w = \tilde{T}w + w_o.$$

This equation can be rewritten by means of (36), (assuming in (36) homogeneous initial conditions), and of (41) to yield

$$w = \int_{\Omega} \tilde{G}(F[w] + p)d\Omega. \tag{47}$$

In (47), the domain of integration has been called Ω for the sake of simplicity. So far, (47) is indeed an *undetermined operator equation* as the term F[w] has not yet been specified. Assume for example, (see (37)),

$$Cu = -cEu \equiv -cu, \; c > 0, \text{ const.}, \tag{48}$$

and in (38)

$$u = Dw = (\sin p)w^2. \tag{49}$$

With (48) and (49), eq. (37) can be changed into

$$\mu \frac{\partial^2 w}{\partial t^2} + \tilde{A}w = p - c(\sin p)w^2, \tag{50}$$

where c is a yet undetermined quantity. Using the inverse of the operator

$$\bar{A}^* w = \mu \frac{\partial^2 w}{\partial t^2} + \tilde{A}w,$$

which for homogeneous initial conditions reads

$$[A^*]^{-1} = \int_\Omega \tilde{G}\ldots d\Omega,$$

as has been determined earlier, eq. (47) assumes the form

$$w = \int_\Omega \tilde{G}[p - c(\sin p)w^2]d\Omega. \tag{51}$$

This is an inhomogeneous, nonlinear integral equation of the Hammerstein type.

In order to remove the underdeterminancy of the problem, which results from the presence of the yet undetermined coefficient c, the following requirements are imposed: (a) the solution of (51) should satisfy certain *performance conditions* , (b) the *solution* of (51) should *exist* and should preferably be *unique*.

Let the performance condition be prescribed as

$$|w| \leq |w_o|_{max} \leq \Omega|\tilde{G}|_{max}|p|_{max}. \tag{52}$$

Let the load p be a bounded function with the norm $\|p\| = \sup|p|$. Moreover, let the load bounds $|p|_{min}$ and $|p|_{max}$ be known. Then, the solutions of (51) will be elements in the Banach space of real bounded functions. Finally, let the coefficient c be assumed in the form

$$c = |p|_{min} (n\Omega^2|K|^2_{max}|p|^2_{max})^{-1} \quad , n > 1. \tag{53}$$

Then, if $|w| < \Omega|\tilde{G}|_{max}|p|_{max}$,

$$|p - c(\sin p)w^2| = |p| - cw^2. \tag{54}$$

Consider the mapping

$$v = \int_\Omega \tilde{G}[p - c(\sin p)w^2]d\Omega, \tag{55}$$

for which the original elements w are supposed to be elements of the closed bounded sphere

$$|w| \leq |w_o|_{max} \leq \Omega|\tilde{G}|_{max}|p|_{max} = |w|_{max}. \tag{56}$$

This is a mapping for which (54) holds. Hence,

$$|v|_{max} \leq \Omega|\tilde{G}|_{max}|p - c(\sin p)w^2|_{max} \leq \Omega|\tilde{G}|_{max}(|p|_{max} - cw^2_{min})$$

$$\leq \Omega|\tilde{G}|_{max}|p|_{max}. \tag{57}$$

Thus,

$$|v|_{max} \leq |w|_{max}, \tag{58}$$

which indicates that v is a mapping of the sphere $|w| \leq |w|_{max}$ into itself. Since the integral operator in the mapping has a kernal $\tilde{G}$ which is uniformly continuous and bounded, the image set of the mapping is compact. Therefore, Schauder's fixed point theorem can be invoked which states that the mapping v has at least one fixed point in the sphere. Consequently, the control problem has at least one solution under the performance condition (51), if the factor c is chosen according to (53).

The uniqueness of the solution can be ensured if the mapping is made a contracting one. For this purpose choose the coefficient n in (53) according to the condition

$$n > 2\frac{|p|_{min}}{|p|_{max}} . \tag{59}$$

From (55) follows

$$|v_1 - v_2| \leq c\,\Omega|\tilde{G}|_{max}\;|w_2^2 - w_1^2| . \tag{60}$$

But,

$$|w_2^2 - w_1^2| = 2|w|\;|w_2 - w_1| \leq 2|w|_{max}|w_2 - w_1|, \tag{61}$$

and by virtue of (56),

$$|w_2^2 - w_1^2| \leq 2\Omega|\tilde{G}|_{max}|p|_{max}|w_2 - w_1|. \tag{62}$$

Substituting (62) in (60) yields

$$|v_1 - v_2| \leq 2c\Omega^2|\tilde{G}|^2_{max}|p|_{max}|w_2 - w_1|. \tag{63}$$

For v to be a contracting mapping, one has to require that

$$2c\Omega^2|\tilde{G}|^2_{max}|p|_{max} < 1,$$

i.e.

$$c < 1/[2\Omega^2|\tilde{G}|^2_{max}|p|_{max}]. \tag{64}$$

This condition for c is obviously satisfied if (59) is used in (53). Hence, for c satisfying (53) and (59), a *unique solution* of the control problem *exists* warranting *performance condition* (52) to be met.

The functional analytical approach outlined here for a beam can of course be extended to more general continua. The treatment of a plate problem by means of this technique has for example been presented in [13].

(ii) Integral Equation Procedure:

As before, it shall be assumed that the generality of the control problem has been reduced by postulating that the operator involved is an integral operator. Then, the system may be described by the equation

$$\int_{\Omega} K[x,\Omega,w(x),\ w(\Omega)]\ \ d\Omega = f(x) \tag{65}$$

where K may be a linear or nonlinear kenrel, w is the system response, f the perturbation and quantitites x and Ω may be "vectors" or scalars.

Use the quadrature formula

$$\int_{\Omega} F(\Omega)d\Omega = \sum_{r=1}^{n} A_r F(\Omega_r) + R, \tag{66}$$

where the Ω_r are chosen pivotal points in , the A_r are appropriate weighing factors and R denotes the remainder term. Let w_j be an approximation for w(xj). Then, using (66) without remainder term in (65), the set of equations (linear or nonlinear ones)

$$\sum_{r=1}^{n} A_r\ K[x_j,\Omega_r,w_j,w_r] = f(x_j),\ j = 1,2,\dots,n, \tag{67}$$

is obtained. It should be noted that the unknowns in this equation are the quantities

$$K_{jr} = K[x_j,\Omega_r,w_j,w_r].$$

Hence, (67) is a highly *undetermined system* as it involves n^2 unknowns K_{jr} but only n equations. It is obvious that the existence of solutions to (67) must be carefully investigated, and an existence of solutions may depend on a correlation between the values which w and f can assume.

Let the problem not be pursued in this generality. Rather, assume that the relation between system response and perturbance is described in terms of a linear, Fredholm integral equation of the second kind, i.e.,

$$w(x) - \lambda \int_{\Omega} K(x,\Omega)w(\Omega)d\Omega = f(x). \tag{68}$$

Applying (66) to (68) without the remainder term yields

$$w(x_j) - \lambda \sum_{r=1}^{n} A_r \, K(x_j,\Omega_r) w(\Omega_r) = f(x_j), \; j = 1,2,\ldots, n. \tag{69}$$

This is a set of linear algebraic equations for the unknowns $K_{jr} = K(x_j,\Omega_r)$. It can be brought into the form

$$\sum_{r=1}^{n} a_{rj} \, K_{jr} = b_j \quad , \; j = 1,2,\ldots, n, \tag{70}$$

where the a_{rj} are elements of the matrix

$$A=\begin{bmatrix} A_1w(\Omega_1) & A_2w(\Omega_2)\ldots\ldots A_nw(\Omega_n) & \ldots\ldots & 0 & \ldots\ldots \\ \ldots & 0 \ldots\ldots & A_1w(\Omega_1)A_2w(\Omega_2)\ldots A_nw(\Omega_n) & \ldots & 0 \ldots \\ & \vdots & \vdots & & \vdots \\ \ldots & \ldots & 0 \ldots\ldots & & A_1w(\Omega_1)A_2w(\Omega_2)\ldots A_nw(\Omega_n) \end{bmatrix} \tag{71}$$

having n rows and n^2 columns, and the b_j are components of the "vector"

$$f = \frac{1}{\lambda}\begin{bmatrix} w(x_1) - f(x_1) \\ w(x_2) - f(x_2) \\ \vdots \\ w(x_n) - f(x_n) \end{bmatrix} \tag{72}$$

hence, (70) is equivalent to

$$A\,X = f, \tag{73}$$

where x is the vector

$$X = \begin{bmatrix} X_1 \equiv K_{11} \\ X_2 \equiv K_{12} \\ \vdots \\ X_{n^2} \equiv K_{nn} \end{bmatrix} . \tag{74}$$

From (71), (72), (73), and (74) it follows that the problem is again underdetermined: n^2 unknowns but only n equations. Thus, for solutions to exist yielding the quantities K_{11}, K_{12}..., K_{nn}, one must require that

$$r_A = r_{AU} \tag{75}$$

holds true. In (75), r_A is the rank of matrix A and r_{AU} is the rank of the *augmented matrix*

$$AU = [A \,\vdots\, f]. \tag{76}$$

Condition (75) is well known in linear algebra and ensures that equation (73) has an infinite number of solution vectors for X. Moreover, it represents the predicted relationship providing the correlation between the values of w and f, which is required for the existence of solutions.

Obviously, the missing uniqueness of the solution X may cause problems for a realization of the control system. Thus, further conditions must be introduced to remove the underdeterminancy of the problem completely. This can be done (if referring to linear problems) by seeking specific solutions satisfying for example a "minimum norm requirement", a solution "optimizing a linear programming", or a solution "optimizing a cost functional" etc. One is thus lead to the idea of *optimal control*. Yet, one should be aware of the fact already mentioned that "optimization" is not superior to any other approach (as the notion "optimization" seems to imply). Since the cost functional is not inherent to the control problem, but is a quantity introduced more or less arbitrarily (and therefore debatable), the optimization achieved with respect to the cost functional is as debatable as this functional itself.

There are still other possibilities to reduce the underdeterminancy of the control problem. These consist in making right away additional assumptions on the nature of the unknown operator. In many cases, such assumptions are indeed indicated by the practical situation in which the control problem arises. For example, assuming that the kernel K in (68) is a symmetric one, would immediately reduce the number of unknowns in (70) from n^2 to $(n+1)n/2$ unknowns. Naturally, also the structure of matrix A in (73) would be different.

In the following, another situation shall be discussed. As a starting point choose (68), i.e.,

$$w(x) - \lambda \int_\Omega K(x,\Omega)w(\Omega)d\Omega = f(x). \tag{68}$$

Assume, that a complete, orthonormal system $\phi_i(x)$ of coordinate functions exists which allows an expansion of w(x) in the form

$$w(x) = f(x) + \sum_i c_i\phi_i(x). \tag{77}$$

Furthermore, assume that the unknown kernel can be written as

$$K(x,\Omega) = \sum_j u_j(\Omega)\phi_j(x), \tag{78}$$

where

$$u_j(\Omega) = \int K(x,\Omega)\phi_j(x)dx \tag{79}$$

is the "Fourier coefficient" of $K(x,\Omega)$ with respect to the system $\phi_i(x)$. The expandability of K in terms of (78) is of course an additional assumption on K, which may be justified by the nature of the problem and verified by the solvability of the problem under assemption (78).

Let now Galerkin's procedure be applied to (68). The result is on the basis of (77)

$$\int_\Omega \left[\sum_i c_i\phi_i(x) - \lambda \int_\Omega K(x,\Omega)w(\Omega)\, d\Omega\right]\phi_k(x)dx = 0, \quad k = 1,2,\ldots . \tag{80}$$

By means of (78) and (79), and observing the orthnormality of the ϕ_i., eq.(80) yields

$$c_k = \lambda \int_\Omega u_k(\Omega)w(\Omega)d\Omega, \; k = 1,2,\ldots \tag{81}$$

This is a result representing a problem which is inverse to the so called "moment problem" in the theory of integral equations: In (81), c_k, λ, and w are prescribed, the functions u_k are to be determined.

Applying (66) to (81), while dropping R, gives

$$c_k = \lambda \sum_{r=1}^{n} A_r\, u_k\,(\Omega_r)w(\Omega_r) \quad , \quad k = 1,2,\ldots, n, \tag{82}$$

which can be written as

$$c_k = \lambda \sum_r a_{rk}\, u_{kr}. \tag{83}$$

Eq. (83) is formally the same underdetermined system of equations for the n^2 unknowns u_{kr} as (70), and has therefore not to be discussed further. All what the preceding deliberations have then shown is that applying Galerkin's method (moment method) does not lead in principal to anything else than to what a direct application of the quadrature formula (66) to the integral equation (68) would have furnished.

Yet, one can give the problem a different twist by proceeding from (81) on differently: assume that the quantities u_k were known or at least anticipated to a certain extent. The simplest assumption were to set

$$u_k = \frac{\phi_k}{\kappa_k}, \qquad \kappa_k = \text{const}, \tag{84}$$

i.e., to assume that the kernel $K(x,\Omega)$ were of the common form

$$K(x,\Omega) = \sum_j \frac{{}_j(\phi)\Omega_j(x)}{\kappa_j} \tag{85}$$

Then, (68) changes into

$$w(x) - \lambda \int_\Omega \sum_j \frac{\phi_j(\Omega)\phi_j(x)}{\kappa_j} w(\Omega)\, d\Omega = f(x). \tag{86}$$

Yet, by virtue of (77),

$$\int_\Omega \phi_j(\Omega)w(\Omega)d\Omega = \int_\Omega f(\Omega)\phi_j(\Omega)d\Omega + \int_\Omega \sum_i c_i\phi_i(\Omega)\phi_j(\Omega)d\Omega. \tag{87}$$

Let

$$\int_\Omega f(\Omega)\phi_j(\Omega)d\Omega = b_j \tag{88}$$

and observe the orthonormality of the ϕ_j. Then, (87) yields

$$\int_\Omega \phi_j(\Omega)w(\Omega)d\Omega = b_j + c_j, \tag{89}$$

and (86) becomes

$$w(x) - \lambda \sum_j \frac{b_j+c_j}{\kappa_j} \phi_j(x) = f(x). \tag{90}$$

Substituting (77) in (90) results in

$$\sum_i c_i\phi_i(x) = \lambda \sum \frac{b_j+c_j}{\kappa_j} \phi_j(x),$$

which can be rewritten as

$$\sum_i \left[c_i - \lambda \frac{b_i+c_i}{\kappa_i}\right]\phi_i(x) = 0. \tag{91}$$

Eq. (90) can only hold true for

$$c_i = \lambda \frac{b_i}{\kappa_i - \lambda} . \tag{92}$$

Relationship (92) is a control condition. In (92), the b_i are prescribed by the prescribed perturbation f. The desired response (77) can be achieved (at least to a certain extent) by making c_i in (77) assume desired values. This can be done by shifting the values of λ and κ_i around, possibly by trial and error. However, it must be noted that $\lambda \neq \kappa_i$ must always be the case.

An example of structural control involving elastoplasticity and contact problems, dealing with optimal control and parameter identification, and containing some of the basic ideas presented here, has been given by B. D. Panagiotopoulos and is very worthy of being read[16].

Finally, it may be mentioned that in some cases, the control problems can be reduced from an operator equation problem to a *functional equation problem.* Examples of such problems have been presented by L. Collatz[17]. Specifically, the example given as 5.3. in Chapter VI on pp. 525-527 of this reference is of some significance. It would obviously become a control problem, if the quantities G and C in that example would be varied.

2. Discretization

In order to avoid the full complexity of control of continuous structures, one may tend to introduce a lumped mass model of the structure. One is then faced with a finite degrees of freedom system to which those methods of modern control theory may be applied which are suitable according to the given circumstances. Wheter such an approach is justifiable, is not so much a matter of mathematics as of physics in the context of model theory. However, one can also carry out discretization of a control problem for continuous structures in a mathematical sense. In a certain way, this has already been forestalled when, in in the preceding examples of section 3 of this paper, an expansion of quantities involved in the calculations had been carried out in terms of a complete set of coordinate functions. The discretization comes about by working with truncated expansions. Such a process of discretization may be preferred to the lumping of masses, as it allows a rigorous mathematical justification.

In order to fixe ideas, let again the beam problem be used as an example. Its control is given by (39) and (23), i.e. by

$$\mu \frac{\partial^2 w}{\partial t^2} + \tilde{A}w = p + Fw, \qquad U[w]_B = 0. \tag{93}$$

Let a complete set of orthonormal coordinate functions $y_i(x)$ be generated by means of the auxilliary problem

$$\tilde{A}y_i = \lambda_i y_i, \qquad U[y_i]_B = 0 \tag{94}$$

Then, apply the expansion theorem [18] which yields

$$w(x,t) = \sum_{i=1}^{\infty} f_i(t)y_i(x). \tag{95}$$

In (95) the $f_i(t)$ are modal amplitudes.

Substituting in (93) w by (95) satisfies the boundary conditions and yields in place of the partial differential equation the expression

$$\mu \sum_i \ddot{f}_i y_i + \sum_i f_i \tilde{A} y_i = p + \sum_i F(F_i y_i). \tag{96}$$

Now, let Galerkins's procedure be applied to (96) which results in

$$\int_\Omega [\mu \sum \ddot{f}_i y_i + \sum_i f_i \lambda_i y_i]\, y_j \, d\Omega = \int_\Omega p\, y_j \, d\Omega + \int_\Omega \sum_i F(f_i y_i) y_j \, d\Omega, \tag{97}$$

if in addition, the first equation in (94) is being used.

Actually, an application of Galerkin's method in this form is equivalent to applying Kantorovich's method, whose mathematical justification has been discussed to some extent in [19]. Assuming that in (97) integration and summation can be interchanged, and that the pertubation p allows an expansion in a Fourier series, i.e.,

$$p = \sum_k P_k(t) y_k(x), \qquad P_k(t) = \int_\Omega p y_k(x) dx, \tag{98}$$

one can derive from (97) the set of ordinary differential equations

$$\mu \ddot{f}_j + \lambda_j f_j = P_j(t) + \sum_i \Phi_{ij}(f_i), \; j = 1,2,3,\ldots, \tag{99}$$

where

$$\Phi_{ij} = \int_\Omega F(f_i y_i) y_j \, d\Omega. \tag{100}$$

By restricting (99) the sum over i to i = 1,2,..., n, and the number of equations to j = 1,2,...,n, where n is supposed to be a finite number, one has obtained with (99) an approximation of the control problem which involves only n degrees of freedom. Thus, to the system (99), the common methods of control as for example discussed in [14] or in [20] can be applied.

It is important to note that the set of eqs. (99) is a coupled one. Although a formulation of the control problem in terms of (99) is already a significant simplification as compared to the one given in terms of differential operators by (39) and (23), the coupling in (99) can still lead to major difficulties: Assume, optimal control theory is being applied to (99). Then, the key to the solution is a matrix Riccati differential equation of high order, if, as usual for structural control, n is a large number. Thus, it will not be easy to solve this Riccati equation. In order to circumvent this difficulty, L. Meirovitch and co-workers have made an interesting proposal: Since one is free to choose the control term $F(w_i)$ within certain limits, let be

$$f(w) = \sum_i y_i(t) F(f_i). \tag{101}$$

With this assumption,

$$\Phi_{ij}(f_i) = \int_\Omega F(f_i)y_iy_j d\Omega = F(f_i)\int_\Omega y_iy_j d\Omega = \delta_{ij}F(f_i) = F(f_j). \tag{102}$$

Using (102) in (99) yields

$$\mu\ddot{f}_j + \lambda_j f_j = P_j(t) + F(f_j), \qquad j = 1,2,\ldots, n. \tag{103}$$

This is an uncoupled problem. Hence, each mode f_j is being controlled individually through (103) and, pursuing optimal control, one is thus led to individual Riccati equations for the n second order equations (103). But, to solve the Riccati equation for a second order equation is an easy task. Hence, the individual optimal modal controls can be determined without great complications. The final control F(w) for the structure as a whole can then be found through (101).

Examples for discretized structural control can be found for instance in [6] and [21].

5. Conclusions

Structural control consists in principle in solving the inverse problems of operator equations, which are essentially improperly posed, underdetermined problems. This underdeterminancy requires introduction of additional conditions, for example performance conditions. The generality of the problem can be essentially reduced by assumptions about the control, making the problem in this way, for example, an inverse integral operator problem or even an inverse functional equation problem. Further simplification of the problems can be achieved through discretization. This may be done by expansions of the relevant quantities involved in terms of elements from a complete system of coordinate functions. In order to justify such a procedure, one should prove that these expansions are uniformly convergent, a requirement which may not always be easy to satisfy. A discretized problem is usually the simplest version of the original control problem and admits treatment by any of the various methods provided by modern control theory as developed for finite degree of freedom systems.

Acknowledgement

This research was carried out with the assistance of NSERC through Grant No. A7297.

References

[1] Roorda, J.

(a) "Tendon Control in Tall Structures", Journal of the Structural Division, ASCE, Vol. 101, No. ST3, Proc. Paper 11537, Sept., 1975, pp. 505-521.

(b) "Active Damping in Structures", Report Aerodynamic No. 8, Cranfield Institute of Technology, Cranfield, England, 1971.

[2] Yang, J. N., and Giannopoulos, F.

(a) "Active Control of Two-Cable Stayed Bridge", Journal of the Engineering Mechanics Division, ASCE, Vol. 105, No. EM5, Proc. Paper 14899, Oct., 1979, pp. 795-810.

(b) "Active Control of Slender Structures", Journal of the Engineering Mechanics Division, ASCE, Vol. 104, No. EM3, Proc. Paper 13836, June, 1978, pp. 551-568.

[3] Yang, J. N., and Yao, J.T.P., "Formulation of Structural Control", Technical Report CE-STR-74-2, Purdue University, West Lafayette, Ind., 1974.

[4] Martin, C.R., and Soong, T.T., "Modal Control of Multistory Structures", Journal of the Engineering Mechanics Division, ASCE, Vol. 102, No. EM4, Proc. Paper 12321, Aug., 1976, pp. 613-623.

[5] Meirovitch, L., Baruh, H., and Oz, H., "A Comparison of Control Techniques for Large Flexible Systems", Paper 81-195 of AAS/AIAA Astrodynamics Specialist Conference, Lake Tahoe, Nevada, Aug. 3-5, 1981.

[6] Abdel-Rohman, M., and Leipholz, H.H.

(a) "Automatic Active Control of Structures", Journal of the Structural Division, ASCE, Vol. 106, No. ST3, Proc. Paper 15260, Mar., 1980, pp. 663-677.

(b) "General Approach to Active Structural Control", Journal of the Engineering Mechanics Division, ASCE, Vol. 105, No. EM6, Proc. Paper 15047, Dec., 1979, pp. 1007-1023.

(c) "Active Control of Flexible Structures", Journal of the Structural Division, ASCE, Vol. 104, No. ST8, Proc. Paper 13964, Aug., 1978, pp. 1251-1266.

[7] Toussi, S., and Yao, J.T.P., "Hysteresis Identification of Multi-Story Buildings", Technical Report CE-STR-81-15, Purdue University, West Lafayette, Ind., 1981.

[8] Petersen, N.R., "Design of Large-Scale Tuned Mass Dampers", ASCE National Convention, Boston, April, 1979.

[9] Zuk, W., "The Past and Future of Active Structural Control Systems", Solid Mechanics Archives, Vol. 5, Issue 1, 1980, pp. 75-90.

[10] Leipholz, H.H.E., ed., "Structural Control", Proc. IUTAM Symp., June, 1979, Univ. of Waterloo, Ont., Canada, North-Holland Pub. Comp., Amsterdam-New York-Oxford, 1980.

[11] Lions, J. L., "Optimal Control of Systems Governed by Partial Differential Equations", Springer-Verlag, Berlin--Heidelberg-New York, 1971.

[12] Payne, L. E., "Improperly Posed Problems in Partial Differential Equations", Soc. for Industrial and Appl. Mathematics, Philadelphia, Pennsylvania 19103, 1975.

[13] Leipholz, H.H.E., "Distributed Control of Elastic Plates", Mechanics Research Communications, Vol. , Issue , , pp. (to appear).

[14] Brogan, W. L., "Modern Control Theory", Quantum Publ. Inc., New York, 1974, pp. 330-343.

[15] Mesarovic, M.D., and Takahara, Y., "General Systems Theory: Mathematical Foundations", Academic Press, New York-San Francisco-London, 1975.

[16] Panagiotopoulos, B. D., "Optimal Control and Parameter Identification of Structures with Convex or Non-Convex Strain Energy Density. Applications to Elastoplasticity and to Contact Problems", Solid Mechanics Archives, Vol. (to appear).

[17] Collatz, L., "The Numerical Treatment of Differential Equations", Second Printing of the Third Edition, Springer-Verlag, Berlin-Heidelberg-New York, 1966.

[18] Meirovitch, L., "Analytical Methods in Vibrations", The Macmillan Co., New York, 1967.

[19] Leipholz, H., "Stability of Elastic Systems", Sijthoff and Noordhoff, Elphen aan den Rign, The Netherlands, 1980.

[20] Schultz, D. G., and Melsa, J.L., "State Functions and Linear Control Systems", McGraw-Hill Book Co., New York-St. Louis-San Francisco-Toronto-London-Sydney, 1967.

[21] Abdel-Rohman, M. and Leipholz, H.H., "Stochastic Control of Structures", Journal of the Structural Division, ASCE, Vol. 107, No. ST7, Proc. Paper 16382, July, 1981, pp. 1313-1325.

ELASTICITY

ON THE ANALYSIS OF A CLASS OF CONTACT PROBLEMS WITH NON-LOCAL FRICTION

J.T. Oden and E.B. Pires

The Texas Institute for Computational Mechanics
The University of Texas
Austin, Texas

ABSTRACT

It is well known that the classical model of Coulomb for static dry friction is neither acceptable from a physical or a mathematical point of view. Physically, contact takes place between two bodies on microscopic set of asperities developed over a film of contaminant, generally oxide, the shear strength of which determines actual resistance to relative sliding. A variety of models of friction have been proposed which provide alternatives to the classical Coulomb law. Among these are so-called non-local models which assert that relative motion at a point occurs when some weighted average of the stresses in the neighborhood of the point reach a critical value. Such non-local laws also produce a more tractable mathematical theory. Indeed, the existence of the solutions to classical boundary-value problems in elasticity with friction has been proved only for very special cases. Duvaut, in particular, has indicated that the use of a non-local law may overcome many of these mathematical difficulties. The present paper deals with the characterization of a non-local friction law and with the study of the Signorini problem with non-local friction. Existence and uniqueness results are developed as well as an approximation theory. Numerical solutions in several representative problems are given.

1. Introduction

We wish to summarize in this note some results that we have obtained recently on the analysis of certain contact problems in elasticity in which a non-local friction law prevails. Complete proofs of all results are to appear in [3,4]; see also the unpublished report [2].

The plausibility of a non-local friction law as an alternative to the classical (local) pointwise Coulomb law of friction is suggested by an examination of both the physics of friction and the mathematics of boundary-value problems in elastostatics with local friction.

From the purely physical side, it has been recognized for many years that Coulomb's law is, at best, a crude approximation of the actual mechanics of friction capable of depicting only gross sliding of one effectively rigid body on another. If one examines carefully the contact surfaces of two metallic blocks pressed together along two apparently flat machined surfaces, one quickly observes that at magnifications of 1000X to 5000X, the contact surfaces exhibit marked deviations from the plane. These irregularities, which are large compared with molecular structures of the metals, are called *asperities* and provide the actual structure through which forces normal to the apparent surfaces are transmitted. As two such bodies are pressed together, the load is transmitted through tips of the asperities; high stresses are developed at the protruding asperities and they quickly yield and fracture to form *junctions* between the bodies in contact. The actual contact area (as opposed to the *apparent* contact area) is the sum of the cross-sections of these junctions formed by crushed, flattened asperities.

Suppose that net opposing normal forces of magnitude N are applied to press two metallic bodies together and that then a tangential shear force T, parallel to the apparent contact plane, is applied so as to cause the sliding of one body relative to another. The normal forces are actually transmitted through a thin film of contaminant and metallic oxide a few angstroms thick, and it is the shear strength of this film that determines the coefficient of friction and not that of the parent metals. As T is increased, the junctions are finally fractured and gross sliding of one body relative to another occurs. Thus, the actual variation of normal stress over the contact surface is not uniformly distributed, but is concentrated in junctions of crushed asperities distributed more or less randomly.

There is also mathematical justification to consider a non-local friction law. Duvaut [1] has shown (see also Duvaut and Lions [6]) that if Coulomb's law is applied pointwise in contact problems involving linearly elastic bodies, then the contact stress σ_n developed normal to the contact surface is ill-defined. Except for some very special cases (see, e.g. Nečas, Jarušek, and Haslinger [8]), the question of the existence of solutions of the friction problem in elastostatics is open. Duvaut [1] observed that the source of difficulty in the proof of existence is the lack of smoothness of the normal contact pressure σ_n. By replacing σ_n by a mollified stress, which might be interpreted as assuming a non-local friction law, he was able to develop a complete existence and uniqueness theory for certain contact problems. Further results in this direction have been obtained by Oden and Pires [3,4] and Demkowicz and Oden [7].

In the present work, the justification of a specific model of non-local friction is given, a variational principle for contact problems with non-local friction is derived, some results on the existence, uniqueness, and approximation of such problems are summarized, and numerical results of some preliminary calculations are given.

2. A Non-local Friction Law

Consider the situation in which a thin ribbon of material A is pressed against a metallic rectangular prism B by a concentrated normal force N as shown in Fig. 1a. At the point of impending motion, the tangential stress distribution σ_T on the contact surface, as predicted by the classical pointwise Coulomb law, is

$$\sigma_T(x) = \nu N\delta \qquad -\ell < x < \ell \tag{2.1}$$

where δ is the Dirac delta. Of course, (2.1) is merely the symbolic representation of the distribution

$$\langle\sigma_T,\phi\rangle = \nu\ \delta\langle\delta,\phi\rangle = \nu N\phi(0) \tag{2.2}$$

for all test functions $\phi \in C_0^\infty(-\ell,\ell)$, where $\langle\cdot,\cdot\rangle$ denotes duality pairing on distributions and test functions (i.e. the action of a distribution q on a test function ϕ is denoted $q(\phi) \equiv \langle q,\phi\rangle$). Alternatively, δ can be interpreted as the limit of a δ-sequence, $\{\omega_\rho\}_{0<\rho}$, $\omega_\rho \in C_0^\infty(-\ell,\ell)$:

$$\phi(0) = \delta(\phi) = \lim_{\rho\to 0}\int_{-\ell}^{\ell}\omega_\rho\phi dx \tag{2.3}$$

As a typical delta sequence, we mention

$$\omega_\rho(x) = \begin{cases} c\ \exp[\rho^2/(x^2-\rho^2)] & |x| \leq \rho \\ 0 & |x| > \rho \end{cases} \tag{2.4}$$

We see that the classical pointwise version of Coulomb's law must be interpreted in the sense of distributions in this situation.

A more realistic model of friction is obtained if we take into account the fact that N is not concentrated at a point but is distributed over a deformed asperity. We shall weigh the normal stress distribution over a circular region of radius ρ_0 on the contact surface using the δ-sequence ω_ρ of (2.4) keeping $\rho = \rho_0$. Since $N = N(x)$ is now a function, we have, instead of (2.2) ,

$$\sigma_T(x) = \nu N(y) * \omega_{\rho_0}(x-y) \tag{2.5}$$

where $*$ denotes the convolution. Thus, we have arrived at a friction law in

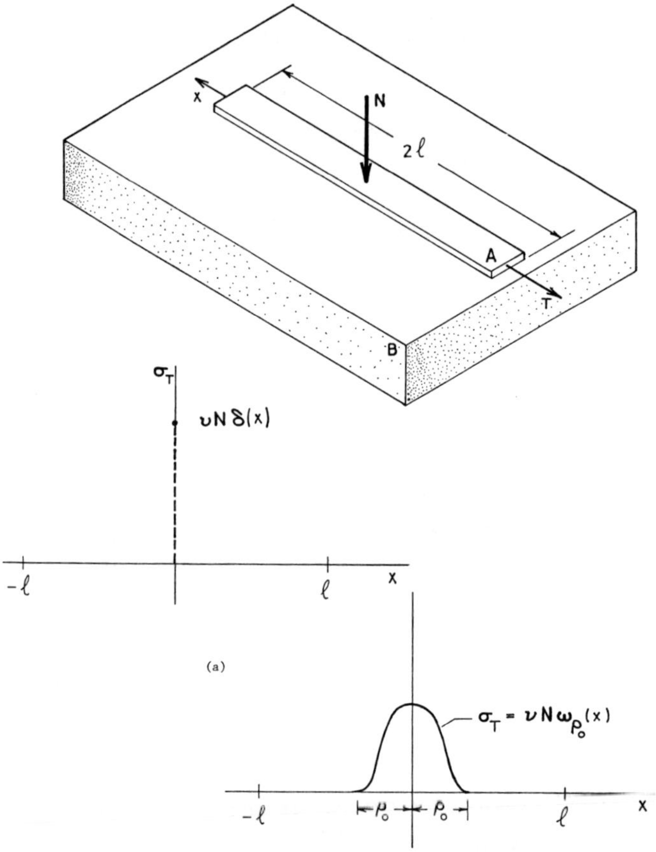

Figure 1. Tangential stress distribution at the point of impending slipping for a) primitive Coulomb law and b) a non-local law of friction.

which impending motion occurs at a point x on the contact surface when the shear stress at that point reaches a value proportional to the weighed average of the normal stress in a neighborhood of the point. If ω_{ρ_0} is used to characterize this weighting function, then the neighborhood is a circular disc of radius ρ_0 centered at x , the maximum weight is given to the stress intensity at the center of the disc (the deformed asperity) and exponentially decreasing weights are assigned to stress intensities as one moves from the center of the neighborhood outward to the periphery of the disc. This is shown in Fig. 1b.

A three-dimensional generalization of this non-local law is immediate: let $\underset{\sim}{u}_T$ be the tangential component of displacement of a point x on the contact surface between two deformable bodies and let $\sigma_n(\underset{\sim}{u})$ and $\underset{\sim}{\sigma}_T(\underset{\sim}{u})$ denote the normal and tangential stresses on the contact surface corresponding to the displacement field $\underset{\sim}{u}$. Then

$$\left.\begin{aligned} &|\sigma_T(\underset{\sim}{u})| < \nu\, S(\sigma_n(\underset{\sim}{u})) \Rightarrow \underset{\sim}{u}_T = \underset{\sim}{0} \\ &|\sigma_T(\underset{\sim}{u})| = \nu\, S(\sigma_n(\underset{\sim}{u})) \Rightarrow \underset{\sim}{u}_T \neq \underset{\sim}{0} \\ &[\text{Specifically, } \exists \lambda \text{ such that } \underset{\sim}{u}_T = -\lambda \underset{\sim}{\sigma}_T(\underset{\sim}{u}),\ \lambda \geq 0] \end{aligned}\right\} \tag{2.6}$$

where S is an operator mollifying the normal stress distribution; e.g.

$$S(\sigma_n(\underset{\sim}{u}))(\underset{\sim}{x}) = \int_{\Gamma_C} \omega_{\rho_0}(|\underset{\sim}{x} - \underset{\sim}{y}|)(-\sigma_n(\underset{\sim}{u}(\underset{\sim}{y})))dy \tag{2.7}$$

where $\underset{\sim}{x}$ and $\underset{\sim}{y}$ are points on the contact surface Γ_C .

3. A Variational Principle for Non-Local Friction

We consider the Signorini problem of contact of a linearly elastic body Ω with a rigid foundation on which the non-local law (2.6) holds. Using standard notations, the governing equations are:

$$\left.\begin{aligned} (E_{ijk\ell}u_{k,\ell})_{,j} + f_i &= 0 \text{ in } \Omega \\ u_i &= 0 \text{ on } \Gamma_D \\ E_{ijk\ell}u_{k,\ell}n_j &= t_i \text{ on } \Gamma_F \\ \underset{\sim}{u}\cdot\underset{\sim}{n} \leq 0,\ \sigma_n(\underset{\sim}{u}) \leq 0,\ &\sigma_n(\underset{\sim}{u})\underset{\sim}{u}\cdot\underset{\sim}{n} = 0 \text{ on } \Gamma_C \end{aligned}\right\} \tag{3.1}$$

$$\left.\begin{array}{l}|\tilde{\sigma}_T(\tilde{u})| \leq \nu\, S(\sigma_n(\tilde{u})) \Rightarrow \tilde{u}_T = \tilde{0} \\ |\tilde{\sigma}_T(\tilde{u})| = \nu\, S(\sigma_n(\tilde{u})) \Rightarrow \exists\, \lambda \geq 0 \text{ s.t.} \\ \qquad \tilde{u}_T = -\lambda \tilde{\sigma}_T \end{array}\right\} \text{ on } \Gamma_c$$

Here $\Gamma = \partial\Omega = \overline{\Gamma}_D \cup \overline{\Gamma}_F \cup \overline{\Gamma}_C$, ν is the coefficient of friction, a positive constant, and f_i, t_i, and Ω are assumed to be smooth (e.g. Ω is Lipschitzian, $f_i \in L^2(\Omega)$, $t_i \in L^2(\Gamma_F)$).

A variational principle for problem (3.1) is embodied in the following variational inequality:

$$\left.\begin{array}{l}\text{Find a displacement field } \tilde{u} \in K \text{ such that} \\ a(\tilde{u},\tilde{v}-\tilde{u}) + \displaystyle\int_{\Gamma_C} \nu\, S(\sigma_n(\tilde{u}))(|\tilde{v}_T| - |\tilde{u}_T|)ds \\ \qquad \geq f(\tilde{v} - \tilde{u}) \quad \forall \tilde{v} \in K \end{array}\right\} \tag{3.2}$$

Here

K = the unilateral constraint set = $\{\tilde{v} = (v_1,v_2,v_3) \in V|$ $\tilde{v} \cdot \tilde{n} \leq 0$ a.e. on $\Gamma_C\}$, with

V = the space of admissible displacements

$= \{\tilde{v} \in (H^1(\Omega))^3 |\ \tilde{v} = \tilde{0} \text{ on } \Gamma_D\}$;

$$\|\tilde{v}\|_V = \|\tilde{v}\|_1 = \{\int_\Omega v_{i,j}v_{i,j}dx\}^{1/2}$$

$a(\tilde{u},\tilde{v})$ = the virtual work of stresses $\sigma_{ij}(\tilde{u})$ on strains $\varepsilon_{ij}(\tilde{v})$

$$= \int_\Omega \sigma_{ij}(u)\varepsilon_{ij}(\tilde{v})dx = \int_\Omega E_{ijk\ell}\, u_{k,\ell}v_{i,j}\, dx$$

[$a(\cdot,\cdot)$ is a symmetric, continuous, V-elliptic, bilinear form on V since we assume $E_{ijk\ell}$ has the usual symmetries and elliptic properties].

$f(\tilde{v})$ = the virtual work of the external forces

$$= \int_\Omega \tilde{f} \cdot \tilde{v}\, dx + \int_{\Gamma_F} \tilde{t} \cdot \tilde{v}\, ds$$

A related auxiliary problem consists of finding $\underset{\sim}{u} \in K$ whenever νS is prescribed as a positive function in $L^2(\Gamma_C)$:

$$\left.\begin{array}{l} \text{Given } \tau \in L^2(\Gamma_C)\ ,\ \tau \geq 0 \text{ a.e., find } \underset{\sim}{u}_\tau \in K \text{ such that} \\ a(\underset{\sim}{u}_\tau,\ \underset{\sim}{v} - \underset{\sim}{u}_\tau) + \int_{\Gamma_C} \tau\left(|\underset{\sim}{v}_T| - |\underset{\sim}{u}_{\tau T}|\right) ds \geq f(\underset{\sim}{v} - \underset{\sim}{u}_\tau) \quad \forall \underset{\sim}{v} \in K \end{array}\right\} \tag{3.3}$$

It can be shown (see Oden and Pires [2]), that

1. $\forall \tau$, there exists a unique solution $\underset{\sim}{u}_\tau$ to (3.3) and the correspondence $\tau \longmapsto \underset{\sim}{u}_\tau$ defines a continuous nonlinear map B from $L^2(\Gamma_C)$ into V ;

$$\underset{\sim}{u}_\tau = B(\tau)$$

2. The normal stress $\sigma_n(\underset{\sim}{u}_\tau)$ is well-defined in the space

$$W' = (H^{1/2}(\Gamma_C))'$$

where $H^{1/2}(\Gamma_C)$ is the space of normal components of displacement on the contact surface, and $\sigma_n : K \to W'$ is continuous.

Returning to the main variational principle (3.2), we note that the following properties of that problem can be established (see Oden and Pires [2] for complete details).

Theorem 1. 1. Every solution of the Signorini problem with non-local friction (3.1) is also a solution of the variational inequality (3.2), conversely, every solution of (3.2) satisfies (3.1) in a weak or distributional sense.

2. Let meas $\Gamma_D > 0$ and the smoothing operator $S:W' \to L^2(\Gamma_C)$ be completely continuous. There exists at least one solution $\underset{\sim}{u} \in K$ of (3.2) for every choice of data $\underset{\sim}{f}$ and $\underset{\sim}{t}$.

3. If the coefficient of friction ν is sufficiently small, then the solution of (3.2) is unique. ◻

One method of proof of items 2 and 3 of this theorem was suggested by Duvaut [1]:

A. Let $T:L^2(\Gamma_C) \to L^2(\Gamma_C)$ denote the operator formed by the composition

$$T = \nu\, S \circ \sigma_n \circ B$$

where $B:L^2(\Gamma_C) \to K$ is the operator inherent in problem (3.3) and $S:W' \to L^2(\Gamma_C)$ is the completely continuous smoothing operator in the non-local friction law.

B. Let ψ^* be a fixed point of T :

$$T(\psi^*) = \psi^*$$

C. Let $\underset{\sim}{u}^*$ be the solution of the auxiliary problem (3.3) corresponding to the choice $\tau = \psi^*$

D. Then, $\forall \underset{\sim}{v} \in K$,

$$a(\underset{\sim}{u}^*, \underset{\sim}{v} - \underset{\sim}{u}^*) + \underbrace{\int_{\Gamma_C} \psi^*(|\underset{\sim}{v}_T| - |\underset{\sim}{u}^*_T|)\, ds}_{\displaystyle \| \atop \displaystyle \underbrace{\int_{\Gamma_C} T(\psi^*)(|\underset{\sim}{v}_T| - |\underset{\sim}{u}^*_T|)\, ds}_{\displaystyle \| \atop \displaystyle \int_{\Gamma_C} \nu\, S(\sigma_n(\underset{\sim}{u}^*))(|\underset{\sim}{v}_T| - |\underset{\sim}{u}^*_T|)\, ds}} \geq f(\underset{\sim}{v} - \underset{\sim}{u}^*)$$

i.e., $\underset{\sim}{u} = B(\psi^*)$ is a solution of the general problem (3.2).

E. Having reduced (3.2) to a fixed point problem, one uses the Schauder-Tychonoff fixed point theorem to show that T has at least one fixed point. For small ν , T becomes a contraction map and the fixed point is unique.

4. Approximations and Numerical Results

Approximations of (3.2) are obtained in the standard way. By constructing a regular series of partitions of $\overline{\Omega}$ into finite element meshes over which piecewise polynomial approximations of the displacements are used, it is usually possible to produce a family $\{V_h\}_{h>0}$ of finite-dimensional subspaces of the space V of admissible displacements, h being the mesh parameter (i.e., the largest diameter of an element in the mesh). Approximations K_h of the unilateral constraint set must generally be constructed so that the contact condition is applied only at nodes on the contact surface Γ^h_C ; thus, in general, $K_h \not\subset K$.

Under these conditions, a Galerkin approximation of (3.2) is characterized as follows:

Find $\underset{\sim}{u}_h \in K_h$ such that

$$a(\underset{\sim}{u}_h, \underset{\sim}{v}_h - \underset{\sim}{u}_h) + \int_{\Gamma_C} \nu\, S(\sigma_n(\underset{\sim}{u}_h))(|\underset{\sim}{v}_{hT}| - |\underset{\sim}{u}_{hT}|)ds \geq f(\underset{\sim}{v}_h - \underset{\sim}{u}_h) \quad \forall \underset{\sim}{v}_h \in K_h \tag{4.1}$$

The following is proved in Oden and Pires [2]:

Theorem 2. Under the conditions of Theorem 1:

1. There exists at least one solution to the approximate problem (4.1); if ν is sufficiently small, the solution is unique.

2. Let A denote the operator $\langle A\underset{\sim}{u},\underset{\sim}{v}\rangle = a(\underset{\sim}{u},\underset{\sim}{v})$ $\forall \underset{\sim}{v} \in V$ and let $A\underset{\sim}{u} - \underset{\sim}{f} \in U$, where U is a Hilbert space densely embedded in the dual V' of V. Then, if ν is sufficiently small, there exists a constant $C > 0$ such that the following error estimate holds,

$$\begin{aligned} \|\underset{\sim}{u} - \underset{\sim}{u}_h\|_1 \leq C\Big\{ &\|\underset{\sim}{u} - \underset{\sim}{v}_h\|_1 + \Big[\|A\underset{\sim}{u} - \underset{\sim}{f}\|_U (\|\underset{\sim}{u} - \underset{\sim}{v}_h\|_{U'} \\ &+ \|\underset{\sim}{u}_h - \underset{\sim}{v}\|_{U'}\Big]^{1/2} + \Big[\nu \|u\|_1 (\|u - \underset{\sim}{v}_h\|_1 \\ &+ \|\underset{\sim}{u}_h - \underset{\sim}{v}\|_1\Big]^{1/2}\Big\} \end{aligned} \tag{4.2}$$

for every $\underset{\sim}{v} \in K$, $\underset{\sim}{v}_h \in K_h$. □

When the solution $\underset{\sim}{u}$ of (3.2) is sufficiently smooth and the spaces V_h exhibit the usual interpolation properties, (4.2) can be used to show that the approximation $\underset{\sim}{u}_h$, satisfying (4.1), converges in V to $\underset{\sim}{u}$ as h tends to zero.

Upon introducing piecewise polynomial approximations $\underset{\sim}{u}_h$ and $\underset{\sim}{v}_h$ into (4.1), we obtain a system of nonlinear algebraic inequalities in the nodal values $\underset{\sim}{u}_i^\alpha$ of $\underset{\sim}{u}_h$. This system can be solved using a number of techniques. We shall now quote results obtained using the following algorithm:

1. Q_2 - (nine-node, biquadratic) polynomial approximations of the displacement field $\underset{\sim}{u}$ are used over each element.

2. An approximation of the auxiliary problem (3.3) is constructed and solved using linear programming techniques (e.g. Uzawa's method, a Newton's method for a regularized problem, standard over-relaxation with projections, etc.).

3. A successive-approximation iterative scheme is then utilized which is based on steps A,B,C,D in the fixed point arguments listed in the previous section.

These steps, of course, describe only in general terms a family of algorithms that could be used for problems of this type. We are currently investigating several of these to determine an efficient scheme for solving friction problems of the type in (3.1) or (3.2).

As a representative numerical example, we present results obtained from a finite element analysis of a rigid-punch problem with non-local friction. The problem is one of plane strain of a block of homogeneous isotropic linearly elastic material indented by a rigid cylindrical punch, as shown in Fig. 2. The mechanical data of the problem is: E = Young's modulus = 1000 (non-dimensional units), μ = Poisson's ratio = 0.3, and the coefficient of friction $\nu = 0.6$. The parameter ρ_0 in the non-local friction law (2.7) was taken to be 0.1. The rectangular mesh of 9-node, Q_2-elements shown in the figure was used in the finite element approximation and the system of nonlinear algebraic inequalities was solved using an algorithm of the type outlined above for a prescribed indentation of $\delta = 0.6$. Other dimensions are given in the figure.

Computed deformed shape and stress profiles are shown in Fig. 3. Observe that the tangential friction stresses do not reach a sharp peak, but are smooth at the point of maximum transverse shear stress. The size of this "boundary-layer" in the neighborhood of the maximum stress, of course, depends upon the magnitude of ρ. As $\rho \to 0$, a cusp is developed at the peak stress corresponding to the case when the classical pointwise version of Coulomb's law holds. For $\rho > 0$, this means that there is no sharp dividing line between full-adhesion (no slip) and sliding areas, but rather a boundary-layer over which a transition from no-slip to slipping occurs. The existence of this boundary-layer is fully consistent with experimental evidence on friction (see Bowden and Tabor [5]).

Acknowledgement. This work was supported by the U.S. Air Force Office of Scientific Research under Contract F-49620-78-C-0083.

References

1. Duvaut, G., "Problèmes Mathématiques de la Mécanique - Équilibre d'un solide élastique avec contact unilatéral de frottement de Coulomb," C.R. Acad. Sc., Paris, t. 290, Série A, pp. 263-265, 1980.

2. Oden, J.T. and Pires, E.B., Contact Problems in Elastostatics with Non-Local Friction Laws, TICOM Report 81-12, Austin, 1981.

3. Oden, J.T. and Pires, E.B., "Non-Local Friction in Contact Problems in Elasticity," (in review).

4. Pires, E.B. and Oden, J.T., "Error Estimates for the Approximation of a Class of Variational Inequalities arising in Unilateral Problems with Friction," J. of Num. Functional Anal. and Optimization, (to appear).

5. Bowden, F.P. and Tabor, T., The Friction and Lubrication of Solids, Part II, Clarendon Press, Oxford, 1964.

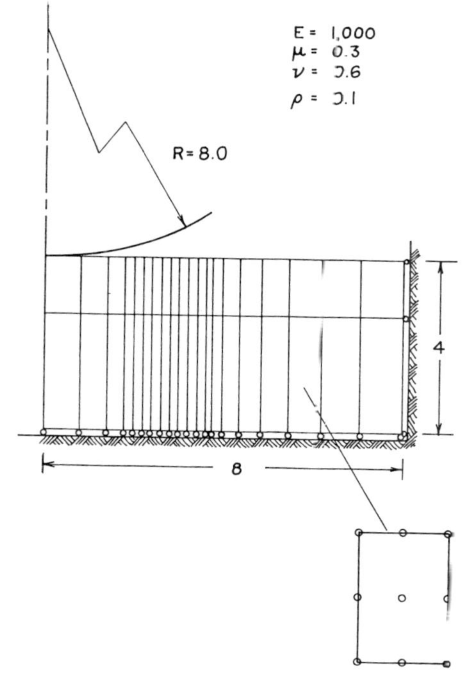

Figure 2. Finite element mesh for the problem of indentation of an elastic block by a rigid cylinder.

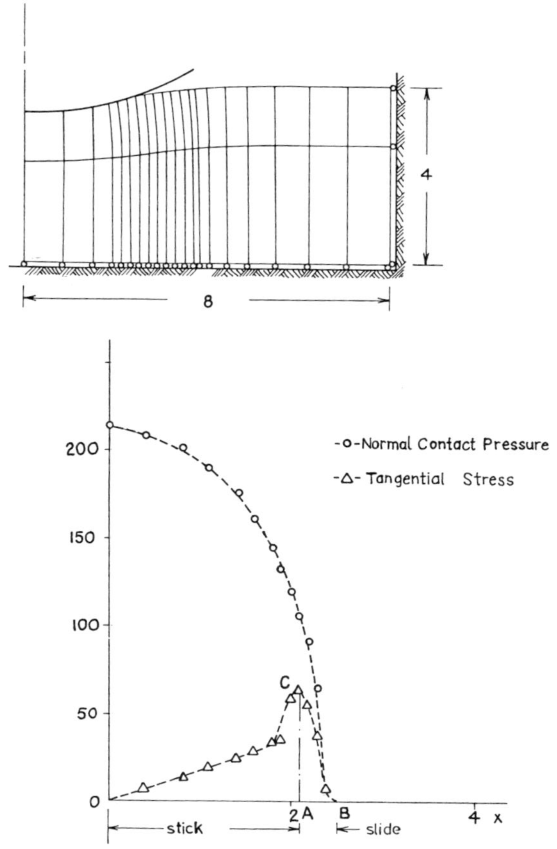

Figure 3. Computed deformed geometry and stresses on the contact surface.

6. Duvaut, G. and Lions, J.L., Inequalities in Mechanics and Physics, Springer-Verlag, New York, 1976.

7. Demkowicz, L. and Oden, J.T., On Some Existence and Uniqueness Results in Contact Problems with Non-Local Friction, TICOM Report 81-10, Austin, 1981.

8. Necas, J., Jarusek, J. and Haslinger, J., "On the Solution of the Variational Inequality to the Signorini Problem with Small Friction," Boll. U.M.I. (5), 17-B, pp. 796-811, 1980.

A UNIFIED FORMULATION FOR A DISLOCATION LYING IN AN INTERFACE

Nicholas J. Salamon
Department of Mechanical and Aerospace Engineering
West Virginia University
Morgantown, WV 26506
USA

ABSTRACT

The elastic fields of a dislocation lying in the interface of an isotropic composite can be written in terms of algebraic relations or Lipschitz-Hankel integrals which reduce to generalized functions. Moreover the interface transmits tractions in the form of Dirac delta functions. It turns out that the limits of Lipschitz-Hankel integrals form a family of generalized representations which provide broader and clearer meaning to physical problems. The development permits a unified formulation for the elastic fields and energy of any shape dislocation loop located near to and in an interface. The special case of a circular loop is discussed in detail.

INTRODUCTION

The problem of a dislocation loop lying in or near and parallel to a bimaterial interface separating two isotropic solids offers an unusual chance to study an intricate, exact solution in the three-dimensional theory of linear elasticity. Yet academics aside, it provides a unique vehicle for study of the mechanical behavior of dislocations falling within the above description. This is because the solution follows from a unified formulation which covers a broad subset of problems and also because the complicated mathematics has been coherently streamlined into a tractable and succinct form.

The general solution is founded upon the continuum theory of dislocations [1] which establishes the formula

$$u_i(x) = \int_S C_{jk\ell m} \left[\frac{\partial}{\partial x_j'} G_{ik}(x,x') \right] b_\ell \, n_m(x') dS(x') \tag{1}$$

for construction of the displacement fields u_i. In (1), C is the tensor of elastic moduli, b is Burgers vector and n is the unit normal to the surface S capping the dislocation loop. G_{ik} is the Green's tensor function for the fundamental solutions describing displacements in the x_i direction occurring at point x due to concentrated unit forces applied at x' and acting in the x_k direction. The objective is to generate the necessary forms of the derivatives of G_{ik} appearing in (1) which satisfy the interface continuity conditions and yield the Burgers displacement discontinuity across the loop. One good way to accomplish this is through use of Papkovich-Neuber potential functions, as reported in Salamon and Dundurs [2], from which the elastic fields may be determined directly.

The utility of the solution hinges upon the nature of the mathematical description, in particular certain integrals over the loop surface. As fortune would have it, those needed for the circular-shaped loop can be cast into convenient forms of the type

$$J(m,n;p) = \int_0^\infty J_m(t) \, J_n(\rho t) \, e^{-|\zeta|t} \, t^p \, dt \tag{2}$$

where $\rho = r/a$, $\zeta = z/a$, a the radius of the loop, and J_k denotes a Bessel function of the first kind and order k. Furthermore r and z are the elements of cylindrical coordinates with $r^2 = x^2 + y^2$ as illustrated in Figure 1. Those integrals necessary to the dislocation problem are discussed in [2]; for computational purposes, transformations to elliptic integrals are provided; for study of the limiting situations when the loop is near to or far from the interface, asymptotic forms are provided. But the intricate behavior of the dislocation loop lying near to and in the interface can only be completely revealed by opening to view the nature of limiting forms of J (m,n;p), in particular

$$J^o(m,n;p) = \lim_{\zeta \to o} J(m,n;p) \tag{3}$$

The details of this study are reported in Salamon and Walter [3]. They show that limits exist in the sense of generalized functions for all values of m, n, p and present a scheme for computing them. For instance, two general results of interest here are as follows:

Lemma. For any integer n > o,

$$1) \quad J^o(n,n;1) = \delta(\rho-1) \tag{4a}$$

where δ is the Dirac delta function;

$$2)\quad J^o\,(n,n-1;o) = \rho^{n-1}\, H(1-\rho) \tag{4b}$$

where H is the Heaviside function.

One result follows from the other. A proof is given in [3].

The purpose of this lecture is to assemble component works and present a unified formulation for the interface dislocation. The displacement fields for a loop of arbitrary shape are given (having been derived directly from [2]) for prismatic and glide configurations denoted by Burgers vectors b_z and b_x respectively. The fields are generated in further detail for the special case of a circular loop and the interface quantities written out. In turn the energy of a circular loop and loop segment are given explicitly while the energy for a loop of arbitrary shape is derived in the form of an integral.

THE ELASTIC FIELD

The material properties enter through the three composite parameters

$$A = \frac{\Gamma(\kappa_1+1)}{\Gamma\ \kappa_1+1},\quad B = \frac{\Gamma(\kappa_1+1)}{\Gamma + \kappa_2},\quad C = \frac{\Gamma}{\Gamma+1} \tag{5}$$

where $\Gamma = G_2/G_1$ and $\kappa = 3-4\nu$. G is the elastic shear modulus and ν is Poisson's ratio with the subscripts 1, 2 denoting properties taken in Regions I, II respectively (Figure 1). It is also convenient to introduce the contractions

$$\begin{aligned} D &= 2\,[A + B + (\kappa_1+1)\,C\,] \\ M &= \kappa_1\,A + B - 2(\kappa_1+1)\,C \\ N &= A + B - (\kappa_1+1)\,C \end{aligned} \tag{6}$$

for subsequent use.

The displacement fields in region I for an arbitrary shaped dislocation lying in the interface are

$$u_i^I = \frac{b_z}{2\pi(\kappa_1+1)}\left[\,\frac{1}{2}\,(\kappa_1 A + B)\frac{\partial}{\partial x_i} - \delta_{3i}\kappa_1 A\frac{\partial}{\partial z} + zA\frac{\partial^2}{\partial x_i \partial z}\,\right]\int_S \frac{dS}{R} \tag{7}$$

for the prismatic loop and

$$u_i^I = \frac{-b_x}{2\pi(\kappa_1+1)} [\delta_{3i} \{\kappa_1 A - (\kappa_1+1) C\} \frac{\partial}{\partial x} + \delta_{i1}(\kappa_1+1) C \frac{\partial}{\partial z} - z A \frac{\partial^2}{\partial x \partial x_i}] \int_S \frac{dS}{R} - \frac{M}{2} \frac{\partial^2}{\partial x \partial x_i} \int_S \ell n\ (R+z)\ dS \tag{8}$$

for the glide loop. Here $x_1 = x$, $x_2 = y$, $x_3 = z$ interchangeably, δ_{ij} is the Kronecker delta, and $R = [(x-x')^2 + (y-y')^2 + z^2]^{1/2}$. The stress fields associated with (7) and (8) are omitted here, but may be generated in the usual manner.

Clearly the fields induced by a finite loop of any shape are readily obtained provided the integrals are available. Some explicit results for the circular shape are given below.

The Circular Prismatic Loop

The elastic fields of a circular prismatic loop together with a discussion on loop interaction with the second phase are given by Dundurs and Salamon [4]. They discovered that such interaction is not restricted to simple repulsion and attraction, but for certain combinations of the elastic constants stable equilibrium positions are possible. Yet despite the use of asymptotic limits, the results did not permit study of the loop in the interface. This was made possible through development of the mathematical machinery given in [3]. Salamon and Comninou [5] then completed the analysis for the interface prismatic loop.

The displacement components in region I are

$$u_\gamma^I = \frac{-b_z \gamma}{2(\kappa_1+1)r} [(\kappa_1 A-B)\ J\ (1,1;0) - 2A\zeta\ J(1,1;1)]$$

$$u_z^I = \frac{b_z}{2(\kappa_1+1)} [(\kappa_1 A+B)\ J\ (1,0;0) + 2A\zeta\ J(1,0;1)] \tag{9}$$

where $\gamma = x$ or y. Region I stress components which transmit interface tractions are

$$\sigma_{zz}^I = \frac{-G_1 b_z}{a(\kappa_1+1)} [(A+B)\ J\ (1,0;1) + 2A\zeta\ J(1,0;2)] \tag{10a}$$

$$\sigma^{I}_{rz} = \frac{-G_1 b_z}{a(\kappa_1+1)} [(B-A)\ J\ (1,1;1) + 2A\zeta\ J\ (1,1;2)] \tag{10b}$$

The cartesian components of shearing traction follow directly from the second of (10) since $\sigma_{xz} = \sigma_{rz} \cos\theta$ and $\sigma_{yz} = \sigma_{rz} \sin\theta$ where $\theta = \arctan(y/x)$ is the angular element of cylindrical coordinates (Figure 1). The remaining stress components may be found in [5].

The Circular Glide Loop

Whereas the fields for the prismatic loop are axially symmetric (making their presentation a simple matter), those for the glide loop are fully three-dimensional, hence considerably more complicated. Salamon and Dundurs [6] discuss the ramifications involving interaction of the glide loop and portions of it with the second phase. Contrary to the prismatic loop, the glide loop is never completely stable. However, if forces are averaged over the entire loop, for certain combinations of the elastic constants, pseudo-equilibrium positions are found. The loop behavior is further complicated by the fact that it cannot be predicted at intermediate distances away from the second phase from knowledge of its asymptotic behavior both near and far from the interface. But it is such intricate behavior that makes the problem more interesting.

Salamon [7] solved the problem for the interface loop. The results were reduced to encompass a subset of problems and provide a sound mathematical basis to some rather subjective ideas. The solution permits clear explanation of intricasies involving, in particular, energy and line tension.

The displacement components for the glide loop in region I are

$$\begin{aligned} u^{I}_{x} = \frac{b_x}{(\kappa_1+1)} \{ & [\kappa_1+1)C + (M/2)\cos^2\theta]\ J\ (1,0;0) \\ & - A\ \zeta\ [J(1,0;1)\cos^2\theta - \rho^{-1} J(1,1;0)\cos 2\theta] \\ & - (1/2)\ M\ \rho^{-1}\ J(1,1;-1)\cos 2\theta\}, \end{aligned} \tag{11}$$

$$u^{I}_{y} = \frac{b_x \sin 2\theta}{4\ (\kappa_1+1)} [M\ J(1,2;0) + 2A\ \zeta\ J(1,2;1)],$$

$$u^{I}_{z} = \frac{b_x \cos\theta}{2(\kappa_1+1)} \{[\kappa_1\ A-B]\ J(1,1;0) + 2A\zeta\ J(1,1;1)\}$$

and the region I stresses which transmit tractions are

$$\sigma^{I}_{xz} = \frac{-\mu_1 b_x}{a(\kappa_1 + 1)} \{[(\kappa_1+1)C + N \cos^2\theta]\ J(1,0;1)$$

$$+ N\rho^{-1} J(1,1;0) \cos 2\theta$$

$$+2A\zeta\ [J(1,0;2) \cos^2\theta - \rho^{-1}\ J(1,1;1) \cos 2\theta]\},$$

$$\sigma^{I}_{yz} = \frac{\mu_1 b_x \sin 2\theta}{2a(\kappa_1 + 1)} \{N\ J(1,2;1) - 2A\ \zeta\ J(1,2;2)\}, \qquad (12)$$

$$\sigma^{I}_{zz} = \frac{\mu_1 b_x \cos\theta}{a(\kappa_1 + 1)} \{[B-A]\ J(1,1;1) - 2A\ \zeta\ J(1,1;2)\}.$$

The remaining stress components may be found in [6].

The Fields in Region II

Since the loop is located in the interface between the two solids, the elastic fields in region II may be expressed in terms of those in region I by interchanging the two materials. When this is carried out, the new composite parameters, denoted by a (*), can be written in the form

$$A^* = \frac{G_1(\kappa_2 + 1)}{G_2(\kappa_1 + 1)} B, \quad B^* = \frac{G_1(\kappa_2 + 1)}{G_2(\kappa_1 + 1)} A, \quad C^* = \frac{G_1}{G_2} C, \qquad (13)$$

with the contractions

$$M^* = \kappa_2 A^* + B^* - 2(\kappa_2 + 1)\ C^* = - \frac{\kappa_2 + 1}{\kappa_1 + 1}\ M,$$

$$N^* = A^* + B^* - (\kappa_2 + 1)\ C^* = \frac{\mu_1(\kappa_2 + 1)}{\mu_2(\kappa_1 + 1)}\ N. \qquad (14)$$

Let P be any point in region I at which the field quantities are defined through the argument list $L = (x,y,z;\ \mu_1,\ \kappa_1;\ A,\ B,\ C,\ M,\ N)$ at most, and P^* its image point at which the respective field quantities are defined through the argument list $L^* = (x,y,\ -z;\ \mu_2,\ \kappa_2\ ;\ A^*,\ B^*,\ C^*,\ M^*,\ N^*)$ at most, then the set of functions E are defined as even if

$$E^{I}(P) = E(L) \rightarrow E^{II}(P^*) = E(L^*) \tag{15}$$

while the set of functions F are defined as odd if

$$F^{I}(P) = F(L) \rightarrow F^{II}(P^*) = -F(L^*). \tag{16}$$

For the prismatic loop, the field quantities which are members of the even set E are u_i, σ_{ij} and σ_{zz} where i,j take on x or y. For the glide loop, the even set members are u_z and σ_{iz}, i = x or y. All other field quantities are members of the odd set of functions F.

INTERFACE DISPLACEMENTS AND TRACTIONS

The field quantities at the interface may be evaluated by first passing to the limit as z approaches zero in (7) and (8). But interesting results are only available if the general integral forms are known. Such results are made possible for the circular loop by exploiting the work of Salamon and Walter [3].

Since continuity of displacements and tractions hold everywhere along the interface except inside the loop where the Burgers discontinuity in displacement occurs, the regional notation (superscripts I,II) may be replaced by 0 or by 0+ and 0- (for the positive and negative side of the plane z = 0) for the discontinuous displacements.

The Prismatic Loop

The interface displacement components are

$$u_{\gamma}^{0} = \frac{-b_z\gamma}{2(\kappa_1+1)r} \; [\kappa_1 A-B] \; J^{0} \; (1,1;0),$$

$$u_{z}^{0+} = \frac{b_z}{2(\kappa_1+1)} \; [\kappa_1 A+B] \; H \; (1-\rho), \tag{17}$$

$$u_{z}^{0-} = \frac{-b_z}{2(\kappa_2+1)} \; [\kappa_2 A^*+B^*] \; H \; (1-\rho).$$

where the aforementioned limit is performed upon Eqs. (9).

With little effort one can show that $u_z^{0+} - u_z^{0-} = b_z$ inside the loop ($\rho<1$); outside the loop, $u_z^{0+} = u_z^{0-} = 0$. It is not obvious, but $J^o\,(1,1;0)$ contains a logarithmic singularity at $\rho = 1$ (see [3]).

The interface tractions follow from (10)

$$\sigma_{zz}^0 = \frac{-G_1\, b_z}{a(\kappa_1 + 1)} \; [A+B]\; J^o\,(1,0;1),$$

$$\sigma_{rz}^0 = \frac{-G_1\, b_z}{a(\kappa_1 + 1)} \; [B-A]\; \delta\,(\rho-1). \qquad (18)$$

The latter result is crystal clear, but the obscured singularity in the former is a Cauchy principal value which varies as $(\rho-1)^{-1}$.

The Glide Loop

The displacement components at the interface follow from (11)

$$u_x^{0+} = \frac{b_x}{4(\kappa_1+1)} \; \{[M + 4(\kappa_1+1)C]\; H(1-\rho) - M\,\rho^{-2}\; H(\rho-1)\; \cos 2\theta\}$$

$$u_x^{0-} = \frac{-b_x}{4(\kappa_2+1)} \; \{[M^* + 4(\kappa_2+1)C^*]\; H(1-\rho) - M^*\rho^{-2}\; H(\rho-1)\; \cos 2\theta\} \qquad (19)$$

$$u_y^0 = \frac{b_x M \sin 2\theta}{4(\kappa_1+1)\,\rho^2} \; H(\rho-1),$$

$$u_z^0 = \frac{b_x \cos\theta}{2(\kappa_1+1)} \; [\kappa_1\, A-B]\; J^o\,(1,1;0).$$

As for the prismatic loop, the Burgers discontinuity is easily verified and the same singular integral appears.

The interface tractions follow from (12)

$$\sigma^0_{xz} = \frac{-G_1 b_x}{a(\kappa_1+1)} \{[(\kappa_1+1)C + N\cos^2\theta]\; J^o(1,0;1) - N\rho^{-1} J^o(1,1;0)\cos 2\theta\}$$

$$\sigma^0_{yz} = \frac{G_1 b_x N \sin 2\theta}{2a(\kappa_1+1)}\; J^o(1,2;1), \qquad (20)$$

$$\sigma^0_{zz} = \frac{G_1 b_x \cos\theta}{a(\kappa_1+1)}\; [B-A]\;\delta(\rho-1)$$

The new integral $J^o(1,2;1)$ which appears also displays a singularity of the Cauchy principal value type which varies as $(\rho-1)^{-1}$.

It is clear that the interface glide loop does not generate interface normal tractions except at the line of the dislocation where the sharp delta singularity appears. In contrast, the prismatic loop follows the same pattern, but for shear tractions instead. Whereas the prismatic loop tends to shear the interface, the glide loop tends to wedge it open. In addition, while the delta singularity associated with the prismatic loop is uniformly distributed around the loop, that for the glide loop is anihilated when θ equals $\pm\,\pi/2$ radians. Thus the screw portions of the loop generate no normal traction; likewise for a straight screw dislocation. When θ equals 0 or π radians, the last equation in (20) and (18) tally with those for a straight edge dislocation given by Comninou [8].

THE ELASTIC ENERGY

The elastic energy of a wedge-shaped loop segment may be computed as the work required to create it. That is

$$\Delta w = -\Delta\theta \int_0^{b_i}\int_0^{a-r_o} \sigma_{iz}\, r\,dr\, db_i \qquad (21)$$

where $\Delta\theta$ is the wedge angle of the segment, r_o is the core radius of the dislocation and i takes on z for the prismatic loop and x for the glide loop. There are two ways to view σ_{iz}; it may either be the stress for a loop already in the interface as given by the first equation of (18) or (20) or it may be the stress for a loop located near the interface which then can be approached in the limit. Although either scheme leads to the same results at the interface, the latter is adopted here because it provides a unified formulation which permits a more general study of interesting interface phenomena.

With the assistance of ref. 3, a limiting process applied to Eq. (9) in ref. 4 yields for a prismatic loop segment in region I the energy

$$\Delta w = \frac{a\,G_1\,b_z^{\,2}\,\Delta\theta}{\pi(\kappa_1+1)}\ \Big\{\ \ell n\,\frac{8}{\varepsilon} - 2 + \frac{1}{2}\,(A+B-2)\ \Big[\ell n\,\frac{4}{(\xi^2+\varepsilon^2/4)^{1/2}} - 2\Big]$$

$$+ \frac{3}{2}\,(A-1)\,\frac{\xi^2}{\xi^2+\varepsilon^2/4}\ \Big\}\ + 0\ (\xi^2\ \ell n\ \xi). \qquad (22)$$

For the glide loop segment in region I, a similar process applied to Eq. (7c) in ref. 6 in conjunction with (21) gives

$$\Delta w = \frac{aG_1\,b_x^{\,2}\,\Delta\theta}{4\pi\,(\kappa_1+1)}\ \Big\{[\kappa_1+1 + (3-\kappa_1)\,\cos^2\theta]\,[\ell n\,\frac{8}{\varepsilon} - 2]$$

$$+ [(2N - 3 + \kappa_1)(\pi-2) - 3 + \kappa_1]\ \ \cos 2\theta$$

$$+ [(\kappa_1 + 1)(2C - 1) + (2N - 3 + \kappa_1)\,\cos^2\theta]\,[\ell n\,\frac{8}{(\varepsilon^2+4\xi^2)^{1/2}} - 2]$$

$$- 16\ (A-1)\,\frac{\xi^2}{\varepsilon^2+4\xi^2}\ \cos^2\theta\ + 0(\xi\ \ \ell n\ \xi)\ \Big\}. \qquad (23)$$

In the above $\varepsilon = r_o/a$ and ξ is the loop-to-interface distance divided by a. The terms containing ξ constitute the distance-dependent energy of interaction between the loop and the interface; the remaining terms are essentially the self-energy of the segment.

The generality of these expressions enable reductions to a variety of cases including the exact solutions for straight dislocations [5],[7]. However what raises the prospects for their utility is their relative simplicity; they are composed of elementary functions. Indeed they may be integrated over θ to obtain the total energy for a loop of any shape.

<u>The Prismatic Dislocation</u>

The energy of the prismatic dislocation lying in the interface is, due to the angular independence of the expressions, quite simple. By setting ξ to zero in (22), one obtains for the loop segment

$$\Delta w_o = \frac{a\, G_1\, b_z^{\,2}\, \Delta\theta}{2\pi(\kappa_1+1)}\, [A+B]\, [\ell n\, \frac{8}{\varepsilon} - 2] \qquad (24)$$

For the total energy of the circular loop, the integration is trivial: multiply (24) by 2π and delete $\Delta\theta$.

Further discussion of the energy and forces acting upon the prismatic interface loop is available in ref. 5.

The Glide Dislocation

The energy of the glide dislocation segment lying in the interface follows by setting ξ to zero in (23):

$$\Delta w_o = \frac{a\, G_1\, b_x^{\,2}\, \Delta\theta}{2\pi\, (\kappa_1+1)}\, \{[(\kappa_1+1)\, C + N \cos^2\theta]\, [\ell n\, \frac{8}{\varepsilon} - 2]$$

$$+ [N(\pi-2) - \pi(3-\kappa_1)/2] \cos 2\theta \}. \qquad (25)$$

For the circular loop, integration over θ from 0 to 2π yields

$$W_o = \frac{a\, G_1\, b_x^{\,2}\, D}{4(\kappa_1+1)}\, [\ell n\, \frac{8}{\varepsilon} - 2] \qquad (26)$$

Further discussion of the energy and forces acting upon the glide interface loop is available in [7].

CLOSURE

It is clear that the classical theory of elasticity applied in concert with modern developments in mathematics provides sufficient power to solve certain nontrivial problems. And in the present case the solutions, though intricate, are quite tractable. Indeed the simplicity of certain limiting cases is rather striking.

But the most important feature of this work is that it provides a unified formulation for the dislocation loop lying in or near to an interface. The three-dimensional character of the solution reveals all facets of loop behavior and, through elementary reductions, the behavior of straight dislocations as well. This is clearly illustrated by the glide loop which is free from the restrictions axial symmetry places upon the behavior of the prismatic loop.

Yet the beauty of these solutions is due to the good fortune of discovering nice integrable potential functions. Yes, good fortune, because so often one is bound by the devil of integration.

REFERENCES

1. Mura, T., "The continum Theory of Dislocations," in Advances in Materials Research, H. Herman, edt., vol. 3, Interscience, New York, 1968.

2. Salamon, N. J. and Dundurs, J., "Elastic Fields of a Dislocation Loop in a Two-Phase Material," J. of Elasticity, vol. 1, 153-164, 1971.

3. Salamon, N. J. and Walter, G G., "Limits of Lipschitz-Hankel Integrals," J. of the Institute for Mathematics and Its Applications, vol. 24, 237-254, 1979.

4. Dundurs, J. and Salamon, N. J., "Circular Prismatic Dislocation Loop in a Two-Phase Material," physica status solidi (b), vol. 50, 125-133, 1972.

5. Salamon, N. J. and Comninou, M., "The Circular Prismatic Dislocation Loop in an Interface," Philosophical Magazine A, vol. 29, 685-691, 1979.

6. Salamon, N. J. and Dundurs, J., "A Circular Glide Dislocation Loop in a Two-Phase Material," J. of Physics C, vol. 10, 497-507, 1977.

7. Salamon, N. J., "The Circular Glide Dislocation Loop Lying in an Interface," J. of the Mechanics and Physics of Solids, vol. 29, 1-11, 1981.

8. Comninou, M., "A Property of Interface Dislocations," Philosophical Magazine A, vol. 36, 1281-1284, 1977.

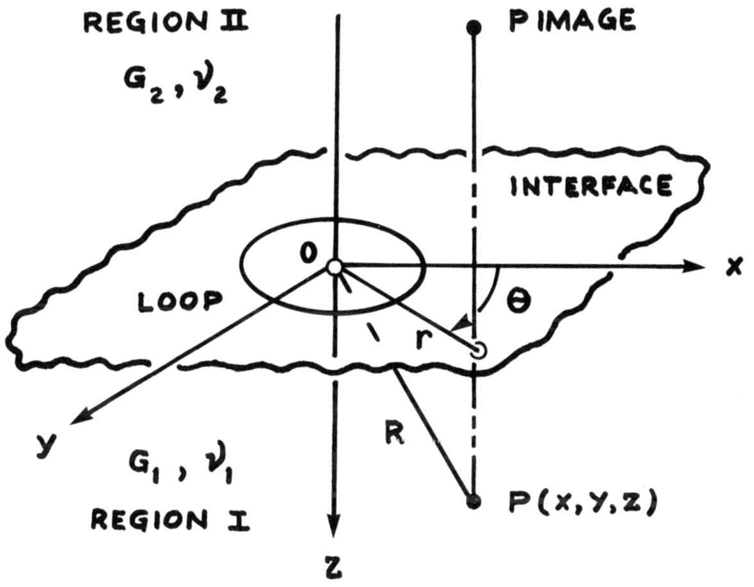

Figure 1. A Dislocation Loop in an Interface.

THE TORSIONAL STIFFNESS OF A RIGID CIRCULAR PUNCH BONDED TO A FINITELY DEFORMED INCOMPRESSIBLE RUBBERLIKE ELASTIC SOLID

A.P.S. Selvadurai

Department of Civil Engineering
Carleton University
Ottawa, Ontario
Canada.

ABSTRACT

In this paper we examine the elastostatic torsional response of a rigid circular punch which is in bonded contact with an initially stressed isotropic incompressible elastic halfspace. The torque-twist relationship for the bonded indentor can be evaluated in exact closed form. This result has potential application to the experimental determination of constitutive parameters of rubberlike elastic solids.

INTRODUCTION

The general problem of infinitesimal deformations superposed on finite deformations of an elastic body has been examined by various investigators and detailed accounts of these developments are given by Green and Zerna [1], Truesdell and Noll [2] and Eringen and Suhubi [3]. The theory of small deformations superposed on finite deformations of an isotropic elastic material developed by Green et al. [4] has received extensive application in the analysis of this class of problems.Special applications of this theory to punch indentation problems related to a halfspace region are given by Green and Zerna [1] and Beatty and Usmani [5]. These general results developed for indentation problems provide a basis for the experimental determination of material properties of rubberlike materials.

The present paper applies the general theory proposed by Green et al. [4] to investigate the superposed torsional response of a rigid circular punch which is bonded to an incompressible isotropic elastic halfspace which is subjected to an initial radial extension or compression. This note extends the classical Reissner-Sagoci problem [6,7] to include effects of an initial finite deformation. Formal analytical results are developed for the incremental displacement and stresses in the finitely deformed halfspace. The relationship between the torque and the rotation for the bonded circular punch is obtained in exact closed form. Numerical results are presented for the torsional stiffness of the rigid circular punch which is in bonded contact with an elastic medium with a strain energy function of the Mooney-Rivlin type.

NOTATION AND FORMULAE

The relevant results of the theory of small deformations superposed on large, developed by Green et al. [4], are summarized here for completeness. The points in the isotropic incompressible elastic halfspace are defined by a general curvilinear coordinate system $\theta_i (\theta_1 = r\ , \theta_2 = \theta,\ \theta_3 = z)$ which moves with the body as it deforms. The covariant and contravariant metric tensors corresponding to the undeformed and deformed states of the body are defined by g_{ij}, G_{ij} and g^{ij}, G^{ij} respectively. For an isotropic incompressible elastic material the contravariant stress tensor τ^{ij}, measured per unit area of the deformed body and referred to θ_i coordinates of the deformed body is given by

$$\tau^{ij} = \Phi g^{ij} + \Psi B^{ij} + pG^{ij}, \tag{1}$$

where

$$B^{ij} = I_1 g^{ij} - g^{ir} g^{js} G_{rs}$$
$$\Phi = 2\frac{\partial W}{\partial I_1}\ ;\quad \Psi = 2\frac{\partial W}{\partial I_2} \tag{2}$$

and $W = W(I_1, I_2)$ is the strain energy function per unit volume of the material. The invariants $I_n (n=1,2)$ are given by

$$I_1 = g^{rs}G_{rs} \; ; \quad I_2 = g_{rs}G^{rs} \; ; \quad I_3 = |G_{ij}|/|g_{ij}| = 1 \tag{3}$$

and p is a scalar pressure which is to be determined by satisfying the traction boundary conditions of the particular problem.

In connection with the solution of the title problem we consider a halfspace region which is subjected to a finite radial stretch μ, with zero surface traction on the bounding plane $z=0$. For this particular finite deformation problem we have

$$\tau^{11} = r^2\tau^{22} = \{\mu^2 - \frac{1}{\mu^4}\}(\Phi + \mu^2\Psi) \tag{4}$$

We now subject the finitely deformed halfspace to a rotationally symmetric deformation which is characterized by the incremental displacement field

$$u_r = 0 \; ; \quad u_\theta = v(r,z) \; ; \quad u_z = 0. \tag{5}$$

The non-zero components of the stress tensor $\tau'^{ij}(r,z)$ governing the superposed deformation can be expressed as

$$\tau'^{11} = r^2\tau'^{22} = \tau'^{33} = p' \tag{6}$$

$$r\tau'^{12} = \alpha_6 \{ \frac{\partial v}{\partial r} - \frac{v}{r}\} \; ; \quad r\tau'^{23} = \alpha_7 \frac{\partial v}{\partial z}$$

where

$$\alpha_6 = \frac{1}{\mu^4}(\Phi + 2\mu^2\Psi) \; ; \quad \alpha_7 = \frac{1}{\mu^4}(\Phi + \mu^2\Psi) \tag{7}$$

In the particular instance where the rotationally symmetric small deformation is superposed on an initial homogeneous finite radial deformation and in the absence of body forces, the equations of equilibrium for the superposed stress field τ'^{ij} reduce to

$$\frac{\partial p'}{\partial r} = \frac{\partial p'}{\partial z} = 0 \,, \tag{8a}$$

$$\frac{\partial \tau'^{12}}{\partial r} + \frac{\partial \tau'^{23}}{\partial z} \frac{3\tau'^{12}}{r} = \frac{\tau^{11}}{r} \left(\frac{\partial^2 v}{\partial r^2} + \frac{1}{r}\frac{\partial v}{\partial r} - \frac{v}{r^2}\right) \tag{8b}$$

The traction boundary conditions corresponding to the superposed stress field are given by

$$\tau'^{ij} n_i + \tau^{ij} n'_i = p'^j \tag{9}$$

where n_i is the covariant component of the unit normal referred to a surface in the finitely deformed body; n'_i and p'^j are, respectively, the covariant component of the unit normal and the contravariant component of the surface force vector referred to a bounding surface in the finitely deformed state.

ROTATIONALLY SYMMETRIC LOADING OF THE HALFSPACE

The method of solution of the title problem employed here is an adaptation of the Hankel transform technique outlined by Sneddon [7,8] for the solution of rotationally symmetric problems in classical elasticity. The first order Hankel transform of the function $s(r)$ is defined as

$$\bar{s}^1(\xi) = H_1 \{s(r);\xi\} \tag{10}$$

where the operator H_1 is defined by the equation

$$H_1\{s(r);\xi\} = \int_0^\infty rs(r)\ J_1(\xi r/a)\ d\xi \tag{11}$$

and a is a typical length parameter in the problem. The appropriate Hankel inversion theorem is

$$s(r) = \frac{1}{a^2} \int_0^\infty \xi\, \bar{s}^1(\xi)\ J_1(\xi r/a)\ d\xi \tag{12}$$

Operating on (8b) with the first order Hankel transform we obtain the following ordinary differential equation for the transformed value of the displacement $v(r,z)$;

$$\left\{\frac{d^2}{dz^2} - \frac{\xi^2}{a^2k^2}\right\} \bar{v}^1(\xi,z) = 0 \tag{13}$$

where

$$k = \left(\frac{\alpha_7}{\alpha_6 + \tau^{11}} \right)^{\frac{1}{2}} \tag{14}$$

The solution of (13) appropriate for the halfspace region is

$$\bar{v}^1(\xi,z) = A(\xi)e^{-\xi z/ak} \tag{15}$$

where $A(\xi)$ is an unknown function. The general solutions for the incremental displacement and stress fields can be obtained by making use of (6), (12) and (15). From (8a) we note that the incremental hydrostatic stress is a constant which should be determined by satisfying the traction boundary conditions of the particular problem.

THE REISSNER-SAGOCI PROBLEM

We consider the problem in which the finitely deformed incompressible elastic halfspace is subjected to a rotationally symmetric incremental deformation which is induced by the torsion of a bonded rigid circular punch of radius a (Fig. 1). In general, the mixed boundary conditions associated with this class of problem are given by

$$\begin{aligned} v(r,0) &= f(r) \quad ; \quad 0 \le \frac{r}{a} \le 1 \\ \tau'^{23}(r,0) &= 0 \quad ; \quad \frac{r}{a} > 1 \end{aligned} \tag{16}$$

where $f(r)$ is a prescribed displacement field. Using the results given in the previous section it can be shown that the boundary conditions (16) reduce to the following pair of dual integral equations

$$\begin{aligned} \int_0^\infty \frac{F(\xi)}{\xi} J_1(\xi\rho)\, d\xi &= g(\rho) \quad ; \quad 0 \le \rho \le 1 \\ \int_0^\infty F(\xi) J_1(\xi\rho)\, d\xi &= 0 \quad ; \quad \rho > 1 \end{aligned} \tag{17}$$

$$F(\xi) = \frac{\xi^2 A(\xi)}{a^3} \quad ; \quad \rho = \frac{r}{a} \quad ; \quad g(r) = \frac{f(r)}{a} \tag{18}$$

The method of solution of the dual system (17) is given by Sneddon [8] and Copson [9] and the details will not be pursued here. The general solution is given by

$$F(\xi) = \left(\frac{2}{\pi}\right)^{\frac{1}{2}} \xi\Big(\xi^{\frac{1}{2}} J_{\frac{1}{2}}(\xi) \int_0^1 \frac{t^2 g(t)\,dt}{\sqrt{1-t^2}} + \int_0^1 \frac{x^2 dx}{\sqrt{1-x^2}} \left\{\int_0^1 g(xt)\ (\xi t)^{3/2} J_{3/2}(\xi t)\ dt\right\}\Big) \tag{19}$$

In the particular case of the rigid circular punch we have $g(\rho) = \Omega\rho$, (where Ω is a constant) and the solution (19) yields

$$A(\xi) = \frac{4\Omega a^3}{\xi\pi}\left\{\frac{\sin\xi}{\xi} - \cos\xi\right\} \tag{20}$$

The components of the incremental displacement and stress fields are given by

$$v(r,z) = \frac{4\Omega a}{\pi}\int_0^\infty \left\{\frac{\sin\xi - \xi\cos\xi}{\xi^2}\right\} e^{-\xi z/ak} J_1(\xi r/a)\, d\xi \tag{21}$$

and

$$r\tau'^{12}(r,z) = \frac{4\Omega\alpha_6}{\pi}\int_0^\infty e^{-\xi z/ak}\left\{\frac{\sin\xi - \xi\cos\xi}{\xi^2}\right\}\left\{\xi J_0(\xi r/a) - \frac{2a}{r} J_1(\xi r/a)\right\}d\xi$$
$$r\tau'^{23}(r,z) = -\frac{4\Omega\alpha_7}{\pi k}\int_0^\infty e^{-\xi z/ak}\left\{\frac{\sin\xi - \xi\cos\xi}{\xi}\right\} J_1(\xi r/a)\, d\xi \tag{22}$$

respectively.

The relationship between the rotation (Ω) and the torque (T) may be derived from the relationship

$$T = \int_0^{2\pi}\int_0^a \{r\tau'^{23}(r,0)\}\, r^2\, dr d\theta \tag{23}$$

Evaluating (22) for $z=0$ we obtain the following expression for the shear stress $\tau'^{23}(r,0)$ in the region $r \le a$;

$$r\tau'^{23}(r,0) = -\frac{4\Omega\alpha_7}{\pi k}\left\{\frac{r}{\sqrt{a^2-r^2}}\right\} \tag{24}$$

With the understanding that the rotation Ω occurs in the direction of the applied torque T we obtain from (23) and (24)

$$T = \frac{16\Omega a^3(\Phi+\mu^2\Psi)}{3\mu^3}\sqrt{\frac{\Psi+\mu^4\Phi}{\Phi+\mu^2\Psi}} \tag{25}$$

It is of interest to note that the superposed rotation Ω is finite for all $\mu(>0)$ provided $\Phi > 0$ and $\Psi > 0$. In this sense, a rotationally symmetric form of the surface instability will not occur at the surface of the half-space, with a traction free surface, which is subjected to all round finite compression. A rigorous proof of this assertion valid for incompressible (or compressible) elastic materials with a general strain energy function requires further investigation.

In the particular case when the incompressible elastic material has a strain energy function of the Mooney-Rivlin type

$$W = C_1(I_1 - 3) + C_2(I_2 - 3) \tag{26}$$

the result (25) reduces to

$$T = \frac{32\Omega(C_1+\mu^2C_2)a^3}{3\mu^3}\sqrt{\frac{\Gamma + \mu^4}{1 + \mu^2\Gamma}} \tag{27}$$

where $\Gamma = C_2/C_1$. Taking the limit of (27) as $\mu \to 1$, and noting that the linear elastic shear modulus $G = 2(C_1 + C_2)$ we obtain

$$T = \frac{16G\Omega a^3}{3} \tag{28}$$

The $T - \Omega$ relationship (28) is in agreement with the result obtained by Reissner and Sagoci [6] and Sneddon [7] for the equivalent problem in classical elasticity. The effect of the initial finite deformation of the torsional rigidity of the bonded rigid circular punch can be examined by comparing (27) with (28). By defining the torsional stiffness as $S(\mu)$ $(= T(\mu)/G\Omega a^3)$ we have

$$\frac{S(\mu)}{S(1)} = \frac{1}{\mu^3} \left\{ \frac{1+\mu^2\Gamma}{1+\Gamma} \right\} \sqrt{\frac{\Gamma+\mu^4}{1+\mu^2\Gamma}} \qquad (29)$$

The effect of μ and Γ on the relative torsional stiffness (as defined by (29)) is illustrated in Fig. 2. As has been observed by Green et al. [4] and others [5, 10, 11] surface instability of a radially compressed incompressible elastic halfspace (with a traction free surface) occurs when μ approaches a value near to 2/3. Accordingly, the minimum value of the radial stretch used in the numerical illustration is taken to be 0.7. The result $\Gamma = 0$ corresponds to a Neo-Hookean material.

REFERENCES

[1] GREEN, A.E. and ZERNA, W., Theoretical Elasticity, Clarendon Press, Oxford (1968).

[2] TRUESDELL, C. and NOLL, W., The Non-linear Field Theories of Mechanics, Hanbuch der Physik(Ed. S. Flugge) Vol. III/3, Springer-Verlag, Berlin (1965).

[3] ERINGEN, A.C. and SUHUBI, E., Elastodynamics, Vol. I, Academic Press, New York (1974).

[4] GREEN, A.E., RIVLIN, R.S. and SHIELD, R.T., Proc. Roy. Soc. Ser. A, 211, 211 (1952).

[5] BEATTY, M.F. and USMANI, S.A., Quart. J. Mech. Appl. Math., 20, 47 (1975).

[6] REISSNER, E. and SAGOCI, H.F., J. Appl. Phys., 15, 652 (1944).

[7] SNEDDON, I.N., J. Appl. Phys., 15, 130 (1944).

[8] SNEDDON, I.N., Fourier Transforms, McGraw-Hill, New York (1951).

[9] COPSON, E.T., Proc. Glasgow Math. Assoc., 5, 21 (1961).

[10] WOO, T.C. and SHIELD, R.T., Arch. Rat. Mech. Anal., 10, 196 (1961).

[11] SELVADURAI, A.P.S., Int. J. Solids Structures, 13, 357 (1977).

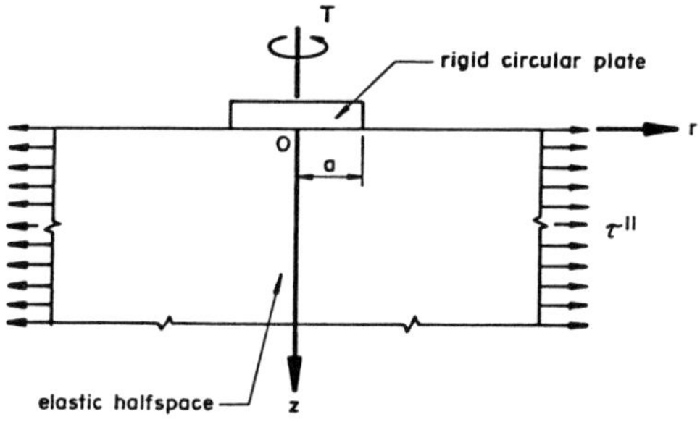

Fig. 1 Torsional loading of the finitely stretched halfspace.

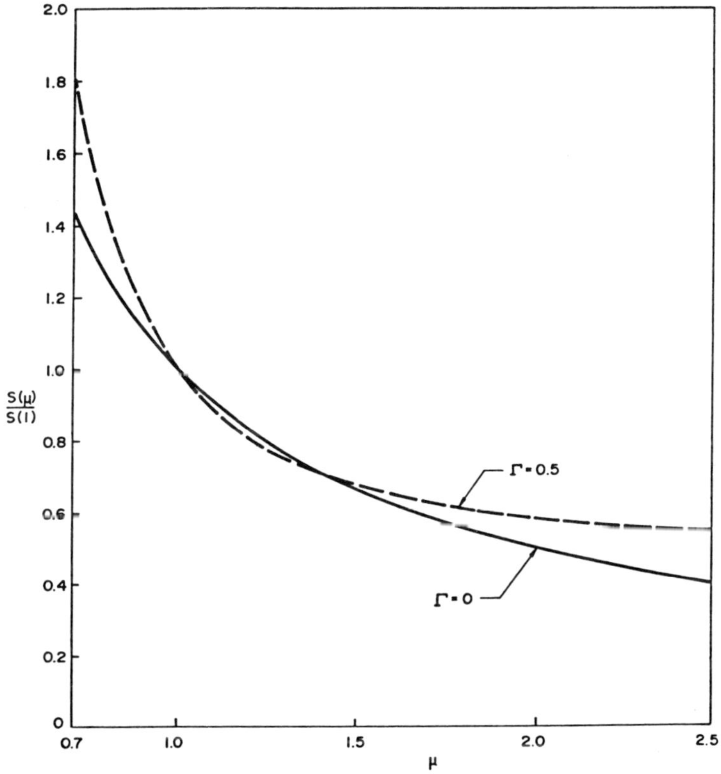

Fig. 2 The variation in the torsional stiffness of the rigid circular punch due to the initial finite deformation.

A NOTE ON THE AXISYMMETRIC OSCILLATIONS OF A RIGID DISC INCLUSION EMBEDDED IN AN ISOTROPIC ELASTIC MEDIUM

A.P.S. Selvadurai and M. Rahman

Department of Civil Engineering
Carleton University
Ottawa, Ontario
Canada.

ABSTRACT

This paper examines the problem of the low frequency oscillation of a rigid circular disc inclusion which is embedded in bonded contact with an isotropic elastic medium of infinite extent. The mathematical analysis of the problem is developed in the context of a mixed boundary value problem associated with an equivalent halfspace region. The system of dual integral equations which are obtained for the description of the mixed boundary value problem are in turn reduced to a single Fredholm integral equation of the second kind. The low frequency dynamic response of the embedded inclusion is derived by considering a series expansion solution of the integral equation expressed in terms of a frequency parameter. This frequency parameter is related to the shear wave velocity of the elastic medium.

INTRODUCTION

The rectilinear and torsional oscillations of a rigid sphere embedded in an unbounded elastic medium was examined by Chadwick and Trowbridge [1]. Kanwal [2] and Williams [3] have examined the application of matched asymptotic expansion techniques to the analysis of vibrations of an inclusion embedded in an unbounded elastic solid. Recently, Datta and Kanwal [4] have used singularity methods to examine the dynamic behaviour of an embedded rigid spheroidal inclusion. This paper shows that the axisymmetric dynamic behaviour of a rigid disc inclusion embedded in an elastic solid can be examined, quite conveniently, by employing techniques developed for the analysis of a rigid foundation resting on a halfspace [5]. It may be noted that the mathematical boundary value problem is formally similar to that which arises when a compressional wave is normally incident on a rigid disc embedded in an infinite solid [6]. The method of analysis outlined here is, however, much simpler and relatively straightforward. The approximate solutions given by Kanwal [2] and Williams [3] are recovered as a lower order approximation of the solution presented.

ANALYSIS

A rigid circular disc inclusion of radius a is embedded in bonded contact with an elastic medium of infinite extent (isentropic Lame constants λ, μ ; equilibrium density ρ_0). The disc is subjected to a periodic load which enforces rectilinear oscillations, about the z-axis, characterized by $\delta e^{i\omega t}$ in the region $r \le a$; $z = 0\pm$. (The notation (+) and (-) refer to the faces of the disc inclusion in contact with the elastic medium occupying the regions $z \ge 0$ and $z \le 0$ respectively.) Considering a Hankel transform development of the governing field equations it can be shown that the axisymmetric displacement field appropriate for the halfspace region $z \ge 0$ is characterized by the displacement components

$$u_r(r,z,t) = \frac{e^{i\omega t}}{a^2}\int_0^\infty \xi\{\frac{A(\xi)}{\alpha}\, e^{-\alpha z/a} + \frac{\beta C(\xi)}{\xi}\, e^{-\beta z/a}\}\, J_1(\xi r/a)\,d\xi$$

$$u_z(r,z,t) = \frac{e^{i\omega t}}{a^2}\int_0^\infty \xi\{A(\xi)\, e^{-\alpha z/a} + C(\xi)\, e^{-\beta z/a}\}\, J_0(\xi r/a)\,d\xi \qquad (1)$$

where

$$\alpha = \begin{cases} [\xi^2 - h^2]^{1/2} & \xi > h \\ -i[h^2 - \xi^2]^{1/2} & ; \quad 0 \leq \xi \leq h \end{cases} \tag{2a}$$

$$\beta = \begin{cases} [\xi^2 - k^2]^{1/2} & ; \quad \xi > k \\ -i[k^2 - \xi^2]^{1/2} & ; \quad 0 \leq \xi \leq k \end{cases} \tag{2b}$$

and

$$h^2 = \frac{\rho_0 \omega^2 a^2}{(\lambda+2\mu)} \quad ; \quad k^2 = \frac{\rho_0 \omega^2 a^2}{\mu} \tag{3}$$

In (1), the arbitrary functions $A(\xi)$ and $C(\xi)$ have to be determined by satisfying the displacement boundary conditions associated with the embedded inclusion problem. Considering the axisymmetric deformation of the infinite space region induced by the displacement of the bonded inclusion we note that the problem is antisymmetric in the normal stress σ_{zz} and in the radial displacement u_r about the plane z=0 and in the region $r \geq a$, for all $t \geq 0$. Also, since the rigid disc inclusion is bonded to the elastic medium $u_r(r,0) = 0$ in the region $r \leq a$. Thus we may restrict our attention to the analysis of a single halfspace region ($z \geq 0$) in which the plane z=0 is subjected to the mixed boundary conditions

$$\begin{aligned} u_r(r,0^+,t) &= 0 \quad ; \quad r \geq 0 \\ u_z(r,0^+,t) &= \delta e^{i\omega t} \quad ; \quad 0 \leq r \leq a \\ \sigma_{zz}(r,0^+,t) &= 0 \quad ; \quad a < r < \infty \end{aligned} \tag{4}$$

Further, by introducing a function $B(\xi)$ such that

$$[A(\xi);\ C(\xi)] = \frac{4(1-\nu)B(\xi)}{(3-4\nu)k^2}\ [-\frac{\alpha}{\xi}\ ;\ \frac{\xi}{\beta}] \tag{5}$$

it can be shown that the boundary conditions (4) are equivalent to a pair of dual integral equations for the unknown function $B(\xi)$. Also, by using a representation for $B(\xi)$ which takes the form

$$B(\xi) = \frac{2\xi\delta a^2}{\pi} \int_0^1 \Theta(\zeta)\ \cos(\xi\zeta)d\zeta \tag{6}$$

these dual integral equations can be reduced to a single Fredholm integral equation of the second kind

$$\Theta(\zeta) + \frac{1}{\pi} \int_0^1 K(\zeta,\tau)\ \ \Theta(\tau)\ d\tau = 1 \tag{7}$$

where

$$K(\zeta,\tau) = 2 \int_0^\infty H(\xi)\ \cos(\xi\zeta)\ \cos(\xi\tau)\ d\xi \tag{8a}$$

and

$$H(\xi) = \frac{4(1-\nu)\xi}{(3-4\nu)k^2}\ \{\frac{\xi^2}{\beta} - \alpha\}\ -1 \tag{8b}$$

The periodic force $P(t)$ $(=2\pi \int_0^a r\ \sigma_{zz}\ dr)$ required to maintain the oscillation $\delta e^{i\omega t}$ is given by

$$P(t) = \frac{32\mu(1-\nu)\delta a e^{i\omega t}}{(3-4\nu)} \int_0^1 \Theta(\zeta)d\zeta \tag{9}$$

Alternatively by denoting the integral in (9) by $\int_0^1 \Theta(\zeta)d\zeta = p_1 + ip_2$ we obtain

$$P(t) = \frac{32\mu(1-\nu)\delta a}{(3-4\nu)}\ \ [p_1^2 + p_2^2]^{1/2}\ \ e^{i(\omega t+\phi)} \tag{10}$$

where $\tan\phi = p_1/p_2$. It may be verified that as $t\to 0$, (10) yields the equivalent statical result [7,8]. The formal analysis can now be extended to include the effect of self weight of the embedded inclusion. We consider the problem where a rigid disc inclusion of mass M is subjected to a periodic displacement $\delta e^{i\omega t}$. The relationship between δ and the maximum amplitude Q^* of the periodic force $Q(t)$ required to maintain this steady oscillation is given by

$$\frac{\delta\mu a}{|Q^*|} = \frac{(3-4\nu)}{32(1-\nu)\ [\{p_1 - \frac{k^2\Delta(3-4\nu)}{8(1-\nu)^2}\}^2 + p_2^2]^{1/2}} \tag{11}$$

where $\Delta = M(1-\nu)/4\rho_0 a^3$ is the mass ratio.

Approximate analysis of the integral equation

The approximate solution of (7) is obtained by making use of the series expansion technique similar to that adopted by Robertson [5] for the analysis of the forced vibration of a rigid disc on an elastic halfspace. The details of this analysis will not be pursued here. It may be noted that $H(\xi)$ has no poles but only branch cuts at $\pm\gamma$ and ± 1 where $\gamma^2 = (1-2\nu)/(2-2\nu)$. It can be shown that, provided $0 \le \nu \le 1/2$, the kernel function $K(t,\tau)$ can be expressed in the following forms: for $t>\tau$

$$K(t,\tau) = \frac{8ik(1-\nu)}{(3-4\nu)} \left[\int_0^{\gamma} \frac{x\{x^2+(\gamma^2-x^2)^{1/2}(1-x^2)^{1/2}\}e^{i(ktx)}}{(1-x^2)^{1/2}} \cos(k\tau x)\,dx + \int_{\gamma}^{2} \frac{x^3\ e^{i(ktx)}}{(1-x^2)^{1/2}} \cos(k\tau x)\,dx \right] \tag{12}$$

If $t<\tau$, the equivalent expression for $K(t,\tau)$ is obtained by interchanging t and τ in (12). It can be shown that $\Theta(\zeta)$ admits a series expansion

$$\Theta(\zeta) = 1 + \sum_{n=1}^{\infty} k^n\ \Theta_n(\zeta) \tag{13}$$

where

$$\Theta_n(\zeta) = -\frac{1}{\pi} \sum_{\gamma=0}^{n-1} \frac{i^{n-\gamma}\ I_{n-\gamma}}{(n-\gamma-1)!} \int_0^1 K_{n-\gamma}(\zeta,\tau)\ \Theta_\gamma(\tau)\,d\tau \tag{14}$$

$$2K_n(\zeta,\tau) = (\zeta+\tau)^{n-1} + |\zeta-\tau|^{n-1} \tag{15}$$

$$I_n = \frac{8(1-\nu)}{(3-4\nu)} \left(\int_0^{\gamma} \frac{x^n\{x^2+(\gamma^2-x^2)^{1/2}(1-x^2)^{1/2}\}dx}{(1-x^2)^{1/2}} + \int_{\gamma}^{1} \frac{x^{n+2}\,dx}{(1-x^2)^{1/2}} \right) \tag{16}$$

The final series expansions for p_1 and p_2 take the forms

$$\begin{aligned} p_1 = 1 &+ k^2\left\{\frac{2I_2}{3\pi} - \frac{I_1^2}{\pi^2}\right\} + k^4\left\{-\frac{2I_4}{15\pi} + \frac{2I_1I_3}{3\pi^2} + \frac{7I_2^2}{15\pi^2} - \frac{2I_1^2I_2}{\pi^3} + \frac{I_1^4}{\pi^4}\right\} + \\ &+ k^6\left\{\frac{4I_6}{315\pi} - \frac{4I_1I_5}{45\pi^2} - \frac{64I_2I_4}{315\pi^2} + \frac{2I_1^2I_4}{5\pi^3} - \frac{2I_3^2}{15\pi^2} + \frac{64I_1I_2I_3}{45\pi^3}\right. \\ &\left. - \frac{4I_1^3I_3}{3\pi^4} + \frac{233I_2^2}{315\pi^3} - \frac{123I_2^2I_1^2}{45\pi^4} + \frac{7I_1^4I_2}{2\pi^5} - \frac{I_1^6}{\pi^6}\right\} \end{aligned} \tag{17a}$$

$$\begin{aligned} p_2 = \frac{kI_1}{\pi} &+ k^3\left\{-\frac{I_3}{3\pi} + \frac{4I_1I_2}{3\pi^2} - \frac{I_1^3}{\pi^3}\right\} \\ &+ k^5\left\{\frac{2I_5}{45\pi} - \frac{4I_1I_4}{15\pi^2} - \frac{22I_2I_3}{45\pi^2} + \frac{I_1^2I_3}{\pi^3} + \frac{62I_1I_2^2}{45\pi^3} - \frac{8I_1^3I_2}{3\pi^4} + \frac{I_1^5}{\pi^5}\right. \end{aligned} \tag{17b}$$

NUMERICAL RESULTS

An approximate analysis of the dynamic displacements induced in an elastic medium of infinite extent, due to the periodic excitation of a massless rigid axisymmetric body of arbitrary shape was considered by Kanwal [2]. The approximate nature of the solution stems directly from the use of matched asymptotic expansion techniques for the analysis of the problem. The problem examined by Kanwal has been extended and corrected by Williams [3]. The result of primary interest to the vibration of the embedded disc inclusion are summarized here. The periodic force P(t) required to maintain the steady oscillation $\delta e^{i\omega t}$ is given by

$$P(t) = P_0\{1 + \frac{i\,k\,P_0(2+\gamma^3)}{12\pi\mu\delta a}\}\,e^{i\omega t} + O(k^2) \tag{18}$$

where $P_0 = 32\mu(1-\nu)\delta a/(3-4\nu)$, is the static load-displacement relationship. This result is valid to order k^2. In the notation employed earlier we note that the Kanwal-Williams solution is equivalent to

$$p_1 = 1 \quad ; \quad p_2 = \frac{8(1-\nu)(2+\gamma^3)k}{3\pi(3-4\nu)} \tag{19}$$

We note that the integrals I_n defined by (16) can be reduced to the form

$$I_n = \frac{8(1-\nu)}{(3-4\nu)}\ \{(1-\gamma^{n+2})\ L_{n+2} + \gamma^{n+2}\ L_n\} \tag{20}$$

where

$$L_n = \frac{(n-1)}{n}\int_0^{\pi/2} \sin^{n-2}\Omega\ d\Omega \ ;\ n>0 \tag{21}$$

with $L_0 = \pi/2$ and $L_1 = 1$. Taking the terms in p_n defined by (17) valid to order k^2 it is evident that the Kanwal-Williams result is contained in (17). A comparison of results for $p_2(k)$ as determined from the series method and the asymptotic expansion technique is presented in Table 1. There is reasonable correlation between the two sets of results for $k \leq 0.2$.

REFERENCES

[1] CHADWICK, P. and TROWBRIDGE, E.A., Proc. Camb. Phil. Soc., 63, 1207 (1965).

[2] KANWAL, R.P., Quart. Appl. Math., 44, 275 (1965).

[3] WILLIAMS, W.E., Q. Jl. Mech. Appl. Math., 19, 413 (1966).

[4] DATTA, S. and KANWAL, R.P., Quart. Appl. Math., 58, 86 (1979).

[5] ROBERTSON, I.A., Proc. Camb. Phil. Soc., 62, 547 (1966).

[6] MAL, A.K., ANG, D.D. and KNOPOFF, L., Proc. Camb. Phil. Soc., 64, 237 (1968).

[7] COLLINS, W.D., Proc. Roy. Soc. Ser. A, 203, 359 (1962).

[8] SELVADURAI, A.P.S. , Geotechnique, 26, 603 (1976).

Table 1 A comparison of results for $p_2(k)$ as derived from the series method and the matched asymptotic expansion (MAE) method.

	$p_2(k)$					
	$\nu = 0$		$\nu = 0.3$		$\nu = 0.5$	
k	Ser.	MAE	Ser.	MAE	Ser.	MAE
0.05	0.033	0.033	0.035	0.035	0.042	0.042
0.10	0.066	0.067	0.070	0.071	0.083	0.085
0.15	0.099	0.100	0.105	0.105	0.125	0.127
0.20	0.132	0.133	0.140	0.140	0.166	0.170
0.25	0.165	0.166	0.176	0.175	0.208	0.212
0.30	0.198	0.199	0.211	0.210	0.249	0.254
0.35	0.232	0.233	0.246	0.245	0.291	0.297
0.40	0.265	0.266	0.282	0.280	0.333	0.340
0.45	0.299	0.299	0.318	0.316	0.375	0.382
0.50	0.333	0.333	0.354	0.351	0.417	0.424

PLASTICITY

FINITE DEFORMATION EFFECTS IN PLASTICITY THEORY

E.H. Lee
Department of Mechanical Engineering,
Aeronautical Engineering & Mechanics
Rensselaer Polytechnic Institute
Troy, New York 12181

Abstract. Plastic strain is defined to be the residual strain at zero stress after a body has been stretched beyond the elastic limit. At zero stress the elastic strain is zero and thus the process of de-stressing uncouples the plastic strain from the elastic strain. In practice measurements of increments of elastic and plastic strain are commonly made by imposing and then removing a stress increment from a specimen already stressed to yield and recording the reversible and residual strain increments. Precise kinematics valid for finite deformation is needed to elucidate the difference between these approaches. Such a rigorous finite deformation theory also permits the influence of the choice of stress-rate definition to be understood. A complete formulation is described using the Jaumann derivative in the case of isotropic hardening.

For kinematic hardening it is shown that a modified Jaumann derivative is needed to avoid large errors in the evaluation of stress distributions. Current practice in finite-element codes considered valid for finite deformation lead to an oscillatory stress for imposed monotonically increasing simple shearing. This arises from the use of the conventional Jaumann derivative in the evolution equation for the back-stress or off-set of the center of the yield surface. A modification of the Jaumann derivative to incorporate the influence of anisotropy eliminates the spurious oscillatory stress behavior.

1. Introduction

Ductile metals can be deformed to extremely large strains without fracturing. Wire, for example, can be extruded from a die which reduces the billet cross-section by a factor of 100. However, the initial development of the theory of plasticity was based on infinitesimal displacement analysis as an outgrowth of the classical theory of elasticity. The physical characteristics were deduced from tensile tests of specimens with cylindrical test sections to generate uniform simple tensile stress σ. Strains were limited to the order of 1% (before the formation of a neck) so that uniform elongation along the gauge length occurred and the longitudinal strain ε would be measured in terms of the increase of the gauge length.

Figure 1 shows a typical stress-strain curve. Plasticity commences at J corresponding to the initial yield stress Y_0. After loading to A, the specimen is unloaded to B following the elastic unloading line AB. The residual strain OB at zero stress is the plastic strain ε^p at B. Since only elastic strain recovery occurs on the unloading path AB, ε^p is also the plastic strain embodied in the material in the state A. The elastic strain at A is BH or ε^e. The elastic strain is reduced to zero in the unstressed state B.

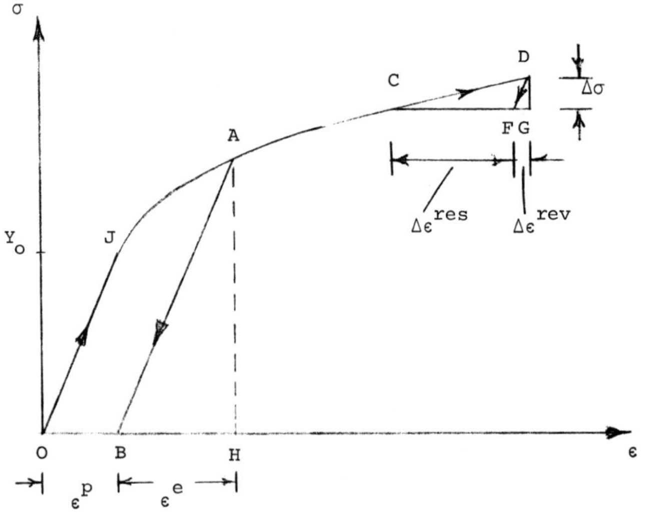

Figure 1 Tension Test

For the purpose of defining stress and strain or increments thereof in such tests, at strains of the order 1% the geometry of the specimen can be considered effectively unchanged by the deformation so that nominal stress (tensile force per unit initial cross-sectional area of the specimen) can be used as the stress variable and nominal strain (extension per unit initial gauge length) as the strain variable. Because of the small strains considered, the difference between these and true stress or Cauchy stress (force per unit current cross-sectional area) and natural strain increment (increment of extension per unit current gauge length) can be considered negligible. Thus, to obtain the total strain at A, the elastic and plastic components are additive just as are the displacements, and the total strain at A is given by

$$\varepsilon = \varepsilon^e + \varepsilon^p \tag{1.1}$$

Plasticity is an incremental or flow type phenomenon in that the constitutive relation involves rate of plastic strain which must be integrated to determine the plastic strain. Thus, in order to build up a constitutive relation for elastic-plastic response to stress, it is necessary to use the time derivative of (1)

$$\dot{\varepsilon} = \dot{\varepsilon}^{e} + \dot{\varepsilon}^{p} \tag{1.2}$$

Combination of the plastic law for $\dot{\varepsilon}^p$ and a derivative of the elastic law for ε^e then permits a constitutive relation for the rate of total strain $\dot{\varepsilon}$ in terms of stress and stress-rate to be deduced.

When somewhat larger but still moderate strains are investigated in the tension test, necking occurs with subsequent plastic deformation being localized in the region of the neck. In order to deduce material characteristics one must then abandon the concept that the specimen configuration is effectively unchanged and analyze instead the response of the material at the bottom of the neck where the major plastic flow is taking place. This can be done to a satisfactory degree of approximation by adopting a strength-of-materials approach, i.e., by considering the approximately cylindrical element at the bottom of the neck to be in simple tension subjected to the true stress σ_t, the tensile force divided by the cross-sectional area at the bottom of the neck. Since at this stage plastic strains dominate elastic strains, the latter can be neglected and the zero volumetric strain associated with plastic flow permits the strain to be expressed in terms of the change in cross-sectional area at the bottom of the neck from A_0 in the undeformed cylindrical specimen to A at the bottom of the neck.

Thus $A_0 \ell_0^e = A\ell^e$, where ℓ_0^e and ℓ^e denote respectively the initial and current lengths of the approximately cylindrical element at the bottom of the neck which contributes the major plastic flow after necking. The local nominal longitudinal strain at the bottom of the neck then becomes

$$\varepsilon_n = \frac{A_o - A}{A} \tag{1.3}$$

and the true or natural strain

$$\varepsilon_t = \ell n\,(A_o/A) \tag{1.4}$$

Thus even to interpret the tension test after the onset of necking one must abandon the concepts of infinitesimal deformation analysis. Summation of strain components (1.1) is no longer obviously valid because the change in geometry of the specimen prevents use of a fixed reference length in expressing strain increments in terms of displacement increments.

As illustrated in Fig.1, the plastic strain at A is the same as that at B after elastic unloading and is defined as the residual strain at zero stress, i.e., when the average or macroscopic elastic strain is reduced to zero. Plastic strain is caused physically by the growth and spread of dislocations and analogous lattice rearrangements through the crystallites in a polycrystalline material. The physical theory of plasticity expresses such plastic strains at zero stress, and hence zero macroscopic elastic strain, in terms of the accumulation of lattice defects and the associated slip along crystallite shear planes. However, it is inconvenient to completely unload after each increment of stress in order to measure the increment of plastic strain uncoupled from the elastic strain. The usual procedure is, having reached a state C on the stress-strain curve, to add a stress increment GD, measure the strain increment CG and then remove the stress increment. This results in a residual strain increment CF and a reversible one FG. Since the process is restricted to small increments of strain superimposed on the state C, these increments are additive since the reference geometry is effectively unchanged. Thus the total increment of strain generated along CD is given by

$$\Delta\varepsilon = \Delta\varepsilon^{rev} + \Delta\varepsilon^{res} \tag{1.4}$$

It has become standard practice to consider the reversible increment to comprise the increment of elastic strain

$$\Delta\varepsilon^{rev} = \Delta\varepsilon^{e} \tag{1.5}$$

and the residual increment to comprise the increment of plastic strain

$$\Delta\varepsilon^{res} = \Delta\varepsilon^{p} \tag{1.6}$$

so that (1.4) yields

$$\Delta\varepsilon = \Delta\varepsilon^{e} + \Delta\varepsilon^{p} \tag{1.7}$$

for small increments, or equivalently in terms of rates

$$\dot{\varepsilon} = \dot{\varepsilon}^{e} + \dot{\varepsilon}^{p} \tag{1.8}$$

This is analogous to (1.2), deduced from (1.1) in infinitesimal displacement theory, but differs since each successive increment of plastic strain is defined in terms of a different reference configuration associated with the state C in Fig.1.before the small increment of stress is applied. Thus the superscript dot in (1.8) should only be considered as a formal notation expressing the rate of strain deduced from the increments (1.7); in general there is no associated summation of strains analogous to (1.1) in the infinitesimal displacement case [1,2]*. In plasticity theory, commonly termed an incremental or flow type structure analogous more to fluid mechanics than solid mechanics, a strain-rate type variable is used and there is no need for this to be the derivative of a finite-strain variable and, indeed, in general it is not. In the case where principal directions of strain increments are fixed in the body, natural or logarithmic strain does fill this role [3,4].

In most elastic-plastic deformation problems the elastic strains are small because the yield stress and traction boundary conditions over at least part of the surface of the body under consideration limit stress magnitudes to be of the order of the yield stress. Exceptions to this circumstance occur where high hydrostatic pressure develops or is prescribed as in explosive generated

---

* Numbers in square brackets refer to the bibliography.

shock problems [5], or in problems where the boundary conditions prescribe appreciable changes in volume of a body. Consideration here will be restricted to the usual circumstance in which the stresses are of the order of the yield stress. Then the elastic strains are small because the yield stress is small compared with the elastic moduli and then Hooke's law expresses the relationship of the elastic strain to the stress.

Although the determination of increments of elastic and plastic strain, by superimposing and then removing an increment of stress and measuring the reversible and residual increments of strain produced is convenient and accurate, it is not exactly equivalent to the definition of plastic and elastic strain presented previously as the residual strain at zero stress and the reversible recovered strain on destressing to zero stress, respectively. As already pointed out, the latter definition uncouples the plastic and elastic components of strain and is in accordance with the physical theory of the phenomena. In incremental loading and unloading from a point on the loading curve (C in Fig.1), the residual strain increment is produced while the body is subjected to the yield stress and is deformed by the associated elastic strain. It is therefore to be expected that this measurement will not express the true increment of plastic strain (residual strain at zero stress) since it will embody some effect of the coupling with the maintained yield-point elastic strain. An exact analysis of this circumstance is present in the next section.

It has already been pointed out in introducing (1.2) that the time derivative of the elasticity law is needed in the development of an elastic-plastic constitutive relation because the plastic strain <u>rate</u> occurs in the plasticity law. Since, as already discussed, we can restrict our consideration to small elastic strains without appreciable loss, Hooke's law is commonly introduced to govern the elastic response, taking the usual form

$$\varepsilon^{e}_{ij} = \frac{1+\nu}{E}\sigma_{ij} \quad \frac{\nu}{E}\sigma_{kk}\delta_{ij} \tag{1.9}$$

The time derivative of this is introduced into (1.8) in the form

$$D^{e}_{ij} = \frac{1+\nu}{E}\frac{d}{dt}(\sigma_{ij}) - \frac{\nu}{E}\frac{d}{dt}(\sigma_{kk})\delta_{ij} \tag{1.10}$$

where d/dt indicates a time derivative. This expression is added to the plastic strain rate term to give the total strain rate D_{ij} which is defined as the symmetrical part of the velocity gradient. Now superimposing a spin Ω_{ij} on the system does not affect D_{ij} and hence d/dt (σ_{ij}) must exhibit the same property, in other words, it must be objective. There is an infinity of such rate definitions to choose from. This lack of definiteness arises since (1.10), with the resulting D^{e}_{ij} defined in terms of the velocity gradient, is not a precise derivative of (1.9) because the symmetric part of the velocity gradient is not in general the derivative of a precisely defined strain. It will be shown in the next section that by adopting a precise kinematics of elastic and plastic strain valid for finite deformation, this difficulty of selecting an appropriate d/dt(σ_{ij}) is resolved as well as establishing the precise connection between the incremental definitions of elastic and plastic strain and measurements made by reducing the stress to zero to uncouple the plastic component.

2. Finite Deformation Kinematics

Figure 2 illustrates the analogue of loading and unloading in a simple tension test (Fig.1) for the case of general deformation. The configuration of the body labeled by the coordinates $\underset{\sim}{X} = (X_1, X_2, X_3)$ represents the initial undeformed state, and material particles after elastic-plastic deformation move to the configuration at time t labeled $\underset{\sim}{x} = (x_1, x_2, x_3)$ according to the mapping

$$\underset{\sim}{x} = \underset{\sim}{x}(\underset{\sim}{X}, t) \tag{2.1}$$

If the body is then destressed, corresponding to the unloaded state represented by point B in Fig.1, it takes on the configuration labeled $\underset{\sim}{p} = (p_1, p_2, p_3)$.

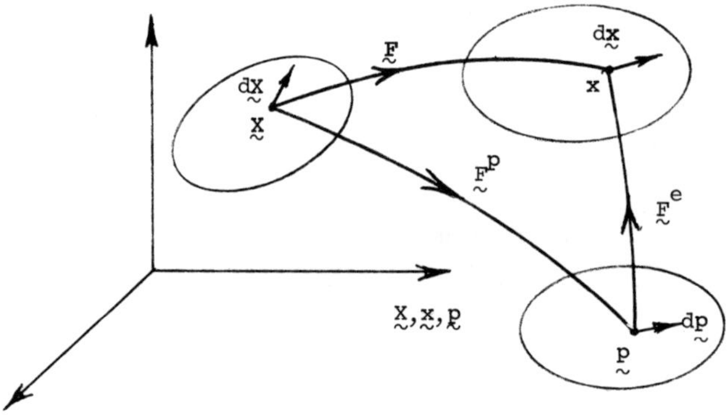

Figure 2 Elastic-plastic Deformation

Since the stress and therefore the elastic strain is zero in the configuration $\underset{\sim}{p}$, the resultant deformation from the initial configuration $\underset{\sim}{X}$ to $\underset{\sim}{p}$ is purely plastic. Because nonzero plastic strain remains when the stress is reduced to zero, if the elastic-plastic deformation is not homogeneous unloading the boundaries commonly leaves the body in a state of residual stress. In order to remove such residual stresses and produce the state $\underset{\sim}{p}$ at zero stress, it would usually be necessary to cut the body into vanishingly small elements, and so the mapping $\underset{\sim}{p}(\underset{\sim}{X}, t)$ expressing the plastic deformation would be noncontinuous. This circumstance does not basically modify the analysis to be described since appropriate definition of the deformation gradients used permits them to represent such discontinuous maps without modifying the structure of the theory [2].

For general stressing, the elastic-plastic deformation corresponding to the state A in Fig.1 for the tension test, is expressed by the deformation gradient matrix of the mapping (2.1)

$$F_{ij} = \partial x_i/\partial X_j \; ; \quad \mathbf{F} = \partial\mathbf{x}/\partial\mathbf{X} \tag{2.2}$$

As already stated the mapping $\mathbf{X} \to \mathbf{p}$ expresses purely plastic deformation with the gradient

$$\mathbf{F}^p = \partial\mathbf{p}/\partial\mathbf{X} \tag{2.3}$$

We consider material which destresses purely elastically from the yield stress at the termination of elastic-plastic deformation to the zero-stress state, so that the plastic strain does not change during this process and the mapping $\mathbf{p}(\mathbf{X},t)$ expresses the plastic strain in the configuration $\mathbf{x}$ (corresponding to A in Fig.1) as well as in the unstressed configuration $\mathbf{p}$. Thus $\mathbf{F}^p$ expresses the deformation gradient of the plastic flow in the material at the yield stress at the initiation of destressing and throughout the destressing process including the unstressed state corresponding to the configuration $\mathbf{p}$. As discussed in [2], the type of analysis described here can be modified to treat the case when plastic flow occurs during destressing as would be the case with a material exhibiting a pronounced Bauschinger effect.

Since elastic deformation is reversible and is zero in the unstressed configuration $\mathbf{p}$, elastic deformation in the configuration $\mathbf{x}$ is expressed by the deformation gradient matrix of the mapping $\mathbf{p} \to \mathbf{x}$

$$\mathbf{F}^e = \partial\mathbf{x}/\partial\mathbf{p} \tag{2.4}$$

Now geometrically the mapping $\mathbf{X} \to \mathbf{x}$ is equivalent to the sequence of mappings $\mathbf{X} \to \mathbf{p} \to \mathbf{x}$ so that the chain rule for partial differentiation gives the matrix product relation

$$F_{ij} = \partial x_i/\partial X_j = (\partial x_i/\partial p_k)(\partial p_k/\partial X_j) = F^e_{ik}F^p_{kj}$$

$$\mathbf{F} = \mathbf{F}^e\mathbf{F}^p \tag{2.5}$$

which is the finite deformation analogue of the strain summation expression (1.1) for the infinitesimal-deformation case. Note that the elastic and plastic terms occur noncommutatively, a physically significant property which is lost in the small strain approximation. Note that the deformation $\mathbf{X} \to \mathbf{p}$ cannot be achieved physically without passing through $\mathbf{x}$ (for example one cannot generate the plastic deformation while remaining at zero stress) but this has no bearing on the formulation of the kinematical relation (2.5).

In the deformation expressed by $\mathbf{F}$, an infinitesimal vector $d\mathbf{X}$ embedded in the material and shown in Fig.2 is deformed into the vector element $d\mathbf{x}$ given by

$$dx_i = (\partial x_i/\partial X_j)dX_j \; ; \quad d\mathbf{x} = \mathbf{F}\,d\mathbf{X} \tag{2.6}$$

The square of the length dS of $d\mathbf{X}$ is given by the scalar product of $d\mathbf{X}$ with itself

$$dS^2 = dX_i dX_i = d\mathbf{X}^T d\mathbf{X} \tag{2.7}$$

in which the superscript T denotes the transpose. An analogous expression for the deformed element using (2.6) yields

$$ds^2 = d\underset{\sim}{x}^T d\underset{\sim}{x} = d\underset{\sim}{X}^T \underset{\sim}{F}^T \underset{\sim}{F}\, d\underset{\sim}{X} \tag{2.8}$$

so that the Lagrange strain is given by

$$\underset{\sim}{E} = (\underset{\sim}{F}^T \underset{\sim}{F} - \underset{\sim}{I})/2 \tag{2.9}$$

By substituting for $\underset{\sim}{F}$ from (2.5)

$$\underset{\sim}{E} = \underset{\sim}{F}^{p^{-1}} \underset{\sim}{E}^e \underset{\sim}{F}^p + \underset{\sim}{E}^p \tag{2.10}$$

where $\underset{\sim}{E}^e$ and $\underset{\sim}{E}^p$ are the elastic and plastic finite strains from their respective reference states

$$\underset{\sim}{E}^e = (\underset{\sim}{F}^{e^T} \underset{\sim}{F}^e - \underset{\sim}{I})/2 \;;\quad \underset{\sim}{E}^p = (\underset{\sim}{F}^{p^T} \underset{\sim}{F}^p - \underset{\sim}{I})/2 \tag{2.11}$$

Since $\underset{\sim}{F}^p$ can be far removed from the unit matrix, it is clear from (2.10) that summability of elastic and plastic strains as defined here, analogous to (1.1) in the infinitesimal deformation case, has no validity.

The particle velocity in the state $\underset{\sim}{x}$ is given by

$$\underset{\sim}{v} = \partial\underset{\sim}{x}/\partial t\Big|_{\underset{\sim}{X}} \tag{2.12}$$

and so $\underset{\sim}{L}$, the gradient of the velocity $\underset{\sim}{v}$ in the space $\underset{\sim}{x}$, is

$$\underset{\sim}{L} = \partial\underset{\sim}{v}/\partial\underset{\sim}{x} = (\partial\underset{\sim}{v}/\partial\underset{\sim}{X})(\partial\underset{\sim}{X}/\partial\underset{\sim}{x}) = \dot{\underset{\sim}{F}}\underset{\sim}{F}^{-1} \tag{2.13}$$

where the superposed dot denotes material derivative, time differentiation at fixed $\underset{\sim}{X}$. Substituting from (2.5) gives

$$\underset{\sim}{L} = \dot{\underset{\sim}{F}}^e \underset{\sim}{F}^{e^{-1}} + \underset{\sim}{F}^e \dot{\underset{\sim}{F}}^p \underset{\sim}{F}^{p^{-1}} \underset{\sim}{F}^{e^{-1}} \tag{2.14}$$

As was pointed out in [4], the configuration in the neighborhood of each material element in the unstressed $\underset{\sim}{p}$ is unique only to within an arbitrary rigid-body rotation. Such rotation does not change the strain so that the stress remains zero and the lack of continuity of the mapping already referred to permits the rotation to be chosen independently for each material element. It is convenient analytically to select the rotation so that for each element destressing from $\underset{\sim}{x}$ occurs without rotation. By the polar decomposition theorem $\underset{\sim}{F}^e$ is then symmetric and is written

$$\underset{\sim}{F}^e = \underset{\sim}{V}^e ;\quad \underset{\sim}{V}^e = \underset{\sim}{V}^{e^T} \tag{2.15}$$

Equation (2.14) then becomes

$$\underset{\sim}{L} = \dot{\underset{\sim}{V}}^e \underset{\sim}{V}^{e^{-1}} + \underset{\sim}{V}^e \dot{\underset{\sim}{F}}^p \underset{\sim}{F}^{p^{-1}} \underset{\sim}{V}^{e^{-1}} \tag{2.16}$$

The symmetric part $\underset{\sim}{D}$ gives the rate of strain about the current configuration and the antisymmetric part gives the spin $\underset{\sim}{W}$, thus

$$\underset{\sim}{D} = \underset{\sim}{D}^e + [\underset{\sim}{V}^e(\underset{\sim}{D}^p + \underset{\sim}{W}^p)\underset{\sim}{V}^{e^{-1}}]_S = \underset{\sim}{D}^e + \underset{\sim}{V}^e\underset{\sim}{D}^p\underset{\sim}{V}^{e^{-1}}|_S + \underset{\sim}{V}^e\underset{\sim}{W}^p\underset{\sim}{V}^{e^{-1}}|_S \quad (2.17)$$

$$\underset{\sim}{W} = \underset{\sim}{W}^e + [\underset{\sim}{V}^e(\underset{\sim}{D}^p + \underset{\sim}{W}^p)\underset{\sim}{V}^{e^{-1}}]_A = \underset{\sim}{W}^e + \underset{\sim}{V}^e\underset{\sim}{D}^p\underset{\sim}{V}^{e^{-1}}|_A + \underset{\sim}{V}^e\underset{\sim}{W}^p\underset{\sim}{V}^{e^{-1}}|_A \quad (2.18)$$

where $\underset{\sim}{D}^e$, $\underset{\sim}{W}^e$, $\underset{\sim}{D}^p$ and $\underset{\sim}{W}^p$ are the symmetric and antisymmetric parts of $\underset{\sim}{\dot{V}}^e\underset{\sim}{V}^{e^{-1}}$ and $\underset{\sim}{\dot{F}}^p\underset{\sim}{F}^{p^{-1}}$, respectively and the subscripts S and A denote the symmetric and antisymmetric parts, respectively.

In the incremental method of measuring elastic and plastic strain, described in the Introduction, an increment of stress is imposed and removed while the stress is maintained. The increment of strain generated by imposing the stress increment is given by $\underset{\sim}{D}$ dt, where $\underset{\sim}{D}$ is given by (2.17). It should perhaps be mentioned that we are concerned with classical time-rate independent plasticity theory appropriate for ambient temperatures and for which the strain rate and spin rate terms at any instant during the deformation increment are proportional to the stress rate applied, so that only the increment of stress influences the resulting increment of strain and not the time it takes to impose the additional stress. Removal of the stress increment is considered to take place by reduction of $\underset{\sim}{V}^e$ without spin.

For a material which exhibits isotropic hardening governed by a Mises type yield condition

$$J_2 = J_2(\bar{\varepsilon}^p) \quad (2.19)$$

where J_2 is the second invariant of the stress deviator tensor and $\bar{\varepsilon}^p$ is the generalised plastic-strain scalar measure (see Hill [6], p.30), normality of the plastic strain-rate vector to the yield surface in the nine-dimensional stress and strain-rate space determines that $\underset{\sim}{D}^p$ has the same principal directions in the current configuration space $\underset{\sim}{x}$ as the stress. Moreover for isotropic elastic response, the elastic pure-deformation matrix $\underset{\sim}{V}^e$ also has the same principal directions as the stress. Thus the three matrices in the product on the right-hand side of (2.17) which expresses the plastic strain-rate contribution can be commuted, the $\underset{\sim}{V}^e$ terms cancel and the contribution to $\underset{\sim}{D}$ simply reduces to an additive $\underset{\sim}{D}^p$. However, the last term in (2.17) prevents the deduction of a simple strain-rate additive law analogous to (1.2) if the spin $\underset{\sim}{W}^p$ is not zero.

The last term in (2.17) involving the matrix product expression $\underset{\sim}{V}^e\underset{\sim}{W}^p\underset{\sim}{V}^{e^{-1}}$ has a clear physical meaning. In a small time increment dt, the spin term $\underset{\sim}{W}^p$ will produce an increment of rotation and the whole expresses the sequence: first a strain due to elastic destressing ($\underset{\sim}{V}^{e^{-1}}$), then the increment of rotation followed by elastic stressing ($\underset{\sim}{V}^e$). The symmetric part expresses the resulting increment of strain which will clearly be nonzero if $\underset{\sim}{W}^p$ is nonzero. This term represents a rate of change of residual strain since on removing the increment of stress this change of strain due to the spin $\underset{\sim}{W}^p$ remains. According to the prescription (1.6) it is therefore considered part of the plastic strain although it is not directly related to the plastic strain rate $\underset{\sim}{D}^p$. Moreover, it can be shown [2] that this component of the residual strain increment is tangential to the yield surface. Clearly to consider it as part of the plastic strain increment is incorrect, and using (1.6) as a definition of plastic strain rate will lead to a violation of the plasticity normality rule unless $\underset{\sim}{W}^p$ is zero. Thus if this "plastic" strain increment is incorporated into a finite element computer code which assumes a Mises type isotropic hardening constitutive relation, the variational principle which governs the calculation will seek a point on the yield surface at which the residual strain increment is normal and this will lead to an error in the stress evaluation.

It is interesting to observe that in (2.17) two terms are added to the elastic strain rate $\underset{\sim}{D}^e$ to give the total strain rate $\underset{\sim}{D}$, one containing $\underset{\sim}{D}^p$ and the other $\underset{\sim}{W}^p$. In the discussion in the Introduction anticipating a shortcoming in the strain rate additive rule (1.8) we were concerned with the potential coupling of the maintained elastic strain with the plastic strain rate. It transpires that for the isotropic Mises strain-hardening law this coupling disappears, but rotation introduces a residual elastic-strain component which vitiates the assumption (1.8). This component is both elastic and residual since the maintained stress combined with material rotation generates a change in elastic strain in the material even though the final stress returns to its original value based on axes fixed in space.

It was pointed out in the Discussion section of [7] that the usual current elastic-plastic finite-deformation computer code is based on the linear kinematics of the summation of elastic and plastic strain rated to give the total strain rate, the incremental law for the plastic strain rate and a derivative of Hooke's law for the elastic strain rate. This combination yields a total strain-rate stress-rate relation, linear in these rates. The Jaumann stress rate based on the total body spin $\underset{\sim}{W}$ is then adopted to ensure compliance of the theory for rigid-body rotation when the strain rate is zero (or equivalently to satisfy the requirement of objectivity). This choice can be considered equivalent to expressing the kinematics relative to axes rotating with the body spin $\underset{\sim}{W}$. The plastic spin $\underset{\sim}{W}^p$ then becomes $(\underset{\sim}{W}^p - \underset{\sim}{W})$ which is small for small elastic strains $(\underset{\sim}{V}^e \sim \underset{\sim}{I})$ according to (2.18). Thus the last term in (2.17) is small and (1.8) is obeyed approximately. Thus for this choice of axes the assumption of linear kinematics for the strain rates provides a good approximation but not an exact analysis. Thus the error in kinematics is largely reduced by selecting an appropriate stress-rate definition.

It was pointed out in the Introduction that the currently accepted approach to finite-deformation elastic-plastic analysis is to use the linear kinematic assumption of additivity of the elastic and plastic strain rates to give the total strain rate and to eliminate the elastic component from this by substituting from Hooke's law. This is effected by transforming the linear Hooke's law into rate form by replacing the strain by the strain rate expressed as the symmetric part of the velocity gradient associated with the elastic deformation, and the stress by an objective stress-rate definition. It was pointed out by Prager [8] that it is then necessary to choose from an infinity of objective stress-rate definitions. Two appear to be popular for elastic-plastic analysis: the Jaumann and Truesdell rates. A problem arises since this transformation of Hooke's law to rate form is not a formal time derivative of each side of the equation by the same operator, nor would such an approach eliminate the difficulty since Hooke's law based on infinitesimal strain is itself not objective under finite rotations. However, use of the exact non-linear kinematics leading to (2.17) combined with the exact elasticity law valid for finite deformation

$$\underset{\sim}{\sigma} = 2\underset{\sim}{F}^e(\partial\psi/\partial\underset{\sim}{C}^e)\underset{\sim}{F}^{e^T}/(\det \underset{\sim}{F}^e); \; \underset{\sim}{C}^e = \underset{\sim}{F}^{e^T}\underset{\sim}{F}^e \tag{2.20}$$

where $\underset{\sim}{\sigma}$ is the Cauchy or true stress and ψ the strain energy function, permits any objective rate operator to be applied to both sides of (2.20). If this permits $\underset{\sim}{D}^e$ to be eliminated from (2.17) and permits the incremental plasticity law for $\underset{\sim}{D}^p$ to be written in terms of the same stress-rate operator, then an elastic-plastic constitutive equation can be generated which relates stress rate with the total strain rate. The form of this relation will depend on the stress-rate definition chosen.

This procedure was carried out for the Jaumann rate in [7]. The resulting operator was introduced into Hill's variational principle [9] commonly used for formulating finite-element elastic-plastic computer codes valid for finite-deformation. When the approximation valid for small elastic strains that $\underset{\sim}{\epsilon}^e = (\underset{\sim}{C}^e - \underset{\sim}{I})/2$ is small so that powers higher than the first can be neglected, the variational principle reduces to that used in [10] which comprises a minor modification the program developed by McMeeking and Rice [11].

Reference [7] presents a rate or incremental formulation of the finite-deformation elastic-plastic theory developed in [4]. The fact that in [4] the elastic constitutive relation was left in total elastic deformation form, rather than in rate form, limited utilization of the theory and application has been restricted to shock-wave analysis in which, by symmetry, principal directions of stress remained fixed in the body. This permits the use of logarithmic or natural strain which generates additivity of elastic and plastic strain as well as of strain rates [5,12]. However this approach is not available for arbitrary loading and the procedure described in [7] removes the restriction on general application of the theory.

3. Stress Analysis for Kinematic Hardening

So far the discussion has been limited to isotropic hardening using a Mises type yield condition. An interesting anomaly has recently arisen in the extension of finite-deformation analysis to materials with anisotropic hardening characteristics, in particular kinematic hardening. This permits properties having the nature of the Bauschinger effect to be incorporated into the theory. In view of the additional complexity introduced and the satisfactory approximation of the elastic and plastic strain-rate additivity hypothesis in the usual case of small elastic strains, this extension of the theory is based on that hypothesis.

In an intriguing paper [13], Nagtegaal and de Jong evaluated the stresses generated by simple shear to large deformation in elastic-plastic and rigid plastic materials which exhibit anisotropic hardening. In conformity with current practice for finite deformation in the case of kinematic hardening, they used an evolution equation for the back stress or shift tensor $\underset{\sim}{\alpha}$ (the current center of the yield surface) which relates the Jaumann derivative of $\underset{\sim}{\alpha}$

to the plastic strain rate. This incorporates effects of finite rotation and ensures objectivity of the evolution equation under superimposed rigid-body rotations. They obtained the unexpected result, for a material which strain hardens monotonically in tension, that the shear traction grows to a maximum value at a shear strain γ of the order unity and then oscillates with increasing strain. The normal traction on the shearing planes exhibited a similar behavior.

In [14] study of the analytical structure of the kinematic hardening law shows that, in the case of simple shear, the use of the conventional Jaumann derivative based on the spin causes the shift tensor $\underset{\sim}{\alpha}$ to rotate continuously and this generates oscillations in the stress field. It is also shown that this analytical structure, which is currently adopted in finite-deformation elastic-plastic codes involving kinematic hardening, is not in accord with the effects of the physical micromechanisms which produce plastic flow. A modified theory consonant with these is developed which yields a monotonically increasing shear traction for the problem under discussion.

Using rectangular Cartesian coordinates for the configuration at time t, simple shearing in the x_1-direction of magnitude, $\gamma = kt$ generates the displacement field

$$u_1 = kt\,x_2\,,\quad u_2 = u_3 = 0 \tag{3.1}$$

and the steady-state velocity field

$$v_1 = kx_2\,,\quad v_2 = v_3 = 0\,. \tag{3.2}$$

The resulting rate of deformation $\underset{\sim}{D}$ (the symmetric part of the velocity gradient) has nonzero components $D_{12} = D_{21} = k/2$ and the spin $\underset{\sim}{W}$ (the skew-symmetric part of the velocity gradient) nonzero components $W_{12} = -W_{21} = k/2$.

A unit square defined by material elements lying initially on the x_1, x_2-axes deforms into a sequence of parallelograms. The sides of the parallelograms originally parallel to the x_2-axis lengthen and rotate towards the x_1-axis, which they approach as $t \to \infty$. Other straight loci of material points move similarly and none ever rotate through an angle greater than π, even though the spin $\underset{\sim}{W}$ is constant corresponding to an angular velocity of magnitude k/2.

The back-stress $\underset{\sim}{\alpha}$ which prescribes the position of the center of the yield surface in stress space provides the anisotropy in the yield function necessary to incorporate such effects as the Bauschinger effect. Since the source of this property is embedded in the material, the tensor $\underset{\sim}{\alpha}$ is considered to rotate with the material so that for finite deformation theory the conventional Jaumann derivative $\overset{\circ}{\underset{\sim}{\alpha}}$ (rate of change of $\underset{\sim}{\alpha}$ with respect to axes rotating with angular velocity equal to the material spin rate) has commonly been used in the evolution equation for $\underset{\sim}{\alpha}$:

$$\overset{\circ}{\underset{\sim}{\alpha}} = c\,\underset{\sim}{D}^p \tag{3.3}$$

This uses a common form for the effect of the current plastic flow on the rate of change of $\underset{\sim}{\alpha}$ inferred from small deformation theory.

In order to examine how $\underset{\sim}{\alpha}$ changes with respect to the axes (x_1, x_2) fixed in space, we need the material derivative $\dot{\underset{\sim}{\alpha}}$ of the shift tensor

$$\dot{\underset{\sim}{\alpha}} = \overset{\circ}{\underset{\sim}{\alpha}} + \underset{\sim}{W}\underset{\sim}{\alpha} - \underset{\sim}{\alpha}\underset{\sim}{W} = C\underset{\sim}{D}^{p} + \underset{\sim}{W}\underset{\sim}{\alpha} - \underset{\sim}{\alpha}\underset{\sim}{W} \ . \tag{3.4}$$

For the case of simple shearing, the spin terms in (3.4) contribute to $\underset{\sim}{\alpha}$ a rotation with constant angular velocity k/2. This rotation contrasts markedly with the rotation of lines of material points in the body for which the total rotation varies with initial orientation but can never exceed π. Using rigid-plastic theory which is appropriate for the large-strain problem under consideration, the plastic flow contribution in (3.4) has fixed principal directions in the (x_1, x_2) plane at angles $\pi/4$ to the axes. Since the hardening mechanisms are embedded in the material, it seems, on physical grounds, implausible that the macroscopic hardening parameter $\underset{\sim}{\alpha}$ could continue to rotate through an unlimited angle while no elements of the material do.

In view of the preceding discussion, it is necessary to determine a direction embedded in the material which characterizes the anisotropy induced by previous plastic flow. As an example appropriate for simple shear, the direction associated with the maximum eigenvalue of the shift tensor $\underset{\sim}{\alpha}$ is taken. Thus a modified Jaumann type derivative is defined which eliminates the effect of rotation of the material lines parallel to the direction of the eigenvector of $\underset{\sim}{\alpha}$ associated with the maximum eigenvalue. This rotation defines a spin $\underset{\sim}{W}^{*}$ readily obtained from the kinematics of simple shearing. Retaining the influence of current plastic flow we obtain the evolution equation analogous to (3.3)

$$\overset{*}{\underset{\sim}{\alpha}} = C\underset{\sim}{D}^{p} \tag{3.5}$$

where the superscript * denotes the modified Jaumann derivative associated with the spin $\underset{\sim}{W}^{*}$. Thus $\dot{\underset{\sim}{\alpha}}$, the material time derivative of $\underset{\sim}{\alpha}$ with respect to fixed axes (x_1, x_2), is given by

$$\dot{\underset{\sim}{\alpha}} = C\underset{\sim}{D}^{p} + \underset{\sim}{W}^{*}\underset{\sim}{\alpha} - \underset{\sim}{\alpha}\underset{\sim}{W}^{*} \tag{3.6}$$

When shearing commences the eigenvector of $\underset{\sim}{\alpha}$ under consideration is initially directed at an angle $\theta = \pi/4$ to the x_1-axis but the motion of the material continually tends to reduce this angle due to the rotation effect of the last two terms in (3.6). As the angle θ decreases, the spin term $\underset{\sim}{W}^{*}$ decreases towards zero. The new evolution equation (3.5) is shown to be objective under superimposed rigid-body rotation.

In formulating the plastic flow law care must be taken since the yield condition will be rotating with the tensor $\underset{\sim}{\alpha}$ and the effect of this must be eliminated from the definition of the stress-rate in the expression for the plastic strain rate. This is treated in a manner analogous to that for the evolution equation be defining a modified Jaumman derivative of the stress based on the rotation of the $\underset{\sim}{\alpha}$ tensor.

Calculations were carried out for both the conventional and modified approaches assuming linear hardening in tension and purely kinematic hardening (although the theory applies to combined kinematic-isotropic hardening). The case of isotropic hardening was also evaluated. Properties appropriate for an aluminum alloy were chosen.

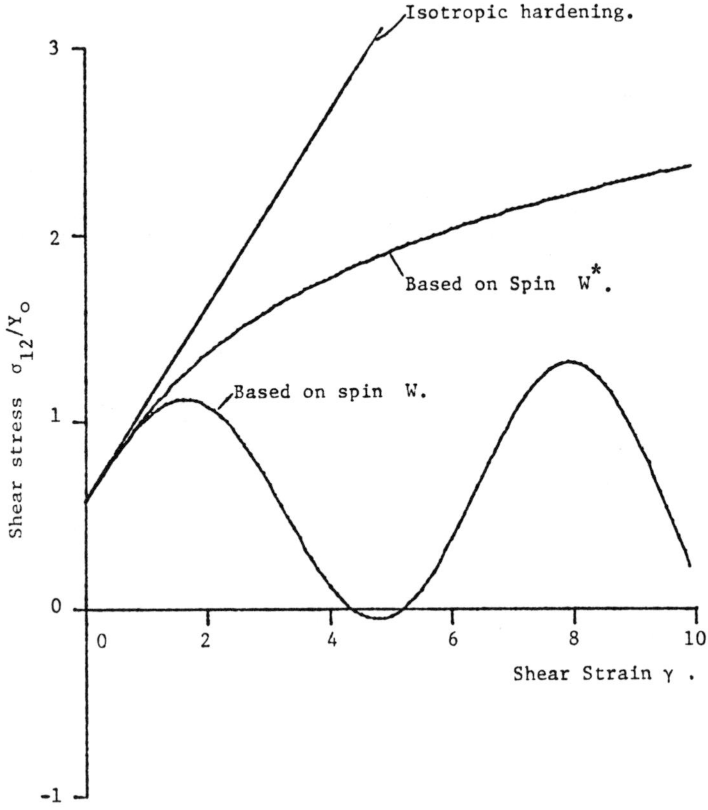

Figure 3 Nondimensional shear stress versus shear strain from rigid-plastic model

Figure 3 shows the shear stress as a function of the shear strain. The strain hardening coefficient is 1.5 Y_0, where Y_0 designates the tensile yield stress. The figure shows the deduced shear stress versus shear strain curves. The use of the conventional Jaumann derivative gives the oscillations presented in [13]. The modified theory is seen to yield a curve which agrees closely with the currently utilized approach up to a shear strain γ of unity, but thereafter the stress increases monotonically with a continuously decreasing modulus. The straight line relation shown is deduced from the tensile behavior on the basis of isotropic hardening with Mises yield condition.

Kinematic hardening, according to the theory presented in this paper, initially gives the same shear stress and hardening modulus as isotropic hardening, but predicts that the modulus decreases monotonically with increasing strain so that at large strains the kinematic hardening curve is well below the isotropic hardening curve.

Comparison of the complete stress analyses with the conventional and the modified Jaumann derivatives indicates that up to strains near 0.5 the difference in the solutions is small. For strains near 2 the difference reaches about 40% and grows rapidly with increasing strain. These results suggest that current codes for stress analysis at large strains which incorporate kinematic hardening should be reassessed if the possibility of serious errors in stress evaluations is to be avoided.

Reference [14] provides a generally applicable formulation of kinematic hardening theory but only a simple example has been evaluated based on a simple hypothesis for the macroscopic influence of the micromechanisms which generate kinematic hardening. The new results and the disparity with those evaluated by the currently accepted method of analysis indicates that we are only in the early stages of development of a satisfactory method of treating anisotropic hardening in the case of finite deformation.

Acknowledgement

This paper was written in the course of research sponsored by the U.S. Army Research Office under Contract No. DAAG 29-82-K-0016 with Rensselaer Polytechnic Institute. The author gratefully acknowledges this support.

References

1. Lee, E.H. and McMeeking, R.M., Concerning Elastic and Plastic Components of Deformation, Int. J.Solids Structures, 16, 715-721, 1980.
2. Lee, E.H., Some Comments on Elastic-Plastic Analysis, Int. J. Solids Structures, 17, 859-872, 1981.
3. Lee, E.H. and Liu, D.T., Finite Strain Elastic-Plastic Theory Particularly for Plane Wave Analysis, J. Appl. Phys., 38, 19-27, 1967.
4. Lee, E.H., Elastic-Plastic Deformation at Finite Strains, J. Appl. Mech., 36, 1-6, 1969.
5. Germain, P. and Lee, E.H., On Shock Waves in Elastic-Plastic Solids, J. Mech. Phys. Solids, 21, 359-382, 1973.
6. Hill, R., The Mathematical Theory of Plasticity, Clarendon Press, Oxford, 1950.
7. Lubarda, V.A. and Lee, E.H., A Correct Definition of Elastic and Plastic Deformation and its Computational Significance, J. Appl. Mech., 48, 35-40, 1981.
8. Prager, W., An Elementary Discussion of Definitions of Stress Rate, Quart. Appl. Math., 18, 403-407, 1961.
9. Hill, R., Some Basic Principles in the Mechanics of Solids without a Natural Time, J. Mech. Phys. Solids, 7, 209-225, 1959.

10. Lee, E.H., Mallett, R.L. and McMeeking, R.M., Stress and Deformation Analysis of Metal-Forming Processes, Numerical Modeling of Manufacturing Processes, PVP-PB-025, eds. R.F. Jones, Jr., H. Armen and J.T. Fong, ASME, 19-33, 1977.
11. McMeeking, R.M. and Rice, J.R., Finite-Element Formulations for Problems of Large Elastic-Plastic Deformation, Int. J. Solids Structures, 11, 601-616, 1975.
12. Lee, E.H. and Wierzbicki, T., Analyses of the Propagation of Plane Elastic-Plastic Waves of Finite Strain, J. Appl. Mech., 34, 931-936, 1967.
13. Nagtegaal, J.C. and de Jong, J.E., Some Aspects of Nonisotropic Work-hardening in Finite Deformation Plasticity, to appear in Proceedings of the Workshop: Plasticity at Finite Deformation, Div. of Appl. Mech., Stanford University, Stanford, California, 1981.
14. Lee, E.H., Mallett, R.L. and Wertheimer, T.B., Stress Analysis for Kinematic Hardening in Finite-Deformation Plasticity, SUDAM Rep. 81-11, Div. of Appl. Mech., Stanford University, Stanford, California, 1981.

YIELDING AND UNLOADING IN PLASTICITY. IDEALIZATIONS AND EXPERIMENTAL RESULTS.

Erhard Krempl
Department of Mechanical Engineering,
Aeronautical Engineering & Mechanics
Rensselaer Polytechnic Institute
Troy, New York 12181

ABSTRACT

Yielding with a stress drop from the upper yield point followed by a yield plateau cannot be considered a material property. It cannot occur if load control with increasing load (stress boundary condition) is imposed in a uniaxial test. Unloading from the plastic range is shown to depend on stress (strain) rate and is not linearly elastic. As zero stress is approached the slope decreases from the elastic one. Yielding and linear elastic unloading cannot be considered distinguishing features of plasticity. History dependence in the sense of plasticity is defined and proposed as a key feature of plasticity.

INTRODUCTION

The representation of actual material behavior in constitutive equations almost always requires that certain key properties be extracted from the vast number of experimental results. This procedure is not necessary when linear theories (elasticity and viscoelasticity), nonlinear elasticity and the theory of nonlinear materials with memory are considered. Mathematical postulates are used to define ideal materials and the properties can be ascertained without the need of experimental information.

The above ideal materials are poor models of the inelastic deformation behavior of metals, i.e. for plasticity and viscoplasticity. This class of materials needs experimental information for the definition of key properties. In the literature the existence of yielding and the elastic unloading which occurs when the load is reduced after plastic deformation are taken as the key properties which are not reproduced by elasticity[1]. On the other hand

[1]There is debate in the literature regarding yielding and hypoelasticity [1-4].

the theory of nonlinear materials with memory is much too general to permit a decision on such specific items.

The yield point and elastic unloading are not mathematical postulates but properties observed in tests. They are therefore subject to the accuracy and the capability of the experimental equipment.

During the last twenty years experimental capabilities have significantly changed through the use of feedback principles, electronics and computers. Servocontrolled mechanical testing machines and properly designed specimens now permit the imposition of slow homogeneous motions during which a stress boundary condition (load control) or a displacement boundary condition (displacement or strain control) can be imposed. Conventional testing machines do not have these capabilities. They impose neither exact load nor displacement control. Stress-strain diagrams obtained with conventional testing machines have in general unknown boundary conditions.

It is the purpose of this paper to examine yielding and unloading behavior on tests performed with the capabilities of the servocontrolled system in mind and to discuss the implication of these findings for the modeling of plasticity (viscoplasticity).

YIELDING

Carbon steel, some low alloy steels and other alloys can exhibit an upper yield point A and lower yield point B as shown in Fig. 1. The plateau following the drop from A is the basis for the elastic perfectly plastic idealization.

In a load controlled test with increasing load the load increment must be greater than zero, $dP > 0$, all the time. It follows that the results of Fig. 1, where $dP \leq 0$ after point A, must have been obtained in other than the load controlled mode. The "typical" stress-strain diagram in Fig. 1 is not a material property since the behavior between A and B cannot occur in load control. This property is demonstrated in Fig. 2 where identical steel specimens are tested in load and displacement control. Yielding is absent in load control.

While the importance of boundary conditions on the appearance of the stress-strain diagram has been recognized by Nadai [5][2], this finding has not been incorporated into theories [1-4].

As shown in Fig. 2 yielding is a special artifact. It disappears after the first loading and is not present in many metals [14]. Even if it is found in other than load control, it has to disappear under stress boundary conditions. It is something special which cannot be taken as a general property which deserves generalization in a theory.

---

[2]To accomplish stress control a special apparatus had to be built; with a servocontrolled machine only a switch has to be activated.

UNLOADING

In applied mechanics the unloading behavior from the plastic range is idealized by linear elastic behavior, see CD in Fig. 1. Not only is this idealization inherent in the yield surface concept, but linear elastic unloading has played a vital role in the development of the endochronic theory [6-8] which does not use a yield surface. It is also an idealization which is necessary to hold for the multiplicative decomposition of the deformation gradient into elastic and plastic parts [9,10].

The type of control can have an influence on the unloading behavior as demonstrated in Fig. 3 where the same input is executed in load (stress) and in displacement (strain) control. For stress control a negative slope is obtained after rate reversal with considerable strain accumulation before positive slope is reached. In strain control the slope changes abruptly to large positive values. In both cases the unloading behavior is not a straight line; rather a slope equal to the elastic modulus is obtained at some stress region below the maximum stress. From then on the slope continues to decrease. For both stress and strain control it is less than the elastic modulus at zero load.

In Fig. 3 viscoplastic behavior is evident by the changes in the flow stress upon jumps in strain rate [11,12]. The steel for which results are shown in Fig. 3 does not correspond to the rate-independent idealization. Rate-dependence was found to be present at room temperature on a stainless steel [11,12] and on a Ti-alloy [13]; it was also found on SAE 1020 steel, A533B steel, copper and brass (unpublished).

From these tests it is not possible to say how a rate (time) - independent material behaves since we have not tested one. However, the tests are on real alloys and represent actual metal behavior which must be considered in modeling.

In Fig. 3 the strain (stress) rate magnitude does not change upon unloading. When the strain (stress) rate magnitude is changed at the point of reversal, various unloading behaviors are possible as shown in Figs. 4 and 5.

The strain rate magnitude is increased (decreased) by two orders of magnitude and the slope leaving the point of reversal is nearly elastic (infinity), see Fig. 4. On the other hand when the stress rate magnitude is increased (decreased), an elastic (small negative) slope is observed in Fig. 5.

Differences caused by the type of control are found right after unloading commences. For every type of control elastic slope is obtained at some positive stress close to the maximum stress. From this point on the slope decreases for every type of control and at zero load the slope is less than the elastic one.

The rate of loading significantly affects the upper portion of the unloading curve but not the part approaching zero load. For the materials we have tested unloading behavior appears to be quite different from the one that is generally assumed in applied mechanics. Specifically the decrease in slope as zero load is approached implies negative plastic work. It is small compared to the plastic work done during loading but significant for thermodynamic considerations.

Published experiments of repeated loadings and unloadings [14] and of hysteresis loops [15,16] do not exhibit elastic unloading and are in accordance with our findings. The unloading behavior is not the primary interest of these publications.

CONCLUSIONS

Yielding and unloading behavior of structural alloys were shown to be substantially different from the common idealizations. The existence of a yield point and linear elastic unloading do not appear to be distinguishing properties which can form the basis of a constitutive idealization. It is for this reason that a new method of defining a key property of plastic deformation was introduced [17,18]. This key property defined as "history dependence in the sense of plasticity" can be ascertained by comparing outputs to the same input of real or thought experiments. It is a general property and does not depend on details of the deformation behavior such as yielding and unloading.

ACKNOWLEDGMENT

This research was supported by a grant from the National Science Foundation. Messrs. V. V. Kallianpur and H. Lu performed the experiments.

REFERENCES

[1] Freudenthal, A. M. and Geiringer, H., "The Inelastic Continuum," in *Handbuch der Physik, Vol. VI,* Springer-Verlag, 1958, specifically p.262.

[2] Truesdell, C., "Second Order Effects in the Mechanics of Materials," *Proc. IUTAM Symp.*, Haifa, Israel, Pergamon Press, 1964, specifically p.17.

[3] Green, A. E., "Hypoelasticity and Plasticity II, *J. Rational Mech. Anal.* *5*, 725-734 (1956).

[4] Hill, R., "Some Basic Principles in the Mechanics of Solids without a Natural Time," *J. Mech. Phys. Solids*, *7*, 209-225 (1959), footnote p.223.

[5] Nadai, A., *Theory of Flow and Fracture of Solids*, McGraw-Hill, 1950, specifically p.309.

[6] Walker, K. P. and Krempl, E., "An Implicit Functional Representation of Stress-Strain Behavior," *Mech. Res. Comm., 5,* 185-190 (1978).

[7] Valanis, K. C., "Fundamental Consequences of a New Intrinsic Time Measure. Plasticity as a Limit of the Endochronic Theory," *Arch. Mech., 32,* 171-191 (1980).

[8] Rivlin, R. S., "Some Comments on the Endochronic Theory of Plasticity, *Int. J. Solids and Structures, 17,* 231-248 (1981).

[9] Lee, E. H., "Elastic-Plastic Deformation at Finite Strain," *Trans. ASME, J. Appl. Mech., 36E,* 1-6 (1969).

[10] Nemat-Nasser, S., "Decomposition of Strain Measures and Their Rates in Finite Deformation Elastoplasticity," *Int. J. Solids and Structures, 15,* 155-166 (1979).

[11] Krempl, E., "An Experimental Study of Room-Temperature Rate Sensitivity, Creep and Relaxation of Type 304 Stainless Steel," *J. Mech. Phys. Solids, 27,* 363-375 (1979).

[12] Kujawski, D., Kallianpur, V. and Krempl, E., "An Experimental Study of Uniaxial Creep, Cyclic Creep and Relaxation of AISI Type 304 Stainless Steel at Room Temperature," *J. Mech. Phys. Solids, 28,* 129-148 (1980).

[13] Kujawski, D. and Krempl, E., "The Rate(Time)-Dependent Behavior of Ti-7Aℓ-2Cb-1Ta Titanium Alloy at Room Temperature under Quasi-Static Monotonic Loading," *Trans. ASME, J. Applied Mech., 48,* 55-63 (1981).

[14] Dalby, W. E., "Researches on the Elastic Properties and the Plastic Extension of Metals," *Phil. Trans. Roy. Soc. London, 221,* 117-138 (1921).

[15] Dolan, T. J., "Nonlinear Response under Cyclic Loading Conditions," *Proc. 9th Midwestern Mechanics Conference,* Univ. of Wisconsin, August 1965, pp. 3-21.

[16] Jhansale, H. R., "A History Dependent Parameter for the Cyclic Stress-Strain Behavior of Metals," Dept. of Theoretical and Appl. Mech. Report, Univ. of Illinois, Urbana, Ill., T & A.M. Rep. No. 383, March 1974.

[17] Krempl, E., "On the Interaction of Rate- and History-Dependent Effects in Structural Metals, *Acta Mechanica, 22,* 53-90 (1975).

[18] Krempl, E., "Plasticity and Variable Heredity," *Arch. Mech., 33,* 289 (1981).

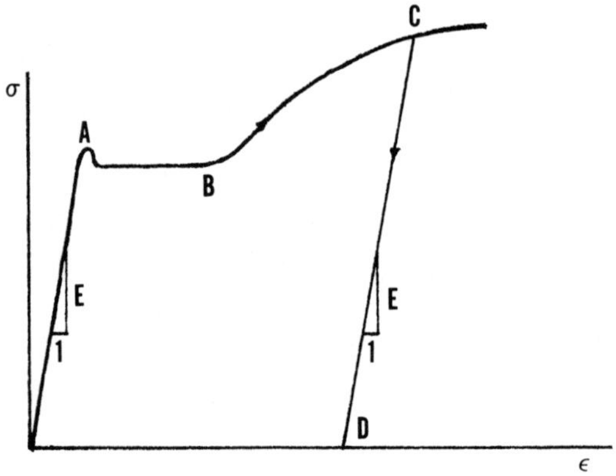

Fig. 1 Schematic showing the stress-strain diagram of a steel with yielding.

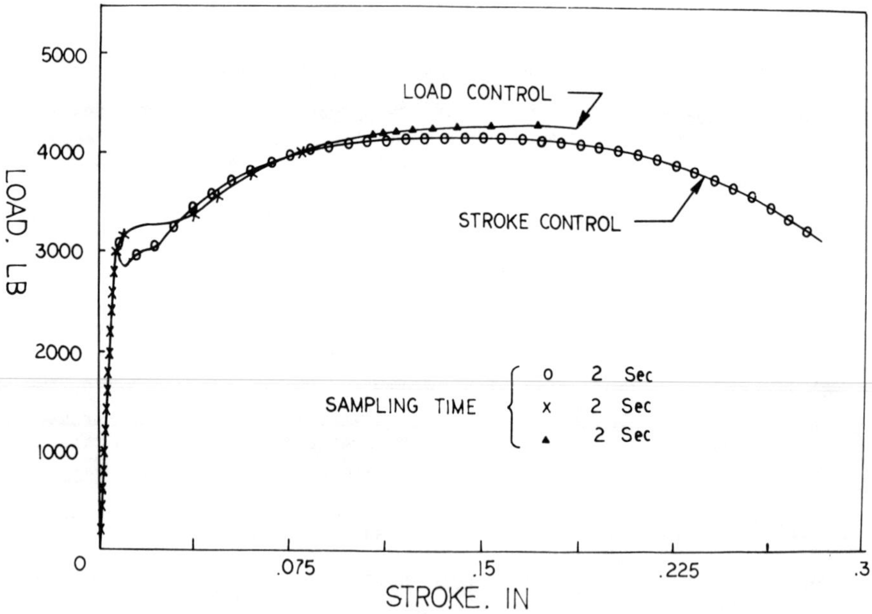

Fig. 2 Load-elongation (stroke) diagram of SAE 1020 steel at room temperature under load and displacement control. Note the absence of a yield point in load control. The two specimens have identical geometry.

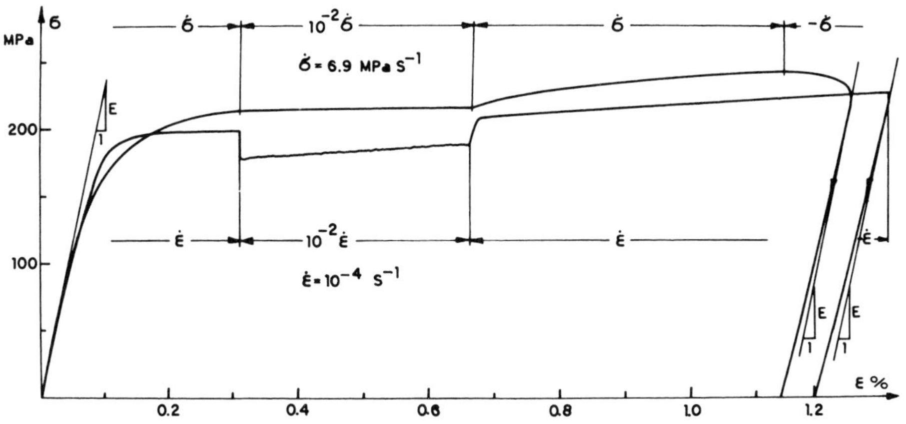

Fig. 3 Influence of control including rate changes and reversal on the stress-strain diagram of annealed AISI Type 304 Stainless Steel at room temperature. Top curve is obtained under load (stress) control, bottom curve for displacement (strain) control, from [12].

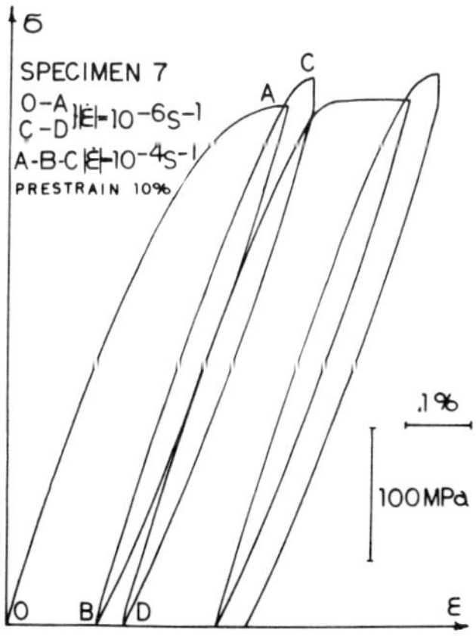

Fig. 4 Behavior of cold-worked AISI Type 304 Stainless Steel under strain control with changes in strain-rate magnitude at points of unloading. At D the program OD is repeated.

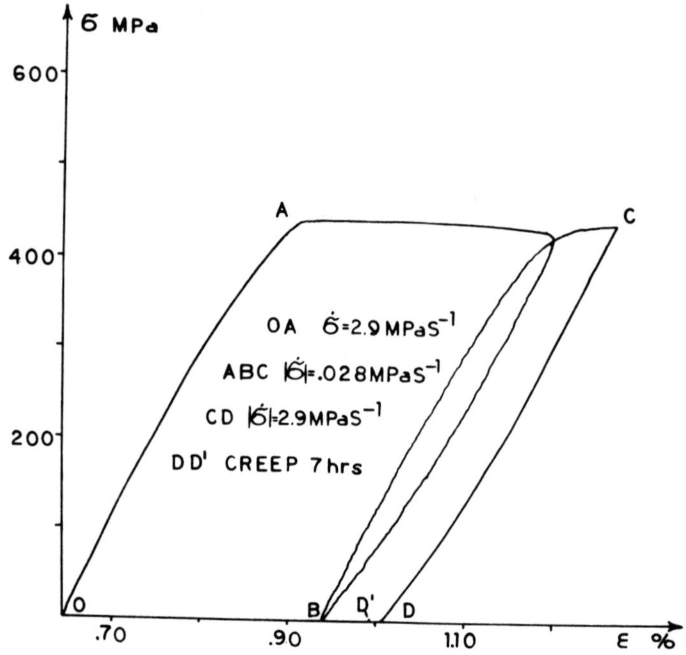

Fig. 5 Behavior of cold-worked AISI Type 304 Stainless Steel under stress control with changes in stress rate at points of unloading.

FLEXURAL MEMBERS

THE CYLINDRICAL SHELL LOADED BY A CONCENTRATED AXIAL FORCE

J. Lyell Sanders, Jr.

Division of Applied Sciences
Harvard University
Cambridge, Massachusetts 02138

ABSTRACT

The paper contains a closed-form solution to the problem according to the semi-membrane equations of cylindrical shells. The results, in combination with the shallow shell solution to the same problem, furnish a solution uniformly accurate over the shell surface.

Introduction

The problem of concentrated forces on infinitely long cylindrical shells has had a rather curious history. An accurate analytical representation in Fourier series form for the solution to the problem of equal opposing normal forces was obtained by Yuan and Ting [1] some years before the correct solution to the problem of a shallow cylindrical shell subjected to a normal force was obtained by Janhanshahi [2]. The complexity of the cylindrical shell equations has less to do with explaining the situation than does the form in which the solutions were represented. In the shallow shell problem attention is focused on the vicinity of the singularity and the results sought are essentially closed-form analytical expressions from which stresses, displacements, etc., can be obtained without any particular numerical difficulties. The analysis required to produce the solution was not elementary. The Fourier series solution to the complete cylindrical shell problem was easier to obtain but it has some serious limitations. The series representations for the various quantities of physical interest are not uniformly convergent. The computation of quantities which become unbounded is especially difficult anywhere near the load point.

Alternative representations of the solution in terms of Fourier integrals also involve difficulties in the same situation. Methods for overcoming the difficulties were devised by Morley [3] in the case of Fourier series and by Lukasiewicz [4] in the case of Fourier integrals. Such methods are effective for getting out numbers in the concentrated load problems as such (provided the shell is not extremely thin), but they are sufficiently complicated to discourage the use of these representations for the kernel functions required in integral equation formulations of boundary-value-problems. They also do not appear to be suited for taking advantage of the known closed-form solution to the corresponding shallow shell problem.

Buchwald [5] has suggested an approach to the problem by means of the method of matched asymptotic expansions. He showed that there is an "outer" solution consisting of a regular part plus a singular part. The regular part is accurate everywhere and the singular part "matches" the shallow shell (or "inner") solution. A suitable combination of inner and outer solutions is uniformly accurate. The solutions in [5] were left in the form of Fourier integrals.

The present author has developed a method for obtaining approximate, but accurate, closed-form outer solutions to a variety of cylindrical shell problems including the concentrated load problems [6,7,8]. The method depends upon making a slight modification in the governing partial differential equation. The small errors thus introduced do not destroy the "matching" property of the solutions. The normal load case was treated in [8]. The axial load case is treated in the present paper with some refinements in the analysis and in somewhat more detail. The remaining tangential load case is left for a later paper.

An interesting finding, which comes out of the analysis in all three load cases, is that the stress and deformation pattern does not at all resemble an elementary or "beam-like" solution within a region extending many diameters away from the point of load application. Other features of the solutions agree qualitatively with those found by previous investigators, all of whom (except Buchwald) considered self-equilibrated sets of loads rather than the single load. Previous results are conveniently gathered together in the book by Lukasiewicz [4]. Extensive calculations due to Bieger [9] are especially to be noted.

Fundamental Equations

Let R be the radius of the cylinder and h its thickness. Let θ be the circumferential angle and zR the axial distance. The fundamental small parameter involved in the theory is defined by

$$\varepsilon^2 = (h/R)[12(1-\nu^2)]^{-1/2} \tag{1}$$

where ν is Poisson's ratio. The way in which the variables involved are scaled and made dimensionless is somewhat arbitrary. What is suitable for one purpose is not suitable for another. To facilitate use of the present results in conjunction with those in [10] the same system will be used here. Let the axial load be P (positive in the direction of increasing z). Dimensionless displacements and stress functions (without an overbar) are given in terms of dimensional variables by

$$(\bar{u},\bar{v}) = (P/Eh)(u,v), \qquad (\bar{\chi}_z,\bar{\chi}_\theta) = PR\varepsilon^2(\chi_z,\chi_\theta)$$
$$\bar{w} = (P/Eh\varepsilon^2)w, \qquad \bar{\psi} = PR\psi \tag{2}$$

where E is Young's modulus.

For membrane and bending stress measures

$$\begin{aligned}(\bar{N}_z,\bar{N}_\theta,\bar{N}_{z\theta}) &= (P/R)(N_z,N_\theta,N_{z\theta})\\ (\bar{M}_z,\bar{M}_\theta,\bar{M}_{z\theta}) &= \varepsilon^2 P(M_z,M_\theta,M_{z\theta})\end{aligned} \tag{3}$$

Complex-valued displacements and stress functions are given in terms of characteristic functions φ and Φ by

$$\begin{aligned} u &= \varepsilon^{-2}\ddot{\Phi}' - \nu\varphi' & \chi_z &= -i\varepsilon^{-2}\ddot{\Phi}' - i\nu\varphi' \\ v &= -\varepsilon^{-2}\dddot{\Phi} - (2+\nu)\dot{\varphi} & \chi_\theta &= i\varepsilon^{-2}\dddot{\Phi} + i(2-\nu)\dot{\varphi} \\ w &= -\ddot{\Phi} + i(1+2i\varepsilon^2)\varphi & \psi &= i\ddot{\Phi} + (1+2i\varepsilon^2)\varphi \end{aligned} \tag{4}$$

where $(\ )^{\bullet} \equiv \frac{\partial}{\partial\theta}$, $(\ )' \equiv \frac{\partial}{\partial z}(\ )$. The characteristic functions Φ and φ, related by $\Phi'' = \varepsilon^2\varphi$ each satisfy the characteristic equation

$$L(\varphi) \equiv \nabla^4\varphi + \ddot{\varphi} - i\varepsilon^{-2}(1+2i\varepsilon^2)\varphi'' = 0 \tag{5}$$

except at $z = \theta = 0$ where there is a singularity. The physical displacements and stress functions are the real parts of the quantities appearing in (4) and similarly for other complex-valued variables which arise. The stress measures are given in terms of φ by

$$\begin{aligned}(N_z,N_\theta,N_{z\theta}) &= (\ddot{\varphi},\varphi'', -\dot{\varphi}')\\ (M_z,M_\theta,M_{z\theta}) &= -i[\varphi'' + \nu\ddot{\varphi},\ddot{\varphi} + \nu\varphi'', (1-\nu)\dot{\varphi}']\end{aligned} \tag{6}$$

Boundary Conditions

The concentrated load is assumed to be applied to the full infinite cylinder and to be equilibrated at infinity, half the load in each direction. By symmetry the analysis can be confined to the half of the cylinder $z > 0$. Two boundary conditions on displacements (at $z = 0$) follow from symmetry. A general derivation of boundary conditions on stress functions in terms of tractions and couples applied on the edge of a shell was given in [11]. In the present case the boundary conditions, holding on $z = 0$ and for $|\theta| < \pi$ are as follows

$$v = 0 \qquad \varepsilon^2 \chi_\theta = \frac{1}{4}\frac{|\theta|}{\theta}(1 - \cos\theta)$$

$$w = 0 \qquad \psi = -\frac{1}{4}\sin|\theta| \qquad . \tag{7}$$

Since the cylinder carries a load the stress-functions are in general multi-valued, and in two distinct ways. One sort of multi-valuedness is associated with complete circuits which enclose the axis of the cylinder. The other sort is associated with complete circuits which enclose the point of load application. The stress-functions are single-valued on the surface of the cylinder with a branch cut on $\theta = \pm\pi$, $-\infty < z < \infty$ and another on $\theta = 0$, $-\infty < z < 0$.

Between equations (4) and (7) two complex boundary conditions on the complete solutions φ_c and Φ_c can be derived. The results are

$$\varphi_c = -\frac{1}{4}(1+i\varepsilon^2\nu)|\theta|$$

$$\ddot{\Phi}_c = \frac{1}{4} i(\sin|\theta| - |\theta|) + \frac{1}{4}\varepsilon^2(2+\nu)|\theta| \tag{8}$$

to hold on $z = 0$ for $|\theta| < \pi$.

The stresses and displacements in the shell at infinity must correspond to an elementary beam-like state (combined bending and tension). Boundary conditions for a "reduced" problem can be obtained by subtracting out these elementary solutions. The solutions to be removed are (for $z \geqslant 0$)

$$8\pi\varphi_B = -(1+i\varepsilon^2\nu)\theta^2 + i[1-i\varepsilon^2(2-\nu)]\varepsilon^2 z^2 + 4\cos\theta + 8\pi B$$

$$8\pi\ddot{\Phi}_B = -i[1+i\varepsilon^2(2+\nu)]\theta^2 + (1+i\varepsilon^2\nu)(2i-\varepsilon^2 z^2) - 2\varepsilon^2 z^2\cos\theta$$

$$- 2i(2\cos\theta - \theta\sin\theta) - 8\pi A\cos\theta + 8\pi i B(1+2i\varepsilon^2) \qquad . \tag{9}$$

The A term here corresponds to a lateral rigid-body-motion. The B terms are a null solution for which displacements and stress-functions vanish identically. Proper choice of A and B makes possible the existence of a unique solution to the reduced problem for which $\varphi \to 0$, $\ddot{\Phi} \to 0$ as $z \to \infty$. The determination of A and B comes presently. Subtracting (9) for $z = 0$ from (8) yields (approximately) the boundary conditions of the reduced problem for φ and $\ddot{\Phi}$.

$$\varphi = -\frac{1}{4}|\theta| + \frac{1}{8\pi}\theta^2 - \frac{1}{2\pi}\cos\theta - B$$

$$\ddot{\Phi} = \frac{1}{4} i(\sin|\theta| - |\theta|) + \frac{1}{8\pi} i\theta^2 + \frac{1}{4\pi} i(2\cos\theta - \theta\sin\theta - 1)$$

$$+ \frac{1}{4}\varepsilon^2(2+\nu)|\theta| + A\cos\theta - iB \qquad . \tag{10}$$

All ε^2 terms except one have been dropped because asymptotic series solutions in powers of ε^2 cannot justifiably be carried beyond the first term according to the theory of thin shells. The remaining ε^2 term enters into the "inner" or shallow shell part of the solution which has boundary conditions corresponding to the *singular* terms in (10). The independent variable θ would be rescaled by a factor of ε and, in effect, the boundary conditions of the shallow shell solution would be

$$\begin{aligned} \varphi &= -\frac{1}{4}|\theta| \\ \ddot{\Phi} &= -\frac{1}{24}i|\theta|^3 + \frac{1}{4}\varepsilon^2(2+\nu)|\theta| \end{aligned} \qquad (11)$$

Since the shallow shell problem has already been completely solved [10] the analysis need not be pursued here. The boundary conditions for the outer solution are those given by (10) with the ε^2 term dropped.

The outer solution is one for which the first term in an asymptotic expansion satisfies the semi-membrane characteristic equation

$$L_{S.M.}(\varphi) \equiv (\ddot{\varphi}+\varphi)^{\cdot\cdot} - i\varepsilon^{-2}\varphi'' = 0 \quad . \qquad (12)$$

Observe that together with (12), $L_{S.M.}(\Phi) = 0$ and $\Phi'' = \varepsilon^2\varphi$ imply

$$\varphi = -i(\ddot{\Phi}+\Phi)^{\cdot\cdot} \quad . \qquad (13)$$

One may verify that the boundary conditions are compatible with this relation. The first condition of (10) becomes superfluous. The reduction in number of boundary conditions comes about because there are no edge effects on $z = 0$, as is apparent on physical grounds.

The condition $\ddot{\Phi} \to 0$ as $z \to \infty$ implies

$$\int_0^{\pi} \ddot{\Phi}(z,\theta)\,d\theta = 0 \,, \qquad \int_0^{\pi} \ddot{\Phi}(z,\theta)\cos\theta\,d\theta = 0 \quad . \qquad (14)$$

These can be proved by a Fourier analysis of the (even) solutions to (12). The conditions (14), applied on $z = 0$, determine A and B to be

$$A = -9i/8\pi \qquad B = -\pi/12 \quad . \qquad (15)$$

The final boundary conditions on the reduced solution are then

$$\begin{aligned} \varphi &= \frac{\pi}{12} - \frac{1}{4}|\theta| + \frac{1}{8\pi}\theta^2 - \frac{1}{2\pi}\cos\theta \\ \ddot{\Phi} &= \frac{i\pi}{12} + \frac{i}{4}(\sin|\theta| - |\theta|) + \frac{i}{8\pi}(\theta^2 - 5\cos\theta - 2\theta\sin\theta - 2) \end{aligned} \qquad (16)$$

to hold on $z = 0$ for $|\theta| < 2\pi$. The right-hand-sides of (16) are (not obviously) even about $\theta = \pm\pi$ and hold in the extended range $|\theta| < 2\pi$, reflecting the physical fact that in the reduced problem there are no resultant forces or couples acting on any cross-section of the cylinder for $z > 0$. In mathematical terms, φ and $\ddot{\Phi}$ are single-valued for $z > 0$ and the boundary conditions (16) can be continued periodically.

A complete and exact solution to the outer problem in Fourier series form could now be obtained trivially from the characteristic equation and (16). The solution in such form is useful (computationally) provided εz is not too small. If εz is $O(1)$ or larger it is the best form to use. However, the series forms, especially those required for stresses, become increasingly unsuitable as εz becomes small.

Outer Solution

The exact outer solution satisfies (12) subject to the stated boundary conditions and conditions at infinity. However, as shown elsewhere [6,8] a very accurate approximate solution can be obtained by basing the analysis upon the following "substitute characteristic equation" in place of (12).

$$L'_{S.M.}(\varphi) \equiv \ddot{\varphi} + \frac{1}{2}\varphi - i^{1/2}\varepsilon^{-1}\varphi' = 0 \quad . \tag{17}$$

The approximation is valid provided φ is periodic and vanishes at infinity, as in the present circumstances. The following asymptotic solution to (17) is accurate provided $\xi = \varepsilon z$ is sufficiently small, about 1/2 or less.

$$\varphi \sim \int_{-\infty}^{\infty} \varphi(0,\eta)K(\xi,\theta-\eta)\, d\eta \tag{18}$$

where

$$K(\xi,\eta) = \frac{i^{1/4}}{\sqrt{4\pi\xi}} \exp\left\{-\frac{1}{2}\, i^{3/2}\xi - \frac{i^{1/2}\eta^2}{4\xi}\right\} .$$

The derivation of (18) was discussed in [8].

For boundary values given by (16) all the integrations in (18) can be carried out in closed form. Results corresponding to the regular terms and singular terms in (16) are given here separately.

$$2\pi\varphi_R = \left(\frac{\pi^2}{6} + \frac{1}{4}\theta^2 - \frac{1}{2}\, i^{3/2}\xi\right)\exp(-i^{3/2}\xi/2) - \cos\theta\,\exp(i^{3/2}\xi/2) \tag{19}$$

$$2\pi\ddot{\Phi}_R = \left(\frac{i\pi^2}{6} - \frac{i}{2} + \frac{i}{4}\theta^2 + \frac{1}{2}i^{1/2}\xi\right)\exp(-i^{3/2}\xi/2)$$

$$- i\left(\frac{5}{4}\cos\theta + \frac{1}{2}\theta\sin\theta - i^{3/2}\xi\cos\theta\right)\exp(i^{3/2}\xi/2) \tag{20}$$

$$\varphi_s = -\frac{1}{4}[\theta\,\mathrm{erf}\,\mu - 2i^{7/4}(\xi/\pi)^{1/2}e^{-\mu^2}]\exp(-i^{3/2}\xi/2) \tag{21}$$

$$\ddot{\Phi}_s = \frac{1}{8}(e^{i\theta}\,\mathrm{erf}\,\mu_1 - e^{-i\theta}\,\mathrm{erf}\,\mu_{-1})\exp(i^{3/2}\xi/2)$$

$$- \frac{i}{4}[\theta\,\mathrm{erf}\,\mu - 2i^{7/4}(\xi/\pi)^{1/2}e^{-\mu^2}]\exp(-i^{3/2}\xi/2) \tag{22}$$

where

$$\mu = i^{1/4}\theta(4\xi)^{-1/2}$$

$$\mu_n = \mu + ni^{3/4}\xi^{1/2} \quad . \tag{23}$$

The semi-membrane version of the beam parts of φ and $\ddot{\Phi}$ are given by

$$2\pi\varphi_B = -\frac{1}{4}(\theta^2 - i\xi^2) + \cos\theta - \pi^2/6 \tag{24}$$

$$2\pi\ddot{\Phi}_B = -\frac{1}{4}\xi^2(1+2\cos\theta) + \frac{i}{4}(2 - \frac{2}{3}\pi^2 + 5\cos\theta + 2\theta\sin\theta - \theta^2) \quad . \tag{25}$$

The results (19) to (25) are valid for $\xi \geqslant 0$.

Displacements and Stress Functions

From (13) it follows that for semi-membrane solutions φ and Φ are the same order and the following simplified formulas for the complex displacements and stress functions hold in place of (4).

$$u = i\chi_z = \varepsilon^{-2}\ddot{\Phi}'$$

$$v = i\chi_\theta = -\varepsilon^{-2}\dddot{\Phi} \quad . \tag{26}$$

$$w = i\psi = -\ddot{\Phi} + i\varphi$$

Results for the beam, regular, and singular parts are as follows (for $\xi \geqslant 0$)

$$2\pi u_B = -\frac{1}{2}\varepsilon^{-1}(1+2\cos\theta)\xi \tag{27}$$

$$2\pi u_R = \frac{1}{4} i^{1/2}\varepsilon^{-1}\left[\left(1+\frac{\pi^2}{3}+\frac{1}{2}\theta^2 - i^{3/2}\xi\right)\exp(-i^{3/2}\xi/2) + \frac{1}{2}(2\theta\sin\theta - 3\cos\theta - 4i^{3/2}\xi\cos\theta)\exp(i^{3/2}\xi/2)\right] \tag{28}$$

$$u_s = -\frac{1}{8} i^{1/2}\varepsilon^{-1}\left\{-\frac{1}{2} i(e^{i\theta}\,\mathrm{erf}\,\mu_1 - e^{-i\theta}\,\mathrm{erf}\,\mu_{-1})\exp(i^{3/2}\xi/2) + [\theta\,\mathrm{erf}\,\mu - 2i^{7/4}(\xi/\pi)^{1/2}e^{-\mu^2}]\exp(-i^{3/2}\xi/2)\right\} \tag{29}$$

$$2\pi v_B = -\frac{1}{2}\varepsilon^{-2}[\xi^2\sin\theta + \frac{i}{2}(2\theta\cos\theta - 2\theta - 3\sin\theta)] \tag{30}$$

$$2\pi v_R = -\frac{i}{4}\varepsilon^{-2}[2\theta\exp(-i^{3/2}\xi/2) + (3\sin\theta - 2\theta\cos\theta - 4i^{3/2}\xi\sin\theta)\exp(i^{3/2}\xi/2)] \tag{31}$$

$$v_s = -\frac{i}{8}\varepsilon^{-2}[(e^{i\theta}\,\mathrm{erf}\,\mu_1 + e^{-i\theta}\,\mathrm{erf}\,\mu_{-1})\exp(i^{3/2}\xi/2) - 2\,\mathrm{erf}\,\mu\,\exp(-i^{3/2}\xi/2)] \tag{32}$$

$$2\pi w_B = \frac{1}{4}[2\xi^2\cos\theta - i(2+\cos\theta + 2\theta\sin\theta)] \tag{33}$$

$$2\pi w_R = \frac{i}{4}[2\exp(-i^{3/2}\xi/2) + (\cos\theta + 2\theta\sin\theta - 4i^{3/2}\xi\cos\theta)\exp(i^{3/2}\xi/2)] \tag{34}$$

$$w_s = -\frac{1}{8}(e^{i\theta}\,\mathrm{erf}\,\mu_1 - e^{-i\theta}\,\mathrm{erf}\,\mu_{-1})\exp(i^{3/2}\xi/2) \tag{35}$$

$$2\pi w'_B = \varepsilon\xi\cos\theta \tag{36}$$

$$2\pi w'_R = \frac{1}{8} i^{1/2}\varepsilon[2\exp(-i^{3/2}\xi/2) + (7\cos\theta - 2\theta\sin\theta + 4i^{3/2}\xi\cos\theta)\exp(i^{3/2}\xi/2)] \tag{37}$$

$$w'_s = -\frac{1}{16} i^{3/4}\varepsilon[4(\pi\xi)^{-1/2}e^{-\mu^2}\exp(-i^{3/2}\xi/2) + i^{3/4}(e^{i\theta}\,\mathrm{erf}\,\mu_1 - e^{-i\theta}\,\mathrm{erf}\,\mu_{-1})\exp(i^{3/2}\xi/2)] \tag{38}$$

For sufficiently small values of ξ the singular parts of the variables have an "active zone" within which $\mathrm{erf}\,\mu_n$ and $e^{-\mu^2}$ vary rapidly with θ. Outside of this zone $\mathrm{erf}\,\mu_n \approx \pm 1$ and $e^{-\mu^2} \approx 0$. The boundary of the active zone is fairly well defined by $|\mu| = 1.5$ or $\theta = \pm 3\sqrt{\xi}$, which is a narrow parabola on the surface of the cylinder centered on the generator $\theta = 0$. The active zone extends completely around the cylinder and overlaps itself at about $\xi = 1$. The asymptotic formula (18) leaves any overlapping out of account, hence the formulas are only accurate up to some value of ξ less than one.

No computational difficulties are encountered with exact Fourier series solutions to (12) for values of $\xi \geqslant 0.1$ $(z \geqslant 0.1\ \varepsilon^{-1})$ so the accuracy of the formulas can be tested numerically. The formula for u_c was tested by this means for several values of ξ. At $\xi = 0.5$ the errors were found to be less than 1% except at $\theta = \pi$ where the error was less than 2%. For smaller values of ξ the errors are less. As ξ becomes larger than 0.5 the accuracy deteriorates the most near $\theta = \pm\pi$, as can be expected because of overlap.

All foregoing results are valid only for $\xi \geqslant 0$ and cannot be continued analytically to negative values of ξ because of the essential singularity in erf μ_n and $e^{-\mu^2}$ at $\xi = 0$. The continuation can, however, be effected by symmetry arguments. The physical displacements and stresses must be either even or odd in ξ. Because of the axial load the real stress functions χ_θ and ψ (which are the imaginary part of v and w) are multi-valued and the symmetry properties for them are more complicated. One can argue that symmetry of the stresses requires the complex v_c and w_c to be odd in ξ *except* for a pure imaginary rigid-body-motion. The complication is that the rigid-body-motion involved in the continuation across $\xi = 0$, $\theta > 0$ differs in sign from that for continuation across $\xi = 0$, $\theta < 0$. First consider the case $\theta > 0$. To the previous expressions for v_c and w_c add an imaginary rigid-body-motion to obtain (for $\xi \geqslant 0$)

$$\begin{aligned} &v_c(\xi,\theta) - \frac{1}{4} i\varepsilon^{-2}(1-\cos\theta) \\ &w_c(\xi,\theta) + \frac{1}{4} i \sin\theta \end{aligned} \quad . \tag{39}$$

These vanish on $\xi = 0$ for $\theta > 0$. Continue them as odd functions of ξ to obtain (for $\xi \leqslant 0$)

$$\begin{aligned} &-v_c(-\xi,\theta) + \frac{1}{4} i\varepsilon^{-2}(1-\cos\theta) \\ &-w_c(-\xi,\theta) - \frac{1}{4} i \sin\theta \end{aligned} \quad . \tag{40}$$

To complete the continuation subtract the same rigid-body-motion to obtain the formulas (for $\xi \leqslant 0$)

$$\begin{aligned} v_c(\xi,\theta) &= \frac{1}{2} i\varepsilon^{-2}(1-\cos\theta) - v_c(-\xi,\theta) \\ w_c(\xi,\theta) &= -\frac{1}{2} i \sin\theta - w_c(-\xi,\theta) \end{aligned} \quad . \tag{41}$$

A similar process applies in the case $\theta < 0$. The complex u_c and w_c' are single-valued and even in ξ. The following results (for $\xi \leqslant 0$) combine all cases.

$$
\begin{aligned}
u_c(\xi,\theta) &= u_c(|\xi|,\theta) \\
v_c(\xi,\theta) &= \frac{1}{2}\, i\varepsilon^{-2}\, \frac{|\theta|}{\theta}\,(1-\cos\theta) - v_c(|\xi|,\theta) \\
w_c(\xi,\theta) &= -\frac{1}{2}\, i\,\sin|\theta| - w_c(|\xi|,\theta) \quad .
\end{aligned}
\tag{42}
$$

These, together with the formulas for $\xi \geqslant 0$ define u_c, v_c, and w_c as single-valued functions on the cylinder cut on $\theta = 0$ from $\xi = -\infty$ to $\xi = 0$ and cut all along $\theta = \pi$. Similar results for φ_c and $\ddot{\Phi}_c$ for negative values of ξ are as follows

$$
\begin{aligned}
\varphi_c(\xi,\theta) &= -\frac{1}{2}\,|\theta| - \varphi_c(|\xi|,\theta) \\
\ddot{\Phi}_c(\xi,\theta) &= \frac{1}{2}\, i(\sin|\theta| - |\theta|) - \ddot{\Phi}_c(|\xi|,\theta)
\end{aligned}
\quad .
\tag{43}
$$

A slight anomaly in the solution needs some discussion. Consider the function u_c, which should be even in ξ. An expansion in powers of ξ for small values of ξ and for $\theta > 0$ contains the following term linear in ξ

$$
\frac{\varepsilon^{-1}}{16\pi}\left[(\pi-\theta)\sin\theta - \frac{5}{2}\cos\theta + \frac{1}{2}(\theta-\pi)^2 - 1 - \frac{\pi^2}{6}\right]\xi \quad . \tag{44}
$$

Such a term would not be present in an exact solution. However, this term is numerically small, amounting to about 3 1/2% (r.m.s.) of the ξ term in the beam part of u in the range $0 \leqslant \theta \leqslant \pi$. Similar remarks apply to unwanted even terms in v_c and w_c.

Further calculations show that for small values of ξ the contribution of the beam part to the solution is nearly annihilated by terms coming from the regular and singular parts. On the other hand, when ξ is the order of one, the regular and singular parts tend to cancel each other, leaving the beam part dominant.

Stress Measures

The stress measures are given by Eqs. (6) in terms of the partial derivations of φ. Results are (for $\xi \geqslant 0$).

$$
2\pi(\ddot{\varphi}_R + \ddot{\varphi}_B) = -\frac{1}{2}\,[1-\exp(-i^{3/2}\xi/2)] - \cos\theta\,[1-\exp(i^{3/2}\xi/2)] \tag{45}
$$

$$
\ddot{\varphi}_s = -\frac{1}{4}\, i^{1/4}(\pi\xi)^{-1/2} e^{-\mu^2} \exp(-i^{3/2}\xi/2) \tag{46}
$$

$$
2\pi(\varphi''_R + \varphi''_B) = \frac{1}{2}\, i\varepsilon^2 [1-\exp(-i^{3/2}\xi/2)] \tag{47}
$$

$$\varphi''_S = \frac{1}{4} i^{5/4}\varepsilon^2(\pi\xi)^{-1/2}[1-(i^{1/2}/2\xi) + (i\theta^2/4\xi^2)]e^{-\mu^2}\exp(-i^{3/2}\xi/2) \tag{48}$$

$$2\pi(\dot{\varphi}'_R + \dot{\varphi}'_B) = \frac{1}{2} i^{3/2}\varepsilon[\sin\theta\exp(i^{3/2}\xi/2) - \frac{1}{2}\theta\exp(-i^{3/2}\xi/2)] \tag{49}$$

$$\dot{\varphi}'_S = \frac{1}{8}\varepsilon(\pi\xi)^{-1/2}[(i^{1/4}\theta/\xi)e^{-\mu^2} + i^{3/2}(\pi\xi)^{1/2}\operatorname{erf}\mu]\exp(-i^{3/2}\xi/2). \tag{50}$$

The expressions for φ''_R and φ''_S were obtained (more accurately) by means of (12) rather than by direct differentiation of (19) and (21). For negative values of ξ, $\ddot{\varphi}$ and φ'' are to be continued as odd functions of ξ, and $\dot{\varphi}'$ as an even function of ξ.

Inner Solution and Matching

The solution to the problem according to the shallow shell equations was given in [10] wherein the coordinates x and y are the present z and θ, and 2λ is the present ε^{-1}. The dependent variables were all scaled the same as in the present paper. In [10] the cut associated with the definition of multi-valued functions was placed on the positive x axis. To make results in [10] correspond to those in the present paper add the rigid-body-motion

$$u = 0\ , \qquad v = \frac{1}{4} i\varepsilon^{-2}y^2, \qquad w = -\frac{1}{2} iy \tag{51}$$

to the expressions given by Eqs. (67) to (69) in that paper and let the polar angle (θ in [10]) vary in the range $(-\pi,\pi)$.

The *singular parts* of the variables in the present semi-membrane solution match the corresponding variables of the shallow shell solution. Put $\theta = \varepsilon^{1/2}s$ (or $y = \varepsilon^{1/2}s$) and calculate the limit z, s, μ fixed, $\varepsilon \to 0$. Matching means that these limits are exactly the same for the two solutions. The limits, called the "common parts" read as follows in terms of ξ and θ (for $\xi \geqslant 0$).

$$u_{CM} = -\frac{1}{4}\varepsilon^{-1}[i^{1/2}\theta\operatorname{erf}\mu + 2i^{1/4}(\xi/\pi)^{1/2}e^{-\mu^2}] \tag{52}$$

$$v_{CM} = \frac{1}{8}\varepsilon^{-2}[i^{1/2}(2\xi + i^{1/2}\theta^2)\operatorname{erf}\mu + 2i^{3/4}\theta(\xi/\pi)^{1/2}e^{-\mu^2}] \tag{53}$$

$$w_{CM} = -\frac{1}{4}[i\theta\operatorname{erf}\mu + 2i^{3/4}(\xi/\pi)^{1/2}e^{-\mu^2}] \tag{54}$$

$$w'_{CM} = -\frac{1}{4} i^{3/4}\varepsilon(\pi\xi)^{-1/2}e^{-\mu^2} \tag{55}$$

$$\ddot{\varphi}_{CM} = -\frac{1}{4}\, i^{1/4}(\pi\xi)^{-1/2} e^{-\mu^2} \tag{56}$$

$$\varphi''_{CM} = \frac{1}{8}\, i^{1/4}\varepsilon^2(\pi\xi^3)^{-1/2}[1 - \frac{1}{2}\, i^{1/2}(\theta^2/\xi)] e^{-\mu^2} \tag{57}$$

$$\dot{\varphi}'_{CM} = \frac{1}{8}\, i^{1/4}\varepsilon(\pi\xi^3)^{-1/2}\theta e^{-\mu^2} \quad . \tag{58}$$

For negative values of ξ, v_{CM} and w_{CM} are given by

$$v_{CM}(\xi,\theta) = \frac{1}{4}\, i\varepsilon^{-2}\theta|\theta| - v_{CM}(|\xi|,\theta) \tag{59}$$

$$w_{CM}(\xi,\theta) = -\frac{1}{2}\, i|\theta| - w_{CM}(|\xi|,\theta) \quad . \tag{60}$$

The common parts of the other variables are continued by symmetry in a manner already indicated.

A uniformly accurate representation of the variables can be formed by a combination of the shallow shell, semi-membrane, and common parts. For w

$$w_U = w_{SS} + w_{SM} - w_{CM} \tag{61}$$

and similarly for other variables. Here w_{SM} is w_c of the present paper for small ξ, or the beam part and a Fourier series representation for larger values of ξ.

Concluding Remarks

The formulas given for the variables of the semi-membrane solution are accurate to within errors of about 1% or less for $\xi \leqslant 0.5$ $(z \leqslant 0.5\ \varepsilon^{-1})$. The range of ξ and the accuracy are adequate for most practical purposes. For shells with a radius to thickness ratio of about thirty or less the errors are no greater than those due to the use of thin shell theory itself. The shallow shell and semi-membrane solutions combined furnish a uniformly accurate solution to the shell problem.

The semi-membrane results are simpler than the shallow shell results and are easy to compute, even on a programmable hand calculator. The way a cylindrical shell reacts to a concentrated axial load can be determined, in a general way, simply by looking at the formulas.

Acknowledgments

The support of the National Science Foundation under Grant CME-80-07477 and the Division of Applied Sciences, Harvard University, is gratefully acknowledged.

References

1. Yuan, S.W. and Ting, L., "On Radial Deflections of a Cylinder Subjected to Equal and Opposite Concentrated Radial Loads," *J. Appl. Mech.* 24 (1957) 278-282.

2. Janhanshahi, A., "Some Notes on Singular Solutions and Green's Functions in the Theory of Plates and Shells," *J. Appl. Mech.* 31 (1964) 441-446.

3. Morley, L.S.D., "The Thin-walled Circular Cylinder Subjected to Concentrated Radial Loads," *Q.J. Mech. Appl. Math.* 13 (1960) 24-37.

4. Lukasiewicz, S., *Local Loads in Plates and Shells*, Sijthoff and Noordhoff, The Netherlands (1979).

5. Buchwald, V.T., "Some Problems of Thin Circular Cylindrical Shells, I. The Equations," *J. Math. and Phys.* 46 (1967) 237-252.

6. Sanders, J.L., Jr., "Closed Form Solution to the Semi-infinite Cylindrical Shell Problem," *Contributions to the Theory of Aircraft Structures*, The Van der Neut Volume, Delft University Press, Rotterdam (1972) 229-237.

7. Chan, A.W. and Sanders, J.L., Jr., "Analysis of a Mitered Joint in a Cylindrical Shell in Bending," *Int. J. Mech. Sci.* 22 (1980) 621-636.

8. Sanders, J.L., Jr., "The Cylindrical Shell Loaded by a Concentrated Normal Force," *Mechanics Today, Vol. 5*, Pergamon Press (1980) 427-438.

9. Bieger, K.W., "Die Kreiszylinderschale unter Konzentrierten Belastungen," *Ing. Arch.* 30 (1961) 57-62.

10. Sanders, J.L., Jr., and Simmonds, J.G., "Concentrated Forces on Shallow Cylindrical Shells," *J. Appl. Mech.* 37 (1970) 367-373.

11. Sanders, J.L., Jr., On Stress Boundary Conditions in Shell Theory," (brief note), *J. Appl. Mech.* 47 (1980) 202-204.

POLYGONAL PLATES UNDER CONCENTRATED LOADS

Thein Wah
Professor
Civil and Mechanical Engineering
Texas A&I University
Kingsville, Texas 78363

Abstract

A general procedure is presented for the solution of elastic plates with rectilinear boundaries under concentrated loads. The method uses eignefunctions for angular regions together with a least squares technique. Excellent agreement is found with the few exact solutions that exist for triangular plates under concentrated loads.

Principal Symbols

a, b, c,	length of sides of triangular plate
$F(\theta)$, $F(\theta,\lambda)$	eigenfunctions
M_r, M_θ, $M_{r\theta}$	bending moments & twisting moment
Q_r, Q_θ	shears
P	concentrated load on plate
w	deflection of plate
r, θ	polar coordinates
α_1, α_2, α_3	plate angles
λ	root of transcendental equation
ν	Poisson's ratio

Introduction

The problem of the elastic plate under concentrated loads is mathematically difficult and few theoretically exact solutions exist. The only solutions which can be classified as exact are for the circular plate under a concentrated load. These solutions have a logarithmic singularity for moments under the load, which is appropriate for the classical (Lagrange) plate theory. The singularity can be avoided only by using a three dimensional theory. For practical purposes the classical theory is still easily the most convenient to use and has the widest applicability. Sufficiently accurate values of bending moments can be obtained close to the concentrated load.

For certain types of rectangular plates, it is possible to obtain a formal solution by expanding the concentrated load in a Fourier series. However, at the point of load application, the series expansions for the bending moments do not converge. Supplementary formulas, exhibiting the required logarithmic singularity have been devised to calculate moment values, but, obviously, such formulas can be obtained only for certain types of boundary conditions.

The advent of electronic computation makes possible a formulation which combines all the main features of an exact solution, namely, the exact satisfaction of all boundary conditions and the expected logarithmic singularity at the concentrated load.

This paper presents such a solution. While it cannot be claimed that the solution is exact throughout the plate, the limited results obtained indicate that it is very nearly so. The solution applies to any polygonal plate whatever the boundary conditions.

It is important to point out that the method is entirely different from the "λ-method" of Quinlan[1]. In Quinlan's solution the boundary conditions are satisfied by solving eight infinite sets of simultaneous equations (for quadrilateral plates) by various Fourier expansions. In our method the boundary conditions are identically satisfied by every term of the series used. Fourier methods are not involved at all. On the other hand there are, in the most general case, four infinite sets of equations to be satisfied. These could possibly be satisfied by expansion in Fourier series. However, it is easier to satisfy them, without expansion, in the sense of least squares, because the equations involve both the coordinate variables.

While the method is general, this paper confines attention to the triangular plate. It turns out that the triangular plate is indeed the basic problem, and application to plates with a greater number of sides is a straightforward logical extension.

Differential Equation

The differential equation governing the deflection w of the plate, for all points except at the load itself is

$$\nabla^4 w = 0 \qquad (1)$$

where $\nabla^4 = (\nabla^2)^2$ is the biharmonic operator. $D = Eh/12(1-\nu^2)$ is defined as the flexural rigidity of the plate. E is the elastic tensile modulus, h is the plate thickness and ν is Poisson's ratio.

Equation (1) is to be solved, for a given position of the concentrated load P, to satisfy the specified boundary conditions on the three edges of the plate shown in Fig. 1. The conventional boundary conditions are simple support, clamped edges and free edges, and may occur in any combination provided that the plate is statically stable. It is also possible to remove one of the supports at the corners of the plate.

Using polar coordinates (r,Θ) (Fig. 1) equation (1) assumes the form

$$\left(\frac{\partial^2}{\partial r^2} + \frac{1}{r}\frac{\partial}{\partial r} + \frac{1}{r^2}\frac{\partial^2}{\partial \Theta^2} \right)^2 w = 0 \qquad (2)$$

The boundary conditions may be stated as follows for the edge $\Theta = \alpha_1$ (Fig. 1).
Simply supported edge

$$w = 0, \quad \nabla^2 w = 0 \quad \text{at } \Theta = \alpha_1 \tag{3}$$

Clamped edge

$$w = 0, \quad \frac{\partial w}{\partial \Theta} = 0 \quad \text{at } \Theta = \alpha_1 \tag{4}$$

Free edge

$$M_\Theta = 0, \quad Q_\Theta + \frac{\partial M_{\Theta r}}{\partial r} = 0 \quad \text{at } \Theta = \alpha_1 \tag{5}$$

together with the following definitions for moments and shears

$$M_r = -D \{(1-\nu) \frac{\partial^2 w}{\partial r^2} + \nu \nabla^2 w\} \tag{6}$$

$$M_\Theta = -D \{\nabla^2 w - (1-\nu) \frac{\partial^2 w}{\partial r^2} \} \tag{7}$$

$$M_{r\Theta} = -M_{\Theta r} = (1-\nu) D \left(\frac{1}{r} \frac{\partial^2 w}{\partial r \partial \Theta} - \frac{1}{r^2} \frac{\partial w}{\partial \Theta}\right) \tag{8}$$

$$Q_r = -D \frac{\partial}{\partial r} \nabla^2 w, \quad Q_\Theta = -D \frac{1}{r} \frac{\partial}{\partial \Theta} \nabla^2 w \tag{10}$$

Eigenfunctions

We assume w in the form

$$w = r^{\lambda+1} F(\Theta,\lambda) \tag{11}$$

Substitution into (2) show that

$$F(\Theta,\lambda) = C_1 \sin(\lambda+1)\Theta + C_2 \cos(\lambda+1)\Theta + C_3 \sin(\lambda-1)\Theta + C_4 \cos(\lambda-1)\Theta \tag{12}$$

The quantities C_1, C_2, C_3, C_4 and λ are constants to be determined from the boundary conditions along the edges AB and AC. Since the boundary conditions are homogenous, the result of substitution of (12) into the boundary conditions is a transcedental equation for determining the value of λ and an eigenfunction $F(\Theta,\lambda)$ which is determined to within a multiplicative constant. The transcendental equations for various boundary conditions and the associated eigenfunctions have been given previously [2] in Tables 1(a) and 2(a) in the Appendix. Roots of all the transcendental equations have been computed for various angles [3] .

Definition of Regions

In Fig. 1, O is the position of the concentrated load. We connect the point O to the vertices of the triangle and extend the lines to meet the opposite sides and complete the triangle A'B'C' as shown in the figure.

For the triangular regions designated 1, 2 and 3 in the figure, the deflection w of the plate may be completely described by three sets of eigenfunctions of the type mentioned above, with undetermined multiplicative constants. These eigenfunctions depend only on the boundary conditions on the edges (AB', AC') for region 1, on the edges (BC', BA') for region 2 and on the edges (CA', CB') for region 3, and of course, on the angles α_1, α_2, and α_3.

The triangle A'B'C', designated region 4 in the figure, contains the concentrated load at O.

The general solution of equation(1)for this region is

$$w' = \frac{P}{8\pi D} R^2 \log R + C_1 + C_2 R^2 + \sum_{n=1}^{\infty} \left(A_n R^n + A_{n+1} R^{n+2} \right) \cos n\Theta_c$$

$$+ \sum_{n=1}^{\infty} \left(B_n R^n + B_{n+1} R^{n+2} \right) \sin n\Theta_c \qquad (13)$$

where A_i, B_i and C_i are unknown constants. R is the radius from the loaded point and Θ_c is defined in Fig. 1. It is easily shown that the integral of the vertical shear taken around the loaded point for any circle, however small, is P, and is given by the logarithmic term, the other terms contributing nothing to the integral of the shear. These functions are well known and have been used for the calculation of circular plates under concentrated load. We designate them circle function for the purposes of this paper.

Continuity Conditions

Consider the region 1 and the triangle O B'C' which is part of region 4.

The deflection function for region 1 is the series of eigenfunctions w of equation (11), while the deflection function for the triangle O'B'C' is w' as given by equation (13). These two functions must satisfy four continuity conditions across C'B'

$$w = w'$$

$$\frac{\partial w}{\partial n} = \frac{\partial w'}{\partial n'} \qquad (14)$$

$$\frac{\partial^2 w}{\partial n^2} = \frac{\partial^2 w'}{\partial n'^2}$$

$$\frac{\partial}{\partial n} \nabla^2 w = - \frac{\partial}{\partial n'} \nabla^2 w'$$

The four equations (14) represent continuity of deflection, slope, curvature and shear, where n and n' define normals to the line B'C'.

The continuity conditions (14) must also be satisfied across the lines C'A' and A'B'.

Least Squares

It is evident that these continuity conditions can never be satisfied exactly. They can, however, be satisfied in the sense of least squares and a procedure identical to that given in reference 2 may be used for deriving the normal equations.

Considering the first of equations (14), we start by writing it at a chosen number of points p along B'C'. We thus obtain the matrix equation

$$\underset{\sim}{A} = \underset{\sim}{B} \tag{15}$$

The matrix $\underset{\sim}{A}$ contains the coefficients of the constants of both the eigenfunctions and the circle functions. The right hand side matrix B consists only of the logarithmic term of the circle functions (13) written for various values of R.

The number of collocation points p must be at least as many as the sum of the unknown constants to be determined. If we designate the number of unknown constants as q, then the matrix $\underset{\sim}{A}$ is of dimension p x q.

If we multiply the ith row of $\underset{\sim}{A}$ by a weighting coefficient w_i we can form an associated weighting matrix $\underset{\sim}{A}^*$. The weighting coefficients may be chosen in accordance with Simpson's rule for approximate integration. If we multiply the equal sides of (15) by the transpose of $\underset{\sim}{A}^*$ there results

$$\underset{\sim}{A}^{*T}\underset{\sim}{A} = \underset{\sim}{A}^{*T}\underset{\sim}{B} \tag{16}$$

This matrix equation constitutes the normal equations of the least squares procedure. It is sufficient to determine the q unknown coefficients, for the L.H.S. is a q x q matrix.

Since however, we have to satisfy four continuity equations across B'C', we must obtain equations similar to (16) for the remaining three continuity equations in (14).

If we designate deflections, slopes, curvatures and shears as "modes", we have four modes for each of the three lines B'C', C'A' and A'B'. It turns out that the most accurate procedure is to first write the collocation equations, for each mode, for all the lines, just as if the three lines were one continuous line, and to repeat this procedure for all the modes. The resulting matrix contains the coefficients of all the unknown constants. Thus if there are p_1 unknown constants for each of the regions 1, 2 and 3 and p_2 unknown constants for the circle functions, the total number of unknowns, and therefore the number of columns in the matrix $\underset{\sim}{A}$ is $3p_1 + p_2$. The number of rows of the matrix depends on the number of collocation points chosen. The procedure adopted was to derive the normal equations for each mode on all the lines and to assemble them into another matrix, and finally obtain the normal equations for all the modes. Very large matrix orders can be avoided, of course, by suitable partitioning.

Classification of Plates

Although there is no difficulty in choosing the eigenfunctions for the regions 1, 2 and 3, the circle functions are different for different plate geometries and load points. For this purpose it is convenient to designate triangular plates as belonging to one of three types. Type 1 is an equilateral plate with identical boundary conditions on all three sides and loaded at the centroid. Reference to Fig. 1 shows that it is sufficient to write the continuity equations along one line only, say, B'C'. The circle functions must be chosen so that they are symmetrical with respect to all three medians. The appropriate function is

$$w' = \frac{PR}{8\pi D} \log R + C_1 + C_2R^2 + \sum_{j=1}^{\infty} (A_j R^{3j} + A_{j+1} R^{3j+2}) \cos 3j\Theta_c \quad (17)$$

If the plate is symmetrical about AA' only, it is designated Type 2, and the appropriate deflection function for region 4 is

$$w' = \frac{P}{8\pi D} R^2 \log R + C_1 + C_2R^2 + \sum_{j=1}^{\infty} (A_j R^j + A_{j+1} R^{j+2}) \cos j\Theta \quad (18)$$

Continuity equations need be satisfied only on 2 lines, B'C' and either C'A' or B'A'.

Where no symmetry exists (Type 3), the corresponding sine functions must be added to equation (18). Continuity must be established along all three lines.

Numerical Results and Discussion

The author has been able to uncover only two solutions for triangular plates which have any claim to being theoretically exact. Both are given in reference 4 .

The first is that of a simply supported isosceles right triangular plate, the solution being obtained by the superposition of two Navier solutions for rectangular plates. The load may be placed anywhere on the triangle. The second solution is for an equilateral triangle, also simply supported, and loaded at the centroid.

Table 1 gives a comparison of the results for the former. We have assumed the load to be placed at the midpoint of the radius at $\Theta_o = 60^o$, which makes it unsymmetrical. The Table shows a listing of deflection and moments along $\Theta = 45^o$ and $\Theta_o = 60^o$. In this and in the Tables that follow Θ_o and R_o are defined in Fig. 1.

We make the following remarks with respect to Table 1. We chose an unsymmetrical problem (Type 3 plate) which involved collocation along all the three lines in Fig. 1. It will be seen that the results are close indeed for deflections as well as moments except near the load point where the bending moments are very different. As noted previously, the Fourier Series solution fails to give the correct values of the moments near the load as the series diverges. Our solution, on the other hand, gives accurate values near the load, but, of course, gives infinite values at R = 0. The tabulated values for moments are at R = .001, R being nondimensionalized by the side AC (Fig. 1).

We have calculated the values for symmetrical cases also. They are even closer than those given in Table 1, and it was felt unnecessary to tabulate them here.

We should add that the numerical results in Table 1 attributed to reference 4 are not given in that reference. They represent values computed by us from the formulas given in the said reference.

Table 2 gives the results obtained for an equilateral, simply supported triangular plate for which reference 4 gives only the deflection under the load and formulas for moments near the load which is at the centroid. These formulas, incidentally, may be considered exact as they are of the logarithmic form. The moments shown are for R = .001. It can be seen that the results are again very close.

Table 3 gives deflection and moments for two very difficult problems: An equilateral triangular plate loaded at the centroid and supported only at the corners, and a right angled isosceles triangular cantilever plate loaded at the mid point of the radius along $\Theta = 45^{o}$.

It would be misleading to suggest that solutions to all triangular plate problems may be obtained routinely by the method given here.

The procedure is very sensitive to the number of roots chosen. It is often necessary to run a problem about three times to test for convergence. While it is generally true that the larger the number of roots used the closer the solution is to the correct one, this is not so beyond a certain number of roots. Raising the radius r to a very large power (say above 25) leads to inaccuracies. Some judgment and comparison with similar solutions is helpful in arriving at the optimum number of roots to be used.

For each eigen root used there are two unknown constants and for each circle function (excluding the logarithmic term) one unknown constant. Usually 5 or 6 eigen roots are sufficient, and, as for the circle functions, the author has found 8, 16, and 22 roots adequate for Types 1, 2 and 3 respectively. He is not able to give more precise guidelines at the present time.

Extension to Polygonal Plates

It is, we believe, rather obvious that the method may be extended to other plates with rectlinear boundaries. One has only to keep in mind that each of the regions at the corners of the plate has to be triangular in order that eigenfunctions may be used, but the interior region, where the circle functions apply, may be of any shape. It will, in fact have as many sides as the polygonal plate under consideration. The number of lines along which continuity is established will equal the number of sides in the polygon in the most general case. In cases of symmetry, the collocation lines will be fewer, and may even be no more than one.

TABLE I

ISOCELES RIGHT TRIANGULAR PLATE

$\Theta = 45^{\circ}$

THIS SOLUTION					REFERENCE (4)			
x/rad	wD/Pb^2	$M_{r/P}$	$M_{\Theta/P}$	$M_{r\Theta/P}$	wD/Pb^2	$M_{r/P}$	$M_{\Theta/P}$	$M_{r\Theta/P}$
0	0	-.0664	.0664	0	0	-.0662	.0662	0
.1	.000233	-.0567	.0694	.00306	.000232	-.0565	.0690	.00341
.2	.000884	-.0277	.0783	.0111	.000880	-.0276	.0775	.0122
.3	.00181	.0205	.0939	.0215	.00180	.0193	.0902	.0240
.4	.00275	.0929	.1021	.0265	.00274	.0943	.1063	.0277
.5	.00335	.1511	.0990	.0041	.00335	..1513	.0999	.0051
.6	.00333	.1211	.0919	-.0195	.00335	.1252	.0998	-.0182
.7	.00285	.0801	.0717	-.0209	.00284	.0731	.0709	-.0206
.8	.00204	.0452	.0484	-.0170	.00203	.0440	.0472	-.0151
.9	.00106	.0203	.0238	-.0132	.00105	.0205	.0230	-.0127
1	0	0	0	-.0119	0	0	0	-.0119

Notes: Simply supported on all sides

$N^{\underline{o}}$ of Eigenroots = 6

$N^{\underline{o}}$ of circle functions = 22

TABLE I (CONTINUED)

ISOCELES RIGHT TRIANGULAR PLATE

$\Theta = 60^{\circ}$

THIS SOLUTION					REFERENCE (4)			
x/AA'	wD/Pb^2	$M_{r/P}$	$M_{\Theta/P}$	$M_{r\Theta/P}$	wD/Pb^2	$M_{r/P}$	$M_{\Theta/P}$	$M_{r\Theta/P}$
0	0	-.0575	.0567	-.0332	0	-.0573	.0573	-.0331
.1	.000218	-.0512	.0632	-.0298	.000217	-.0514	.0632	-.0295
.2	.000844	-.0297	.0804	-.0207	.000843	-.0310	.0812	-.0196
.3	.00178	.0148	.1113	-.00988	.00179	.0115	.1139	-.00695
.4	.00287	.0946	.1800	.00302	.00285	.0889	.1810	.00487
.5	.00363*	.0552*	.619*	.0119*	.00361*	.2976*	.3094*	.0132*
.6	.00342	.1215	.1736	.0171	.00344	.1174	.1741	.0172
.7	.00276	.0621	.1025	.0192	.00278	.0643	.1001	.0184
.8	.00191	.0313	.0624	.0196	.00191	.0325	.0606	.0198
.9	.00097	.0137	.0336	.0202	.00095	.00866	.0332	.0209
1	0	-.01213	.01213	.0210	0	-.0129	.0129	.0224

$\Theta_o = 60^{\circ}$

$R_{o/b} = .366$ *AT LOAD

$\nu = .3$

TABLE 2

EQUILATERAL PLATE

$\theta = 30^\circ$

x/AA'	wD/Pb^2	M_r/P	M_θ/P	wD/Pb^2	M_r/P	M_θ/P
.6	.00393	.1453	.2136			
C.G	.00428	.571*	.628*	.00431	.571*	.625*
.7	.00417	.226	.258			

Notes:

Simply Supported on all sides, Loaded at C.G, *Moments calculated at R/b = .001, $\nu = .3$, $R_o/b = .5774$, N$^{\underline{o}}$ of Eigen Roots = 5, N$^{\underline{o}}$ of Circle Functions = 8

TABLE 3

	CORNER SUPPORTED PLATE, $\theta = 30^\circ$			CANTILEVER PLATE $\theta = 45^\circ$		
x/AA'	wD/Pb^2	M_r/P	M_θ/P	wD/Pb^2	M_r/P	M_θ/P
0	0	.2957	-.0986	.0341	0	0
.1	.0142	.2951	-.0749	.0301	.0010	.0472
.2	.0258	.2940	-.2987	.0262	.0061	.0780
.3	.0349	.2924	.0301	.0224	.0338	.0893
.4	.0417	.2919	.1045	.0188	.0572	.1937
.5	.0463	.2999	.1988	.0150*	.4712*	.6078*
.6	.0489	.3489	.3412	.0177	.00148	.1428
.7	.0494	.3731	.4606	.0067	-.1064	.0473
.8	.0478	.1823	.3653	.0033	-.1947	-.0189
.9	.0456	.0787	.3581	.00092	-.2866	-.0741
1	.0436	0	.3733	0	-.4050	-.1215
C.G	.0495*	.737*	.793*			

Notes:

Equilateral plate free on all sides and loaded at C.G
$\nu = .3$
* AT LOAD
N$^{\underline{o}}$ of Eigenroots = 6
N$^{\underline{o}}$ of Circle Functions = 8

Notes:

90° Isosceles Cantilever
$\alpha_1 = 90^\circ$, unsupported at A
$\nu = .3$
* AT LOAD
N$^{\underline{o}}$ of Eigenroots = 6
N$^{\underline{o}}$ of Circle Functions = 16

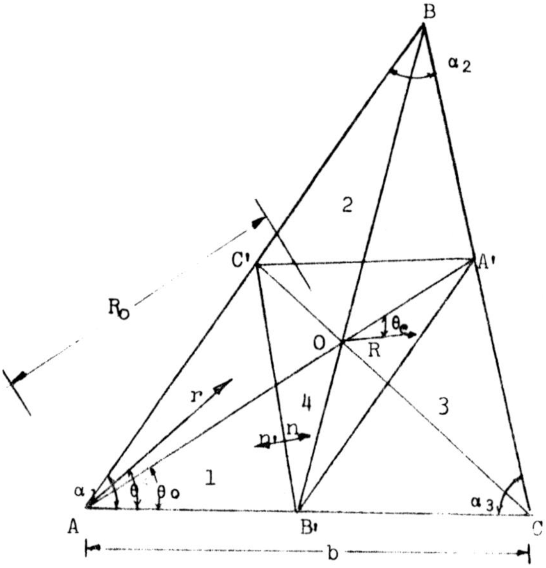

Fig. 1

DEFINITION SKETCH

REFERENCES

1. Quinlan, P.M., The λ-Method for Rectangular Plates, Proceedings of the Royal Society, Series A, 288, 1965, pp 371-395.
2. Wah, Thein, Elastic Quadrilateral Plates, Journal of Computers and Structures, Vol. 10, pp. 457-466, 1979.
3. Wah, Thein, Roots of Transcendental Equations, National Technical Information Service, Accession No. PB272929/AS (1977), U.S. Department of Commerce, Springfield, VA22161.
4. Timoshenko, S., and Woinowsky-Krieger, S. , "Theory of Plates and Shells", 2nd Edition, McGraw Hill, New York (1959).

ON MINIMIZING THE MEAN SQUARE DEFLECTION OF THIN ELASTIC BEAMS

Robert Reiss
and
Tiruvadi S. Raviguruurajan
Department of Mechanical Engineering
Howard University
Washington, D.C. 20059

ABSTRACT. The problem of minimizing the mean square deflection of thin elastic beams subject to a fixed cost constraint is formulated. For specificity and simplicity, a linear relationship between specific cost and bending stiffness-typified by sandwich structures - is assumed. The derived optimality conditions require a constant value for an appropriate mutual strain energy. An iteration technique is presented and shown, through the use of specific examples, to provide the optimum design after only a few iterations.

1. INTRODUCTION

A number of recent papers have treated the minimization of some measure of the elastic deflection of structures under static load. Shield and Prager [1] and Huang [2] elegantly generalized Barnett's [3] pioneering effort on the minimum deflection design at a specified point for statically determinate beams. The more interesting problem of minimizing the maximum deflection [4-6] has been solved only for relatively simple loads. The complications resulting from the unknown number and location of local maxima of the deflection [7] may be avoided by optimizing, instead, an integral measure of the displacement. The most common of these measures is external work [8-10]. Recently, Banichuk et. al. [11] proposed minimizing the p -th root of the integral of the absolute deflection to the p-th power. As $p \to \infty$, this problem approaches the minimum-maximum problem. On the other hand, for p=2, the subject of the present study, Banichuk's integral reduces to the mean square deflection.

2. PROBLEM FORMULATION

Consider a thin beam of length L and bending stiffness S which is a function of the centroidal axial coordinate X. The kinematic support conditions and applied pressure P(X) (force/unit length) are prescribed. Any applied concentrated forces and/or couples are symbolically included in P through generalized functions. For a fixed amount of cost, it is desired to determine the stiffness distribution S(X) so that the mean square deflection

$$\int_0^L W^2(X)\, dX \tag{1}$$

is minimized. In (1), W represents the deflection of the beam. If shear deformation is neglected, the only relevant field equation is

$$(SW'')'' = P , \tag{2}$$

where $(\ )' = d(\ )/dX$.

The discussion in this study shall be restricted to beams whose bending stiffness is proportional to the specific cost. Thus the prescribed cost constraint may be expressed

$$\int_0^L \{S(X) - S_o\}\, dX = 0, \tag{3}$$

where S_o is a given constant. The optimization problem may now be stated: Minimize the functional (1) subject to the subsidiary constraints (2) and (3) and the given kinematic and static boundary conditions.

The side constraints may be relieved in the usual way by introducing the Lagrange multipliers w(X) and k and defining a new functional J*(W,S)

$$J^* = \frac{1}{2}\int_0^L W^2(X)dX - \int_0^L w(X)[(SW'')''-P]dX$$

$$+ k\int_0^L (S-S_o)dX . \tag{4}$$

3. THE OPTIMALITY CONDITIONS

The desired optimality conditions are determined by requiring that the variation δJ^* vanish for all infinitesimal and sufficiently regular variations δW and δS compatible with the static and kinematic boundary conditions. It is now a straight forward matter to show that

$$\delta J^* = \int_0^L (k-W''w'')\ \delta S\ dX$$

$$+ \int_0^L [W-(Sw'')''] \, \delta w \, dX$$

$$+ \{w\delta Q - w'\delta M - Sw''\delta W' + (Sw'')'\delta W\}\Big|_0^L , \qquad (5)$$

where Q and M are, respectively, the internal shear force and bending moment associated with the load P. Consequently, the Euler conditions necessary for optimality are immediately determined to be

$$(Sw'')'' = W,$$

$$W''w'' = k\ . \qquad (6)$$

The natural boundary conditions associated with the augmented functional J* may be determined by considering each classical boundary condition separately.

Clamped Edge. Here $W=W'=0$ and consequently $\delta W=\delta W'=0$. It is now evident from (5) that $w\delta Q-w'\delta M$ must also vanish at the clamped edge. Since δQ and δM are treated as independent variations, the natural boundary conditions become $w=w'=0$.

Simply Supported or Pinned Edge. Here M, W, δM and δW necessarily vanish. Therefore the natural boundary conditions become $w = Sw'' = 0$.

Free Edge. Since $M=Q = \delta M = \delta Q = 0$, the natural boundary conditions are $Sw'' = (Sw'')'=0$.

It is apparant that w(X) obeys the same boundary conditions as the actual deflection W. Therefore, in view of the first of (6), the Lagrange multiplier w may be interpreted as the deflection of a beam whose design is S(X) when acted upon by the load W(X).

4. STATICALLY DETERMINATE BEAMS

It is convient to introduce m(X), the bending moment associated with the load W. Thus

$$SW'' = -M\ , \qquad m'' = -W\ , \qquad (7)$$

$$\int_0^L (S-S_o)dX = 0, \quad kS^2 = Mm \qquad (8)$$

defines the desired solution. It may be shown from (8) and Castigliano's Theorem that the optimal value of the mean square deflection is kS_oL.

For statically determinate beams, the solutions to (7) may be explicitly expressed in terms of an appropriate Green's function. Thus,

$$W(X) = -\int_0^L \frac{G(X,Y)M(Y)}{S(Y)}\,dY\ ,$$

$$m(X) = -\int_0^L G^*(X,Y)\ W(Y)dY, \qquad (7')$$

where $G^*(X,Y) = G(Y,X)$ is the adjoint Green's function. It should be observed that (7') provides the solution W and m of (7) once S is known, and that S and k may be determined from (8) once m is known. This suggests an examination of the following iterative approach.

Iteration Method I

1. Arbitrarily select $S \geq 0$, say, $S \equiv S_o$.

2. Calculate W and m from (7').

3. Calculate k from (8), i.e.

$$k^{\frac{1}{2}} = (S_oL)^{-1}\int_0^L \sqrt{Mm}\ \ dX.$$

4. Set $S = \sqrt{Mm/k}$.

5. Go back to Step 2.

The utility of the foregoing method depends, of cause, upon whether it converges to the optimal design S(X), and upon whether the convergence is dependent upon the trial design (Step 1). Secondary concerns are the effects of the number of intervals N selected for the numerical integrations in steps 2 and 3. A theoretical discussion of these considerations is beyond the scope and intent of this paper; however, closed form solutions which may be obtained for carefully selected loads may be used to validate Iteration Method I.

<u>Example 1</u>. For the special case of a simply supported beam loaded by

$$P(X) = P_o(X/L)(1-X/L), \qquad (9)$$

the exact solution may be shown to be

$$S = 5\ S_o[(X/L)^4 - 2(X/L)^3 + X/L]\ ,$$

$$k = P_o{}^2L^8/432000\ S_o^3 = 2.3148\text{x}10^{-6}\ P_o^2\ L^8/S_o^3\ . \qquad (10)$$

Figure 1 shows that the computer generated plots of the exact design (10) and the approximate design S obtained after the first iteration are virtually indistinguishable. Comparable results were obtained on the second iteration for W. The numerical integration was based upon N=30 intervals. The value k converged to $2.3154 \times 10^{-6}\ P_o^2 L^8/S_o^3$ in 3 iterations; Subsequent iterations did not change the value of k. On the other hand, with N=100, the value k converged nicely to $2.3148 \times 10^{-6}\ P_o^2 L^8/S_o^3$ within just 3 iterations.

Example 2. For a simply supported beam uniformly loaded, the bending moment is

$$M = P_o X(L-X)/2.$$

In this case an exact solution is unavailable; however, Figure 2 shows the first four approximate designs (N=30) are virtually indistinguishable and consequently it may be inferred that, as in the prior example, just one iteration is necessary to adequately approximate the optimum design. The values for

$$kS_o^3 \times 10^5/P_o^2 L^8$$

converged to 5.7826 and 5.7811 for N=30 and 100, respectively. For comparison purposes, it is noted that the mean square deflection for uniform stiffness ($S \equiv S_o$) exceeds the optimum value by 48%.

Example 3. Consider now a cantilevered beam at X=0 and free at X=L. For a uniformly distributed load, the bending moment is

$$M = -P_o(X-L)^2/2.$$

Once again, Iteration Method I was applied to obtain the solution. Results for the four iterations (N=30) are shown in Fig. 3. The values $kS_o^3 \times 10^3/P_o^2 L^8$ converged for N=30 and 100 to 1.3759 and 1.3758, respectively. Once again, it is observed that the optimal design provides a 66% reduction for mean square stiffness compared to the uniform stiffness design.

5. STATICALLY INDETERMINATE BEAMS

The curvature of the deflected indeterminate beam will change sign as many times as the number redundant support conditions. At these inflection points M and, as a consequence of (8), S both vanish. Furthermore, it is necessary to allow for a jump in slope of the deflection wherever the stiffness vanishes.

It should be observed that once the inflection points, say X_i, are known, the beam becomes statically determinate. Thus a procedure anologous to Iteration Method I could be applied to determine k for each set X_i. It is clear that X_i must be chosen so that k is minimized. One approach is to extremize k numerically; another is to develop additional optimality criterion by considering small variations in the location X_i analogous to Masur's method [12]. If the latter approach is adopted, the condition

$$Q[w'] + q[W'] = 0 \tag{11}$$

results, where q is the shear force associated with m, $[\cdot]$ denotes the jump in the indicated function, and all functions are evaluated at the inflection point X_i.

The remainder of the discussion will be restricted to a beam clamped at X=0 and pinned at X=L. There is one redundancy and hence one singular point X_o. In order to cast equations (7),(8) and (11) in a form appropriate for an iterative solution, (7) is first integrated for the particular boundary conditions under discussion to provide

$$W(X) = \int_0^X (Y-X)\frac{M(Y;X_o)}{S(Y;X_o)}dY + [W']\, H(X-X_o)(X-X_o) ,$$

$$m(X) = \int_{X_o}^X (Y-X)\, W(Y)dY + \frac{X-X_o}{L-X_o}\int_{X_o}^L (L-Y)\, W(Y)\, dY, \qquad (12)$$

where H is the Heaviside function. Furthermore, it may easily be shown that

$$[W'] = \frac{1}{L-X_o}\int_0^L (L-Y)\frac{M(Y;X_o)}{S(Y;X_o)}\, dY ,$$

$$[w'] = \frac{1}{L-X_o}\int_0^L (L-Y)\frac{m(Y;X_o)}{S(Y;X_o)}\, dY. \qquad (13)$$

In terms of the dimensionless parameters

$\bar{W} = \zeta W/L$, $x = X/L$, $y = Y/L$, $x_o = X_o/L$, $\bar{k} = k\zeta^2 S_o/L^2$, $s=S/S_o|x-x_o|$,
$\bar{M} = \zeta LM/S_o(x-x_o)$, $\bar{m} = \zeta m/L^3\ (x-x_o)$,

where ζ is a non-dimensional constant, Eqs. (12), (8) and (11), respectively, become

$$\bar{W} = \begin{cases} \int_0^x F(x,y;x_o)dy\ , & x < x_o \\ \left[\int_0^{x_o} F(x_o,y;x_o)dy - \int_{x_o}^1 F(x_o,y;x_o)dy\right]\frac{1-x}{1-x_o} + \\ \int_x^1 F(x,y;x_o)dy\ , & x > x_o\ , \end{cases} \qquad (14a)$$

$$\bar{m} = (1-x_o)^{-1}\int_{x_o}^1 (1-y)\,\bar{W}(y)dy + (x-x_o)^{-1}\int_{x_o}^x (y-x)\bar{W}(y)dy, \qquad (14b)$$

$$\bar{k}^{\frac{1}{2}} = \int_0^1 |\,x-x_o|\ \{\bar{M}\ \bar{m}\}^{\frac{1}{2}}\ dx, \qquad (15a)$$

$$\bar{k}\, s^2 = \bar{M}\, \bar{m}\ , \tag{15b}$$

$$\left[\int_0^{x_o} F(1,x;x_o)dx - \int_{x_o}^{1} F(1,x;x_o)dx\right]\int_{x_o}^{1} (1-x)\ \bar{W}(x)dx$$

$$+\left[\int_0^{x_o} G(1,x;x_o)dx - \int_{x_o}^{1} G(1,x;x_o)dx\right](1-x_o)\bar{M}(x_o) = 0, \tag{16}$$

where $F(x,y;x_o) = (x-y)\ \bar{M}(y;x_o)/s(y;x_o)$ and $G(x,y;x_o) = (x-y)\ \bar{m}(y;x_o)/s(y;x_o)$.

Since $\bar{M}\ (x;x_o)$ is a specified function, a slight variation in Iteration Method I should be expected to produce the statically indeterminate solution.

Iteration Method II

1. Arbitrarily choose x_o and s, say, $s \equiv 1$.

2. Compute $\bar{W}$, and then $\bar{m}$, from (14).

3. Compute $\bar{k}$ from (15a).

4. Set $s = \{\bar{M}\ \bar{m}/\bar{k}\}^{1/2}$.

5. Calculate x_o from (16).

6. Go back to Step 2.

Example 4. It may be shown that if the load

$$P(X)=\begin{cases} P_o x^2 & x < y_o \equiv 1 - \sqrt{2}/2 \\ P_o(1-x)(1+x-4y_o) & x > y_o \end{cases}$$

is applied to the indeterminate beam, then an exact solution may be obtained. With $\zeta = 12S_o/P_oL^3$, the results are

$$x_o = y_o\ , \quad k = 1.4093\text{x}10^{-7}\ P_o^{\ 2}L^8/S_o^{\ 3}.$$

Iteration Method II was then applied to check the results. The starting value for x_o was selected as 0.2, and the value of $kS_o^{\ 3}\text{x}10^7/P_o^{\ 2}L^8$ converged to 1.4094 and 1.4092 after 4 iterations for N=50 and 100, respectively. In each case x_o converged to 0.292893. Different choices for the initial value of x_o had no effect upon the results.

Example 5. A uniform load P_o was assumed for the clamped-pinned beam. Since $M = P_o(L-X)(X-X_o)/2$, the choice $\zeta = 2S_o/P_oL^3$ yields

$$\bar{M} = 1-x.$$

As shown in Fig. 4, the optimal design converged after the third iteration (N=50) starting with a trial value $x_o = 0.2$. The results were $x_o = 0.2926$ (N=50) and 0.2903 (N=100). The corresponding values for k are $k = 7.7784 \times 10^{-6} P_o^2 L^8/S_o^3$ (N=50) and $k = 7.7782 \times 10^{-6} P_o^2 L^8/S_o^3$ (N=100).

As a general rule the starting value of x_o had virtually no effect on the number of iterations required; indeed, in each trial run, the value x_o converged before k did.

6. SUMMARY

It has been shown that iteration techniques are remarkably efficient in solving the non-linear optimality conditions obtained by minimizing the mean square deflection of beams. For statically indeterminate beams, the location of the inflection (singular) points, which are treated merely as additional design variables, are also readily obtained through the iteration process. These results support the contention of Khan et. al [13] that iteration techniques applied to optimality criterion in design, when available, are generally superior to mathematical programming methods.

The methods of this paper apply to other structures such as plates and shells as well. This work is currently in progress and will be reported in the near future.

REFERENCES

[1]. Shield, R.T. and Prager, W., "Optimal Structural Design for Given Deflection," ZAMP, Vol. 21, 1970, pp. 513-523.

[2]. Huang, N.C., "On the Principle of Stationary Mutual Complementary Energy and Its Application to Structural Design," ZAMP, Vol. 22, 1971, pp. 608-620.

[3]. Barnett, R.L., "Minimum Weight Design of Beams for Deflection," ASCE Journal of Engineering Mechanics, Vol. 87, EM2, 1961, pp. 75-109.

[4]. Huang, N.C., "Optimal Design of Elastic Beams for Minimum-Maximum Deflection," ASME Journal of Applied Mechanics, Vol. 38, 1971, pp. 1078-1081.

[5]. Komkov, V. and Coleman, N.P., "An Analytic Approach to Some Problems of Optimal Design of Beams and Plates," Archives of Mechanics, Vol. 27, No.4, 1975, pp. 565-575.

[6]. Cinquini, C., "Optimal Elastic Design for Prescribed Maximum Deflection," Journal of Structural Mechanics, Vol. 7, No. 1, 1979, pp. 21-34.

[7]. Reiss, R., "Optimum Design for Minimum-Maximum Deflection," Developments in Mechanics, Vol. 11, Proc. 17th Midw. Mech. Conf., May 1981, pp. 161-162 (Abstract).

[8]. Prager, W. and Taylor, J.E., "Problems of Optimal Structural Design," ASME Journal of Applied Mechanics, Vol. 35, 1968, pp. 102-106.

[9]. Huang, N.C., "Optimal Design of Elastic Structures for Maximum Stiffness," International Journal of Solids and Structures, Vol. 4, 1968, pp. 689-700.

[10]. Masur, E.F., "Optimum Stiffness and Strength of Elastic Structures," ASCE Journal of Engineering Mechanics, Vol. 95, EM5, 1970, pp. 621-640.

[11]. Banichuk, N.V., Karelishvili, V.M. and Moronov, A.A., "Optimization Problems with Local Performance Criteria in the Theory of Plate Bending," MTT, Vol. 13, No. 1, 1978, pp. 116-122.

[12]. Masur, E.F., "Optimality in the Presence of Discreteness and Discontinuity," Optimization in Structural Design, (Edit. A. Sawczuk and Z. Mröz), Springer-Verlag, 1975, New York, pp. 441-453.

[13]. Khan, M.R., Thornton, W.A. and Willmert, K.D., "Optimality Criterion Techniques Applied to Mechanical Design," ASME Journal of Mechanical Design, Vol. 100, 1978, pp. 319-327.

Acknowledgment

This research was supported by NSF grant CME-7921128

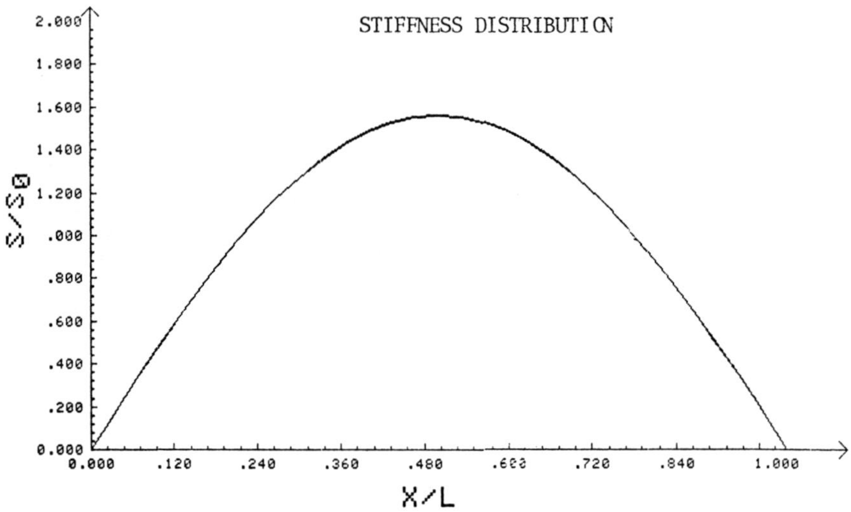

Fig. 1. Comparison of the exact and first approximation to the stiffness distribution of Example 1.

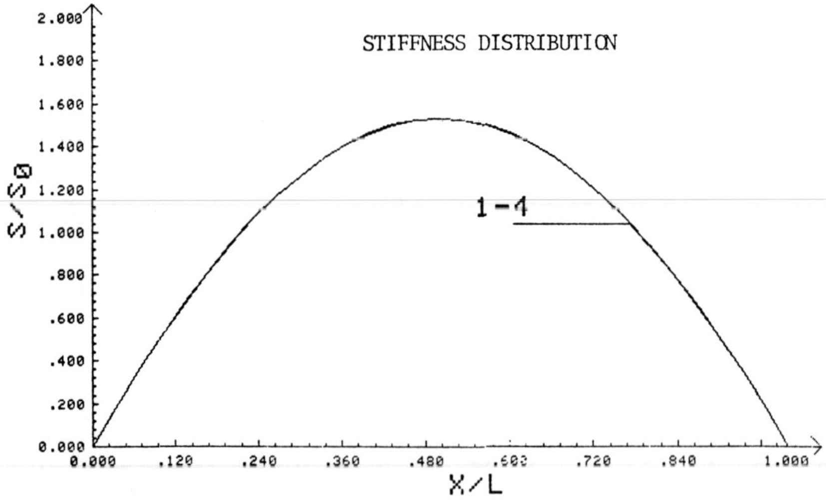

Fig. 2. Stiffness distributions for the first 4 iterations of a uniformly loaded simply supported beam.

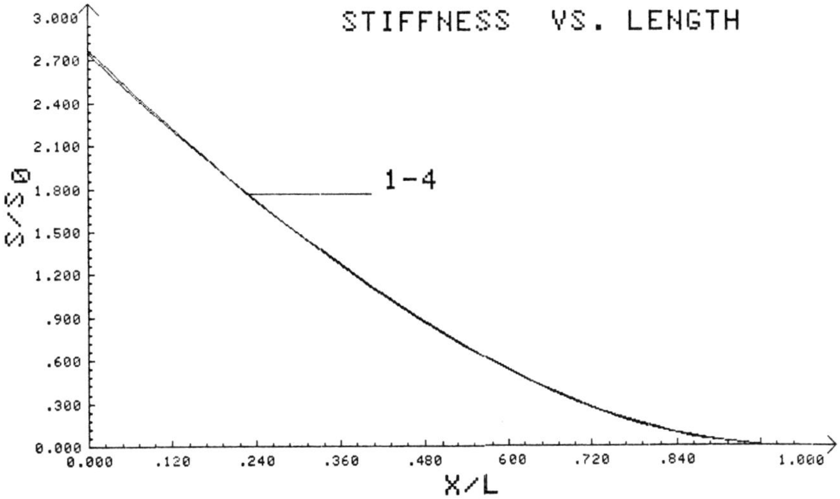

Fig. 3. Stiffness distributions for the first 4 iterations of the cantilevered beam.

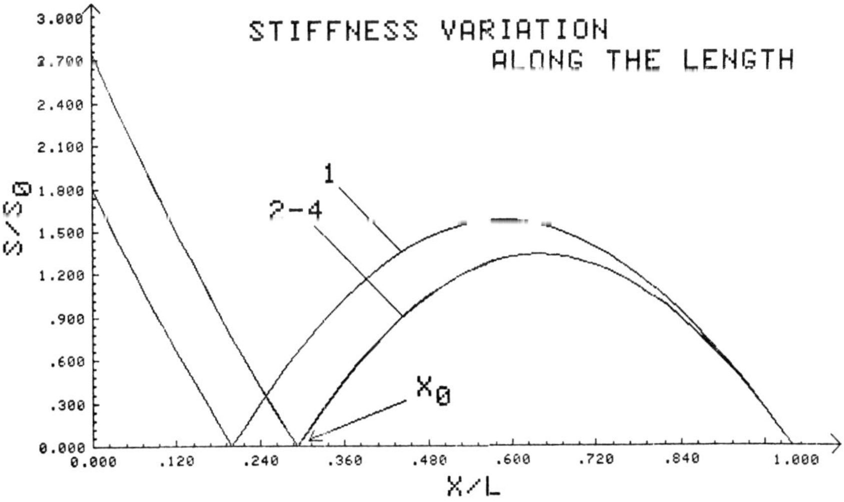

Fig. 4. Stiffness distributions for the first 4 iterations of the uniformly loaded clamped-pinned beam.

A CONSISTENT HIGHER ORDER BEAM THEORY

William B. Bickford

Arizona State University

ABSTRACT

A consistent theory for the static and dynamic behavior of straight elastic prismatic bars which generalizes the usual ad hoc assumptions on displacement beyond the elementary and Timoshenko theories is presented. With these assumptions for the displacements it is shown that in order for the vectorial and variational approaches to coincide, it is necessary to generalize the concepts of shear force and bending moment. Resulting solutions to classical static problems are generated and shown to exhibit a boundary layer character. These solutions are then compared to those of the theory of elasticity. Travelling wave solutions are investigated and compared to those of Pochhammer and Timoshenko.

I. INTRODUCTION

A multitude of papers over the years have addressed the problem of analyzing the static and dynamic behavior of beams. The extension of the elementary Bernoulli-Euler theory to include shear deformation is popularly attributed to Timoshenko, [1] and [2], although as pointed out in [3], the shear effect and rotatory inertia were given by Bresse in [4]. Since then many papers have dealt with the shear coefficient of the Timoshenko theory.

A deficiency of the usual shear deformation theory is that it contains a contradiction regarding the distribution of shear over the cross-section. This difficulty is considered by Levinson [5] who includes additional coefficients in a power series expansion of the displacements in terms of in plane coordinates in order to satisfy identically the boundary conditions on the top and bottom surfaces of a beam of rectangular cross-section. He then uses the usual definitions for the bending moment and the shear force to generate force resultant-displacement relations and vectorial equations of motion. The resulting equations of motion in terms of the displacements are exactly the same as those of the Timoshenko theory when the shear coefficient in that theory is chosen as 5/6. These displacement equations of motion do not however agree with those obtained using Hamilton's Principle starting with the same assumptions for the displacements.

It is the purpose of this effort to show that the approach taken by Levinson [5] can, by introducing appropriate generalizations of the shear force and bending moment, be made to agree with the above mentioned results via Hamilton's Principle and thus produce a consistent higher order theory. In addition it is demonstrated that this consistent theory produces a 'boundary layer effect.'

As mentioned above, the basic idea, presented in [5] is to generalize the usual plane sections remain plane assumption by representing the displacements as

$$u(x,z,t) = z\psi(x,t)+z^3\phi(x,t) \tag{1}$$

$$w(x,z,t) = w(x,t)$$

We form

$$\gamma_{xz} = \frac{\partial w}{\partial x} + \frac{\partial u}{\partial t} = \frac{\partial w}{\partial x} + \psi + 3z^2\phi$$

and require

$$\gamma_{xz}\Big|_{z=\pm\frac{h}{2}} = 0$$

which yields

$$u(x,z,t) = -\frac{4z^3}{3h^2}\cdot\frac{\partial w}{\partial x} + \left(z - \frac{4z^3}{3h^2}\right)\psi$$

$$w(x,z,t) = w(x,t) \tag{2}$$

Within the spirit of the approximate nature of the analysis we take

$$\sigma_x = E\varepsilon_x = E\left[\left(z - \frac{4z^3}{3h^2}\right)\frac{\partial\psi}{\partial x} - \frac{4z^3}{3h^2}\frac{\partial^2 w}{\partial x^2}\right]$$

$$\tau_{xz} = G\gamma_{xz} = G\left(\frac{\partial w}{\partial x} + \psi\right)\left(1 - \frac{4z^2}{h^2}\right) \tag{3}$$

as the only non-vanishing stress components. Equations 2) and 3) then embody the basic assumptions which will be used in constructing the governing equations of motion.

II. VARIATIONAL FORMULATION

As is well known in mechanics, there are two basic approaches to generating equations of motion, the vectorial and variational approaches. These two paths should lead when applied in a consistent fashion to the same results. To this end, consider first the variational approach. We form using equations 2) the Kinetic and Potential energies, i.e.,

$$2T = \int_0^{\ell} \frac{\rho I}{105} (5\dot{w}'^2 + 68\dot{\psi}^2 - 32\dot{\psi}\dot{w}') + \rho A\dot{w}^2 \; dx$$

$$2V = \int_0^{\ell} \frac{EI}{105} (5w''^2 + 68\psi'^2 - 32\psi' w'') \qquad (4)$$

$$(8AG/15)(w'+\psi)^2 - p(x,t)w] \; dx$$

and use Hamilton's Principle

$$\delta \int_{t_1}^{t_2} (T-V)dt = 0$$

to generate the equations of motion,

$$-\alpha(w'+\psi) + \frac{\partial}{\partial t}(\gamma(16\dot{w}'-68\dot{\psi})) - \frac{\partial}{\partial x}(\beta \; (16w''-68\psi')) = 0 \qquad (5a)$$

and

$$\frac{\partial}{\partial x}(\alpha(w'+\psi)) - \frac{\partial^2}{\partial x^2}(\beta \; (5w''-16\psi'))$$

$$+ \frac{\partial^2}{\partial x \partial t}(\gamma(5\dot{w}'-16\dot{\psi})) - \rho A\ddot{w} + p(x,t) = 0 \qquad (5b)$$

and boundary conditions

$$[-\alpha(w'+\psi) + \frac{\partial}{\partial x}(\beta(5w''-16\psi')) + \frac{\partial}{\partial t}(\gamma(5\dot{w}'-16\dot{\psi}))]\delta w|_0^{\ell} = 0 \qquad (6a)$$

$$\beta(16\psi'-5w'')\delta w'|_0^{\ell} = 0 \qquad (6b)$$

$$\beta(16w''-68\psi')\delta\psi|_0^{\ell} = 0 \qquad (6c)$$

where

$$\alpha = \frac{8AG}{15} \quad \beta = \frac{EI}{105} \quad \gamma = \frac{\rho I}{105} \quad ' = \frac{\partial}{\partial x} \quad \dot{} = \frac{\partial}{\partial t}$$

These are obviously not the same as those presented in [5] which, as mentioned previously, are essentially the same as those of Timoshenko.

We observe that these constitute a 6th order system in x and as such there are now 3 boundary conditions which must be satisfied at each end. In particular, both w' and ψ can be specified at a boundary as opposed to the usual shear theory. This will be discussed in section V.

III. CONSISTENT VECTORIAL FORMULATION

Consider the two dimensional equations of elasticity in the form

$$\frac{\partial \sigma_x}{\partial x} + \frac{\partial \tau_{xz}}{\partial z} = \rho \ddot{u} \tag{7a}$$

$$\frac{\partial \tau_{xz}}{\partial x} + \frac{\partial \sigma_z}{\partial z} = \rho \ddot{w} \tag{7b}$$

In order to produce the results in (5), we define

$$M_1 = -\int z\sigma_x dA$$

$$V_1 = \int \tau_{xz} dA$$

which upon using (3) yields

$$M_1 = -\frac{EI}{5}(4\psi' - w'') \tag{8a}$$

$$V_1 = \frac{2}{3} AG(w' + \psi) \tag{8b}$$

Then multiply (7a) by zdA and integrate over the area to obtain

$$-\frac{\partial M_1}{\partial x} - V_1 = \frac{\partial^2}{\partial t^2}\left(\frac{EI}{5}(4\psi - w')\right) \tag{9}$$

Next multiply (7b) by dA and integrate over the area to obtain

$$\frac{\partial V_1}{\partial x} + p = \rho A\ddot{w} \tag{10}$$

Equations (8), (9) and (10) are those developed by Levinson [5] and are essentially equivalent to the Timoshenko theory. They incorporate the traditional definitions for shear and bending moment which are used in the elementary and shear deformation beam theories.

In order to develop Equations (5) vectorially it is necessary to generalize the ordinary definition for the bending moment. To this end define

$$M_3 = -\int z^3\sigma_x dA = \frac{EIh^2}{140}(5w''-16\psi') \tag{11}$$

which can be thought of as a higher order moment resultant. We then multiply eqn (7a) by $z-4z^3/3h^2$ and integrate over the area. This yields using (2), (8), and (11),

$$-\alpha(w'+\psi) + \frac{\partial}{\partial t}(\gamma(16\dot{w}'-68\dot{\psi})) - \frac{\partial}{\partial x}(\beta(16w''-68\psi')) = 0$$

which is (5a). In order to produce the 2nd variational eqn. we proceed as follows. Form, from equations 7,

$$z\left(\frac{\partial^2\sigma_x}{\partial x^2} + \frac{\partial^2\tau_{xz}}{\partial x\partial z}\right) + \frac{\partial\tau_{xz}}{\partial x} + \frac{\partial\sigma_z}{\partial z} = \rho z\frac{\partial\ddot{u}}{\partial x} + \rho\ddot{w}$$

i.e., compute the derivative of 7a) with respect to x, multiply by z, and add to 7b). Integration over the area yields again with (2), (8), and (11),

$$\frac{\partial}{\partial x}(\alpha(w'+\psi)) - \frac{\partial^2}{\partial x^2}(\beta(5w''-16\psi'))$$
$$+ \frac{\partial^2}{\partial x\partial t}(\gamma(5\dot{w}'-16\dot{\psi})) - \rho A\ddot{w} + p(x,t) = 0$$

which is (5b).

Thus it is seen that when the displacements are generalized beyond the traditional plane sections remain plane assumption, the force resultants must also be generalized in order for the vectorial and variational formulations to coincide. It remains to be seen whether or not there can be developed a rational method for constructing the vectorial equations without knowing, a priori, the variational equations.

Finally, we note that the equations of motion and boundary conditions can be expressed in terms of the force resultants as

$$\frac{\partial}{\partial x}\left(\frac{4V_1}{5} - \frac{4}{3h^2}\frac{\partial M_3}{\partial x}\right) + p(x,t) = \rho A\ddot{w} - \frac{\partial^2}{\partial x \partial t}(\gamma(5\dot{w}'-16\dot{\psi}))$$

$$\left(\frac{4V_1}{5} - \frac{4}{3h^2}\frac{\partial M_3}{\partial x}\right) + \frac{\partial M_1}{\partial x} = \frac{\partial}{\partial t}(\gamma(16\dot{w}' - 68\dot{\psi}))$$

and

$$\left[\frac{4}{5}V_1 - \frac{4}{3h^2}M'_3 + \frac{\partial}{\partial t}(\gamma(5w'-16\psi))\right]\delta w\Big|_0^{\ell} = F_w \delta w\Big|_0^{\ell} = 0$$

$$\frac{4}{3h^2}M_3\delta w'\Big|_0^{\ell} = F_{w'}\delta w'\Big|_0^{\ell} = 0$$

$$\left(M_1 - \frac{4}{3h^2}M_3\right)\delta\psi\Big|_0^{\ell} = F_\psi\,\delta\psi\Big|_0^{\ell} = 0$$

respectively, where F_w, $F_{w'}$, and F_ψ are the generalized forces. The time independent parts of the equations of motion, namely,

$$\frac{d}{dx}\left(\frac{4V_1}{5} - \frac{4}{3h^2}\frac{dM_3}{dx}\right) + p = 0$$

and

$$\frac{4V_1}{5} - \frac{4}{3h^2}\frac{dM_3}{dx} + \frac{dM_1}{dx} = 0$$

clearly indicate that $F_w = 4V_1/5 - 4M_3'/3h^2$ is the appropriate transverse shear force resultant for this theory.

IV. STATIC SOLUTIONS

Consider the classical problem of the cantilever beam loaded as shown. Deleting the time dependent terms in equations (5) there results, with $p = 0$,

$$\alpha w''-5\beta w^{iv} + \alpha\psi' + 16\beta\psi''' = 0$$

$$\alpha w'+16\beta w''' + \alpha\psi - 68\beta\psi'' = 0$$

Eliminating ψ yields the 6th order equation

$$\frac{d^6 w}{dx^6} - \lambda^2 \frac{d^2 w}{dx^2} = 0$$

with $\lambda^2 = 420/(1+\nu)h^2$. The non-essential boundary conditions at $x = 0$ are

$$(M_1 - \frac{4}{3h^2} M_3)\Big|_{x=0} = 0$$

$$\frac{4}{3h^2} M_3 \Big|_{x=0} = 0$$

$$(\frac{4}{5} V_1 - \frac{4}{3h^2} M_3')\Big|_{x=0} = P$$

The choice as to what combination of boundary conditions to impose at $x = L$ is of interest in terms of which is 'appropriate.' For the usual shear theory it is possible to require that either of ψ, the rotation, or w' the bending slope, vanish. The difference in these solutions is usually referred to as a 'deflection due to shear.' For this theory it is obviously possible to require that both ψ and w' vanish at a 'clamped' end. It is also possible to still require that either ψ or w' vanish. On order to determine the relative merits of the different choices we consider

Case 1 $\quad w'\Big|_{x=L} = 0$

$$(M_1 - \frac{4}{3h^2} M_3)\Big|_{x=L} = 0$$

Case 2 $\quad \psi\Big|_{x=L} = 0$

$$\frac{4}{3h^2} M_3\Big|_{x=L} = 0$$

Case 3 $\quad w'\Big|_{x=L} = 0$

$$\psi\Big|_{x=L} = 0$$

where in each case $w\Big|_{x=L} = 0$ is also enforced.

The 3 solutions respectively are

$$w_1 = w_E - \frac{84PL^3}{EI}\frac{1}{\lambda L}\frac{\cosh\lambda L}{\sinh\lambda L}(1-\frac{x}{L}) - \frac{PL^3}{5EI}(1+\nu)(\frac{h}{L})^2(1-\frac{\sinh\lambda x}{\sinh\lambda L})$$

$$w_2 = w_E - \frac{21}{4}\frac{PL^3}{EI}\frac{1}{\lambda L}\frac{\cosh\lambda L}{\sinh\lambda L}(1-\frac{x}{L}) - \frac{PL^3}{20EI}(1+\nu)(\frac{h}{L})^2$$

$$(1-\frac{\sinh\lambda x}{\sinh\lambda L} + 5\,(1-\frac{x}{L}))$$

$$w_3 = W_E - \frac{1}{5}\frac{PL^3}{EI}(1+\nu)(\frac{h}{L})^2(1-\frac{x}{L}-\frac{\sinh\lambda L-\sinh\lambda x}{\lambda L\cosh\lambda L})$$

where

$$w_E = -\frac{PL^3}{3EI}(1-\frac{3}{2}\frac{x}{L}+\frac{1}{2}(\frac{x}{L})^3)$$

is the elementary solution. We note that all 3 have exponential terms which give an edge or boundary layer character to the solution. In particular consider each of the 3 solutions evaluated at x = 0 namely

$$w_1\Big|_{x=0} = -\frac{PL^3}{3EI}(1+252\,\frac{1+\nu}{420}\frac{h}{L}\frac{\cosh\lambda L}{\sinh\lambda L}-\frac{3}{5}(1+\nu)(\frac{h}{L})^2)$$

$$w_2\Big|_{x=0} = -\frac{PL^3}{3EI}(1+\frac{63}{4}\frac{1+\nu}{420}\frac{h}{L}\frac{\cosh\lambda L}{\sinh\lambda L}+\frac{9}{10}(1+\nu)(\frac{h}{L})^2)$$

$$w_3\Big|_{x=0} = -\frac{PL^3}{3EI}(1+\frac{3}{5}(1+\nu)(\frac{h}{L})^2-\frac{3}{5}(1+\nu)\frac{1+\nu}{420}(\frac{h}{L})^3\frac{\sinh\lambda L}{\cosh\lambda L}$$

The first two of these expressions each contain a term of order h/L which does not compare favorably with the corresponding elasticity results [6] namely

$$w\Big|_{x=0} = -\frac{PL^3}{3EI}(1+\frac{3}{4}(1+\nu)(\frac{h}{L})^2)$$

The third expression however has the usual $(h/L)^2$ shear correction term plus a higher order $(h/L)^3$ shear correction term. The appearance of the (h/L) terms for the first two combination of B,C's make it clear that for this theory it is appropriate to impose $w = w' = \psi = 0$ at a clamped end.

It is also of interest to note that the shear force transmitted along the length of the beam is from eqns (11)

$$F_w = \frac{4}{5} V_1 - \frac{4}{3h^2} M_3'$$

For this problem

$$\frac{4}{5} V_1 = \frac{4P}{5} \left(1 - \frac{\cosh \lambda x}{\cosh \lambda L}\right)$$

$$\frac{4}{3h^2} M_3' = -\frac{P}{5} \left(1 + 4 \frac{\cosh \lambda x}{\cosh \lambda L}\right)$$

The relative values of each of these is indicated in Fig. 1, with h/L = .1

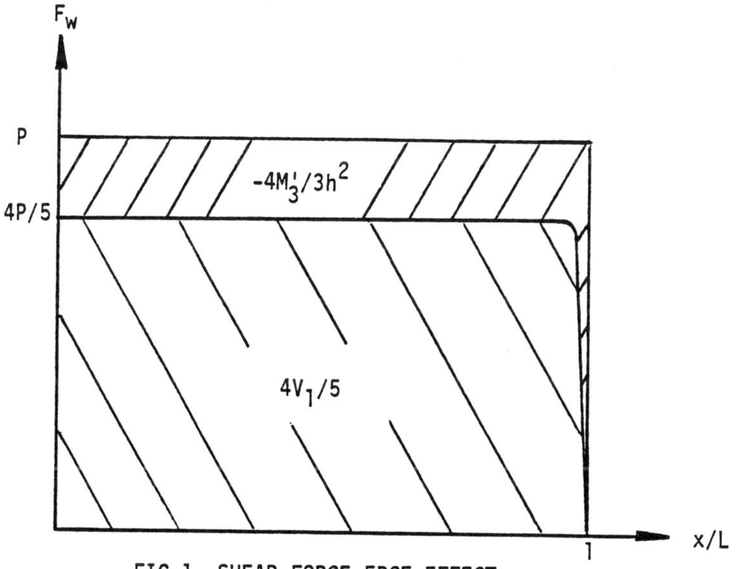

FIG 1. SHEAR FORCE EDGE EFFECT

Here the previously mentioned boundary layer character of the solution is clearly evident in terms of the higher order resultant M'_3 which contributes 1/5 of the total transverse load except in the neighborhood of x = L where it increases very rapidly to supply the total.

The problem of the simply supported uniformly loaded beam of length L can easily be solved to yield

$$w = \frac{p_0 L^4}{24EI}\left(\frac{x}{L} - 2\left(\frac{x}{L}\right)^3 + \left(\frac{x}{L}\right)^4\right) + \frac{p_0 L^4}{10EI}(1+\nu)\left(\frac{h}{L}\right)^2\left(\frac{x}{L} - \left(\frac{x}{L}\right)^2\right)$$

$$+ \frac{p_0 L^4}{2100EI}(1+\nu)^2\left(\frac{h}{L}\right)^4\left(\frac{\sinh\lambda x - \sinh\lambda L + \sinh\lambda(L-x)}{\sinh\lambda L}\right)$$

Here the first and second terms represent the elementary and usual sort of shear correction terms respectively. The third term has a hyperbolic factor which again exhibits a boundary layer character at the ends. It contributes to the displacement over essentially the entire length of the beam as indicated in Fig. 2.

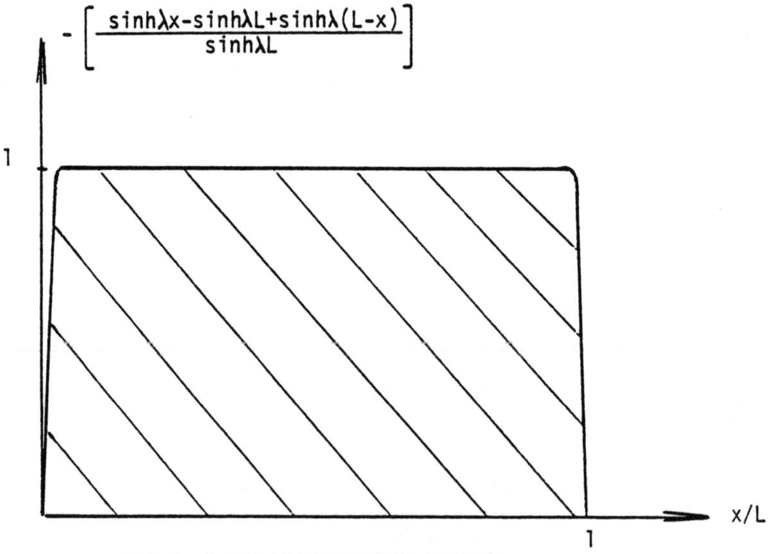

FIG 2. DISPLACEMENT EDGE EFFECT

The maximum displacement at the center can be written as

$$w\Big|_{x=\frac{L}{2}} = \frac{5p_0L^4}{384EI}\left(1+\frac{48}{25}(1+\nu)\left(\frac{h}{L}\right)^2-\frac{32}{875}(1+\nu)^2\left(\frac{h}{L}\right)^4\right)$$

The first two terms coincide exactly with the elasticity solution in [6] and the third term provides an additional higher order shear correction term.

V. DYNAMIC SOLUTIONS

One of the major tests for any dynamic beam theory is its ability to reasonably approximate the results of the Pochhammer-Chree theory. To this end equations (5) can be put in the form

$$\begin{aligned}&\frac{\partial^6 w}{\partial x^6}-\frac{70b^2}{R^2}\frac{\partial^4 w}{\partial x^4}-\frac{2}{C_0^2}\frac{\partial^6 w}{\partial x^4\partial t^2}+\left(\frac{70b^2+85}{C_0^2R^2}\right)\frac{\partial^4 w}{\partial x^2\partial t^2}\\&+\frac{1}{C_0^4}\frac{\partial^6 w}{\partial x^2\partial t^4}-\frac{70b^2}{C_0^2R^2}\frac{\partial^2 w}{\partial t^2}-\frac{85}{R^2C_0^4}\frac{\partial^4 w}{\partial t^4}=0\end{aligned} \tag{12}$$

where

$$R^2 = I/A \quad C_0^2 = E/\rho \quad b^2 = G/E$$

and p(x,t) has been set to zero.

Seeking solutions of the form

$$w(x,t) = w_0 \sin\frac{2\pi(x-ct)}{\lambda}$$

results in

$$Ay^4 - By^2 + C = 0 \tag{13}$$

where

$$A = \frac{255}{4\pi^2e^2}+1$$

$$B = \frac{630}{16\pi^4e^4}+\frac{255+210b^2}{4\pi^2e^2}+2$$

$$C = \frac{210b^2}{4\pi^2 e^2} + 1$$

$$y = c/c_0$$

$$e = h/2\lambda$$

For small e (low frequencies) the two branches of the frequency curve (13) are displayed in Fig. 3 along with those of the Timoshenko theory. Also included for comparison are the results from the Pochhammer theory [7] for a circular bar. From this we conclude that for low frequencies both branches of the present theory closely approximate the exact results. For high frequencies ($e \to \infty$) the frequency equation (13) becomes

$$y^4 - 2y^2 + 1 = 0 \quad ,$$

i.e., both branches of the frequency curve approach 1. For the Timoshenko theory the corresponding frequency equation (as $e \to \infty$) becomes

$$y^4 - (1+b^2)y^2 + b^2 = 0$$

with roots b^2 and 1, the lower branch being approximately correct. In this regard the predictions of the first branch of present theory do not measure up to the less sophisticated 'plane sections remain plane' theory of Timoshenko.

VI. DISCUSSION AND CONCLUSIONS

A consistent higher order beam theory which generalizes the usual plane sections remain plane assumptions used in the elementary and Timoshenko theories has been presented. The definitions of the usual bending moment and transverse shear have been generalized so as to obtain coincidence between the vectorial and variational formulations.

The boundary value problem which results is 6th order in space and 4th order in time, compared to previous theories which were 4th order in each variable. This makes it necessary to impose 3, instead of the usual 2, boundary conditions at each end. For a free end there are 3 natural b.c.'s, for the simply supported end, one forced and two natural b.c.'s and at a clamped end it turns out to be appropriate to prescribe zero values for all of w, w', and ψ.

The solution to the classical problem of the end loaded cantilever exhibits a usual $(h/L)^2$ type correction term and from the present theory an additional $(h/L)^3$ type correction term. For the simply supported uniformly loaded beam, the usual $(h/L)^2$ term and an additional $(h/L)^4$ term obtain. Both of the usual $(h/L)^2$ type terms compare favorably with the elasticity results.

Traveling wave solutions to the dynamic equations are investigated with less than satisfying results. Both branches of the frequency curve approach one for large values of wave length as opposed to either the Pochhammer theory or the more usual Timoshenko theory where the lower root is asymptotic to the ratio of the shear and dilatational speeds.

In conclusion, a consistent formulation for the vectorial and variational approaches has been demonstrated. Solutions to the static equations are shown to exhibit a boundary or edge effect not previously obtained for beam equations. These solutions compare favorably to those of elasticity theory. Traveling wave solutions to the equations of motion compare favorably to the Pochhammer theory for large wave length (lower frequencies) but are asymptotically incorrect for very high frequencies.

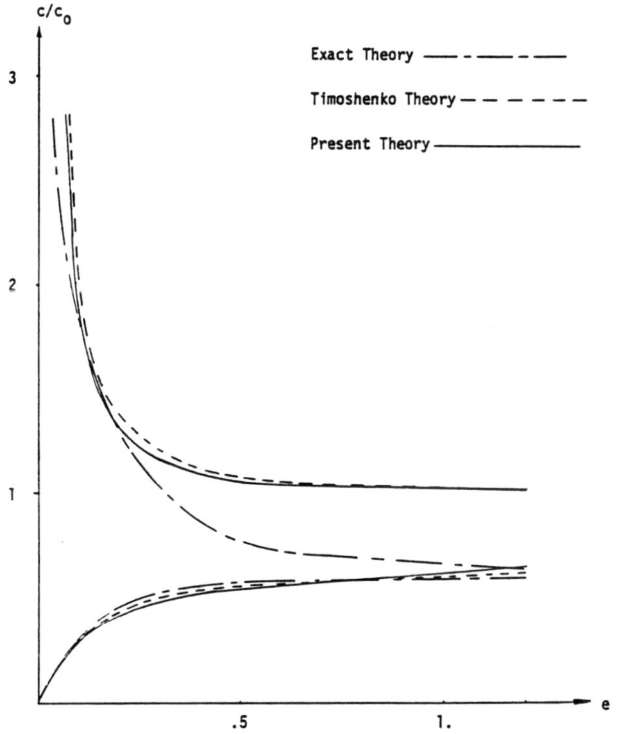

FIG 3. PHASE VELOCITY CURVES

REFERENCES

1. Timoshenko, S. P. Philosophical Magazine 41, 1921, pp. 744-746. On the correction for shear of the differential equation for transverse vibrations of prismatic bars.

2. Timoshenko, S. P. Philosophical Magazine 43, 1922, pp. 125-131. On the transverse vibration of bars of uniform cross-section.

3. Mindlin, R. D., and Deresiewkz, H. Proceedings of the Second U. S. National Congress of Applied Mechanics, pp. 175-178. Timoshenko's shear coefficient for flexural vibrations of beams.

4. Breese, M. Cours de Mec'Anique Appliquee, Mallet-Bachelier, Paris, 1859.

5. Levinson, M. Journal of Sound and Vibration, 74, 1981, pp. 81-87. A new rectangular beam theory.

6. Timoshenko, S. P. Theory of Elasticity, Third Edition, McGraw-Hill, pp. 41-50.

7. Pochhammer, L. J. F. Math (Crelle) Bd 81, 1876, p. 324.

LARGE AMPLITUDE VIBRATION OF MODERATELY THICK CIRCULAR PLATES

M. Sathyamoorthy
Department of Mechanical and Industrial Engineering
Clarkson College of Technology
Potsdam, New York 13676

ABSTRACT

This paper presents the results of an analytical investigation of nonlinear flexural vibration of circular plates which are moderately thick. Transverse shear deformation and rotatory inertia effects are incorporated into the governing differential equations which are derived by applying the well-known Berger approximation. The system of equations given here readily reduce to the dynamic Berger-type equations applicable for thin plates when transverse shear and rotatory inertia effects are neglected. By employing a multiple-mode approach, solutions to the governing equations are obtained by means of Galerkin's technique and Runge-Kutta numerical integration procedure. Numerical results are reported for period ratios considering isotropic and orthotropic circular plates at different amplitudes and radius-to-thickness ratios. Comparisons made with existing results for all special cases indicate good agreement for both static and dynamic problems. Results indicate significant influences of transverse shear deformation and rotatory inertia on the large amplitude vibration behavior of moderately thick circular plates. The method of analysis given here is a simple and interesting alternative to a much more complicated von Kármán-type approach to the solution of the title problem.

NOMENCLATURE

a	radius of plate
F	non-dimensional amplitude, A_1-A_2
$E_x, E_y, \nu_{xy}, \nu_{yx}$	orthotropic material constants
h	plate thickness
$q(x,y)$	lateral load per unit area
q_o	uniformly distributed load per unit area
R	radius-to-thickness ratio
RI	tracing constant for rotatory inertia
T	nonlinear period with transverse shear and rotatory inertia effects
T_o	linear period with no transverse shear and rotatory inertia effects
TS	tracing constant for transverse shear
u, v, w	displacement components
x, y	rectangular cartesian coordinates
ρ	mass density
τ	nondimensional time, $t(\frac{E_x}{\rho a^2})^{1/2}$

INTRODUCTION

Nonlinear behavior of elastic plates has received considerable attention in recent years. A review of literature indicates that most of the investigations are concerned with rectangular plate geometries [1]. Using Galerkin's method, Nowinski [2] investigated the nonlinear transverse vibrations of rectilinearly orthotropic circular plates with clamped and stress-free boundaries. Yamaki [3] used a similar technique to study isotropic circular plates with clamped immovable edges. Axisymmetric nonlinear vibrations of thin circular plates have also been reported by Bulkelay [4], Srinivasan [5] and Kung and Pao [6]. These investigations are based on the most widely used dynamic von Kármán-type theory given by Herrmann [7] and a single-mode approach. Furthermore, in these analyses the effects of transverse shear deformation and rotatory inertia have been neglected. A fairly simple theory which was originally proposed by Berger [8] for the nonlinear static analysis of thin plates based upon the neglect of the second invariant of the middle surface strains in the expression for the extensional strain energy of the plate was subsequently extended by Nash and Modeer [9] for the dynamic study of plate problems. This approximation which was later known as the Berger's approximation was widely used for plates of various geometries with immovable boundaries. Iwinski and Nowinski [10], Basuli [11], Pal [12], Datta [13] and Banerjee [14] used this approxmation to study the static nonlinear problems concerning isotropic and orthotropic circular plates. Wah [15] investigated the nonlinear dynamic behavior of isotropic circular plates by means of Berger-type equations suggested by Nash and Modeer [9]. Further work in this area based on the extension of Berger's approximation can be found in References [16-19]. None of these papers, however, consider the effects of transverse shear deformation or rotatory inertia on the nonlinear behavior.

A review of literature indicates that there are several nonlinear theories for orthotropic and anisotropic plates which include the effects of transverse shear deformation and rotatory inertia. Solutions to these nonlinear theories, however, cannot be found in the literature. In view of this, Wu and Vinson [20] derived a simplified set of equations using the Berger's approximation and presented results for the nonlinear vibration of composite rectangular plates. The combined effects of transverse shear and rotatory inertia on the nonlinear vibration of orthotropic rectangular plates have also been discussed by the author [21] based on an improved version of the Berger-type theory and on a generalization of the classical von Kármán-type plate theory [22,23]. From these investigations it may be concluded that the effects of transverse shear deformation and rotatory inertia are very important for moderately thick plates, particularly for non-isotropic plates.

This paper is concerned with the large amplitude flexural vibration of moderately thick rectilinearly orthotropic circular plates with rigidly clamped edges. Effects of transverse shear deformation and rotatory inertia have been incorporated into the governing nonlinear equations which are obtained by using the Berger's approximation [20,24]. The nonlinear equations thus obtained are expressed in terms of the first invariant of the middle surface strain e, and lateral displace w [24]. These equations are independent of slope functions α and β which are considered as additional variables in the equations given by Wu and Vinson [20]. The resulting Berger-type equations are solved here by following a multiple-mode analysis approach. A solution for w is carefully chosen in the polynomial form to satisfy the appropriate boundary conditions and substituted into the first of the two governing equations which is in a quasilinear form and integrated in conjunction with the immovable boundary conditions. Using the results of this integration and the transverse displacement assumed earlier, the equation of transverse motion is approximately satisfied by integrating the error function over the area of the plate. This procedure results in a set of two nonlinear time-differential equations whose solutions are obtained by applying the Runge-Kutta numerical integration technique. If the transverse shear and/or rotatory inertia effects are neglected at this stage in the analysis, the nonlinear differential equations reduce to a set of Duffing-type equations. Numerical results are graphically presented for different orthotropic and isotropic circular plates including the effects of transverse shear deformation and rotatory inertia. Present results for all special cases are in good agreement with existing solutions for nonlinear vibration and bending problems. The analysis approach used here is substantially simple and yet yields reasonably accurate results comparable with those obtained by a more tedious non Kármán-type approach to the solution of the problem.

GOVERNING EQUATIONS

A circular plate of radius a and thickness h is considered with origin at the center of the plate and x and y axes along horizontal and vertical directions respectively (see Fig. 1). For a moderately thick rectilinearly orthotropic plate the governing Berger-type equations in order to account for transverse shear deformation and rotatory inertia are [20,21]:

$$\varepsilon_x^o + k\varepsilon_y^o = \frac{\delta^2 h^2}{12} = e \tag{1}$$

$$N(I_o) + U(w) = 0 \tag{2}$$

where

$$\varepsilon_x^o = u_{,x} + \frac{1}{2} w^2_{,x} \quad , \qquad \varepsilon_y^o = v_{,y} + \frac{1}{2} w^2_{,y}$$

$$I_o = q(x,y) - \rho h w_{,tt} + C_1 \frac{\delta^2 h^2}{12} (w_{,xx} + k w_{,yy}) \tag{3}$$

$$C_1 = \frac{E_x h}{\mu} \quad , \qquad \mu = 1 - \nu_{xy}\nu_{yx} \quad , \qquad k = \left(\frac{E_y}{E_x}\right)^{1/2}$$

A comma denotes partial differentiation with respect to the corresponding coordinate. The sixth order differential operators N and U in equation (2) are defined in [21]. These operators are written in terms of various coefficients a_i (a_1 to a_{22}) which depend upon the orthotropic plate material constants as well as tracing constants TS and RI that are useful in identifying the effects of transverse shear deformation and rotatory inertia, respectively. If these effects are included in the analysis, TS = RI = 1 and TS = RI = 0 if these effects are ignored. Thus equations (1) and (2) constitute a set of two equations governing the large amplitude flexural vibrations of rectilinearly orthotropic moderately thick plates wherein the transverse shear and rotatory inertia effects are included. When TS = RI = 0, these equations readily reduce to those given by Nowinski [2] for thin plates.

Solutions to the system of equations (1) and (2) are presented below for the nonlinear flexural vibrations of orthotropic circular plates with clamped immovable boundaries.

METHOD OF SOLUTION

The boundary of the circular plate is defined by the equation $x^2 + y^2 = a^2$ and is assumed to be immovably clamped. Therefore, the boundary conditions are:

$$u = v = w = w_{,x} = w_{,y} = 0 \qquad \text{along } x^2 + y^2 = a^2 \tag{4}$$

In order to satisfy the boundary conditions on w, a multiple-mode expression for w is chosen as follows:

$$w(x,y,\tau) = \frac{h}{a^4}(x^2 + y^2 - a^2)^2 [A_1(\tau) + \frac{1}{a^2} A_2(\tau)(x^2 + y^2 - a^2)] \tag{5}$$

In equation (5) $A_1(\tau)$ and $A_2(\tau)$ are unknown functions of nondimensional time τ defined in Nomenclature. Substituting the expressions for ϵ_x^o and ϵ_y^o from equation (3) in equation (1) and integrating with the conditions that u=v=o along $x^2 + y^2 = a^2$, the following relationship is obtained:

$$e = -\frac{1}{2\pi a^2} \iint w\,(w_{,xx} + k\, w_{,yy})\, dx\, dy \tag{6}$$

Since w is already known from equation (5) the integration in equation (6) can be readily performed to yield

$$\begin{aligned} e = -\frac{h^2}{\pi a^8} \Big[& 2a^2 (3 + k) \{ b_1 (A_1^2 - 3A_1A_2) + c_1 (A_1A_2 - 3A_2^2)\} \\ & + 2a^2(3k+1) \{b_2 (A_1^2 - 3A_1 A_2) + C_2 (A_1A_2 - 3A_2^2)\} + 3 (5+k) \\ & (b_3 A_1A_2 + C_3 A_2^2) + 3 (5k + 1) (b_4 A_1A_2 + C_4 A_2^2) + 18 (1+k) \\ & (b_5 A_1A_2 + C_5 A_2^2) + 18 (1+k) (b_5 A_1A_2 + C_5A_2^2) + a^4 (1+k) (3b_6 A_1A_2 \\ & - 2C_6 A_1A_2 + 3 C_6A_2^2 - 2b_6 A_1^2) \Big] \end{aligned} \tag{7}$$

where b_i (b_1 to b_6) are double integral values obtained by multiplying $\frac{1}{a^4}(x^2 + y^2 - a^2)^2$ with x^2, y^2, x^4, y^4, x^2y^2 and unity. Similarly c_i (c_1 to c_6) are the corresponding double integral values of the same expressions with $\frac{1}{a^4}(x^2 + y^2 - a^2)^2$ replaced by $\frac{1}{a^6}(x^2 + y^2 - a^2)^3$.

Equation (7) is rewritten for convenience in the following form:

$$e = d_1 A_1^2 + d_2 A_1 A_2 + d_3 A_2^2 \tag{8}$$

the coefficients b_i and c_i are not defined here for the sake of saving space but can be easily obtained by performing the integrations mentioned earlier.

Having solved equation (1), attention is now turned to the solution of equation (2) which represents the dynamic equation of equilibrium in the transverse direction of the vibrating plate. Substituting equations (5) and (8) in equation (2) and integrating the resulting expression over the area of the plate, the following two coupled nonlinear time-differential equations in A_1 (τ) and A_2 (τ) are obtained.

$$\begin{aligned}
&B_1(A_1^3),_{\tau\tau\tau\tau} + B_2(A_1^2A_2),_{\tau\tau\tau\tau} + B_3(A_1\,A_2^2),_{\tau\tau\tau\tau} + B_4(A_3^3),_{\tau\tau\tau\tau} \\
&+ B_5(A_1),_{\tau\tau\tau\tau\tau\tau} + B_6(A_2),_{\tau\tau\tau\tau\tau\tau} + B_7(A_1^3),_{\tau\tau} + B_8(A_1^2A_2),_{\tau\tau} \\
&+ B_9(A_1A_2^2),_{\tau\tau} + B_{10}(A_2^3),_{\tau\tau} + B_{11}(A_1),_{\tau\tau\tau\tau} + B_{12}(A_2),_{\tau\tau\tau\tau} \\
&+ B_{13}(A_1^3) + B_{14}(A_1^2A_2) + B_{15}(A_1A_2^2) + B_{16}(A_2^3) + B_{17}(A_1),_{\tau\tau} \\
&+ B_{18}(A_2),_{\tau\tau} + B_{19}A_1 + B_{20}A_2 = q_o^* \qquad (9)
\end{aligned}$$

$$\begin{aligned}
&C_1(A_1^3),_{\tau\tau\tau\tau} + C_2(A_1^2A_2),_{\tau\tau\tau\tau} + C_3(A_1A_2^2),_{\tau\tau\tau\tau} + C_4(A_2^3),_{\tau\tau\tau\tau} \\
&+ C_5(A_1),_{\tau\tau\tau\tau\tau\tau} + C_6(A_2),_{\tau\tau\tau\tau\tau\tau} + C_7(A_1^3),_{\tau\tau} + C_8(A_1^2A_2),_{\tau\tau} \\
&+ C_9(A_1A_2^2),_{\tau\tau} + C_{10}(A_2^3),_{\tau\tau} + C_{11}(A_1),_{\tau\tau\tau\tau} + C_{12}(A_2),_{\tau\tau\tau\tau} \\
&+ C_{13}(A_1^3) + C_{14}(A_1^2A_2) + C_{15}(A_1A_2^2) + C_{16}(A_2^3) + C_{17}(A_1),_{\tau\tau} \\
&+ C_{18}(A_2),_{\tau\tau} + C_{19}A_1 + C_{20}A_2 = q_0^* \qquad (10)
\end{aligned}$$

where q_o^* is the nondimensional load defined in terms of the uniformly distributed load q_o as $q_o^* = \dfrac{q_o a^4}{E_x h^4}$. B_1 to B_{20} and C_1 to C_{20} are nondimensional coefficients which depend upon plate material constants, geometric plate parameters of the plate as well as tracing constants TS and RI. Equations (9) and (10) are applicable to moderately thick circular plates including the effects of transverse shear deformation and rotatory inertia i.e., TS = RI = 1. If either or both of these effects are neglected i.e., TS = 1, RI = 0; TS = 0, RI = 1; TS = RI = 0 then the time-differential equations (9) and (10) simplify as follows:

$$\begin{aligned}
&B_{13}A_1^3 + B_{14}A_1^2A_2 + B_{15}A_1A_2^2 + B_{16}A_2^3 + B_{17}A_1,_{\tau\tau} \\
&\quad + B_{18}A_2,_{\tau\tau} + B_{19}A_1 + B_{20}A_2 = q_0^* \qquad (11)
\end{aligned}$$

$$C_{13}A_1^3 + C_{14}A_1^2A_2 + C_{15}A_1A_2^2 + C_{16}A_2^3 + C_{17}A_{1,\tau\tau}$$

$$+C_{18}A_{2,\tau\tau}+C_{19}A_1+C_{20}A_2 = q_0^* \qquad (12)$$

Equations (9-12) can be used to investigate the static and dynamic nonlinear behavior of circular plates with or without the effects of transverse shear deformation and rotatory inertia. For static nonlinear problems, numerical solutions for equations (11) and (12) can be obtained by means of Newton's method with TS=1 or TS=0 (RI=0). In the case of static problems $A_1(\tau)$ and $A_2(\tau)$ are independent of time and therefore equations (11) and (12) reduce to a system of two nonlinear algebraic equations whose numerical solutions by Newton's method will give the load-deflection relationship. In the case of nonlinear dynamic problems, equations (9) and (10) are used for moderately thick plates whereas equations (11) and (12) are used for all other combinations of TS and RI. Numerical solutions for these sets of equations are obtained by means of an IMSL Subroutine which uses the Runge-Kutta-Verner fifth and sixth order integration method.

NUMERICAL RESULTS

Numerical results are reported for the static and dynamic cases of some high-modulus composite and isotropic circular plates. The elastic constants with respect to x and y directions of glass-epoxy (GE), boron-epoxy (BE) and graphite-epoxy (GRE) plates are taken from [23]. The nonlinear period of vibration T which includes the effects of transverse shear and rotatory inertia and the linear period T_o which does not include these effects are computed for various plate parameters.

The fundamental linear frequencies for isotropic circular plates obtained from the present analysis considering both single and multiple terms in w are presented in Table 1. These results show excellent agreement with those available in the literature [25, 26]. It can be seen that the natural frequency calculated here by using a two-term mode shape for w is almost equal to the exact natural frequency. In the case of static problems the nonlinear load-deflection relationship obtained from the numerical solution of equations (11) and (12) are presented in Table 2. Comparisons are also made with the results of Yamaki [3]. Again, there is a remarkable agreement between the present and reported results in the case of single-mode solutions. The effect of transverse shear deformation has been ignored in the results given in Table 2. However, by taking TS=1 similar results could be readily obtained for moderately thick plates. In the case of dynamic problems, the values of period ratios at different nondimensional amplitudes for thin plates (TS=RI=0) obtained from the solutions of equations (11) and (12) are presented in Table 3 with the corresponding available values for comparison. Similar results are also presented for moderately thick isotropic plates in Table 3. Any observed deviations in the numerical values between single-term solutions in Tables 2 and 3 are due the fact that Yamaki [3] used the von Kármán-type field equations for thin plates whereas here Berger-type equations have been used. For nonlinear static problems q_o^* evaluated at any given deflection on the basis of a two-term solution is higher than that obtained by a single-term solution. Similarly, the period ratio at any given amplitude is lower than that predicted by a single-mode solution. These variations could be attributed to an equivalent increase in the flexural rigidity of the plate and therefore clearly demonstrate the nonlinear coupling between modes.

Turning the attention now to the numerical results for moderately thick plates, it is required to solve equations (9) and (10) which are nonlinear and highly coupled. In order to make the numerical procedure manageable, the forty coefficients in these two equations are evaluated for different plate parameters. A comparison of the coefficients indicate that some higher order derivatives in equations (9) and (10) can be neglected in the final numerical solution. Following this procedure, numerical results are obtained by means of Runge-Kutta-Verner numerical integration technique. These results are graphically presented in Figures 1 to 4. For easy comparison, period amplitude curves for thin plates (TS = RI = 0) are also presented. For all the cases investigated here, the period ratio decreases with increasing amplitude of vibration and therefore the period-amplitude behavior is of the hardening type. The effects of transverse shear and rotatory inertia are shown by an increase in the period ratio at any amplitude of vibration with maximum occurring at small amplitudes. It is also seen that the effects of transverse shear and rotatory inertia are considerably significant for composite plates whereas these effects are not so important for isotropic plates. It is observed that the effect of transverse shear is more than the effect of rotatory inertia whether the plate material is isotropic or composite.

In conclusion, it is to be mentioned that Berger-type field equations are rather easier to deal with particularly when using multiple mode approach to the solution of moderately thick plates. The numerical results of both isotropic and composite plates are reasonably accurate and can be accepted for all practical purposes. Small deviations in the numerical results are due to the approximate nature of the Berger-type field equations. The efforts involved in obtaining a reasonable solution to the problem by means of Berger-type theory is far less than that is needed when more accurate theories are used. Finally, multiple-mode solutions seem to exhibit the same type of general qualitative behavior as given by the single-mode solutions. At large amplitudes, the effect of coupling of vibrating modes becomes all the more important for composite plates than isotropic plates.

ACKNOWLEDGEMENT

The results presented in this paper were obtained in the course of research partially supported by Clarkson College of Technology, Potsdam, New York.

REFERENCES

1. Sathyamoorthy, M. and Pandalai, K.A.V., Large Amplitude Vibrations of Certain Deformable Bodies - Part II, Journal of Aeronautical Society of India, Vol. 25, 1973, pp. 1-10.

2. Nowinski, J.L., Nonlinear Vibrations of Elastic Circular Plates Exhibiting Rectilinear Orthotropy, ZAMP, Vol. 14, 1963, pp. 113-124.

3. Yamaki, N., Influence of Large Amplitude on Flexural Vibrations of Elastic Plates, ZAMM, Vol. 41, 1961, pp. 501-510.

4. Bulkelay, P.Z., An Axisymmetric Nonlinear Vibration of Circular Plates, ASME Journal of Applied Mechanics, Vol. 30, 1963, pp. 630-631.

5. Srinivasan, A.V., Large Amplitude Free Oscillations of Beams and Plates, AIAA, Vol. 3, 1965, pp. 1951-1953.

6. Kung, G.C. and Pao, Y.H., Nonlinear Flexural Vibrations of a Clamped Circular Plate, ASME Journal of Applied Mechanics, Vol. 39, 1972, pp. 1050-1054.

7. Herrmann, G., Influence of Large Amplitudes on Flexural Motions of Elastic Plates, NACA TN-3578, 1955.

8. Berger, H.M., A New Approach to the Analysis of Large Deflection of Plates, ASME Journal of Applied Mechanics, 1955, pp. 465-472.

9. Nash, W.A. and Modeer, J.R., Certain Approximate Analysis of the Nonlinear Behavior of Plates and Shells, Proceedings of the Symposium on the Theory of Thin Elastic Shells, 1960, pp. 331-354.

10. Iwinski, T. and Nowinski, J., The Problem of Large Deflections of Orthotropic Plates, Arch. Mech. Stos, Vol. 9, 1957, pp. 593-603.

11. Basuli, S., Note on the Large Deflection of a Circular Plate Under a Concentrated Load, ZAMP, Vol. 12, 1961, pp. 357-362.

12. Pal, M.C., Large Deflections of Heated Circular Plates, Acta Mechanica, Vol. 8, 1969, pp. 82-103.

13. Datta, S., Large Deflection of a Circular Plate on Elastic Foundation Under a Concentrated Load at the Center, ASME Journal of Applied Mechanics, Vol. 42, 1975, pp. 503-505.

14. Banerjee, M.M., Note on the Large Deflection of an Orthotropic Circular Plate with Clamped Edge Under Symmetrical Load, Journal of the Indian Institute of Science, Vol. 58, 1976, pp. 175-180.

15. Wah, T., Vibration of Circular Plates at Large Amplitudes, ASCE Journal of Engineering Mechanics Division, Vol. 89, 1963, pp. 1-15.

16. Srinivasan, A.V., Nonlinear Vibrations of Beams and Plates, International Journal of Nonlinear Mechanics, Vol. 1, 1966, pp. 179-191.

17. Pal, M.C., Large Amplitude Vibration of Circular Plates Subjected to Aerodynamic Heating, International Journal of Solids and Structures, Vol. 6, 1970, pp. 301-313.

18. Chiang, D.C. and Chen, S.S.H., Large Amplitude Vibration of a Circular Plate with Concentrated Rigid Mass, ASME Journal of Applied Mechanics, Vol. 39, 1972, pp. 577-583.

19. Huang, C.L. and Al-Khattat, I.M., Finite Amplitude Vibrations of a Circular Plate, International Journal of Nonlinear Mechanics, Vol. 12, 1977, pp. 297-306.

20. Wu, C.I. and Vinson, J.R., On the Nonlinear Oscillations of Plates Composed of Composite Materials, Journal of Composite Materials, Vol. 3, 1969, pp. 548-561.

21. Sathyamoorthy, M., Vibration of Plates Considering Shear and Rotatory Inertia, AIAA, Vol. 16, 1978, pp. 285-286.

22. Sathyamoorthy, M., Effects of Large Amplitude, Shear and Rotatory Inertia on Vibration of Rectangular Plates, Journal of Sound and Vibration, Vol. 63, 1979, pp. 161-167.

23. Sathyamoorthy, M. and Chia, C.Y., Nonlinear Vibration of Orthotropic Circular Plates Including Transverse Shear and Rotatory Inertia, Proceedings of ASME Winter Annual Meeting, New York, 1979, pp. 357-372.

24. Sathyamoorthy, M., Transverse Shear and Rotatory Inertia Effects on Nonlinear Vibration of Orthotropic Circular Plates, Computers and Structures (to appear).

25. McNitt, R.P., Free Vibration of Clamped Elliptical Plates, Journal of Aerospace Science, Vol. 29, 1962, pp. 1124-1125.

26. Ng, S.F. and Sharma, A., Vibration of Thin Plates Using the Ritz Method, Canadian Aeronautics and Space Journal, Vol. 25, 1979, pp. 372-381.

Single-mode	Multiple-mode	[25,26]	Exact
10.328	10.217	10.217	10.216

Table 1. Nondimensional frequency parameter $\lambda^2 = \omega a^2 (\frac{\rho}{D})^{1/2}$ for isotropic circular plates.

$\frac{w_{max}}{h}$	Nondimensional load q_o^*		
	Single-term	Multiple-term	Yamaki [3]
0	0	0	0
0.5	3.297	3.343	3.276
1.0	8.791	9.295	8.623
1.5	18.681	21.148	18.113

Table 2. Load-deflection values for isotropic circular plates, TS = 0

Amplitude	Thin plates, TS = RI = 0			TS = RI = 1, R = 10	
A_1-A_2	Single-mode	Multiple-mode	Yamaki [3]	Single-mode	Multiple-mod
0	10000	10000	10000	10159	10468
0.5	9563	9633	9584	9691	10014
1.0	8541	8287	8586	8606	8255
1.5	7397	6175	7460	7413	6157
2.0	6373	5284	6510	6360	5327

Table 3. Values of $(T/T_o)\ 10^4$ for isotropic circular plates

i	B_i	C_i
1	0.000020	0.000024
2	-0.000053	-0.000067
3	0.000050	0.000064
4	-0.000016	-0.000021
5	0.000406	0.000451
6	-0.000338	-0.000386
7	0.015709	0.018756
8	-0.042343	-0.052611
9	0.039458	0.050810
10	-0.012660	-0.016965
11	0.316889	0.352531
12	-0.264398	-0.303002
13	3.111219	3.697289
14	-8.377919	-10.393260
15	7.799994	10.056240
16	-2.499931	-3.364326
17	61.914580	68.956250
18	-51.717190	-59.439740
19	5.860757	5.860752
20	-4.606479	-7.243792

Table 4. Coefficients of equations (9) and (10) for TS = RI = 1, R = 10.

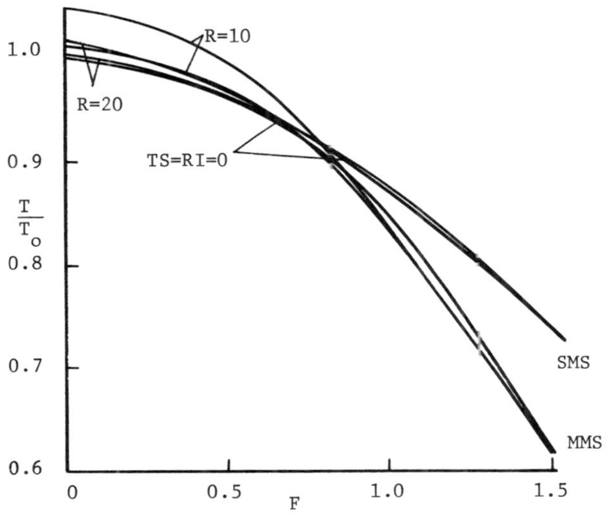

Fig. 1 Amplitude-period response curves for isotropic circular plates, SMS - Single Mode Solution, MMS - Multiple Mode Solution

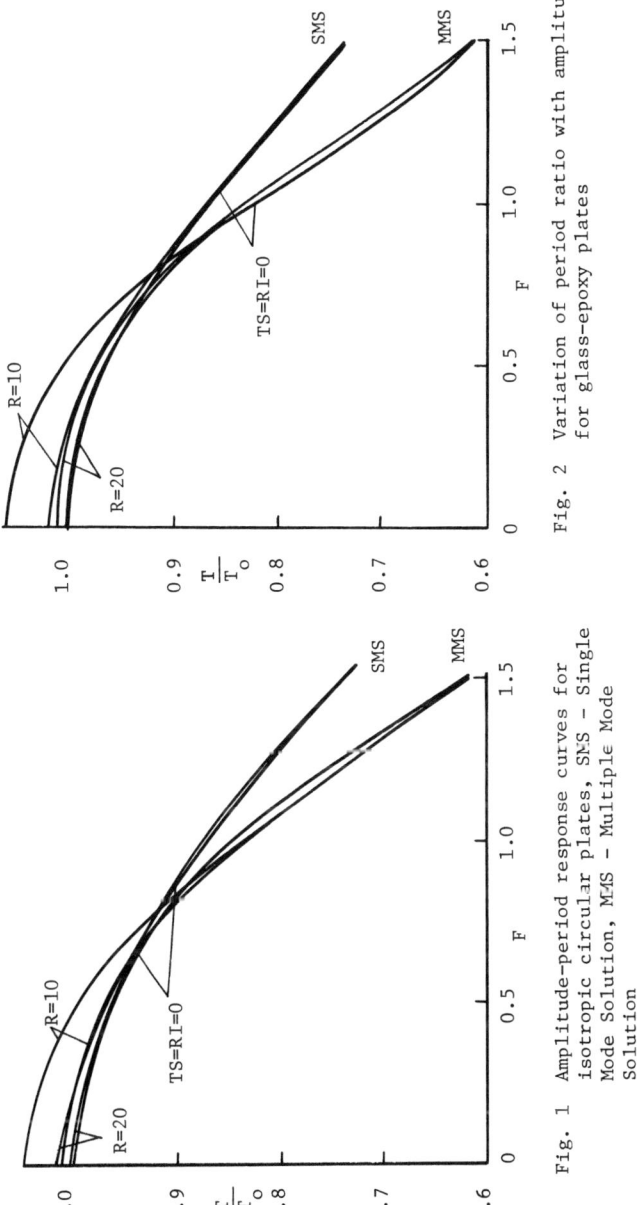

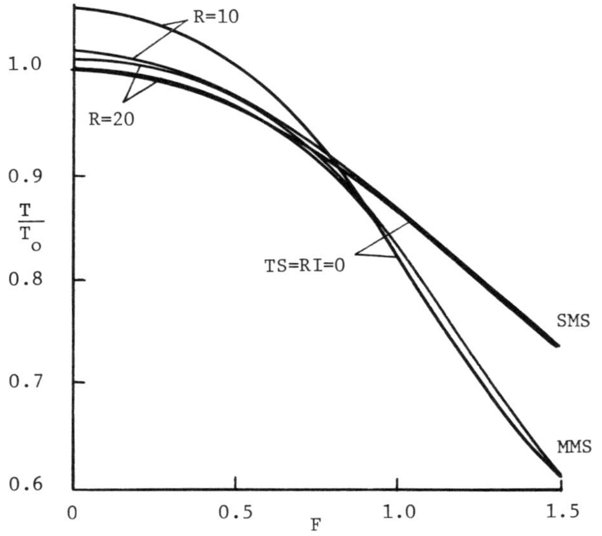

Fig. 2 Variation of period ratio with amplitude for glass-epoxy plates

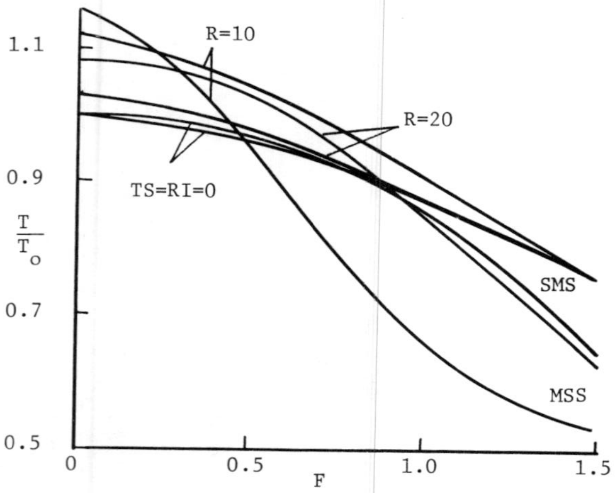

Fig. 3 Variation of period ratio with amplitude for graphite-epoxy plates

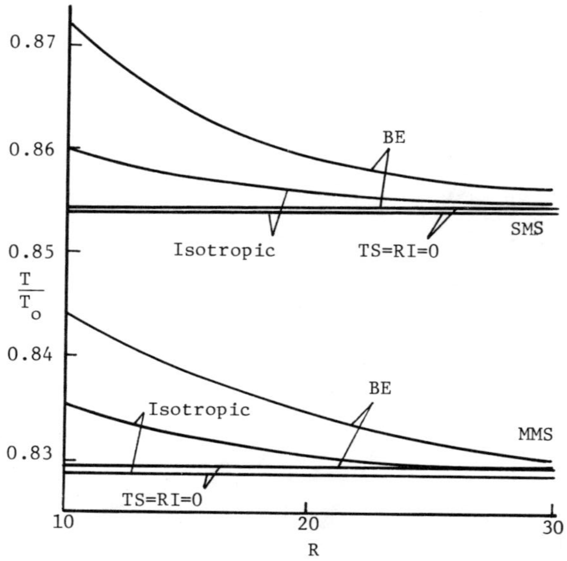

Fig. 4 Variation of period ratio with radius-to-thickness ratio for unit nondimensional amplitude

EXPERIMENTAL MECHANICS

PREDICTION OF SUB-CRITICAL CRACK GROWTH DATA FROM MODEL EXPERIMENTS

C. W. Smith, D. Post and G. Nicoletto
Department of Engineering Science and Mechanics
Virginia Polytechnic Institute and State University
Blacksburg, VA 24061

ABSTRACT

The majority of in-service brittle fractures result from the sub-critical growth of small cracks to critical size under fatigue loads. Such cracks typically originate in regions of high triaxial constraint, (i.e. high local stress gradients). Such constraints are caused by complex boundary shapes (notches, grooves, keyways, inclusions, etc.). Consequently the sub-critical crack growth regime includes an extremely complex three dimensional cracked body problem which typically involves non-planar flaws as well as curved crack fronts and nonuniform stress distributions which defy analytical tractability except by approximate numerical methods.

In order to provide code verification for such solutions the first author and his colleagues undertook an effort, beginning over a decade ago, to develop a cost effective experimental technique for modelling such problems. The technique is based upon an idea of G. R. Irwin, and currently utilizes a marriage between the optical methods of "frozen stress" photoelasticity and moiré interferometry and the near field equations of linear elastic fracture mechanics.

After briefly reviewing the foundations of the method, the present paper presents results from application of the method to several problems of current interest in order to illustrate the importance of flaw shape upon stress intensity values. It is suggested that, within certain constraints, the method can successfully predict crack shapes and stress intensity distributions to within engineering accuracy where none of these quantities are known a priori.

INTRODUCTION

The most frequent chronology of events leading to service fractures in metal parts begins with a defect or tiny flaw, usually on an exposed surface near a re-entrant corner, which enlarges under the action of repeated (vibratory, fatigue, etc.) loading until it reaches a critical size, after which catastrophic fracture results. The problem geometry which exists during this sub-critical flaw growth period is often quite complex, involving irregular body boundaries and often non-planar cracks and/or curved crack fronts. Such problems can readily be identified as three dimensional (3D) cracked, finite body problems where the stress intensity factor (SIF) varies along the flaw border. Such problems have resisted efforts of applied mathematicians to render them tractable in the closed form sense. However much progress has been made recently in modelling such problems numerically [1]-[4]. A major difficulty has been lack of knowledge of the crack shape during sub-critical flaw growth and its influence upon the SIF distribution.

Over a decade ago, the first author and his associates began to study experimental modelling techniques for use in simulating the growth of natural cracks and from which SIF distributions could be estimated. Originally, the method, involving a marriage between near field equations of linear elastic fracture mechanics with the frozen stress photoelastic technique, was applied only to Mode I problems [5]. Subsequently, it was extended to include mixed mode analysis [6]-[8]. In the process of modelling nozzle corner cracks in flat plates [9]-[10], it was noticed that, when cracks were grown from tiny starter flaws in the photoelastic models under monotonic loads, the resulting shapes would, under certain conditions, overlay those produced by tension-tension fatigue in geometrically similar models, even though the shapes were not simple. Figure 1 provides such an example for a deep nozzle-corner flaw.

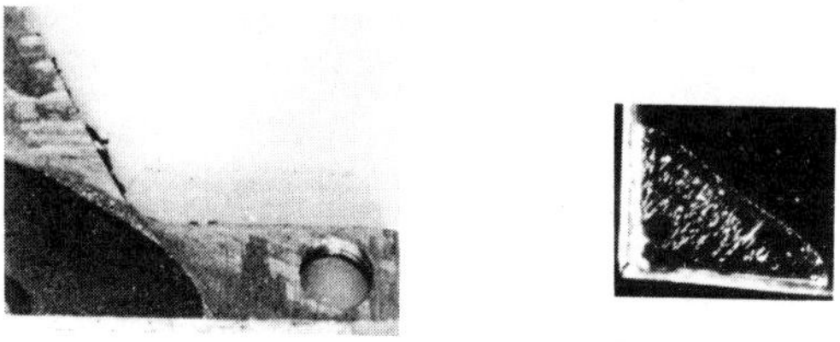

(a) (b)

Figure 1 Photo of Deep Cracks in Fatigued Reactor Steel (a) and Photoelastic (b) Models. (a) Courtesy M. G. J. Broekhoven - Delft University of Technology.

Such observations suggest that, under the proper conditions, the frozen stress method, when utilizing natural cracks, can provide predictions of both flaw shape and SIF distributions where neither are known a priori. Substantial work has been done (and is continuing) towards the goal of delineating the conditions necessary to produce the desired similarity in model behavior. Important considerations include:

i) Absence of fatigue overloads in prototype.

ii) Similarity in starter crack shapes for model and prototype.
iii) Assessment of Poisson's Ratio Effect.
iv) Assessment of compressive part of load spectrum.

By properly controlling these influences, the potential for reasonable correlation appears promising.

Since any experimental technique carries with it a certain amount of error, and since most 3D problems lack exact solutions, it was decided recently to integrate the frozen stress method with an independent alternate determination of the SIF distribution from the in-plane displacement field. Because of its high sensitivity the moiré interferometric technique, developed by the second author, [19], appeared to suit the purpose.

After including a brief review of the analytical foundations of the experimental method for Mode I analysis, this paper focuses upon the current status of the correlation of contemporary fracture analysis with modelling of surface flaws in flat plates. Results from other problems where flaw shapes deviate from simple geometrical shapes are also included in order to demonstrate the influence of flaw shapes upon SIF distributions. Comparisons with numerical results are included.

REVIEW OF THE EXPERIMENTAL METHOD

Although the method has been extended to include shear mode, we shall focus on the Mode I case here since our studies suggest an absence of shear modes in growing cracks.

A) Analytical Foundations

For the case of Mode I loading, one begins with equations of the form:

$$\sigma_{ij} = \frac{K_I}{r^{1/2}} f_{ij}(\theta) + \sigma^0_{ij}(r,\theta) \qquad (i,j = n,z) \qquad (1)$$

for the stresses in a plane mutually orthogonal to the flaw surface and the flaw border referred to a set of local rectangular cartesian coordinates as pictured in Fig. 2, where the terms containing K_1, the SIF, are identical to Irwin's Equations for the plane case and σ^0_{ij} are normally taken to be constant, for a given point along the flaw border, within the measurement zone but may vary from point to point. Eqs. 1 need not satisfy boundary conditions on the crack surface since they are only applied in a region near $\theta = \pm \pi/2$. This approach is often used in connection with hybrid finite element analysis. Observing that isochromatic fringes tend to spread approximately normal to the flaw surface (Fig. 3), Eqs. (1) are evaluated along $\theta = \pi/2$ (Fig. 2) and the maximum in-plane shearing stress is computed as:

$$\tau^{nz}_{max} = 1/2\, [(\sigma_{nn} - \sigma_{zz})^2 + 4\sigma^2_{nz}]^{1/2} \qquad (2)$$

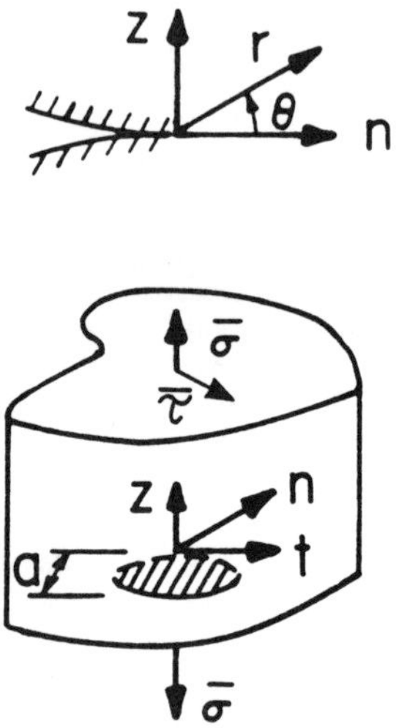

Figure 2 Mode I Problem Geometry and Notation.

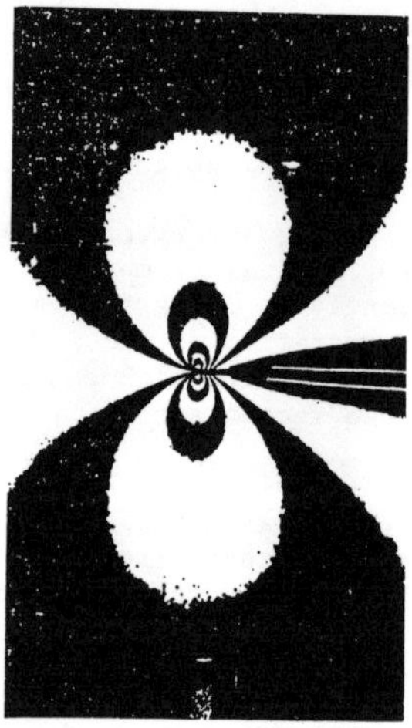

Figure 3 Photo of Mode I Stress Fringe Pattern.

which, when truncated to the same order as Eqs.(1), leads to the two parameter equation:

$$\tau_{max}^{nz} = \frac{A}{r^{1/2}} + B \quad \text{where} \quad A = K_1(8\pi)^{1/2}, \quad B = f(\sigma_{ij}^{0}) \qquad (3)$$

which can be rearranged into the normalized form

$$\frac{K_{AP}}{q(\pi a)^{1/2}} = \frac{K_I}{q(\pi a)^{1/2}} + \frac{f(\sigma_{ij}^{0})(8)^{1/2}}{q}\left\{\frac{r}{a}\right\}^{1/2} \qquad (4)$$

where $K_{AP} = \tau_{max}^{nz}(8\pi r)^{1/2}$ and, from the Stress-Optic Law, $\tau_{nz} = Nf/2t'$ where N is the stress fringe order, f the material fringe value and t' the slice thickness in the t direction, q is the remote loading parameter such as uniform stress (σ), or pressure (p), and a the characteristic flaw depth. Eqs. (1) with σ_{ij} as described above, require that a linear relation exist between the normalized apparent SIF and the square root of the normalized distance from the crack tip. Thus, one needs only to locate the linear zone in a set of photoelastic data and extrapolate across a very near field non-linear zone to the crack tip in order to obtain the SIF. An example of this approach using data from cracked plate tests is given in Fig. 4. Since the method is 3D, several slices are always removed from a given flaw border. The proper location for the linear zone must be common to all such slices and lie approximately in the region located between 0.1 and 1.0 mm from the crack plane.

Once the photoelastic data have been collected a linear diffraction grating is replicated on the surface of the slices according to the technique developed by Post, [11]. Assuming reversibility of the displacement field in a stress-freezing material, the slices are annealed. The deformed active grating is interrogated by superposition of a reference grating of 1200 lines per millimeter. The resulting moiré fringes describe the "negative" of the in-plane displacement field locked in the model during the stress freezing cycle. The moiré relationship gives

$$u_i = g_i N \qquad (i = n,z) \qquad (5)$$

where g_i is the pitch of the grating having lines perpendicular to the i direction and N is the moire fringe order.

According to LEFM the in-plane displacement field eqs. are

$$u_z = \frac{2(1+\nu)}{E} K_{AP} \left\{\frac{r}{2\pi}\right\}^{1/2} \sin\frac{\theta}{2}\left[2 - 2\nu - \cos^2\frac{\theta}{2}\right]$$

$$u_n = \frac{2(1+\nu)}{E} K_{AP} \left\{\frac{r}{2\pi}\right\}^{1/2} \cos\frac{\theta}{2}\left[1 - 2\nu + \sin^2\frac{\theta}{2}\right] \qquad (6)$$

When only a linear grating normal to the z-axis is present and the measurements are taken along a line defined by $\theta = \frac{\pi}{2}$ in Fig. 2.

$$K_{AP} = \frac{CN}{r^{1/2}} \qquad (7)$$

A plot of the normalized K_{AP} vs. $(r/a)^{1/2}$ allows one to determine a region in which the data fall along a straight line. Its intersection with the y-axis yields an estimate of K_I by moiré interferometry. A typical set of moiré data is shown in Fig. 5.

B. Experimental Procedure

In order to obtain flaw shapes and two independent estimates of the SIF for problems where neither are known a priori, we begin with fringe free cast sections of the part to be examined. We hold a sharp blade normal to the surface of the model part at the desired point and strike the blade with a hammer. A starter crack will propagate dynamically to a short distance from the blade tip and will then arrest itself. The parts of the model are then glued together, taking care that flaw borders are not to be located near glue lines, and the model is heated to critical temperature. Sufficient live load is applied above critical temperature to cause the crack to grow slowly. During growth, the crack is monitored through a glass oven port and when it reaches its approximatly desired size, the load is reduced to terminate flaw growth, and the model is then cooled under reduced load to room temperature. Upon unloading at room temperature, negligible recovery occurs and slices are removed with a band saw at intervals along the flaw border parallel to the nz

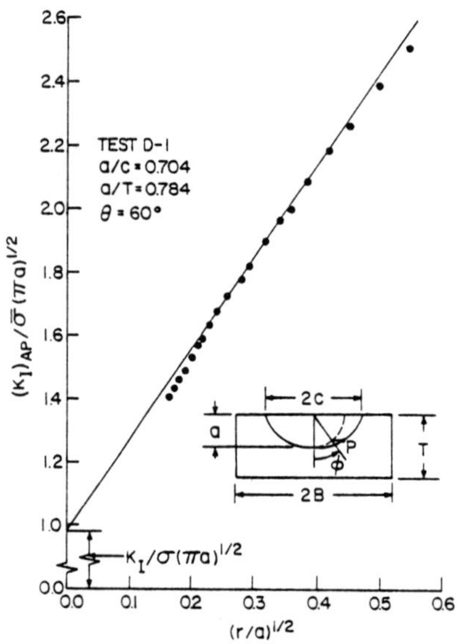

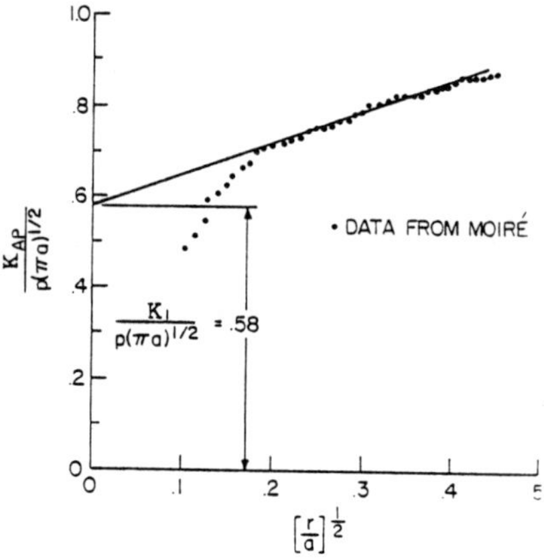

Figure 5 Estimation of K_1 from Moiré Data.

Figure 4 Estimation of K_1 from Raw Photoelastic Data from a Stress Frozen Slice.

plane (Fig. 2), coated with matching index fluid, and viewed through a crossed circular polariscope using white light and reading tint of passage isochromatics. The Tardy method is used to obtain fractional fringe orders. Optical data are then fed into a simple least squares computer program for obtaining the straight line of best fit (Fig. 4) for use in estimating the normalized K_1 value. Linear diffraction gratings are then replicated on the slices which are subsequently annealed.

The annealed slices are next introduced in the optical set-up, shown in Fig. 6a, where the deformed grating is analyzed by superposition of a virtual grating, [11].

The virtual grating, which is a three-dimensional array of walls of constructive and destructive interference, is generated whenever two collimated beams of light intersect and its frequency is a function of the angle of intersection as well as of the wavelength of the light.

The configuration of the optical elements used in this study to create the virtual grating is pictured in Fig. 6b and the virtual grating itself is denoted as a hatched zone.

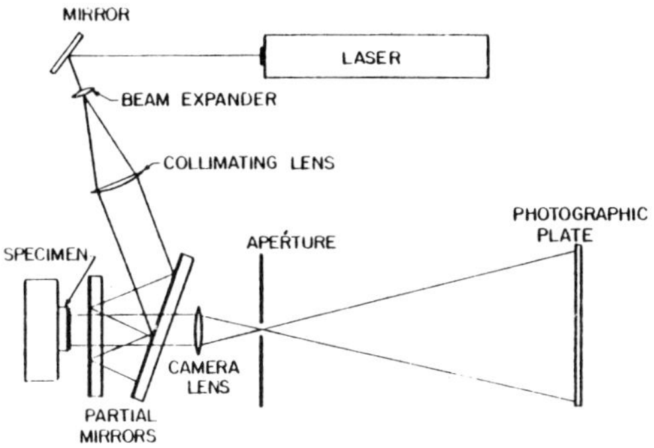

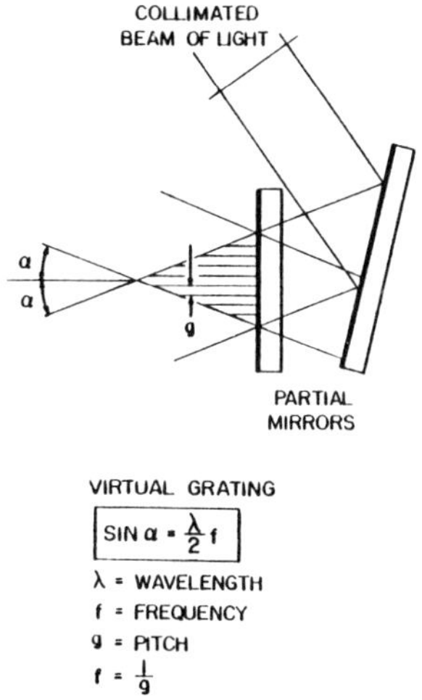

Figure 6 Experimental Setup for Moiré Fringe Measurement.

APPLICATIONS

A) <u>Surface Flaws in Flat Plates Under Mode I Loading</u>. This class of problems has received considerable attention in the literature since 1962, when Irwin [12] first classified the characteristic shape for such a flaw as being semi-elliptic (Fig. 7). Analytically, these studies have led to the development of a number of numerical models (op. cit. [1]-[3]). Moreover, a number of such models have recently been improved and compared (op. cit. [4]). Experimentally, extensive studies have been conducted, some recent results of which are summarized in Figure 8. These studies suggest that starter cracks at the surface which emanate from point defects exhibit aspect ratios (a/c) of approximately 0.9. As they grow, their aspect ratios decrease. A second class of surface flaws are those created by thermal shock, which initially exhibit very low starter crack aspect ratios, say a/c ≈ 0.1 to 0.2, and their aspect ratios increase as they grow.

Recently, Newman and Raju [13] have incorporated a crack growth rate model into a finite element model for semi-elliptic cracks so as to be able to predict aspect ratio versus relative flaw depth (a/t) curves such as those shown in Fig. 8. The finite element model utilized isoparametric linear strain elements away from the crack tip and pentahedron shaped singular elements at the crack tip with square root displacement terms which provided the stress singularity at the crack tip. SIF's were calculated using a nodal force method so as to avoid the assumption of plane stress or plane strain. Convergence studies indicated SIF determinations to within one or two percent of closed form 3D solutions. The models had 4300 to 4800 degrees of freedom. They analyzed both tensile and bending loads. Their results compared quite favorably with the data shown in Fig. 8 for tension and also compared favorably with similar curves for bending. However, the bending data were restricted to $a/t \leqslant 0.5$.

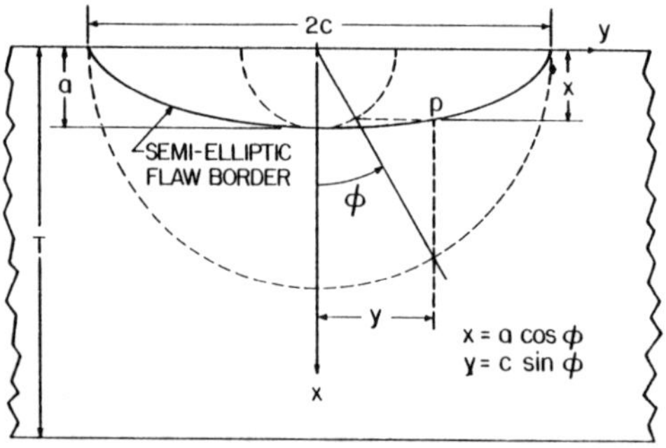

Figure 7 Semi-Elliptic Surface Flaw Notaton

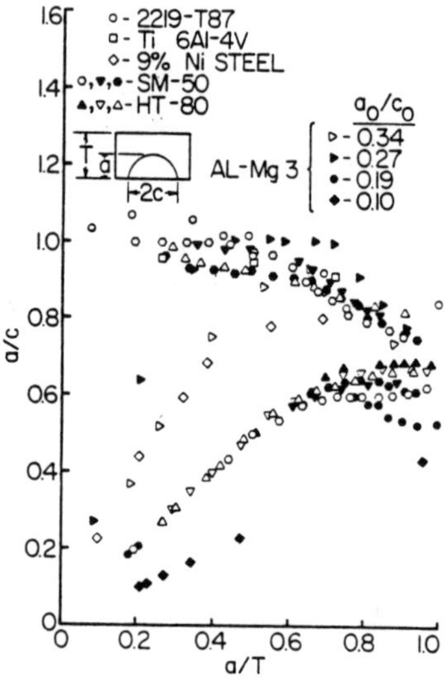

Figure 8 Experimental Results from Tests on Surface Flaws in Tension, Ref. (13).

Newman and Raju prescribed independent ΔK coefficients in their Paris-type crack growth rate equations for growth in the surface and maximum flaw depth directions in order to facilitate a variation in the aspect ratio to occur during flaw growth. However, they then prescribed a relation between the two coefficients in order to obtain semi-circular shallow flaws.

Due to the good correlation achieved between the Newman-Raju model and the variable aspect ratio fatigue test data on metals, it was decided to apply the authors' experimental modeling technique to check the SIF distributions from the Newman-Raju model and to determine if the natural cracks grown in the photoelastic models deviated from the assumed semi-elliptic shapes.

For the case of remote tension, Fig. 9 presents a comparison between photoelastic results and the Newman-Raju model for both shallow and deep flaws. The analytical SIF's were evaluated for $\nu = 0.45$ since the photoelastic results were for $\nu = 0.50$. It was also found that the flaw shapes remained semi-elliptic in the photoelastic experiments even for the deep ($a/t \approx 0.8$) flaws.

In other studies [14][15], the first author and his associates have found that, in bending, the flaw shape retains it's semi-elliptic shape up to mid-depth of the plate. However, for deeper flaws in bending distortions of the nature pictured in Fig. 10 have been observed (op. cit. [15]). In Fig. 11, we compare the experimental results obtained from tensile loading of a flaw previously distorted by bending with the average of six numerical solutions (op. cit. [4]) for a semi-elliptic crack of the same aspect ratio. After taking into account the various inevitable experimental inaccuracies, the authors believe that the rather significant increase in the SIF at midflaw is

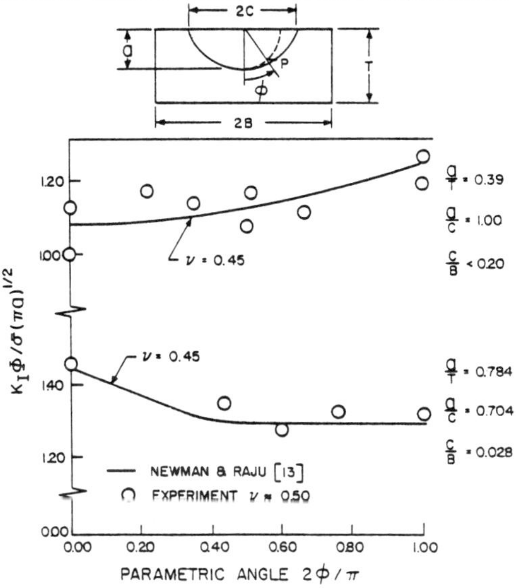

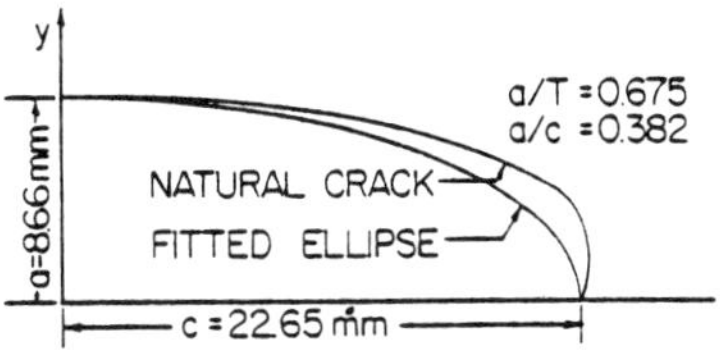

Figure 10 Distortions in Surface Flaws Due to Bending.

Figure 9 Analytical and Experimental Results for Semi-Elliptic Surface Flaws in Tension.

due to the distortion in shape of the flaw. This result suggests that in the case of bending loads surface flaws which penetrate beyond about two thirds of the plate depth may not be adequately represented by a semi-elliptic shape. On the other hand, for uniform tension, the semi-elliptic shape appears adequate for flaw depths up to a/t ≈ 0.80.

B) Cracks Emanating from Holes in Uniform Tensile Fields. Another illustration can be offered in order to show how distortion of the crack front can alter the SIF distribution. Consider the hole-crack problem pictured in Fig. 12. When loaded in uniform Mode I tension, a quarter circular starter crack will grow into a quarter elliptic shape with most of the growth along the direction of a. If however, the plate is flexed to maintain growth along c equal to a, then as the flaw enlarges, the flaw border tends to flatten in the central region in the neighborhood point I. The most current numerical

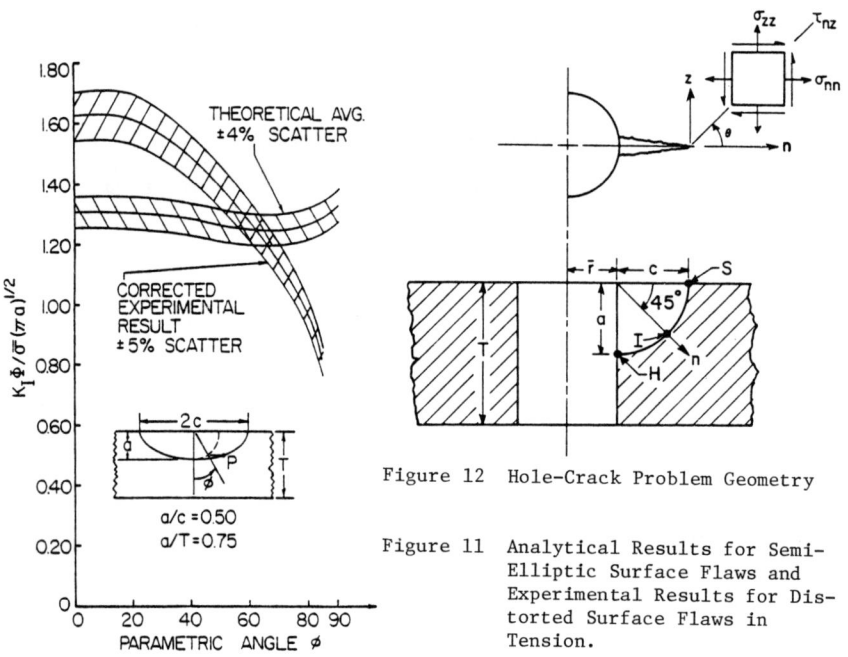

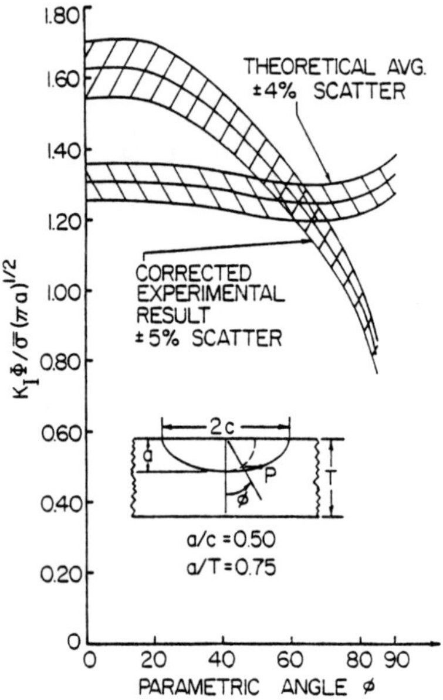

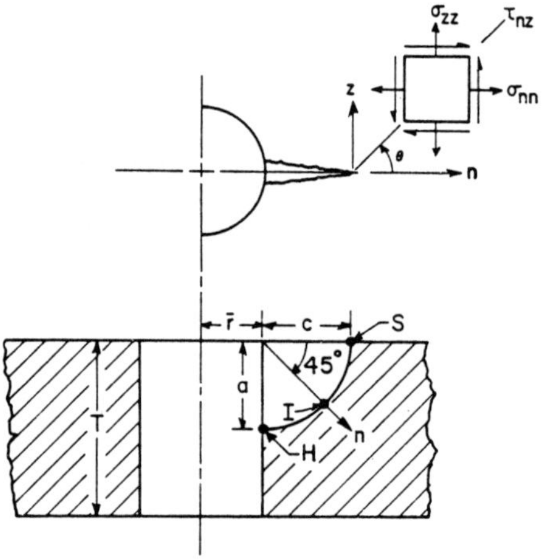

Figure 12 Hole-Crack Problem Geometry

Figure 11 Analytical Results for Semi-Elliptic Surface Flaws and Experimental Results for Distorted Surface Flaws in Tension.

solution for the SIF distribution in this problem is a hybrid finite element - alternating method developed by Smith and Kullgren for quarter elliptic holes [16]. The alternating method iterates between two solutions:

i) Stress distribution in a finite body due to surface tractions using a 3D finite element method.

ii) Stress distribution in an infinite body due to an arbitrary loading applied to an embedded ellipse.

The procedures employed lead to direct calculation of the SIF's and do not require extrapolation of stresses or fitting crack opening displacements as required of some models.

In Fig. 13, we first compare experimental (1) and analytical (2) results when the flaw is beginning to flatten in the central region. As the flaw size increases, the central flattening increases (3) and the difference in the experimental SIF distribution (3) from that predicted for a quarter elliptic flaw (4) is significant. The experimental results were taken from [17].

Similar behavior has been observed for nozzle corner cracks in nuclear reactors [18].

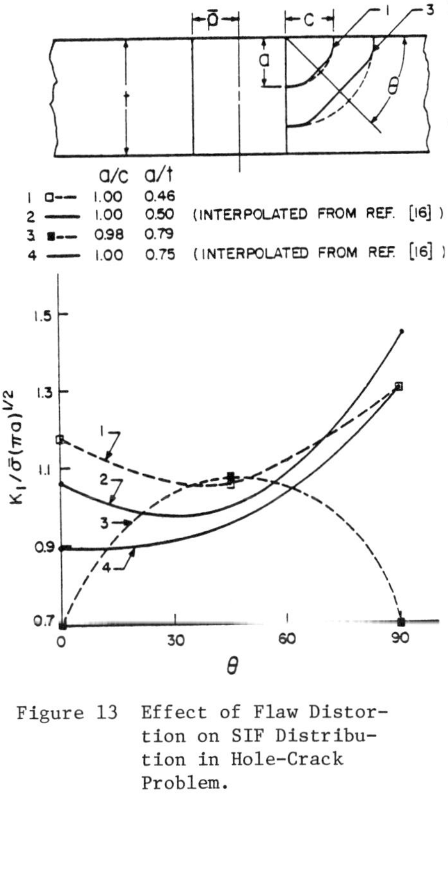

Figure 13 Effect of Flaw Distortion on SIF Distribution in Hole-Crack Problem.

SUMMARY

An experimental modelling method which integrates equations of LEFM with the optical analysis methods of frozen stress photoelasticity and moiré interferometry for estimating flaw shapes and stress intensity distributions in 3D cracked body problems was reviewed for Mode I analysis.

The method was then applied to two classes of problems for which approximate solutions were available. Results suggest that the method can provide reasonable computer code verification where flaw shapes are known, and can provide estimates of the influence of flaw shape distortions on the SIF distributions. The accuracy of the method is currently believed to lie between ± 2% and ± 10% for a broad class of 3D cracked body problems.

Acknowledgements

The authors are pleased to acknowledge the contributions of their colleagues as cited in the references, particularly F. W. Smith, T. E. Kullgren, J. C. Newman, Jr. and I. S. Raju. They are also indebted to the National Science Foundation for support of this work under Eng. 76-20824-1.

REFERENCES

[1] Swedlow, J. L., ed., The Surface Crack: Physical Problems and Computational Solutions, Applied Mechanics Division of ASME Special Publication, 1972.

[2] Rybicki, E. F. and Benzley, S. E., Computational Fracture Mechanics, Pressure Vessels and Piping Division of ASCE, Special Publication, 1975.

[3] Hulbert, L. E., "Benchmark Problems for Three Dimensional Fracture Analysis", International Journal of Fracture, Vol. 13, pp. 87-91 (1977).

[4] McGowan, J. J., ed., "A Critical Evaluation of Numerical Solutions to the 'Benchmark' Surface Flaw Problem", Fracture Committee of SESA Monograph, 1980.

[5] Smith, C. W., "Use of Three Dimensional Photoelasticity and Progress in Related Areas", Experimental Techniques in Fracture Mechanics, Vol. 2, SESA Monograph, A. S. Kobayashi, Ed., pp. 3-58, 1975.

[6] Smith, C. W., Peters, W. H. and Andonian, A. T., "Mixed Mode Stress Intensity Distributions for Part Circular Surface Flaws", Journal of Engineering Fracture Mechanics, Vol. 13, pp.615-629, 1979.

[7] Smith, C. W. and Hardrath, W. T., "Photoelastic Determination of Stress Intensities for Mode III", Recent Advances in Engineering Science, Vol. 15, pp. 195-200.

[8] Smith, C. W. and Hardrath, W. T., "A Method for Measuring K_3 in Three Dimensional Cracked Body Problems", Developments in Mechanics, Vol. 10, pp. 225-230, 1979.

[9] Smith, C. W. and Peters, W. H., "Experimental Observations of 3D Geometric Effects in Cracked Bodies", Developments in Theoretical and Applied Mechanics, Vol. 9, pp. 225-234, 1978.

[10] Smith, C. W. and Peters, W. H., "Prediction of Flaw Shapes and Stress Intensity Distributions in 3D Problems by the Frozen Stress Method", Preprints from Sixth International Congress on Experimental Stress Analysis, pp. 861-865, 1978.

[11] Post, D. and Baracat, W. A., "High-Sensitivity Moiré Interferometry - A Simplified Approach", J. of Experimental Mechanics, Vol. 21, No. 3, pp. 100-104, March 1981.

[12] Irwin, G. R., "Crack Extension Force for a Part Through Crack in a Plate", Journal of Applied Mechanics, Vol. 29, Trans. ASME Series E, pp. 651-654, Dec. 1962.

[13] Newman, J. C., Jr., and Raju, I. S., "Analyses of Surface Cracks in Finite Plates Under Tension or Bending Loads", NASA TP No. 1578, Dec. 1979.

[14] Schroedl, M. A. and Smith, C. W., "Local Stresses Near Deep Surface Flaws Under Cylindrical Bending Fields", Progress in Flaw Growth and Fracture Toughness Testing, ASTM STP 536, pp. 45-63, 1973.

[15] Smith, C. W., Peters, W. H., Kirby, G. C. and Andonian, A., "Stress Intensity Distributions for Natural Flaw Shapes Approximating 'Benchmark' Geometries", VPI-E-80-17, Feb. 1980 (In Press), Fracture Mechanics, ASTM STP 743.

[16] Smith, F. W. and Kullgren, T. E., "Theoretical and Experimetnal Analyses of Surface Cracks Emanating from Fastener Holes", U.S. Air Force Flight Dynamics Laboratory-TR-76-104, 180 pp., Feb. 1977.

[17] Smith, C. W., Peters, W. H. and Gou, S. F., "Influence of Flaw Geometries on Hole-Crack Stress Intensities", Fracture Mechanics, ASTM-STP No. 677, pp. 431-445, 1979.

[18] Smith, C. W., Peters, W. H., Hardrath, W. T. and Fleischman, T. S., "Stress Intensity Distributions in Nozzle Corner Cracks of Complex Geometry", Paper No. G4/4, Transactions of the Fifth International Conference on Structural Mechanics in Reactor Technology, Vol. G, 1979.

[19] Post, D., "Optical Interference for Deformation Measurements--Classical, Holographic and Moire Interferometry", Mechanics of Nondestructive Testing, W. W. Stinchcomb ed., Plenum Publishing Co., New York, 1980.

THE EXPERIMENTAL BOUNDARY INTEGRAL METHOD IN PHOTOELASTICITY

I. Umeagukwu, W. Peters and W. Ranson
College of Engineering
University of South Carolina
Columbia, South Carolina 29208

ABSTRACT

In the stress analysis of structures as an integral and necessary part of the design process, analysts often assume that critical points of interest lie on the boundary of the member being analyzed. However, the possibility exists that critical areas of interest do not occur on a boundary of the region of interest. The purpose of this paper is to present a hybrid experimental analytical solution procedure which simplifies quantification of stresses in the interior region of a finite two-dimensional member subjected to mechanical loading in its plane. The coupling of photoelasticity with the boundary element potential solution is applied to the optimization of a symmetrically notched tensile bar as an illustrative example.

INTRODUCTION

Many engineers and scientists are well acquainted with methods such as finite differences of finite elements. These methods discretize the domain of the problem under consideration into a number of elements or cells. The governing equations of the problem are then approximated over the region by function which fully or partially satisfy the boundary conditions. An alternative to these techniques is called the boundary methods which has as its foundation the theory of the potential [1]. These methods use approximating functions that satisfy the governing equations in the domain but not the boundary conditions. They are becoming more popular because they offer an elegant and economic alternative to the other two methods. They can also be combined with the other two to obtain a better representation of the boundary conditions in a finite element or finite difference program. Another important advantage of the boundary methods is that they are usually able to represent regions of stress concentration in a better manner than finite elements, although this will depend on the type of approximating function used. One of the most interesting features of these methods is that much smaller systems of equations and considerable reduction in data is required to solve a problem.

Boundary methods can be of many types as the "domain" methods [finite elements and finite differences are called the domain methods], ranging from simple techniques such as the so called indirect method to the more versatile direct techniques.

The least sophisticated boundary technique is the indirect method of analysis. This technique starts with a solution that satisfies the governing equations in the domain but which has some unknown coefficients. These coefficients are then determined by enforcing the boundary conditions at a number of points or subregions. Although this is a crude technique, it can give good results in many practical applications. This technique has been used for many years due to its simplicity.

Direct methods, on the other hand, are generally presented as based on Green's [2] identity. They are more versatile and general than indirect methods, which may be presented as a special case of direct formulations. Development of the coupling of these direct boundary methods with the photoelastic method is presented in detail in a later section.

The optimization of a tension load bar with symmetrically located holes (see Figure 1) was chosen as an illustrative example for the application of this hybrid experiment-analytical solution procedure. The general problem of the reduction of stress concentration around holes and filets is of particular interest to designers and has been studied quite extensively by Durelli [3],[4]. The optimization process embodies the removal of material (by filing or machining) from the low stress zone of the hole boundary of a photoelastic model until an isochromatic fringe is coincident with the hole boundary. The results of this procedure are illustrated in a later section. Of particular interest, however, will be the application of the hybrid analysis procedure for the determination of the stresses in the interior region of both the original geometry and the final optimized geometry.

PROBLEM FORMULATION

The main objective in this problem is to obtain stress components at any point in a two-dimensional stressed model. Solutions to

$$\nabla^2(\sigma_1 + \sigma_2) = \nabla^2(I_1) = 0 \tag{1}$$

where σ_1 and σ_2 are the principal and $I_1 = \sigma_1 + \sigma_2$ is the first stress invariant are needed. This information is then coupled with the photoelastic data

$$(\sigma_1 - \sigma_2) = \frac{nf}{t} \tag{2}$$

where n is the fringe number, f is the photoelastic fringe constant and t is the thickness of the model. Thus, if $\sigma_1 + \sigma_2$ and $\sigma_1 - \sigma_2$ are known at any desired point, the principal stress components can easily be separated. Although the experimental data does not yield I_1, points which occur on a free boundary can be found where I_1 can be calculated. On the free boundary at point A, as shown in Figure 1, $\sigma_n \equiv 0$, and for a tension boundary $\sigma_T = \sigma_1$. Also, from the photoelastic data σ_1 can be calculated. Therefore, $I_1 = \sigma_1$ at each point on the boundary can be measured. The main objective, then, is to obtain solutions to $\nabla^2 I_1 = 0$ at any point in the model.

Illustration of the potential solution known as the Boundary Element Formulation is shown in Figure 2. Boundary conditions imposed on this problem are two types:

A. Essential Conditions, such as:

$$I_1 = \sigma_n + \sigma_s \text{ on } \partial R_1 \tag{3}$$

B. Natural Conditions of the type:

$$\bar{I}'_1 = \frac{\partial}{\partial n}(\sigma_n + \sigma_s) \text{ on } \partial R_2 \tag{4}$$

The total boundary is $\partial R = \partial R_1 + \partial R_2$. For problems in photoelasticity we specify boundary condition A at all points on the boundary and calculate $\partial/\partial n\ (\sigma_n + \sigma_s)$ at all points on the boundary.

If the solution $\nabla^2 I_1$ is transformed to the boundary, an equation of the form

$$\int_R (\nabla^2 I_1)\ u^*\ dR = \int_{\partial R_2} \left(\frac{\partial I_1}{\partial n} - \frac{\partial \bar{I}_1}{\partial n} \right) u^*\ d(\partial R_2) - \int_{\partial R_1} (I_1 - \bar{I}_1)\ \frac{\partial u^*}{\partial n}\ d(\partial R_1) \tag{5}$$

is obtained. If this equation is integrated by parts, the starting point for the Boundary Integral Equation results

$$\int_R I_1(\nabla^2 u^*)\, dR = -\int_{\partial R_2} \frac{\partial \bar{I}_1}{\partial n} u^*\, d(\partial R_2) - \int_{\partial R_1} \frac{\partial I_1}{\partial n} u^*\, d(\partial R_1) - \int_{\partial R_2} I_1 \frac{\partial u^*}{\partial n}\, d(\partial R_2) + \int_{\partial R_1} \bar{I}_1 \frac{\partial u^*}{\partial n}\, d(\partial R_1) \tag{6}$$

The aim now is to find a solution satisfying the Laplace Equation. Assume that a concentrated load is acting at a point 'i', the governing equation is

$$\nabla^2 u^* + \Delta^i = 0 \tag{7}$$

where Δ^i is Dirac delta function. The solution of this equation is called the fundamental solution with the property that

$$\int_R I(\nabla^2 u^* + \Delta^i)\, dR = \int_R I\, \nabla^2 u^*\, dR + I^i \tag{8}$$

where I^i represents the value of the unknown function I at the point of the application of the charge. If Equation (5) is satisfied by the fundamental solution,

$$I^i = -\int_R I(\nabla^2 u^*)\, dR \tag{9}$$

then Equation (4) becomes

$$I^i + \int_{\partial R_2} I \frac{\partial u^*}{\partial n}\, d(\partial R_2) + \int_{\partial R_1} \bar{I}_1 \frac{\partial u^*}{\partial n}\, d(\partial R_1) = \int_{\partial R_2} \frac{\partial \bar{I}^i}{\partial n} u^*\, d(\partial R_2) + \int_{\partial R_1} \frac{\partial I_1}{\partial n} u^*\, d(\partial R_1) \tag{10}$$

For an isotropic three-dimensional medium the fundamental solution of Equation (7) is

$$u^* = \frac{1}{4\pi r} \tag{11}$$

where r is the distance from the point of application of the unit potential to the point under consideration.

The three-dimensional Laplace Equation (7) in polar coordinates and taking symmetry into consideration becomes,

$$\frac{\partial^2 u^*}{\partial r^2} + \frac{2}{r} \frac{\partial u^*}{\partial r} = \Delta^i \tag{12}$$

When Equation (11) is substituted into Equation (12) it is seen that it is satisfied for all values of r except zero. The integration,

$$\int_R \nabla^2 u^* \, dR = -\int_R \Delta^i \, dR = -1 \tag{13}$$

in a sphere surrounding the point where the load is applied must be performed to study the case of $r \equiv 0$.

It can easily be proved that the left hand side of Equation (13) is also equal to minus one by writing the following expression

$$\int_R \Delta^2 u^* \, dR = \int_{\partial R} \frac{\partial u^*}{\partial n} d(\partial R) = \int_{\partial R} \frac{\partial u^*}{\partial r} d(\partial R) \tag{14}$$

Substituting the fundamental solution (11) into (14) gives

$$\begin{aligned} \int_{\partial R} \frac{\partial u^*}{\partial n} d(\partial R) &= \frac{1}{4\pi} \int_{\partial R} (-\frac{1}{r^2}) \, d(\partial R) \\ &= -\frac{1}{4\pi} \left(\frac{4\pi r^2}{r^2} \right)_{\partial R} = -1 \end{aligned} \tag{15}$$

The above result is independent of r and as such shows that when $r \rightarrow 0$ the left hand side of Equation (13) is also equal to -1. For two dimensions the fundamental solution for the isotropic case is

$$u^* = \frac{1}{2\pi} \ln \frac{1}{r} \tag{16}$$

Equation (10) is valid for any point in the domain, but in order to formulate the problem as a boundary technique it must be taken to the boundary. The final result for a boundary point 'i' is

$$\begin{aligned} \frac{1}{2} I_1^i + \int_{\partial R_2} I_1 \frac{\partial u^*}{\partial n} d(\partial R) + \int_{\partial R_1} \bar{I}_1 \frac{\partial u^*}{\partial n} d(\partial R) \\ = \int_{\partial R_2} \frac{\partial \bar{I}_1}{\partial n} u^* \, d(\partial R) + \int_{\partial R_1} \frac{\partial I_1}{\partial n} u^* \, d(\partial R) \end{aligned} \tag{17}$$

In general, Equation (17) can be written as

$$\frac{1}{2} I^i + \int_{\partial R} \bar{I}_1 \frac{\partial u^*}{\partial n} d(\partial R) = \int_{\partial R} \frac{\partial \bar{I}_1}{\partial n} u^* d(\partial R) \qquad (18)$$

This equation applies to a particular node 'i' [2].

In general, Equation (18) must be solved numerically. For the hybrid case, I_1, is known at discrete points in the boundary. Thus $\partial I/\partial n$ represents the unknown quantities to be determined. This solution procedure is to assume constant value of I_1 and $\partial I/\partial n$ for each segment on the boundary and reduce (18) to a set of algabraic equations to be solved.

Once I_1 and $\partial I_1/\partial n$ are known on the boundary, then Equation (10) can be solved for values of I_1 at any point in the domain.

EXPERIMENTAL PROCEDURE

A standard diffuse light circular polariscope was employed in this study. The model was constructed from 1/8 in. thick photoelastic material (PSM1, Manufactured by Photoelastic, Inc., Raleigh, N.C.) to the specification shown in Figure 3. The model was then loaded with a tensile load of 310 lbs. and the resulting whole order fringe pattern is shown in Figure 4. The fringe pattern exhibited a uniform fringe order of 4 at a location of 2.5 in. from the filet location (origin of x-y coordinates). In order to generate input data for the potential solution, sixty coordinate points were established around the boundary of the specimen as shown in Figure 3. As mentioned previously, one of the two principal stresses is zero on the free boundary (see Figure 1), therefore, $I_1 = \sigma_1 = nf/t$. In order to obtain sufficient data around the boundary, the method of Tardy Compensation [5] was used to obtain nine fractional fringe orders between each whole order fringe. This data was used as input for the potential solution as shown in Equation (3). The potential solution was employed to calculate $(\sigma_1 + \sigma_2)$ at selected interior points along the dashed lines of symmetry shown in Figure 3. Separation of the principal stresses along these lines was accomplished by combining the results of the potential solution with photoelastic data $(\sigma_1 - \sigma_2)$ at the same selected points.

After sufficient data was obtained to calculate individual stresses at the desired interior points for the geometry specified in Figure 3, the optimization process discussed previously was carried out. This process involved viewing the model in a diffuse light polariscope while simultaneously removing material with a file in the low stress region around the filets. Dimensions of the optimized filet along the horizontal line of symmetry and the edges of the specimen were kept equal to the dimensions of the original geometry since a change in these dimensions would invalidate a comparison of the optimized shape with the original semicircular notches. The filing was carried out until a photoelastic fringe was coincident with the boundary of the optimized shape as shown in Figure 5. The boundary of this optimized geometry was then described in a fashion similar to the original specimen and photoelastic data was used as input to the potential solution. The results of this solution, $(\sigma_1 + \sigma_2)$, when combined with the photoelastic data $(\sigma_1 - \sigma_2)$ generated principal stress information along the same lines of symmetry as shown in Figure 3 for the original geometry.

RESULTS

It can be observed from the photoelastic fringe patterns in Figures 4 and 5 that the fringe order at the filet is n = 12 for the original geometry and n = 10 for the optimized geometry. The stress concentration factor can be easily calculated by dividing the filet fringe order by the fringe order n = 4 at the point where the cross sectional stress is uniform. The stress concentration for the optimized shape (10/4 = 2.5) has been substantially reduced when compared with the original unoptimized shape (12/4 = 3). The reduction accomplished by this optimization compares favorably when compared with other values presented in the literature [3].

The results of the application of the hybrid experimental-analytical method for the estimation of stresses interior to the boundary are shown in Figures 6 and 7 for the unoptimized geometry. In Figure 6 the stresses σ_1 and σ_2 have been separated along the horizontal line of symmetry (dashed horizontal line in Figure 3). In Figure 7, the stresses σ_1 and σ_2 are shown along the vertical line of symmetry (vertical dashed line in Figure 3). Confidence in this solution can be obtained by considering the area under the σ_1 vs horizontal distance curve. The result of this integration times the model thickness should be equal to one-half the applied load (the integration is performed only on one-half of the specimen). The actual value was one hundred and fifty pounds which compared well with the half-load value of one hundred and fifty-five pounds. Along the vertical axis, equilibrium requires that the area under the σ_2 vs vertical distance curve must be zero. This integration yielded a near zero result.

The interior principal stresses are shown in Figures 8 and 9 for the optimized filet. Equilibrium checks for the σ_1 vs horizontal distance data and for the σ_2 vs vertical distance data demonstrated results similar to those for the unoptimized filet.

The acquisition of interior principal stress values yields a better understanding of the filet optimization problem. Logical reasoning dictates that a stress redistribution must occur across the horizontal cross section if the stress concentration factor is reduced. The hybrid experimental-analytical method allows us to verify this redistribution of maximum principal stress at this location. A comparison of Figure 8 with Figure 6 shows that the severe stress gradient (in σ_1) present in the unoptimized filet substantially has been reduced. The optimized filet region causes the maximum principal stress to be more uniformily distributed throughout the cross section.

SUMMARY

The hybrid experimental-analytical method has the capability of uniquely quantifying principal stresses in the interior regions of two dimensional solid bodies. This procedure should provide useful information in optimization problems where a knowledge of interior stress distributions is required. The additional capability of ascertaining the location of critical points of interest should provide insight into problems with complicated geometry and/or load where these critical points do not lie on the boundary of the object being analyzed. The procedures developed herein should provide a very useful tool in the analysis and optimization of problems not amenable to analytical solutions.

REFERENCES

1. Kellogg, O.D., Foundations of Potential Theory, Dover Publications Inc., 1953.

2. Brebbia, C.A., The Boundary Element Method for Engineers, Pentech Press, 1978, pp.2, 46-51 and 146.

3. Durelli, A.J., Brown, K. and Yee, P., "Optimization of Geometric Discontinuities in Stress Fields," Experimental Mechanics, Vol. 19, No. 8, August 1978, pp. 303-308.

4. Durelli, A.J. and Rajiah, K., "Lighter and Stronger," Experimental Mechanics, Vol. 20, No. 11, November 1980, pp. 371-375.

5. Tardy, M.H.L., "Methode Pratique d'Exame de Mesure de la Birefringenge des Verres d'Optique," Rev. D'Optique, Vol. 18, pp. 59-69, 1929.

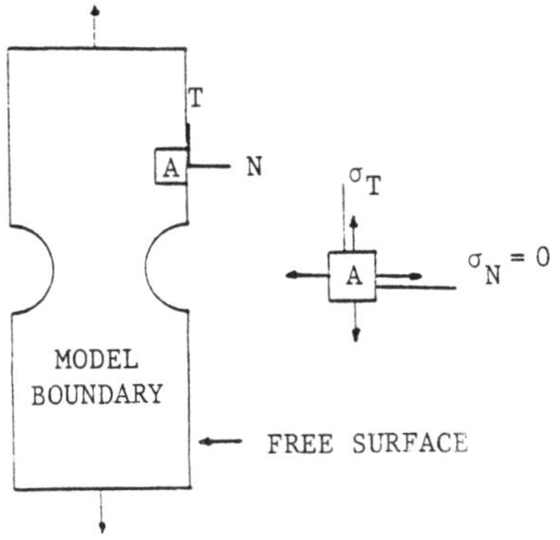

FIGURE 1. PROBLEM GEOMETRY

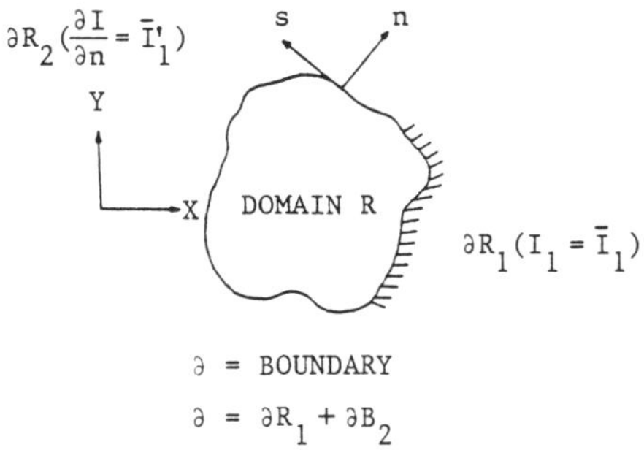

FIGURE 2. BOUNDARY ELEMENT FORMULATION

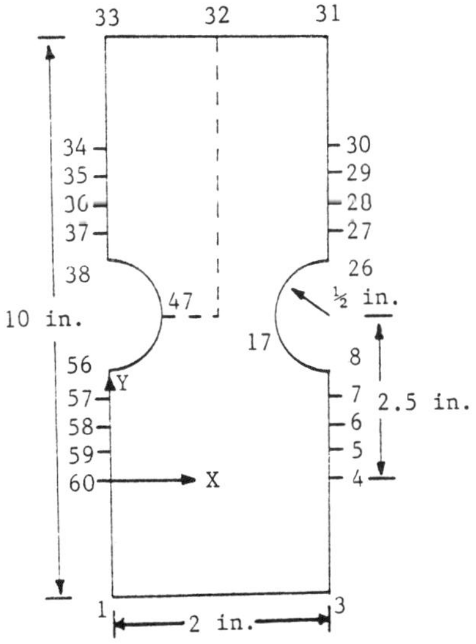

FIGURE 3. DEFINITION OF BOUNDARY POINTS

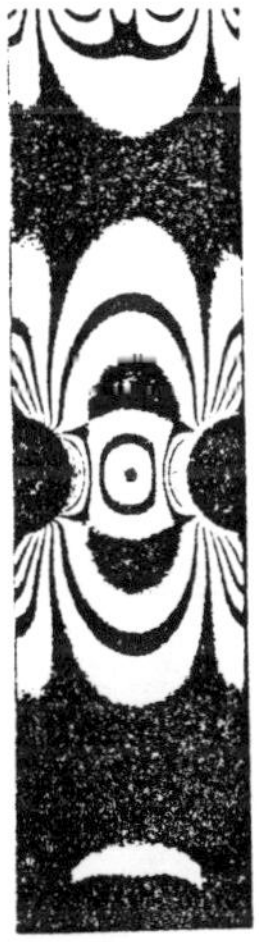

FIGURE 4. UNOPTIMIZED GEOMETRY

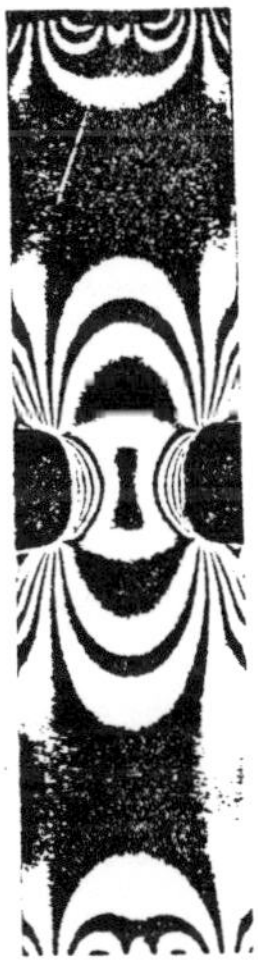

FIGURE 5. OPTIMIZED GEOMETRY

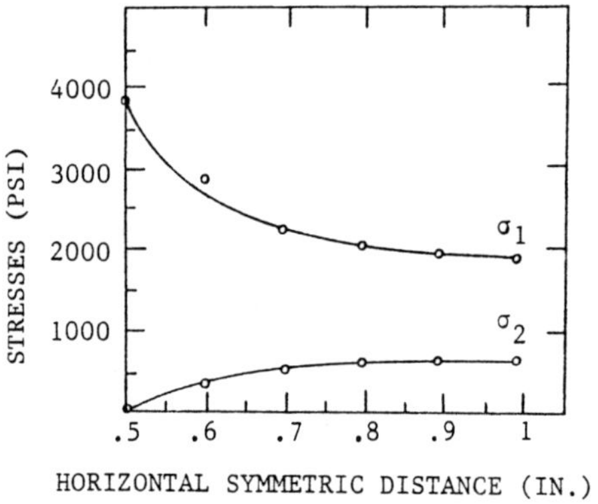

FIGURE 6. PRINCIPAL STRESSES ALONG HORIZONTAL SYMMETRIC DISTANCE (UNOPTIMIZED BOUNDARY)

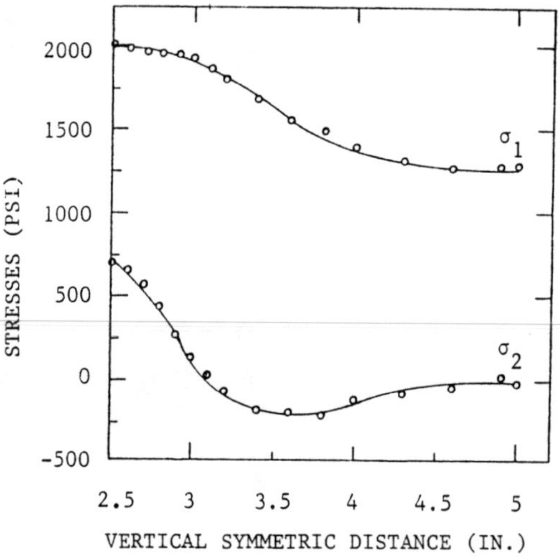

FIGURE 7. PRINCIPAL STRESSES ALONG VERTICAL SYMMETRIC DISTANCE (UNOPTIMIZED BOUNDARY)

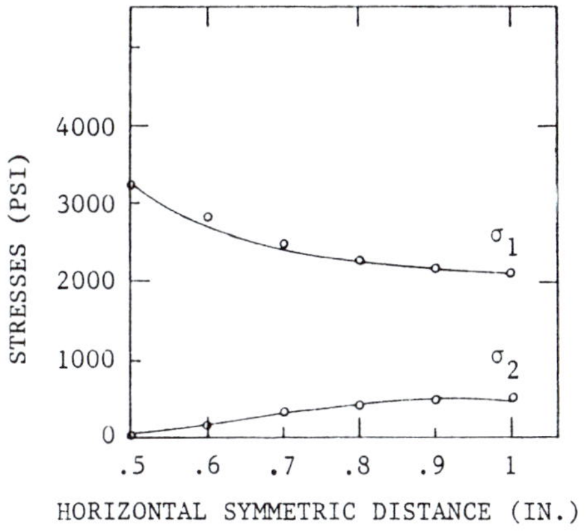

FIGURE 8. PRINCIPAL STRESSES ALONG HORIZONTAL SYMMETRIC DISTANCE (OPTIMIZED BOUNDARY)

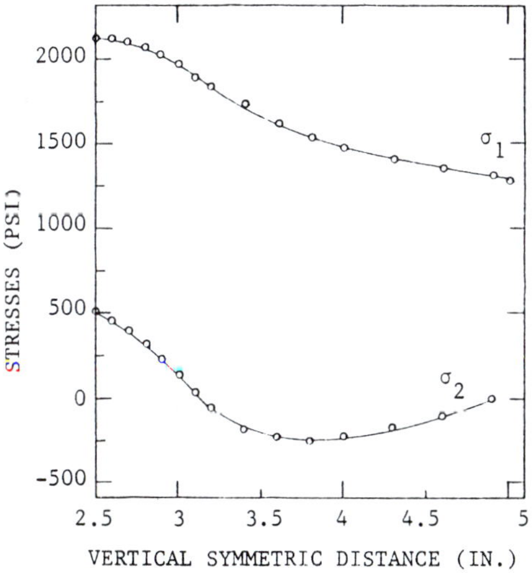

FIGURE 9. PRINCIPAL STRESSES ALONG VERTICAL SYMMETRIC DISTANCE (OPTIMIZED BOUNDARY)

RESISTANCE OF SHOT-PEENED MILD STEEL THIN TARGET PLATES TO PENETRATION BY PROJECTILES AT ORDNANCE VELOCITIES

A.M. Eleiche and M.S. Abdel-Kader
Dept. of Mechanical Design & Production
Faculty of Engineering, Cairo University
Guiza, Egypt

ABSTRACT

Ballistic tests with 7.62 mm rifle lead ammunition as projectiles were performed on as-received and shot-peened mild steel plates of different thicknesses up to 4 mm, in order to separately investigate the effects of projectile impact velocity, target plate thickness and target configuration on resistance to penetration. A special setup was used in the tests allowing the measurement of impact and post-perforation velocities. It was found that the process of shot peening generally improves target plate resistance to penetration, the extent of improvement being mainly dependent on other test conditions.

INTRODUCTION

To properly describe the interaction between striker and target during penetration, many parameters are found to be involved. These are classified, as shown in Fig. 1, into: impact conditions, projectile characteristics and target characteristics. The role of various penetration parameters and a description of the resulting failure patterns and mechanisms are reviewed in great detail elsewhere (1, 2).

The present work was conducted to investigate the capability of standard 7.62 mm rifle lead ammunition to penetrate at normal incidence mechanically-treated brass and mild steel target plates of various thicknesses, and in the single or laminated configuration. The type of treatment chosen was shot peening, rather than the classical methods of surface treatment such as carburizing, in order to determine its feasibility as a substitute method for increasing penetration resistance.

The present paper is concerned with the behavior of mild steel target plates, whereas the detailed results on brass targets are being presented elsewhere (3).

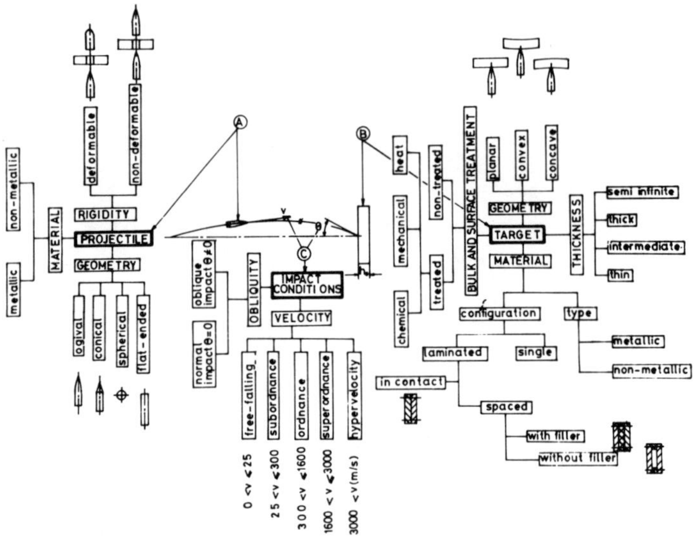

Fig. 1

EXPERIMENTAL PROCEDURE

Mild steel plates were secured in the form of sheets of 1, 2, 3 and 4 mm thickness. For ease of shot peening, strips of one meter by 20 cm of different thicknesses were prepared. Some of these strips were peened on one side only, while others were treated on both sides, one at a time. During the process of shot peening, diagrammatically illustrated in Fig. 2, the workpiece is exposed to a blast of shot consisting of carbon steel abrasives of bianitic structure and a hardness of 410 to 420 HV, having a mean diameter of 0.8 to 1.25 mm. Peening conditions were so adjusted such that the maximum possible effect could be attained; this effect being evaluated in terms of the change in hardness, toughness and ductility of the plates.

The hardness gradient induced throughout the peened plates was determined by microhardness measurements taken at 5 to 9 equidistant points along the thickness. These measurements were repeated at 2 to 3 different places and average values calculated.

Besides hardness, the strength, ductility and toughness of the target material affect its penetration resistance (4). For this purpose, quasi-static testing was performed on standard specimens cut in two perpendicular directions in the plane of each target plate. Load-displacement curves were autographically recorded for each test, thus allowing the engineering stress-strain

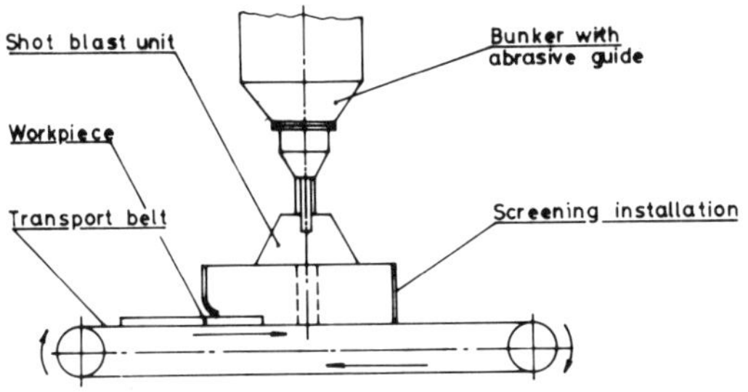

Fig. 2

curves to be deduced, and consequently values of strength, ductility and toughness to be calculated. These values should reflect, at least to a first approximation, the general characteristics of the plate material. As discussed in detail elsewhere (5), more exact evaluation of properties should be determined at dynamic shear and compressive strain rates in a direction parallel to the projectile motion.

In the ballistic tests performed, measurements were mainly concerned with the determination of the projectile impact and post-perforation velocities. A schematic of the setup used is shown in Fig. 3, which is self explanatory. Details of various particularities can be found in (5).

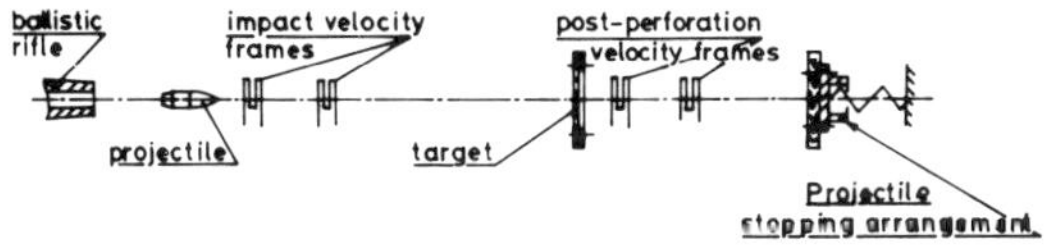

Fig. 3

RESULTS AND DISCUSSION

The target marking code consisted of letters and numbers separated by dashes, to specifically describe the material type, plate thickness, surface treatment and spacing, if any, as follows: (a) material of plate: S; (b) plate thickness: by a number representing the thickness in tenths of mm. For a laminated target, two or three numbers are used giving the thickness of the first, second and third plate respectively, starting from the firing side; (c) surface treatment: 0P for as-received plates, 1P for plates peened on the firing-side surface only, and 2P for plates peened on both sides; (d) spacing: 0S for zero spacing and xS for a spacing of x mm. Thus: S-10+20-2P+0P-3S refers to a composite steel target consisting of two plates: the firing-side lamina is 1 mm thick and peened on both sides, and the second lamina is 2 mm thick and unpeened; the two laminae being spaced by 3 mm.

Results of the microhardness measurements on the 1, 3 and 4 mm-thick plates are plotted in Figs. 4-6 versus through-thickness distance from the peened side if any. Results for the 2 mm-thick plates are not shown, since they were

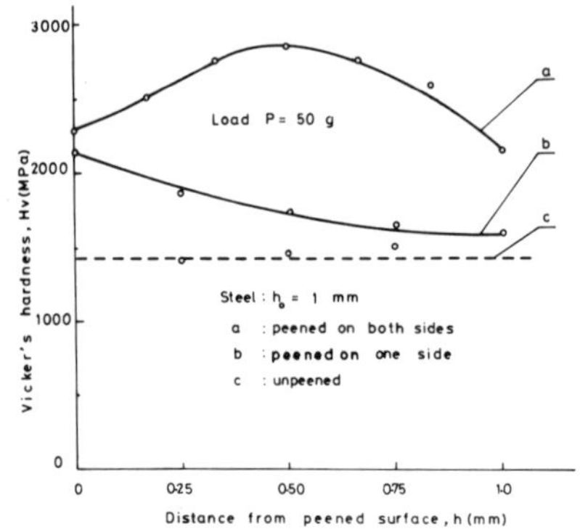

Fig. 4

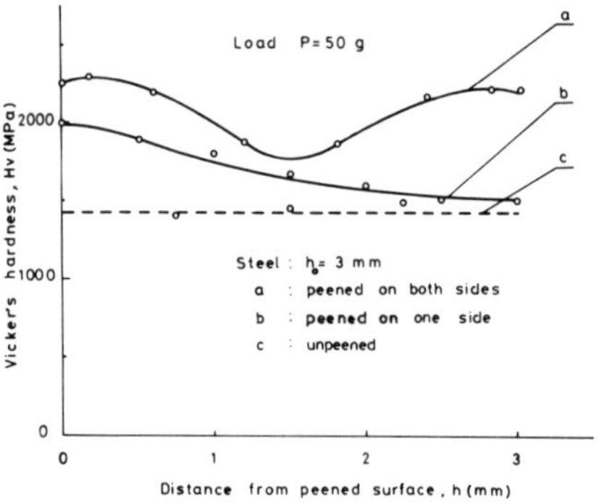

Fig. 5

approximately of the same pattern as those for the 3 mm-thick plates. On the peened side, where direct impact with abrasives took place, the hardness exhibits a maximum value, then decreases to a minimum at the other side for plates peened on one side only. For plates peened on both sides, the minimum hardness value is actually reached midway through the thickness. However, the minimum hardness values of the treated plates, whether peened on one or both sides, are greater than the initial hardness of the unpeened plates; the difference being greater the thinner the plate. This result should be expected, since the effect of peening on one side of any of the used plates, within the range of thickness chosen, extended to the other side, the extent being dependent on the plate thickness.

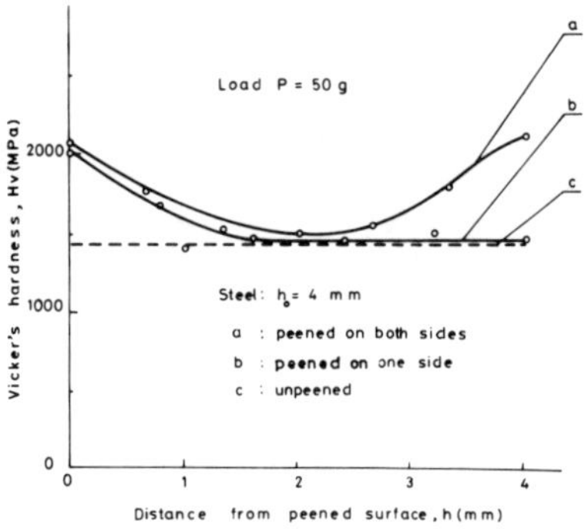

Fig. 6

For the particular case of 1 mm-thick plates peened on both sides, the situation is somewhat different, since the maximum hardness is obtained midway through the thickness. This is probably because the maximum compressive stress induced by peening of each side of the plate is not achieved exactly at the surface, but at a point underneath (6). As the plate is relatively thin in this case, this approximately occurs in the middle of the plate thickness.

Representative load-displacement traces obtained during the testing of 3 mm-thick plates are shown in Fig. 7. Engineering stress-strain curves were deduced from such traces and values of the yield and ultimate stresses as well as strain to fracture determined. A general increase in strength as induced by shot peening is found to take place; the increase being greater for plates

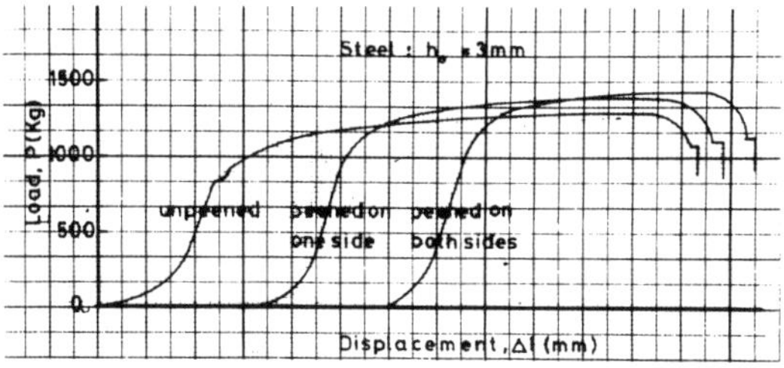

Fig. 7

peened on both sides. However, a corresponding reduction in ductility as expressed by strain to fracture also occurs.

For each firing test performed, impact and post-perforation velocities were measured and important quantities calculated. In all cases, the average of at least two firing test results, collected under similar testing conditions, whose values differed by no more than 3 %, was taken as the basis for analysis.

In many previous investigations (7, 8), either the projectile velocity drop or the loss in projectile energy was used to represent the target plate resistance to penetration. In the present study, the "specific energy loss", e, defined as the loss in projectile energy divided by the total thickness of target was adopted as a more relevant parameter. A wide variety of behavior was displayed depending on impact velocity, plate thickness, surface treatment and configuration. This is now discussed in greater detail.

Effect of Projectile Impact Velocity

This was only studied for the case of non-treated plates. For this purpose, rounds were specially filled with propellant charges ranging in weight from 1.0 to 1.6 g and differing by 0.1 g each, and projectiles so prepared were fired against as-received plates.

The relationship between impact velocity and specific energy loss for different plate thicknesses are presented in Fig. 8. With the increase in impact velocity, the specific energy loss first decreases to a minimum, the value of which depends on the plate thickness and then increases. This behavior was previously observed and discussed by many investigators such as Recht and Ipson (8) and Osborn and Maj (9). At velocities near the ballistic limit, thermal effects do lower the material strength and consequently the plate resistance to penetration, whereas at higher velocities the effect of strain rate on raising the material fracture resistance is predominating. It is also interesting to note that with the increase in plate thickness, the amount of energy absorbed, both overall and specific, increases.

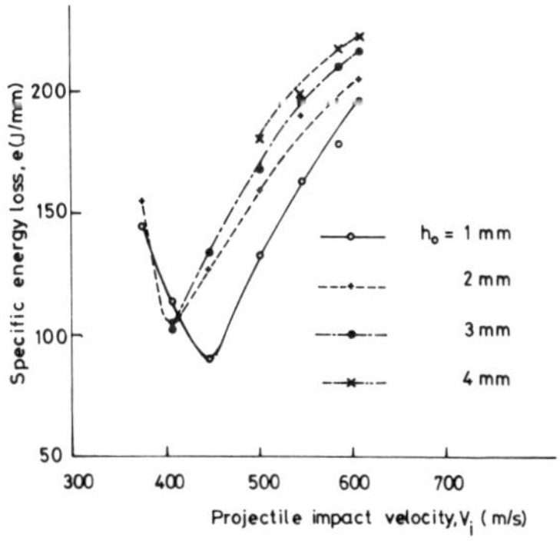

Fig. 8

Effect of Target Thickness

First and most importantly, a suitable projectile impact velocity had to be chosen anf fixed. For this purpose, use was made of the results of Fig. 8. A velocity of 544 m/s was found effective in providing the projectile with a sufficient amount of energy to successfully perforate the different varieties of target plates at each and every impact. On the other hand, this velocity was greater than the ballistic limits of all plates tested, and in turn fell far from the zone of mixed results, thus providing repeatability and accuracy.

For the single-plate configuration, three target varieties were tested, namely unpeened, peened on one side only and peened on both sides. Results are shown in Fig. 9 where it is noted that the specific energy loss increases with the increase in plate thickness. This is true for both treated and as-received plates.

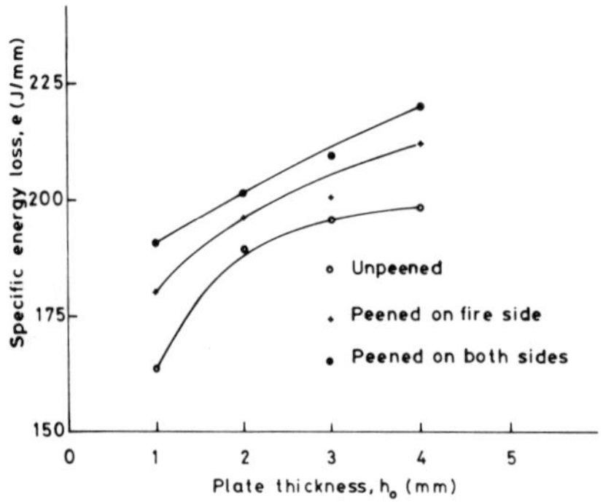

Fig. 9

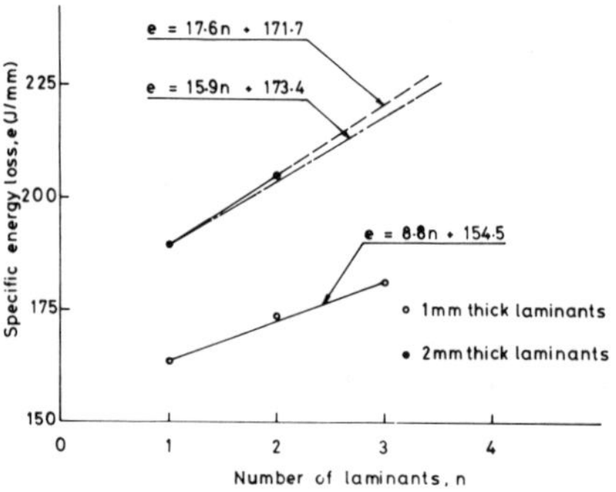

Fig. 10

Composite targets built-up using both 1mm and 2mm-thick laminae were also tested. Values of the specific energy loss were calculated for different in-contact laminated targets of total thickness ranging from 1 mm for a single plate to 3 mm for a three-lamina target and of total thickness ranging from 2 mm for a single plate to 4 mm for a two-lamina target. For 1mm-thick-lamina composite targets, the relationship between e and the number of laminae, n, building up the target, as deduced from Fig. 10, was expressed by:

$$e = 8.8\, n + 154.5 \tag{1}$$

If a simple relationship of this sort is assumed valid for other lamina thicknesses as well, a set of straight lines should be obtained in Fig. 10. For instance, with targets built-up using 2mm-thick laminae, expression (1) will take the form

$$e = 17.6\, n + 171.7 \; ; \tag{2}$$

use being made of the test result for the 2 mm-thick single plate. However, the corresponding expression based on the two test results known for the 2mm-thick-lamina composite targets, namely 2 mm-thick single plate and 4 mm-thick two-lamina target, is found to be:

$$e = 15.9\,n + 173.4 \quad . \qquad (3)$$

The difference between expressions (2) and (3) is well within the accuracy of testing. This suggests the validity of the simple linear relationship between e and n. However, generalizations to other configurations should still need further experimental confirmations.

It is obvious from the previous results that whether the target plate is single or laminated, e increases with the increase of total plate thickness. For composite targets, a further increase is actually obtained with the increase in lamina thickness. It is usually agreed that the major parameter affecting the amount of energy absorbed during projectile penetration seems to be the ratio of the target thickness to the projectile diameter (10). This explains in part the present observed behavior. Furthermore, this may also be related to the fact that the energy required to cause target failure can be considered as consisting of two parts: that initiating failure, whatever the mechanism may be, and that propagating it throughout the thickness. For relatively thin plates, the second portion may be negligible, whereas for thicker plates its magnitude is appreciable and improves target resistance considerably.

Effect of Surface Treatment

Results of firing tests have shown that plates peened on both the firing and back sides were more resistant than those peened on the firing side alone. This is due to the change in mechanical properties induced, which is not localised at the surface but extends to varying depths depending on plate thickness and peening conditions.

As detailed elsewhere (5), the hardness gradient throughout the thickness of single targets can be accounted for by a mean value, whereas the overall hardness and toughness of laminated targets can be expressed by the mean of the values of these two quantities in individual laminae. On the other hand, since both hardness and toughness seem to play competitive roles in affecting penetration resistance (4), a non-dimensional parameter, ε_a, which can be thought of as "apparent ductility" of the material, is adopted as representative for the overall effect of peening. ε_a is defined by:

$$\varepsilon_a = U / (HV)_m \quad , \qquad (4)$$

where U and $(HV)_m$ are the overall toughness and hardness of the target. Values of mean hardness, toughness and apparent ductility were calculated for various single and laminated plates. Results have shown that the increase in target mean hardness is actually accompanied by corresponding decrease in toughness as seen in Fig. 11 for single plates; the amount of change being substantially dependent on plate thickness.

The specific energy loss for different plates tested is plotted as function of mean hardness, toughness and apparent ductility in Figs. 12-14 for single plates and in Figs. 15-17 for laminated ones. For single plates, e increases

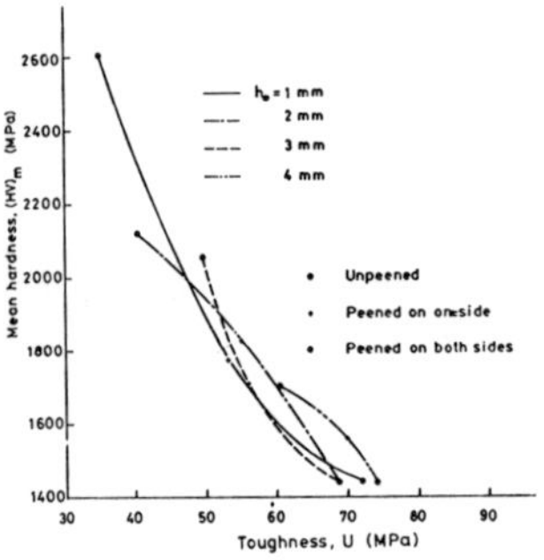

Fig. 11

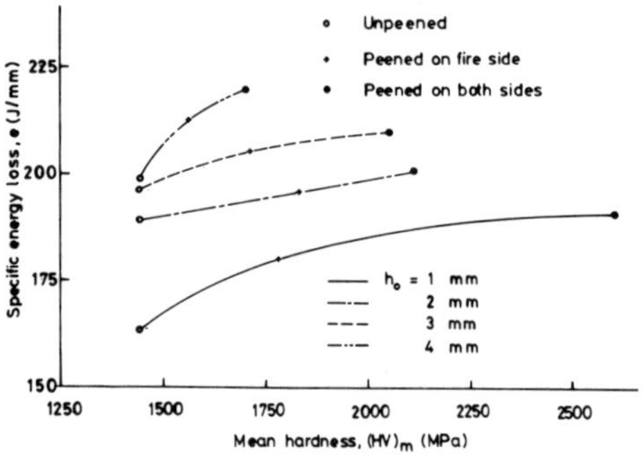

Fig. 12

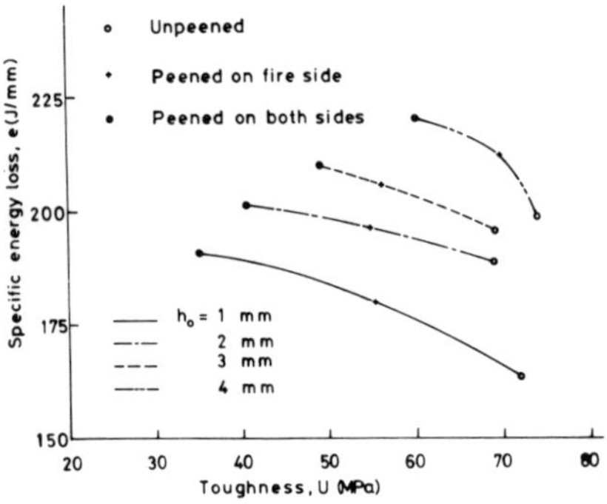

Fig. 13

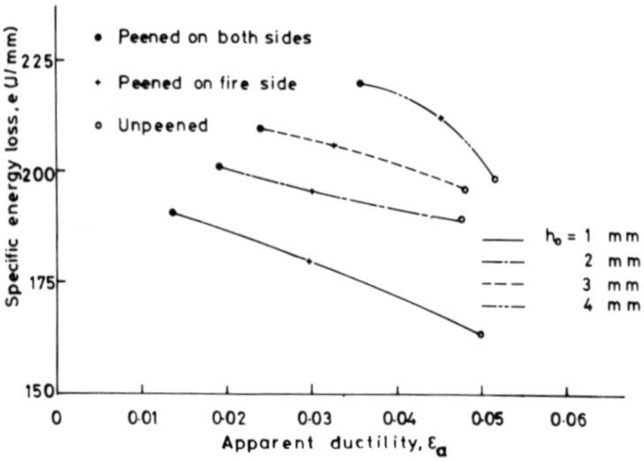

Fig. 14

with the increase in mean hardness of target plate and corresponding decrease in toughness and apparent ductility. Results for laminated plates reveal a different and more interesting type of behavior. Thus, e first increases with the decrease in apparent ductility, ε_a, until a maximum is reached, then decreases with further decrease of ε_a. It can be concluded, therefore, that for the present test conditions maximum resistance to penetration is actually obtained at a certain critical value of apparent ductility; further decrease in ε_a would diminish target resistance. Consequently, surface treatment of thin plates should be thought of as a double-sided process in terms of its effect on overall target resistance, and optimum conditions for surface treatment should always be determined for each case of lamination. In this respect, the plate thickness is expected to play a substantial role and the present results cannot be generalized for all thicknesses.

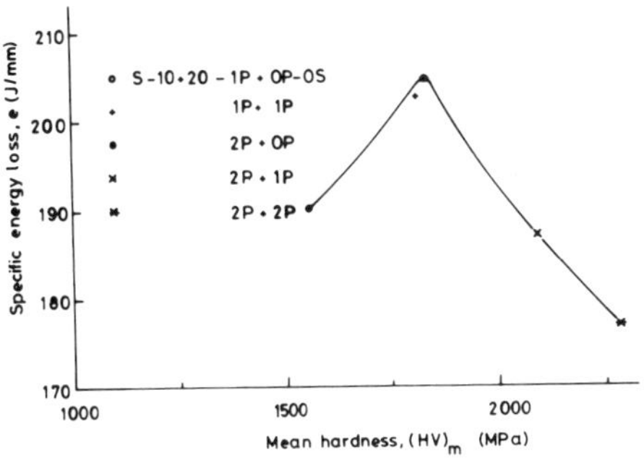

Fig. 15

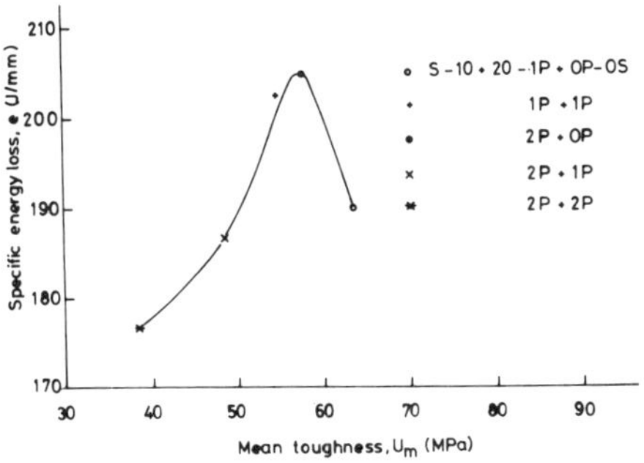

Fig. 16

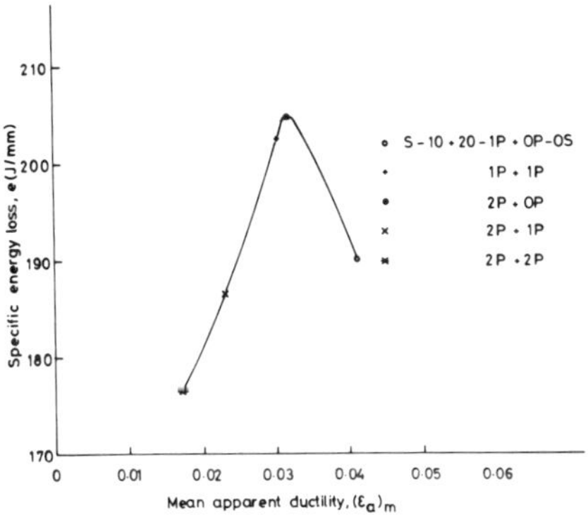

Fig. 17

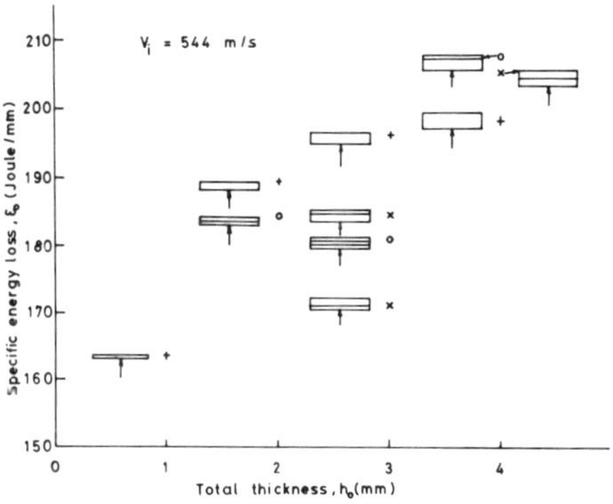

Fig. 18

Effect of Target Configuration

In this study, tests were largely confined to laminated in-contact target plates of 3 mm total thickness, though two laminated target varieties of 4 mm total thickness were also tested. Results are shown in Fig. 18 together with those for single plates of equivalent thicknesses for comparison. It seems that, whatever the configuration is, a laminated plate of up to 3 mm total thickness is less resistant than a single plate of equivalent thickness. For greater plate thickness, however, laminated targets are apparently more effective than single ones since laminated target plates of 4 mm total thickness exhibited greater resistance than that shown by single plates of the same thickness. This typical behavior of relatively thick plates was also revealed by other investigators such as Zaid and Travis (10) who found that, on impact with a flat-ended projectile of 9.375 mm in diameter, a laminated steel target plate is more resistant than a single plate of equivalent thickness when the total thickness is greater than 6.25 mm, i.e. for plate thickness-to-projectile diameter ratio greater than 0.67. For the present test conditions, this ratio is 0.525. Defining thin plates as those having a thickness-to-projectile diameter ratio less than 0.5 (8), it may than be concluded that for thin plates, single targets are more effective than laminated ones.

It is clear from Fig. 18 that for the laminated target of 3 mm total thickness the maximum effectiveness was exhibited in case when the backup plate is of greater thickness than the firing-side one. Based on these results, 4 mm-thick laminated targets were tested; different lamination varieties providing equally thick laminae or thicker backup plate being only considered. As seen in Fig. 18, results were in general agreement with this observation.

CONCLUSIONS

(a) Shot peening, in general, improves the resistance to penetration of steel thin target plates. The "apparent ductility", a property that combines both hardness and toughness effects on plate resistance to penetration - thus useful for comparison of behavior - seems to have a certain critical value at which the target exhibits maximum resistance. Plates peened on both sides are also found to be more effective than those peened on the firing side alone.

(b) Resistance of untreated targets decreases with the increase in projectile impact velocity up to a critical value at the neighborhood of the limit velocity. Above this critical value, the penetration resistance increases with further increase in impact velocity.

(c) Increasing plate thickness increases penetration resistance, both overall and specific.

(d) With laminated targets, a linear relationship seems to exist between the specific energy loss and the number of laminae. Also, penetration resistance is improved by increasing the thickness of the backup plate.

REFERENCES

1. Backman, M. and Goldsmith, W., Int. J. Eng. Sci. 16, 1-99 (1978)
2. Abdel-Kader, M.S. and Eleiche, A.M., Sci. Eng. Bull. Fac. Eng. Cairo Univ., In print (1982)
3. Eleiche, A.M. and Abdel-Kader, M.S., SESA/JSME Joint Conf. Exp. Mech. (1982)
4. Rinehart, J.S. and Pearson, J., Behaviour of Metals Under Impulsive Loads, Dover Publications, Inc., New York (1965)
5. Eleiche, A.M. and Abdel-Kader, M.S., Int. J. Sol. Struct., Submitted for publication
6. Hertzberg, R.W., Deformation and Fracture Mechanisms of Engineering Materials, John Wiley, New York (1976)
7. Awerbuch, J. and Bodner, S.R., Int. J. Sol. Struct. 10, 685-699 (1974)
8. Recht, R.F. and Ipson, T.W., Experimental Mechanics 15, 249-257 (1975)
9. Maj, S. and Osborn, C.J., Proc. Int. Conf. on Fracture, Vol. 3, pp. 617-620, Pergamon Press, New York (1977)
10. Zaid, A.I.O. and Travis, F.W., Proc. Conf. on Mech. Prop. of Mat. at High Rates of Strain, pp. 417-428, Institute of Physics, London (1974)

FURTHER STUDIES ON DYNAMIC CRACK CURVING

Y.-J. Sun*, M. Ramulu, A. S. Kobayashi and B. S.-J. Kang

University of Washington
Department of Mechanical Engineering
Seattle, Washington 98195

The elasto-dynamic stress field surrounding rapidly propagating cracks in thin polycarbonate, double edged crack tension specimens were analyzed by dynamic photoelasticity using a 16-spark gap Cranz-Schardin camera system. Crack curving was observed in two slanted double edged crack specimens and in two offset parallel double edged crack specimens. In another test, the crack ran straight between two symmetrically located twin cracks. Results of these five tests were used to verify the dynamic crack curving criterion by Ramulu et al. in which a reference distance of

$$r_o = \frac{1}{128\pi}\left[\frac{K_I}{\sigma_{ox}} V(c, c_1, c_2)\right]^2$$

from the crack tip is incorporated into the maximum circumferential stress or minimum strain energy density criteria. The critical material property for crack curving in this thin polycarbonate sheet was found to be about r_c= 0.5 mm.

*On leave from Northern Jioa-Tong University, Beijing, Peoples Republic of China.

INTRODUCTION

In a paper in 1963, Erdogan and Sih [1] used the orientation of maximum circumferential stress to predict crack extension of inclined cracks in tension specimens. This mixed mode crack extension criterion, commonly referred to as the maximum circumferential stress criterion, was later advanced among others by Williams and Ewing [2] and Finnie and Saith [3]. More recently, Streit and Finnie [4] incorporated the second order term, σ_{ox}, in the crack tip stress field and proposed a crack curving criterion based on the directional stability of a mode I crack propagation. This stability criterion introduced another critical material parameter, r_o, which is the radial distance from the crack tip. The second order term of σ_{ox} was also used by Cotterell and Rice [5] for predicting the crack curving direction of a slightly curved crack. Historically,Yoffe was the first to use the maximum circumferential stress theory to explain surface roughening and crack branching of a rapidly propagating crack in 1951 [6].

As a natural extension of Griffith's energy release rate, Hussain et al. [7], Palaniswamy and Knauss [8] among others predicted the direction of crack kinking based on the maximum energy release criterion. Focusing directly on energy, Sih [9] predicted that the crack would kink in the direction of the minimum strain energy density factor, S. In an early critique of 1976, Swedlow [10] concluded that the difference between the crack kinking angle predicted by the maximum circumferential stress criterion and the minimum strain energy density criterion "are modest at most". Theocaris and Andrianopoulos [11] recently modified the S theory and designated the mean value of S, a critical material value for crack extension. Sih [12] also applied the minimum S theory to predict crack kinking of a dynamic crack.

Ramulu et al., in a recent paper [13], incorporated the second order term of σ_{ox} in the dynamic crack tip stress field and then derived the dynamic counterpart of the crack stability model based on the maximum circumferential stress criteron as well as the minimum strain energy density factor. This dynamic crack curving criterion was used to evaluate nine dynamic photoelasticity tests involving curved as well as straight propagating cracks in fracturing Homalite-100 specimens. The critical material property of r_c was found to be 1.3 mm for the Homalite-100 specimens investigated. More importantly, the crack curving directions predicted by either the maximum circumferential stress theory or the minimum strain energy density theory were generally within 5 degrees of each other for the relatively low crack velocities observed in these tests. The sign of σ_{ox} was also found to influence significantly the crack angle of a running crack. The purpose of this paper is to provide further evidence in support of the dynamic crack curving criterion advanced by Ramulu et al.

DYNAMIC CRACK CURVING CRITERIA

Maximum Circumferential Stress Criterion

The maximum circumferential stress criterion, as modified by Ramulu et al. [12], assumes that the crack will extended towards the maximum circumferential stress which reached its critical value at a critical distance, r_c, away from the rapidlly propagating crack tip. r_c is a characteristic distance derivable from the crack stability criterion involving the second order term of σ_{ox} in the dynamic crack tip stress field. This characteristic distance is [13]

$$r_0 = \frac{1}{128\pi}\left[\frac{K_I}{\sigma_{ox}} V(c,c_1,c_2)\right]^2$$

where K_I is the dynamic stress intensity factor.

σ_{ox} is the second order term in the dynamic stress field and is often referred to as the remote stress component.

c, c_1 and c_2 are the crack velocity, dilatational wave velocity and the shear wave velocity, respectively.

$V(c,c_1,c_2)$ is a dynamic correction factor to the static crack instability criterion and is given in Reference [13].

This crack curving criterion which is the dynamic extension of that by Streit and Finnie [3] can be used to predict the crack curving of a crack propagating under pure mode I as well as mixed mode, i.e. modes I and II, conditions.

Minimum Strain Energy Density Criterion

The minimum strain energy density criterion, as modified by Ramulu et al. [13], also incorporates the characteristic distance of r_0 and thus the second order term of σ_{ox} in the strain energy density factor, S, of Sih [9]. Unlike the maximum circumferential stress criterion, the minimum S condition yields a relation between the crack curving angle and r_0 in terms of the given σ_{ox} and modes I and II stress intensity factors, K_I and K_{II}. Given r_0, however, the extended minimum strain energy density criterion with the σ_{ox} term can be used to predict crack curving of a mode I static crack or a crack propagating at a low crack velocity.

Homalite-100 Fracture Specimens

The validity of the above two dynamic crack curving criteria were assessed through re-evaluated dynamic photoelasticity results of Homalite-100 fracture specimens [14-18]. r_0 for the minimum S criterion was equated to rc which was found to be about 1.3 mm for the Homalite-100 data used in evaluating the maximum circumferential stress criterion. For the relatively low crack curving angles of -20 to 25 degrees observed in the nine tests, both the maximum circumferential stress and the minimum strain energy density criteria predicted the fracture angles within 1 degree for most of the 81 data points considered in this investigation [13].

POLYCARBONATE FRACTURE SPECIMENS

In order to further verify the above dynamic crack curving criteria, a series of dynamic photoelastic fracture experiments involving thin polycarbonate fracture experiments were conducted. Specimen configurations and the crack paths in five double edged crack tension specimens with either offset parallel cracks, offset slanted cracks and symmetically located twin cracks, used in this investigation are shown in Figure 1. The annealed thin polycarbonate specimens with blunt starter cracks exhibited brittle fracture with shear lips less than 10 percent of the thickness and an apparent crack tip yield zone of less than 1.5 mm. The dynamic isochromatics surrounding the propagating crack were recorded with a 16 spark gap Cranz-Schardin camera system.

The isochromatic data were reduced by least square fitting to the recorded dynamic isochromatics a theoretical mixed-mode, dynamic crack tip stress field with disposable parameters of K_I, K_{II} and $_{ox}$ [19]. The characteristic crack tip distance, r_o, was then computed by Equation (1). With r_o known, the predicted crack curving angle can be computed by the maximizing condition for the maxmimum circumferential stress criterion or the minimizing condition for the minimum S criterion. Details of this data reduction procedure can be found in Reference [20].

RESULTS

As shown in Figure 1, crack always propagated from the longer left crack and curved towards the shorter stationary right crack except for Specimen S2-810518 which involved a symmetrically located twin crack. Also the eccentric loading of Specimens S15-810727 and the longer initial crack length of Specimen S5-810530 caused the rapidly propagating upper crack to intersect with the stationary lower crack at its midcrack length.

Figure 2 shows two typical dynamic isochromatic patterns in a fracturing offset, slanted, double edged crack tension specimen. The right edge crack did not propagate during the entire fracture event. Both frames in Figure 2 show the expanding shear wave front which emanated from the original crack tip of the left edged crack when it started to propagate. Figure 3 shows the variations in K_I, K_{II}, σ_{ox} and crack velocity of the upper crack in Figure 2. While the crack velocity and K_I remained essentially constant through the relatively straight propagation of the upper crack, σ_{ox} oscillated consistent with previous observations [13].

Figure 4 shows two typical dynamic isochromatic patterns in a fracturing offset, parallel double edged crack specimen. Although the upper crack had cut the specimen in half, the stationary lower crack continued to show a high mode II crack tip stress field in Frame No. 10 of this figure. A similar high mode II crack tip stress field was observed in all stationary cracks during the latter part of crack propagation history. Figure 5 shows the variations in K_I, K_{II}, σ_{ox} and crack velocity of the upper crack in Figure 4. The larger excursion in K_I in this figure is associated with the curved crack shown in Figure 4. Figure 6 shows the K_I, K_{II} and σ_{ox} of the lower stationary crack in Figure 4. While K_I=3 Mpa$\sqrt{m}$, of this crack was close to the estimated fracture toughness of polycarbonate, the small difference in the crack tip bluntness probably prevented crack propagation of the lower right crack. K_{II} of the stationary crack varied from -8.5 to 0.6 MPa$\sqrt{m}$. and is about half of the K_I value. This high value, not commonly observed in previous dynamic photoelastic experiments, is due to the load redistribution caused by the decreasing remaining ligament in the fracturing specimen.

Figure 7 shows typical isochromatics of an edge crack propagating between two symmetrically located twin cracks. High mode II crack tip stress fields are noted in the stationary twin cracks in Frame No. 12 of this figure. Figure 8 shows the variations in K_I, K_{II}, σ_{ox} and crack velocity of the propagating left crack. Oscillations in σ_{ox} are smaller in this straight crack which is propagating at approximately the same crack velocity of $0.2c_1$ as the other cracks.

Figure 9 shows the variations in characteristic distance, r_o, which was computed by Equation (1), for the propagating cracks in the the five tests. r_o for the curved portion of the rapidly propagating crack is larger than the r_o of the straight crack. Also r_o, within the scatter band indicated in Figure

9, always dropped to a minimum value at the onset of crack curving. The scatter band of the minimum value of r_0 yield an average r_c=0.5 mm. This value is consistent with the r_c value estimated by Theocaris [21].

Having estimated the r_c for this thin polycarbonate fracture specimens, the crack curving angle was then estimated either by using the maximum circumferential stress or the minimum S criteria. The results, as summarized in Table 1, show that the predicted and experimentally observed crack curving angles for these five tests were mostly within 1 degree of each other regardless of the crack curving criterion used.

CONCLUSIONS

1. Dynamic crack curving angle under pure mode I and mixed mode I and II conditions can be predicted by using either the extended maximum circumferential stress or the minimum strain energy density criteria.

2. The critical characteristic crack tip distance is r_c=0.5 mm for the thin polycarbonate fracture specimens considered in this investigation.

ACKNOWLEDGEMENT

The results reported in this paper were obtained through ONR Contract No. 00014-76-C-0600 NR 64-478. The authors wish to thank Drs. N. Perrone and Y. Rajapske, ONR for their support during the course of this investigation.

REFERENCES

1. Erdogan, F. and Sih, G. C., "On the crack extension in plates under plane loading and transverse shear", Trans. ASME J. Basic Engrg., 85(D), 1963, pp. 519-527.

2. Williams, J. G. and Ewing, P.D., "Fracture under complex stress - The angled crack problem", Int. J. Fracture Mech. 8, 1972, pp. 441-446.

3. Finnie, I. and Saith, A., "A note on the angled crack problem and the directional stability of crack", Int. J. Fracture 9, 1973, pp. 484-486.

4. Streit, R. and Finnie, I., "An Experimental Investigation of Crackpath Directional Stability", Experimental Mechanics, Vol. 20, No. 1, January 1980, pp. 17-23.

5. Cotterell, B. and Rice, J. R., "Slightly Curved or Kinked Cracks", Int. J. Frac., Vol. 11, No. 2, 1981, pp. 155-164.

6. Yoffe, E. H., "The Moving Griffith Crack", Philosophical Magazine, Vol. 42, 1951, pp. 739-750.

7. Hussain, M. A., Pu, S. L. and Underwood, J., "Strain-Energy-Release Rate for a Crack Under Combined Mode I and Mode II", ASTM STP 560, 1974, pp. 2-28.

8. Palaniswamy, K. and Knauss, W. G., "On the Problem of Crack Extension in Brittle Solids Under General Loading", Mechanics Today, edited by S. Nemat-Nasser, Pergamon Press, 1978, pp. 87-148.

9. Sih, G.C., "A Special Theory of Crack Propagation", Methods of Analysis and Solutions of Crack Problems, Vol. 1, pp. 21-45, edited by G. C. Sih, Noordhoff International Publishing, Leyden, 1973.

10. Swedlow, J. L., "Criteria for growth of the angled crack", ASTM STP 601, 1976, pp. 506-521.

11. Theocaris, R. S. and Andrianopoulos, N. P., "A Modified Strain Energy Density Criterion Applied to Crack Propagation", to be published in the ASME Journal of Applied Mechanics.

12. Sih, G. C., "Dynamic Crack Problems: Strain Energy Density Fracture Theory", Elastodynamic Crack Problems, Vol. 4, pp. 17-37, edited by G. C. Sih, Noordhoff International Publishing, Leyden, 1977.

13. Ramulu, M. and Kobayashi, A. S., "Dynamic Crack Curving - A Photoelastic Evaluation", submitted to Experimental Mechanics.

14. Bradley, W. B. and Kobayashi, A. S., "Fracture Mechanics - A Photoelastic Investigation", Engineering Fracture Mechanics, Vol. 3, 1971, pp. 317-322.

15. Wade, B. G. and Kobayashi, A. S., "Photoelastic Investigation on the Crack Arrest Capability of Pretensioned Stiffened Panels", Exp. Mechanics, 15, 1, 1975, pp. 1-9.

16. Kobayashi, A. S., Mall, S. and Lee, M. H., "Fracture Dynamics of Wedge-Loaded Double Cantilever Beam Specimen", Cracks and Fracture, ASTM STP, 601, June 1976, pp. 274-290.

17. Kobayashi, A. S. and Chan, C. F., "A Dynamic Photoelastic Analysis of Dynamic-tear-test Specimen", Experimental Mechanics, Vol. 16, No. 5, May 1976, pp. 176-181.

18. Kobayashi, A. S. and Mall, S., "Dynamic Fracture Toughness of Homalite-100, Experimental Mechanics, 18, 1, Jan. 1978, pp. 11-18.

19. Kobayashi, A. S. and Ramulu, M., "Dynamic Stress Intensity Factor for Unsymmetric Dynamic Isochromatics", Experimental Mechanics, Vol. 21, Jan. 1981, pp. 41-48.

20. Ramulu, M. "Dynamic Crack Curving and Branching", a PhD submitted to the University of Washington, March 1982.

21. Theocaris, P. S., "The Caustic as a Means to Define the Core Region in Brittle Fracture", Engineering Fracture Mechanics, Vol. 12, 1980, pp. 235-242.

TABLE 1

SUMMARY OF EXPERIMENTAL AND THEORETICAL RESULTS

Total Number of Experiments:	5
Type of Fracture Specimens:	Double Edged Crack Specimen
Number of Data Points:	114
Crack Velocity, c/c_1:	About 0.2
K_I (MPa $\sqrt{m}$):	1.5 to 3.2
K_{II} (MPa$\sqrt{m}$):	-0.5 to 0.6
σ_{ox}/K_I:	-11.1 to 2.5
r_o (mm):	0.25 to 0.75
Measured Crack Curving Angle:	-20 to 3 degrees
Predicted Crack Curving Angle:	
Maximum Circumferential Stress:	-19 to 5 degrees
Minimum S:	-18 to 5 degrees

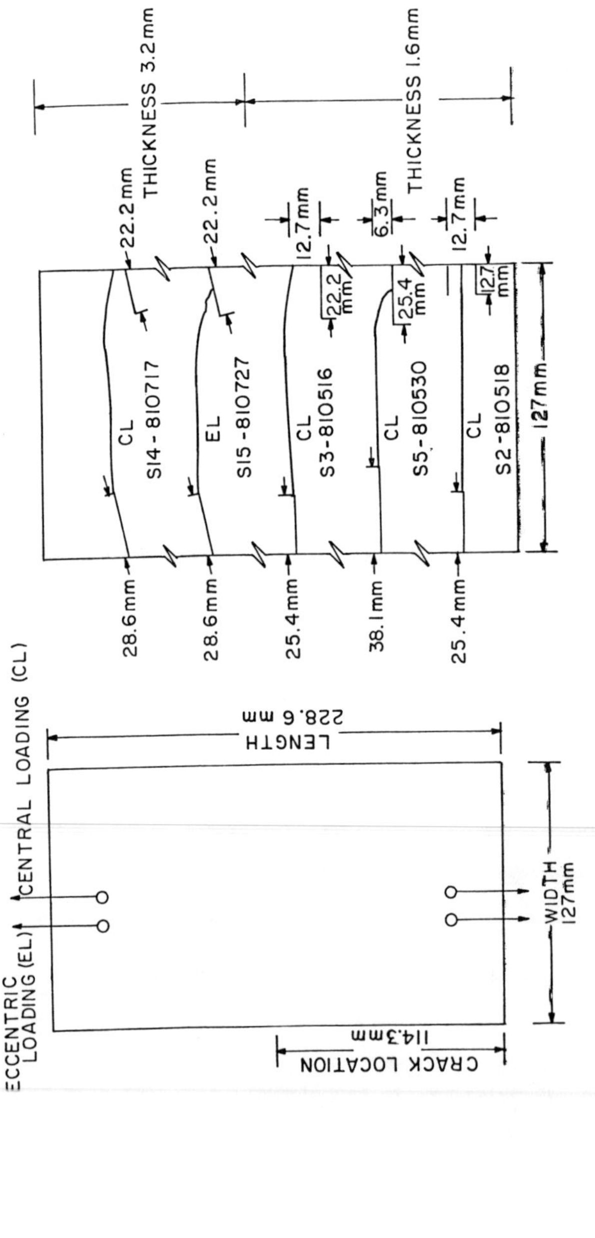

(a) SPECIMEN DIMENSIONS

(b) CRACK PATHS

FIG. 1. POLYCARBONATE DOUBLE EDGED CRACK TENSION SPECIMEN.

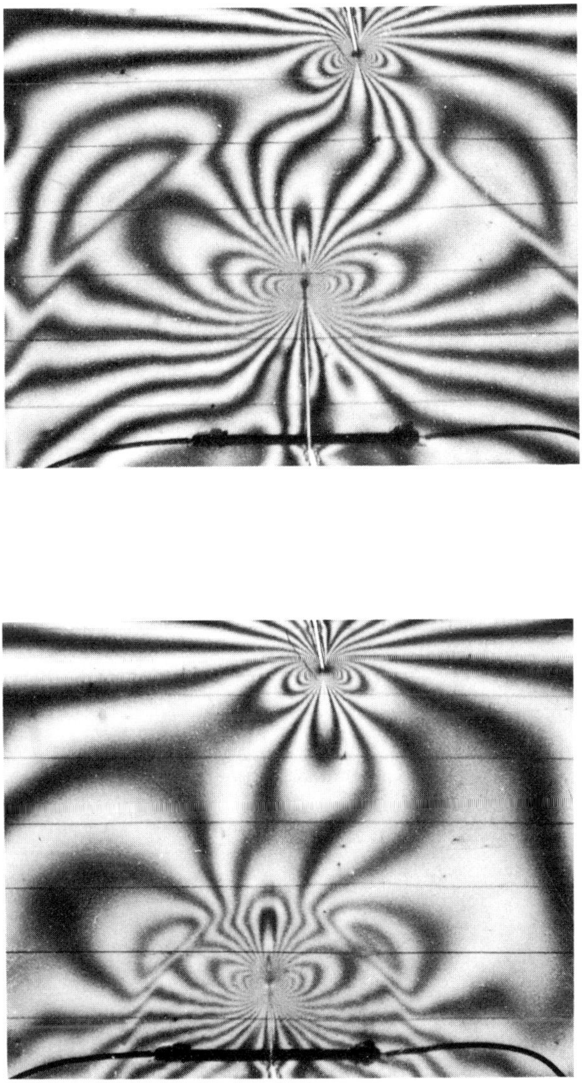

FIRST FRAME
$6\,\mu$ SECONDS

SEVENTH FRAME
$28\,\mu$ SECONDS

FIG. 2. TYPICAL DYNAMIC PHOTOELASTIC PATTERNS IN POLYCARBONATE SLANTED DOUBLE EDGED CRACK TENSION SPECIMEN. SPECIMEN NO. SI4-810717.

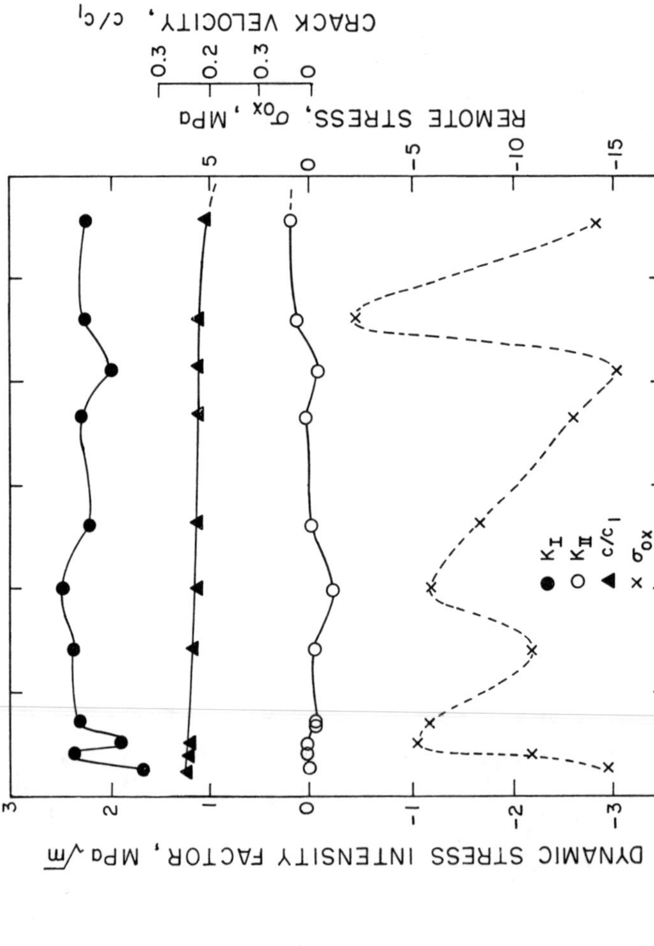

FIG. 3. DYNAMIC STRESS INTENSITY FACTOR, CRACK VELOCITY AND σ_{ox} OF UPPER CRACK OF SPECIMEN NO. S14-810717.

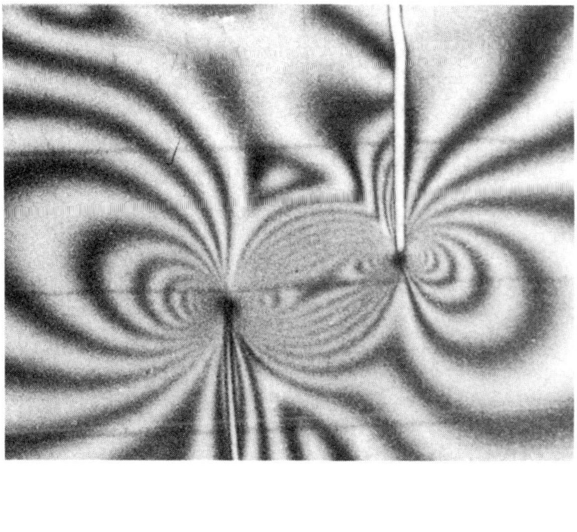

SEVENTH FRAME

150 μ SECONDS

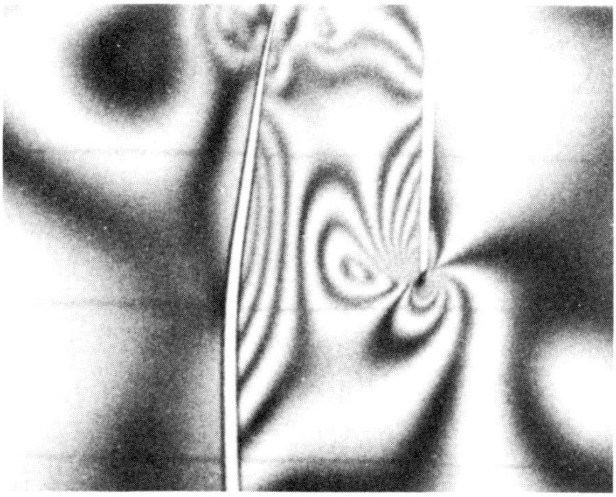

TENTH FRAME

225 μ SECONDS

FIG. 4. TYPICAL DYNAMIC ISOCHROMATICS OF A CURVED CRACK IN A POLYCARBONATE PARALLEL DOUBLE EDGED OFFSET CRACK TENSION SPECIMEN. SPECIMEN NO. S3-810526.

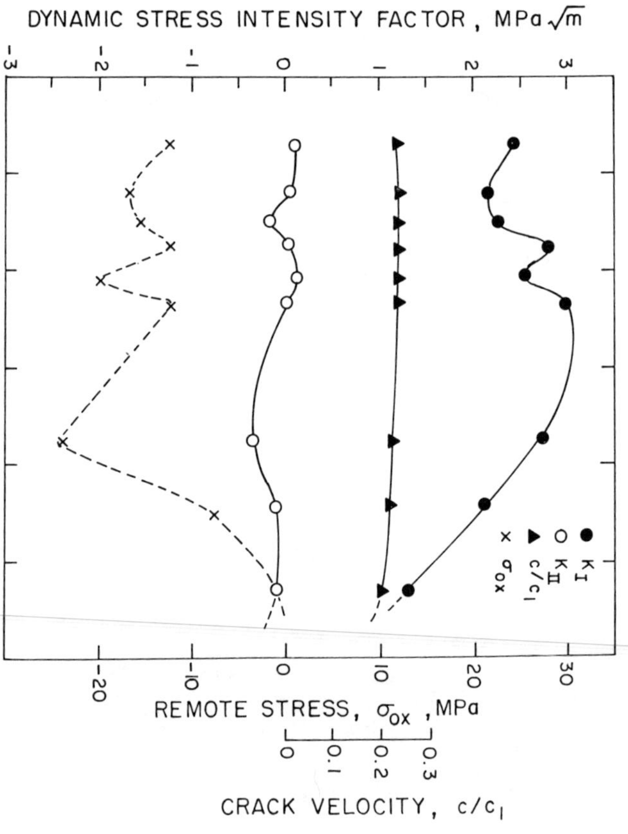

FIG. 5. DYNAMIC STRESS INTENSITY FACTOR CRACK VELOCITY AND σ_{ox}

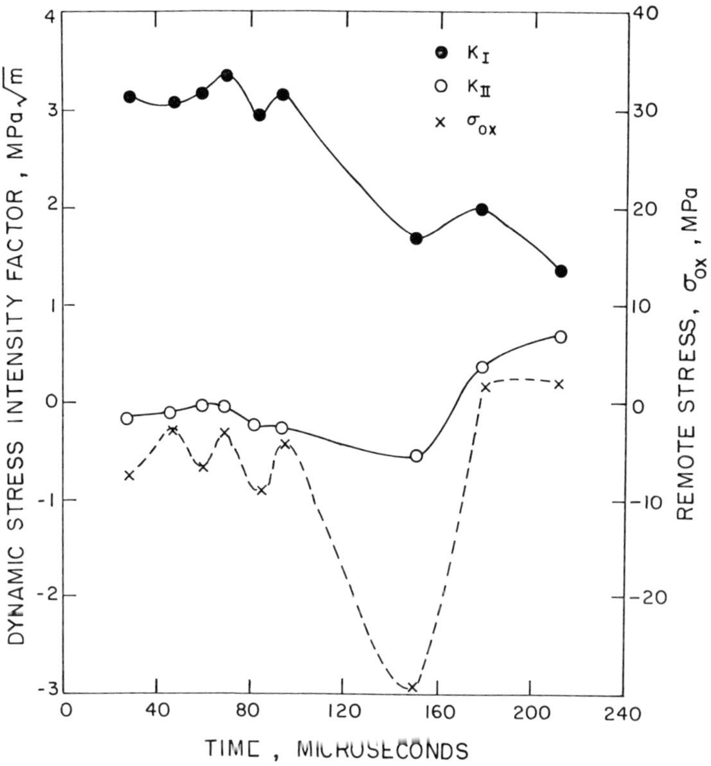

FIG. 6. DYNAMIC STRESS INTENSITY FACTOR AND σ_{ox} OF THE STATIONARY LOWER CRACK OF SPECIMEN NO. S3-810526.

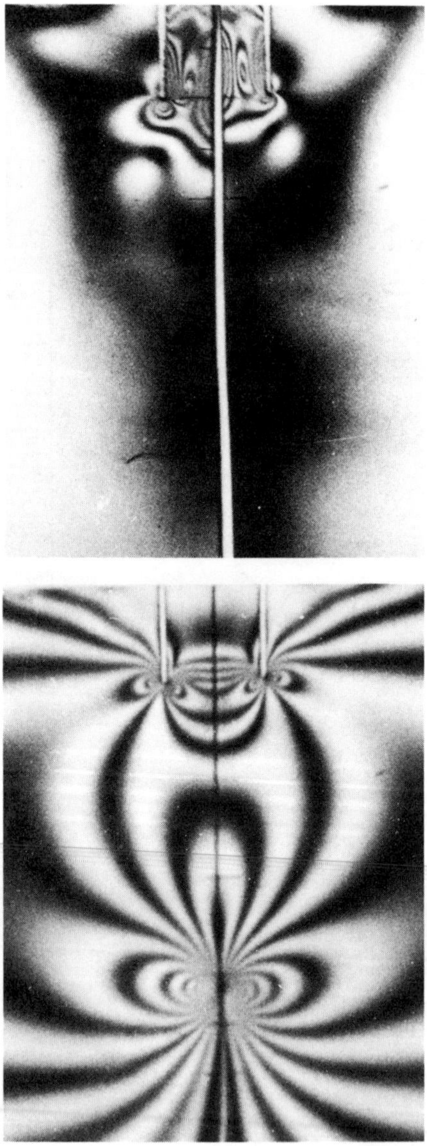

SIXTH FRAME
100 μ SECONDS

TWELFTH FRAME
262 μ SECONDS

FIG. 7. TYPICAL DYNAMIC ISOCHROMATICS OF A STRAIGHT CRACK PASSING THROUGH SYMMETRICALLY LOCATED TWIN CRACKS. SPECIMEN NO. S2-810518.

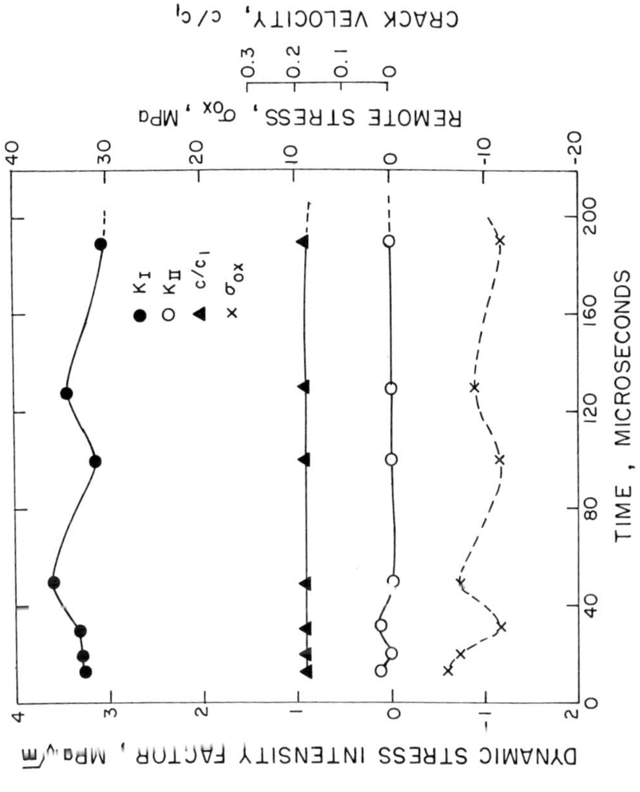

FIG. 8. DYNAMIC STRESS INTENSITY FACTORS, CRACK VELOCITY AND σ_{ox} OF THE RUNNING CRACK OF SPECIMEN NO. S2-810518.

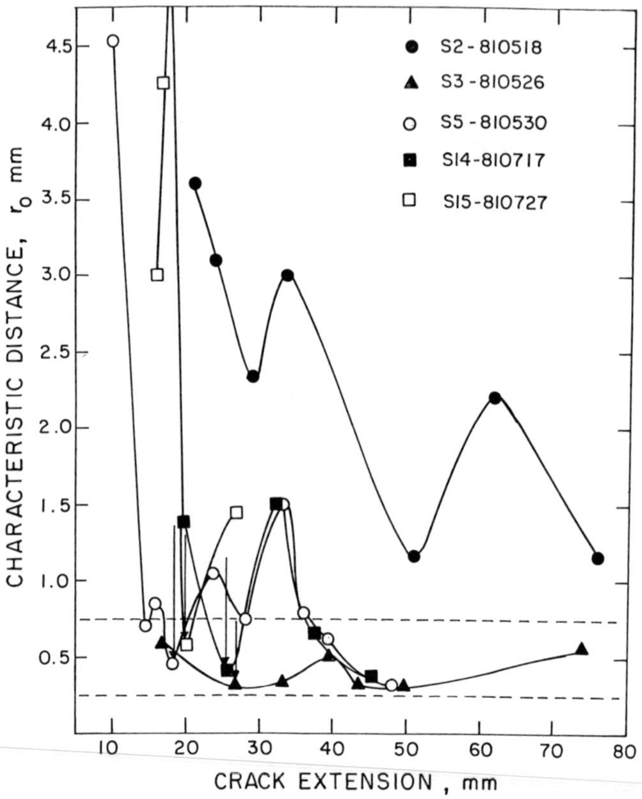

FIG. 9. CHARACTERISTIC DISTANCE r_0 OF PROPAGATING CRACK IN THE POLYCARBONATE DOUBLE EDGED CRACK TENSION SPECIMENS.

COMPUTER AIDED OPTICAL NONDESTRUCTIVE FLAW DETECTION SYSTEM FOR COMPOSITE MATERIALS

John A. Schaeffel and Bobby R. Mullinix
Ground Equipment and Missile Structures Directorate
Technology Laboratory, U.S. Army Missile Research and Development Command, Redstone Arsenal, Alabama

W. F. Ranson
Mechanical Engineering Department
University South Carolina
Columbia, South Carolina

W. F. Swinson
Mechanical Engineering Department
Auburn University
Auburn University, Alabama

ABSTRACT

This paper records the use of speckle interferometry to locate flaws in thin walled pressure vessels made of composite materials. Speckle interferometry is used to generate Young's fringes caused by illuminating the test piece with a laser. A double exposed negative records the speckle pattern before and after deformation. The negative is interrogated with a laser beam to yield Young's fringes. A computer is used to assist in data collection and reduction. Two variations of data collection are used to transmit information to the computer and compared. One variation is a rotating mirror system and the other variation is a vidicon camera system.

INTRODUCTION

Growth in the use of composite materials for structural applications has generated an urgent need for the detection and evaluation of flaws inadvertently induced in such structures by manufacturing procedures. Such detection and evaluation procedures can be grouped, quite naturally, with other nondestructive test (NDT) methods used in the inspection and evaluation of homogeneous structures. It appears that recent advances in the fields of optical interferometry may permit applications in the NDT field.

In recent years, filament-wrapped composite structures have been utilized in small solid-propellant rocket motors and launch tubes with great success. Use of composite materials in this application has merit due to the high strength-to-weight ratio, the low cost of manufacture, and the ease of assembly of the rocket motor case and grain components. Due to the size and nature of such rocket motors and launch tubes, production would occur in an automated fashion at very high rates. Obviously, such high production rates will require an automated system of quality control.

It has been found by experience that flaws are likely to occur in the manufacture of small rocket motor cases in the form of unbonded surfaces where bonds should be. These unbonded regions can occur for several reasons. For example, if the region is contaminated with a foreign substance such as water, grease, etc., the bond will not form. Also, failure to apply adhesive properly to the entire region will prevent the bond from forming. Such conditions are bound to occur occasionally in a mass-production-type operation. The problem is to detect such flaws, assess their significance, and discard the faulty motor case if conditions warrant.

At the present time, experimental techniques for detecting flaws have been demonstrated qualitatively. Assessment of flaw sizes and shapes have been demonstrated.

The objective of this work was to determine the magnitude of the surface displacement and surface strains of three hollow composite cylindrical pressure vessels in the vicinity of three typical flaws (or flaw simulations).

DEFINITION OF THE PROBLEM

To aid in the task of developing an experimental technique for flaw detection in composite structures, three test structures were designed with simulated flaws placed in specific locations and with specific orientations.

The simulated flaws were created by placing thin strips of Teflon plastic between the two helical wraps of filaments at the center of the walls of the hollow cylindrical structures. The Teflon strips were intended to inhibit bonding of the resin in the composite between layers at the location of the strips. Thus, the two center layers of the composite structure would be free to move relative to each other. The purpose of the three models designed for study was to permit a comparison to be made of experimental and analytical predictions of the displacements on the surface of the cylinders in the vicinity of the flaws. All models were manufactured with 60° helix wraps and details of each type specimen are shown in Figures 1, 2 and 3.

SPECKLE INTERFEROMETRIC ANALYSIS USING YOUNG'S FRINGES

Several investigators have shown that inplane displacement data can be determined using speckle interferometric data to produce Young's fringes [1-6]*. The general arrangement for data collection is shown in Figure 4 and data is retrieved with the arrangement shown in Figure 5. Reference [6] notes that interpretation of the fringe data can be mathematically modeled and maxima given from

$$u = m_1 \frac{\underline{\lambda} f}{\underline{X}} \quad \text{for } m_1 = 0, 1, 2, \ldots \tag{1}$$

where

u = inplane displacement,
m_1 = fringe order,
$\underline{\lambda}$ = wave length of the laser source,
$\underline{X}$ = coordinate distance on the screen, and
f = lens focal length or if lens is omitted the distance from film to screen.

In application a double exposed film record was made of the speckles on the composite flawed cylinders. One exposure with cylinder pressurized to 120 psi and the second exposure at atmospheric pressure. The film was processed and placed in a holder in front of the laser beam. A coordinate system was choosen as illustrated in Figure 6. Displacement data was calculated at coordinate intersections at 0.1 inch intervals for the full field of the model. From this strain was calculated as

$$E_y = \frac{\Delta u_y}{\Delta y} \tag{2}$$

where

E_y = normal strain in the y-direction,
Δu_y = change in displacement in the y-direction, and
Δy = change in y coordinate.

*Numbers in [ ] refer to references

The effect of the flaws in each cylinder was clearly evident by abrupt changes in displacements, which indicated high strain. Figure 7, 8 and 9 illustrate the flaw effects in each flawed cylinder. The flaw size and location are evident and offer quantitative values for accepting or rejecting a cylinder. In addition this technique does not require the stringent stability that holography does and further offers a comparatively straight (compared to holography) way of calculating strain results. The applied loads are well below maximums possible in such cylinders thus will not cause structure damage to the cylinders as full pressure testing has the potential of doing.

COMPUTER AIDED DATA REDUCTION SYSTEMS AND SOFTWARE

A computer aided data reduction system was developed for applications in the speckle interferometric data analysis. However, other applications of optical data reduction are amenable to this type of fringe analysis such as reflecting grid and Moire'. The basic system consists of two parts which are an optical recording of interference data and the computer hardware used in the numerical analysis. Optical recording of the data can be obtained by several methods depending on the required results. If only one-dimensional displacement data using speckle photography are desired, then a rotating mirror system can be employed. Two-dimensional data analysis utilizes a vidicon TV camera system which is also used in whole field data analysis such as the reflecting grid and Moire'. An x-y film translation stage is common to both systems along with some of the computer interfaces. This section will discuss each data recording system. Examples are selected which illustrate the data analysis procedures. A typical fringe pattern used for analysis is shown in Figure 10. Interference fringes are a measure of the object motion or surface deformation of the body. Because only a small area is illuminated by the laser, the information in the data analysis yields the displacement at a point. Therefore, to obtain a complete map of the surface, each point on the film will have to be illuminated. This is accomplished by an x-y translation stage in which the transparency has been mounted. The stage has the capacity to translate 6 in. in each direction in 0.001-in. increments with a 0.0001-in. repositioning accuracy. Synchronous stepping motors provide the control for the translation directions. A photographic transparency is mounted in a 4 x 5 in. window as shown in Figure 11. The laser beam passes through a point on the film and the film is then translated relative to a stationary laser beam. The beam direction mirrors are attached to a post separate from the translation stage.

Location of a point in the window is controlled by the stepping motors which are in turn controlled by the computer. Thus, the computer can control the translation distance with a minimum interval of 0.001 in. One way to process this data is to image the diffraction pattern on a vidicon camera, Figure 10. The image on the camera can then be viewed on a TV monitor or stored in memory in the computer.

The computer and interface system is shown schematically in Figure 12. A PDP 11/40 minicomputer is used to control the motion of the translation table and process the video signal. Operation of the system begins with a command from the computer to move the table to a specified position, the computer sends a command through the video interface to store the vidicon image in the scan conversion memory. The scan conversion memory operates in either a read, write, or erase mode depending on the command from the PDP 11/40. Upon receipt of an erase command, the scan conversion memory switches to the erase mode and the election beam is blanked and is raster-scanned. A write command is then given to the scan conversion memory and the image on the vidicon camera is then stored in memory. A command from the computer then converts the information stored in the scan conversion memory into digital form. The complete image is not processed but only selected lines of video information are processed. This process completes the data acquisition at a point on the film and the computer then locates a new coordinate position and the process is repeated.

Typical information stored in the scan conversion memory is shown in Figure 13. Data analysis consists of determining the locus of points of minimum intensity which is referred to as a fringe. However, this analysis is complicated by the fact that the film transparency is illuminated by a coherent light source which produces laser speckle. The presence of this speckle is to superimpose a noise-like signal on the spatial information which degrades the image. In Figure 13, the speckle appears as bright spots on the photography. Therefore the effect of laser speckle is to superimpose bright spots on the film which complicates the location of the fringes.

In the x-y read mode, the computer receives a voltage from the scan conversion memory corresponding to an intensity at points along a scan line. This signal is stored in memory in the PDP 11/40 for processing of fringe information. A digitized single scan line of the image in Figure 13(a) is shown in 13(b). This signal illustrates the effect of the laser speckle on the image. The image should appear as a sine wave without the presence of speckle; therefore, methods of reducing speckle noise are necessary to locate the position of a fringe.

A speckle averaging technique was developed which reduces the effect of noise in locating points of minimum intensity. Fringes which appear as parallel lines on the camera simplify the speckle averaging technique. The procedure is to digitize a single scan line and store into memory the location and value of intensity along this line. Another scan line is also digitized and stored into memory. These two scan lines, which are separtated by a small distance, are then added and the result is the average of the two signals. The best scan line separation appears to be approximately 10 lines apart. The average of these two lines had the effect of reducing the speckle noise superimposed on the image.

Fringe separation is determined by first locating the point of maximum intensity from the scan average. Then the first minimum point is located to each side of this maximum and the difference in location determines the fringe spacing.

An example of a cantilever beam with a transverse end load was chosen as a test for the accuracy of fringe spacing and scan line separation. This example was selected because the theoretical solution is known and fringe spacing varies from infinite at the fixed end to a small spacing at the free end. Therefore, a wide range of fringe spacing conditions can be examined from this one example.

Figure 14 is a graphical display of measured values made by an observer and a computer scan with 10-line separation. Both sets of measured values are compared to the analytical solution to assess the accuracy of the measurements. Compared to the theoretical solution, both sets of measured values are within 2% of the analytical solution.

Another way of processing the data is with a rotating mirror which increases the speed of the results. A schematic diagram of the basic system is shown in Figure 15. This system differs from the vidicon camera system in that only a single direction can be scanned by the mirror. However, the diffraction halo projected on the screen (Figure 10) is the same as the halo projected on the rotating mirror (Figure 15). As the mirror rotates, the fringe pattern is swept across the aperture opening in the photodiode. The aperture size is sufficiently small such that only a small beam of light passes through the opening. Therefore as the mirror rotates, the fringe pattern is swept across the face of the photodiode. The optical performance of the system was evaluated to determine some experimental parameters of the rotating mirror. The film translation stage is the same as discussed in the previous section.

The photodetector consists of a SGD-100 photodiode whose output is amplified by an operational amplifier. The output is then sent to an analog-to-digital (A/D) converter for processing.

Figure 16 shows the basic block schematic of the rotating mirror analyzer. The output of the photodetector is filtered with a low pass filter to remove excess noise. The filtered signal is monitored on an oscilloscope and is converted into digital form by a Digital AR-11 A/D converter whose output is processed by a PDP 11-40 computer processing unit. The computer interfaces with a digital DR11-C digital input/output interface which controls the scan converter interface. The x-y translation stage is controlled by the scan converter interface. The motor assembly of the rotating mirror (2 in. diameter) rotates at 39 rpm. Filtering by the low pass filter is accomplished with a 0-10K ohm 1.0 MFD resistor-capacitor circuit. The AR-11 digitizes data with a conversion time of 32 sec.

In operation, the x-y translation stage is advanced to a position where data are to be gathered. The rotating mirror then rotates the fringe pattern image in front of the photodetector. The photodetector converts the fringe intensity information into electrical form which is then filtered to remove noise and be digitized with the AR-11 A/D converter. The digitized intensity information is then processed to determine fringe spacing and localized displacement information. Because the A/D converter and computer software routines require a fixed and predetermined amount of time to perform the digitization function, it is not necessary to use a line clock to keep track of the location of the fringe pattern.

A typical oscilloscope trace of an unfiltered fringe pattern from the rotating mirror system is presented in Figure 17, which shows that the presence of laser speckle produces a high frequency noise on the signal. This signal is filtered by a low pass filter to reduce the speckle effect on the signal. The filtered signal then served as an analog input to the PDP 11/40 computer. Fringe spacing is then determined in this case by locating the maxima of the wave form. The maximum points are located by sampling the data and determining when the slope between three successive data point samples changes from positive to negative.

To test the accuracy of the rotating mirror system, the cantilever beam example was utilized. A graphical display of the results is given in Figure 18. Compared to the theoretical solution the measured values are within 2% of the theoretical solution.

In conclusion Young's fringe information from speckle interferometry can be effectively used to locate flaws and determine strains for nondestructive structural evaluation of thin composite structures. The data can be processed with computer systems for relatively fast and accurate results.

REFERENCES

1. Leendertz, J., "Interferometric Displacement Measurement on Scattering Surfaces Utilizing Speckle Effect", Journal of Physics E, Vol. 3, 1970, p. 214.

2. Butters, J., "Laser Holography and Speckle Patterns in Engineering Metrology", Symposium on Advanced Experimental Technology in the Mechanics of Materials, September 1970.

3. Butter, J., and Leenderts, J., "A Double Exposure Technique for Speckle Pattern Interferometry", Journal of Physics E, Vol. 4 1971, p. 277.

4. Hung, Y., Der Hovanesian, J., "Full-field Surface-strain and Displacement Analysis of Three-dimensional Objects by Speckle Interferometry", Experimental Mechanics, Vol. 12, No. 10, October 1972, p. 454.

5. Cloud, G. L., "Quantitative Speckle-Moire Interferometry", paper presented at the 1973 Fall Meeting of the Society for Experimental Stress Analysis.

6. Kinariwala, V. R., Peters, W. H., Ranson, W. F., Swinson, W. F., "Speckle Patterns for Measuring Displacements Described as Diffraction Gratings" Proceedings 12 Annual Meeting, Society of Engineering Science, 1975.

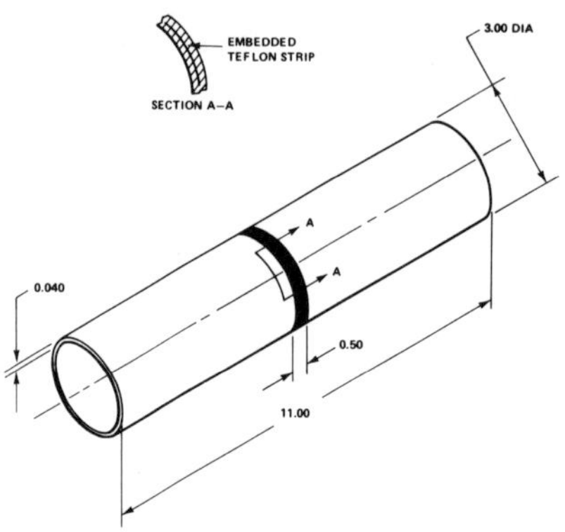

Figure 1. Composite cylinder with simulated circumferential flaw.

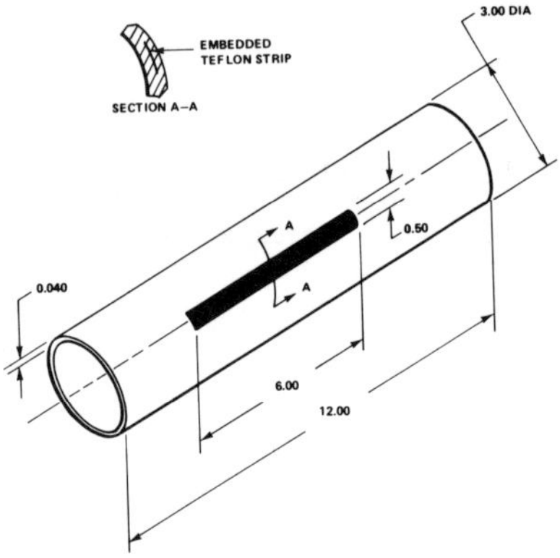

Figure 2. Composite cylinder with simulated longitudinal flaw.

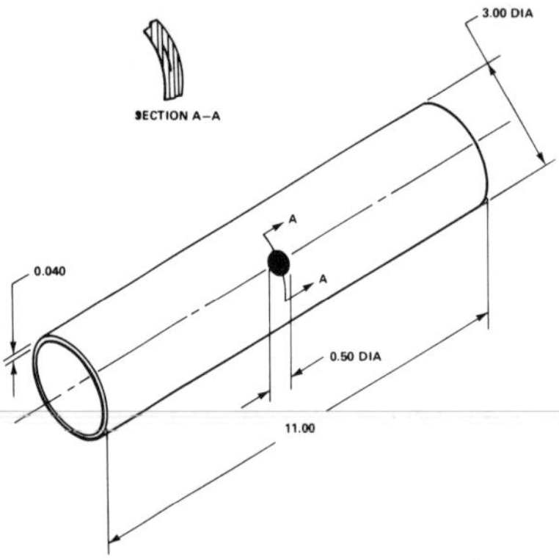

Figure 3. Composite cylinder with simulated spot flaw.

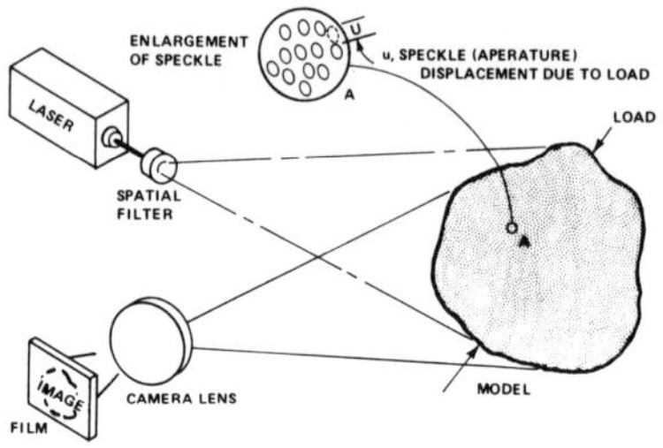

Figure 4. Collection of speckle interferometric data.

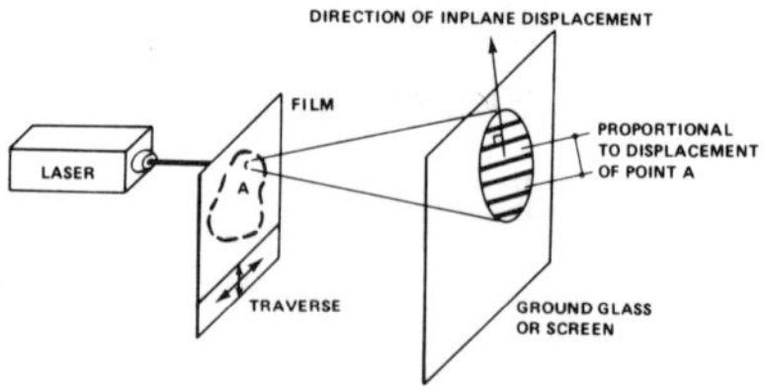

Figure 5. Data analysis

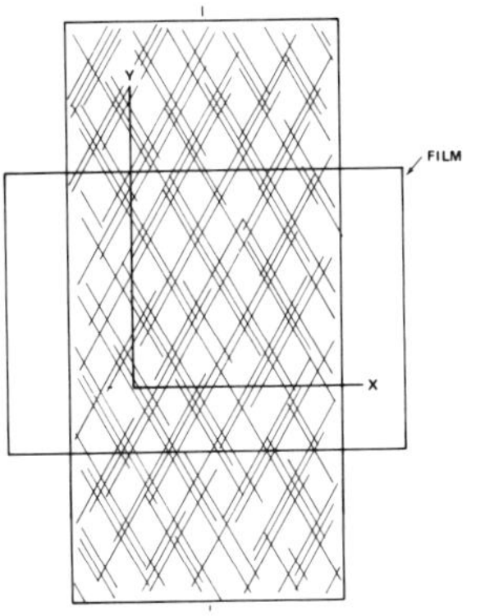

Figure 6. Coordinate axis

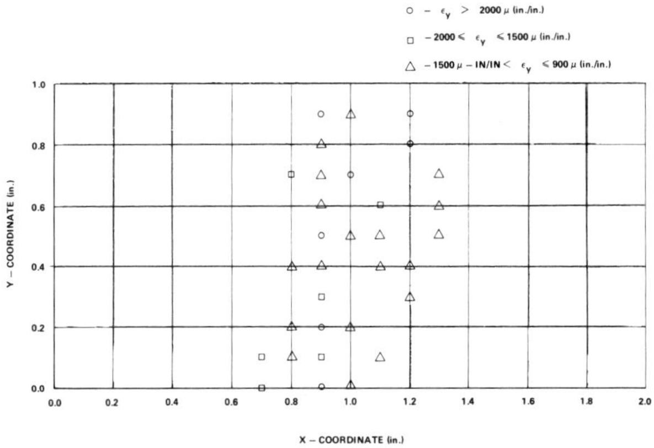

Figure 7. Strain distribution for longitudinal flaw cylinder.

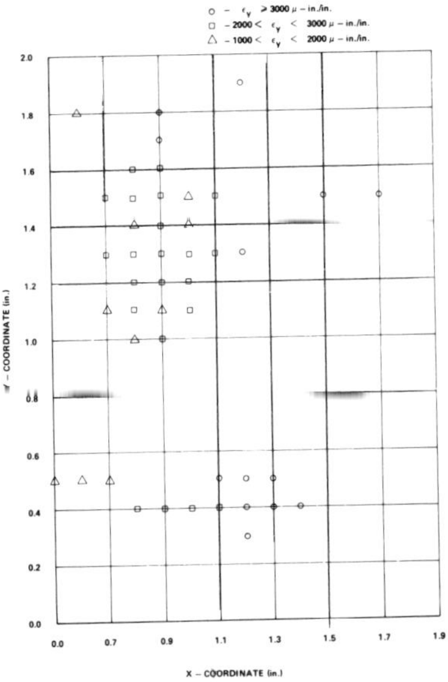

Figure 8. Strain distribution for spot flaw cylinder.

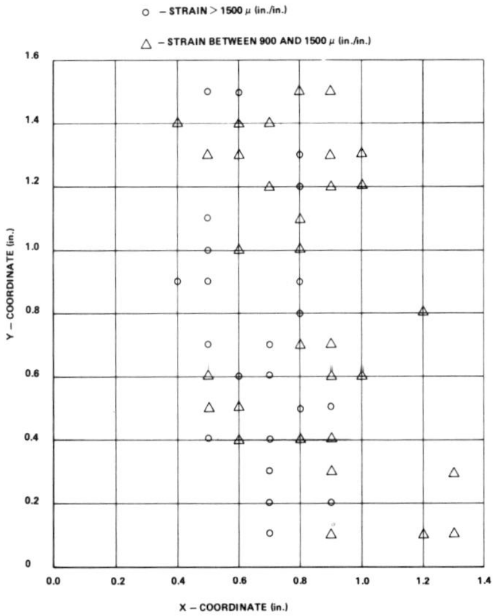

Figure 9. Strain distribution for circumferential flaw cylinder.

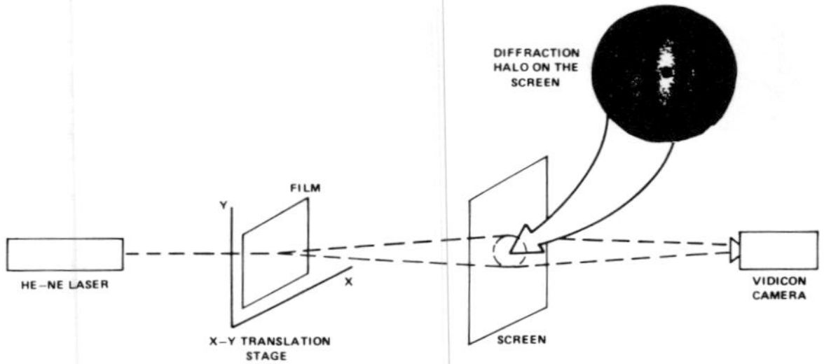

Figure 10. Optical system used for computer aided data reduction.

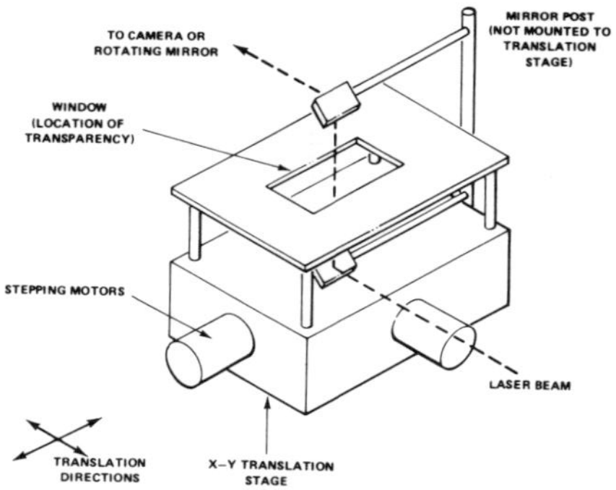

Figure 11. Translation table and laser beam path.

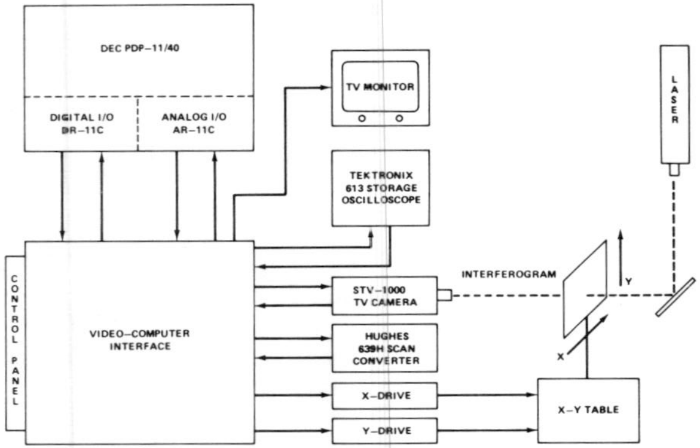

Figure 12. Computer-aided speckle analyzer.

Figure 13. (a) Photograph of diffraction pattern, (b) digitized scan of a horizontal line.

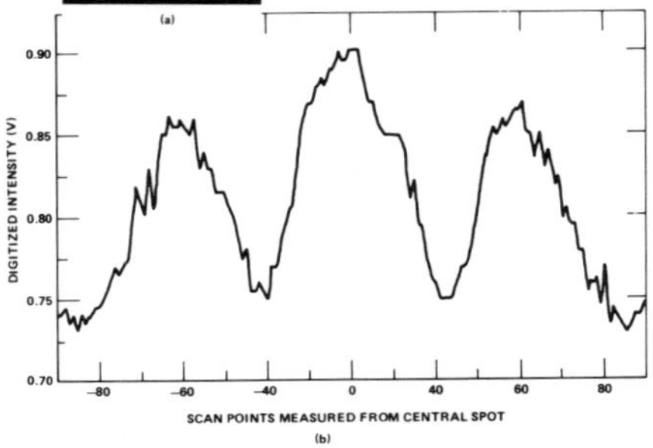

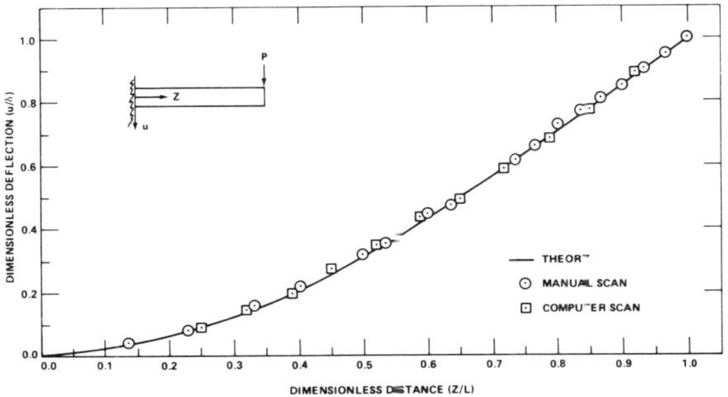

Figure 14. Comparison of fringe spacing values measured by an observer and computer scan measurement.

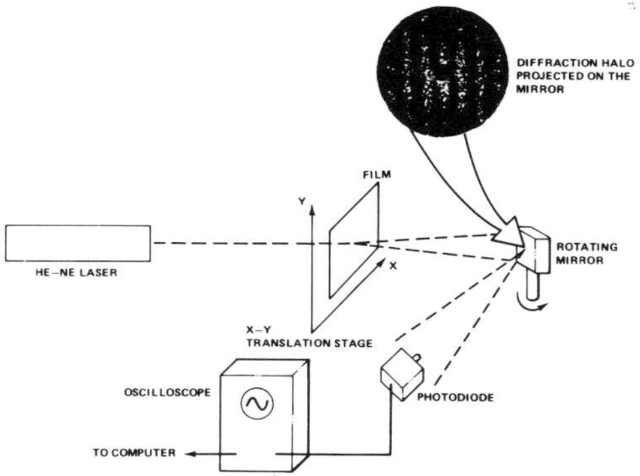

Figure 15. Rotating mirror system used for computer aided data reduction.

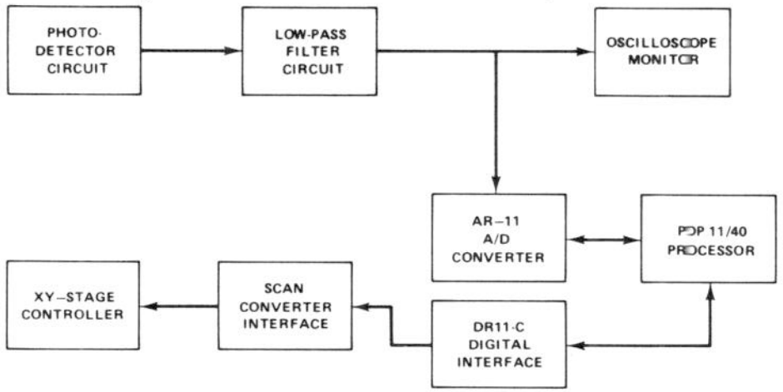

Figure 16. Rotating mirror system block diagram.

(a) (b)

Figure 17. (a) Unfiltered scope trace of the photodiode output, (b) Filtered trace of the same signal (horizontal sweep 5 msec/cm, vertical 0.5 V/cm).

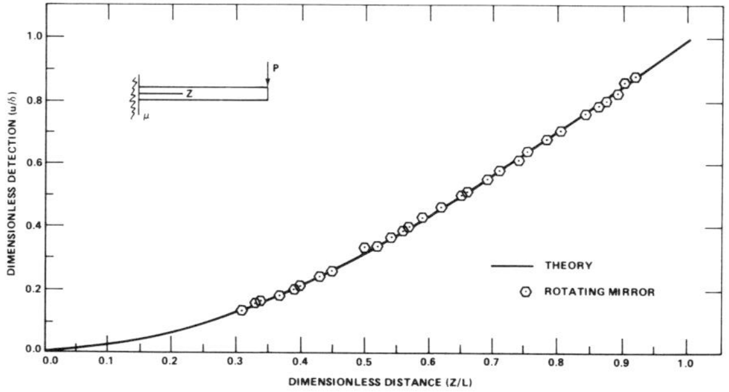

Figure 18. Comparison of fringe spacing values measured by rotating mirror system.

OBLIQUE PENETRATION OF MONOLITHIC TARGETS

Aaron D. Gupta, John J. Misey, John D. Wortman
Ballistic Research Laboratory
Aberdeen Proving Ground, Maryland

ABSTRACT

The response of a kinetic energy penetrator during the oblique penetration of an aluminum and a steel target was computed and compared with experimental data at specific gage positions on the rod surface. The predictions from two codes, i.e., REPSIL and EPIC-3, showed satisfactory correlation with the experimental records during the loading phase.

INTRODUCTION

Most investigations on projectile performation are limited to normal penetration of projectiles and relatively little has been reported on oblique penetration. Zener and Peterson [1] studied the effect of the angle of impact on the projectile's residual velocity. Recht and Ipson [2] developed an energy analysis to examine the total angular change of the projectile. Summers [3] studied the effect of the angle of impact on the depth of penetration.

Goldsmith, et al [4], carried out oblique impact experiments on steel beams to study the crater geometry and crater depth. Awerbuch and Bodner [5] conducted an experimental study of yawed rod impacts by using reverse ballistic techniques.

Bertholf, et al [6] analyzed the oblique impact of a high velocity, tungsten alloy projectile with a multi-layerd steel target using two-dimensional stress wave propagation computer programs. Recently, G. E. Hauver [7] developed a forward ballistic technique for hard target penetration wherein a long rod, instrumented with strain gages was impacted against a stationary target at velocities greater than

1000 meters per second. The experimental results were in good agreement with simulated results from two dimensional analysis [8] at normal impact. This technique was extended to solve oblique penetration problems. The measurements from tests employing the forward as well as reverse ballistics techniques were ultimately intended to serve as a comparison with predictions from a computer code, to provide input data of material parameters for more exact computations, and to improve the simulation of the penetration process.

This paper complements the work of Mr. Hauver by taking his oblique penetration measurements on a steel rod penetrating a relatively soft aluminum as well as a hard steel target and performing numerical simulations of the penetrator response.

Numerical Methods

The first method involved simulation of oblique penetration into a relatively soft aluminum target using a finite-difference structural response type code, i.e. REPSIL [9,10] developed at the Ballistic Research Laboratory. The second method employed a finite element formulation in the three dimensional Lagrangian code, EPIC-3, developed by Dr. Gordon Johnson [11] of Honeywell and involved a steel rod penetrating a hard RHA (Rolled Homogenous Armor) target at an ordnance velocity.

A beam bending version of the REPSIL code was used to predict strain-time response at specific gage locations. REPSIL is a finite-difference code which calculates the large deflection transient motion of thin Kirchoff shells. The program marches out the solution timewise by cyclically solving an explicit, time-centered, difference formula for displacements. The modified version was extended to include free-edge boundary conditions and allow the bending moment and transverse load to be superposed at the tip of the hemispherical nose of the rod. To minimize instability in the numerical procedure symmetric differential equations were used wherever feasible. However instability could occur if the time step at any cycle exceeds the wave-speed-transit time which effectively limits the maximum allowable time step. The code does not currently include a material failure criterion and is not capable of simulating hard target penetration unless an erosion rate is known a-priori at the tip of the rod as described by Gupta et.al [12].

The EPIC-3 code performs Elastic-Plastic Impact Computations in three dimensions involving complex geometry and loading conditions. It is based on a Lagrangian finite element, lumped mass formulation. The equations of motion are integrated directly, rather than through the traditional stiffness matrix approach. Nonlinear material strength and compressibility effects are included to account for elastic-plastic flow and wave propagation. The code has material descriptions which include strain hardening, strain rate effects, thermal softening and failure. Mesh generators are included to produce quickly configurations such as flat plates, spheres, and rods with blunt, ogival, or conical nose shapes. Complex shapes can also be represented simply by providing the adequate assemblage of elements to represent the desired geometry. The elements are tetrahedral in shape and are well suited to represent severe distortions generally occuring during high velocity impact.

Material failure is currently dependent on the equivalent plastic strain and the volumetric strain. The equivalent plastic strain, $\bar{\varepsilon}p$, is obtained by integrating the equivalent strain rate, $\dot{\bar{\varepsilon}}p$ with respect to time during plastic flow such that

$$\bar{\varepsilon}p(t+\Delta t) = \bar{\varepsilon}p(t) + \dot{\bar{\varepsilon}}p(t)\Delta t$$

where Δt is the integration time increment. The volumetric strain ε_V is obtained by observing the current and initial volume of the element in the following manner

$$\varepsilon_V = V/Vo-1$$

where Vo is the initial volume of the element and V is the current volume.

When the failure criterion has been met for these two strains the equivalent tensile stress is set to zero, and no tensile or shear stress is allowed to develop in the failed element. The net result is that the failed elements acts like a liquid inasmuch as it can develop hydrostatic compression with no shear or tensile stress. Another option is available wherein the element fails totally and all stresses and pressures are set equal to zero.

Oblique Impact Simulation

The projectile-target configuration selected for the oblique impact simulation is similar to that used by Mr. Hauver et.al [13] in the experimental phase. A schematic drawing of the experimental set up for the soft target penetration is shown in Figure 1.

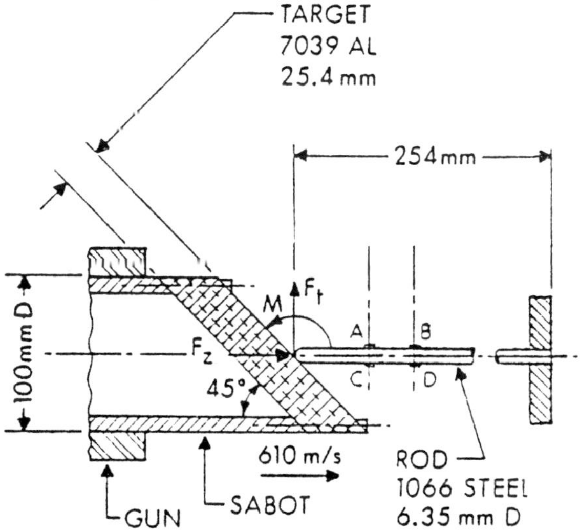

Figure 1. Oblique impact configuration of a steel projectile for soft target penetration.

A reverse ballistic technique is used in this case. The stationary projectile is a 1066 steel rod with an L/D ratio of 10. Type EP high elongation foil strain gages are located at 38.1 mm and 63.5 mm from the tip of the rod. The target is the relatively soft 7039 aluminum, 25.4 mm thick, 100 mm in diameter and is launched at 45° obliquity from a light gas gun at 610 meters/second.

For the case of a hard target penetration a forward ballistic technique was employed. The schematic of the target-projectile configuration with associated data is shown in Figure 2. The projectile is a S-7 (VIMVAR) steel rod 254 mm long, 8.1 mm in diameter, with strain gages located 20, 40, and 60 mm from the tip of the rod. Launched at 1,000 meters/second the projectile impacts a rolled homogeneous armor target at 45° obliquity which has a line-of-sight thickness of 12.7 mm.

In the EPIC-3 code the rod portion is configured with 1848 nodes and 7020 elements consisting of composite bricks with six tetrahedral elements in each brick. The positions at which strain calculations are made is determined by selecting adjacent nodes in an element that is located on the plane of symmetry and at the approximate 20 mm, 40 mm, and 60 mm positions on opposite sides of the projectile. The strain calculation is accomplished by noting the change in length between these nodes and comparing this change with its original length. The positions of the gages are shown in Figure 2.

In the REPSIL code the cross-section of the rod is approximated with six slices of varying widths but identical thickness to represent the circular section area. Twenty segments of equal lengths are used to represent the rod in the axial direction.

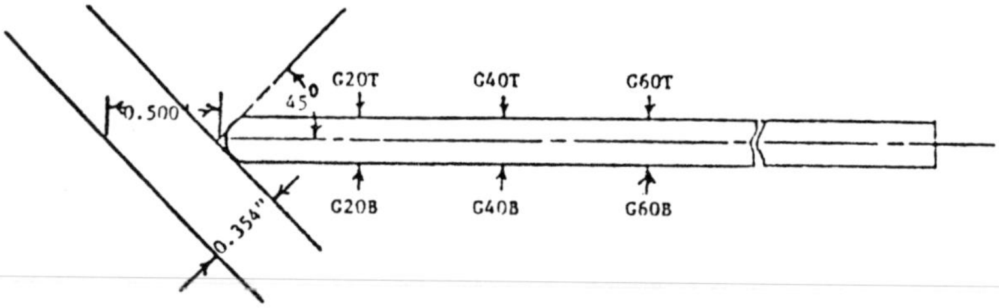

ROD: VIMVAR W/O MOLY, OHLER MFG. #54, 10" LONG, 0.32" DIAM., RC = 48

TARGET: RHA(from 4" stock), 0.500" LINE-OF-SIGHT THICKNESS, RC = 26-27

IMPACT VELOCITY: 1,000 METERS/SEC. = 3,281 FEET/SEC.

GAGE LOCATIONS:

G20T = 20.16 MM. = 0.794"	G20B = 20.12 MM. = 0.792"
G40T = 40.09 MM. = 1.578"	G40B = 40.03 MM. = 1.576"
G60T = 60.02 MM. = 2.363"	G60B = 60.11 MM. = 2.366"

Figure 2. Schematic for oblique impact of a steel projectile for hard target penetration.

Material Properties

The material properties for the computations were obtained from the Solid Mechanics Branch of the Laboratory. These properties are shown in Table 1.

Table 1. Material Properties

Material	E (GPa)	ν	τ_y (GPa)	τ_u (GPa)	ρ (Mg/m^3)
S7 Steel	206..8	0.3	1.44	2.68	7.8
1066 Steel (Rockwell C54)	212.5	0.3	1.7	2.60	7.8
RHA	206.8	0.3	0.96	1.14	7.8
7039Al	69.7	0.3	0.45	0.80	2.74

E - elastic modulus
ν - Poission's ratio
τ_y - yield strength
τ_u - ultimate strength
ρ - density

Dynamic and quasi-static tests were run to determine the elastic modulus and yield strength [7] for S7 steel. The results are shown in Figure 3. The ultimate strength was taken from data derived by Bell [13]

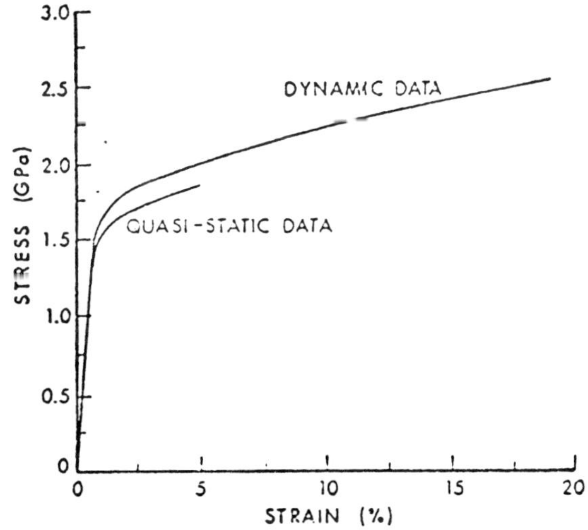

Figure 3. Dynamic and Quasi-Static Relations between stress and strain for S7 steel.

Only quasi-static stress-strain data for 1066 steel in uniaxial compression were available in [15] and are reproduced in Figure 4. The results show

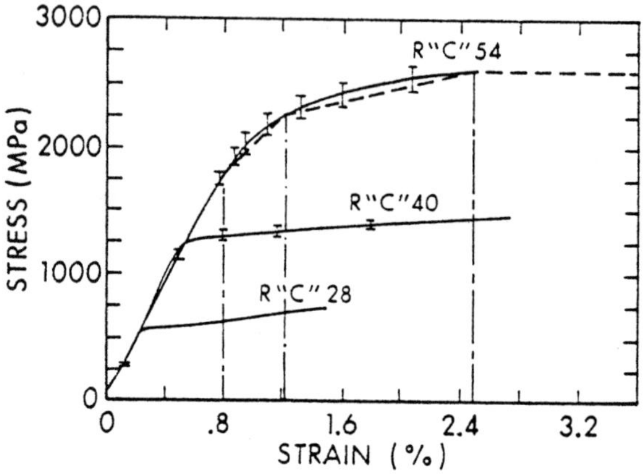

Figure 4. Quasi-static stress-strain curves for 1066 steel loaded in uniaxial compression.

the influence of hardness. The continuous curves represent test data with the experimental uncertainty shown by superimposed error bars. The dotted lines indicate the trilinear representation at low strains and the perfectly plastic behavior at higher strains which approximated the stress-strain characteristics of the rod material in the REPSIL code as a polygon.

In EPIC-3 the failure criterion was considered where only failure in the projectile was permitted in shear at a true strain of 40% and total failure at a true strain of 100%. These values for strain to failure were extrapolated from data for S7 steel obtained from Oak Ridge by Dr. E. W. Bloore [16]. The data indicated elongation at failure of 9.4 to 10.5% at a strain rate of .033 m/m/s and 20 to 22% at a strain rate of 280.0 m/m/s.

Estimation of Projectile Loads

Unlike the EPIC code where projectile loads are self-generated in the code due to target-projectile interaction, the structural response code, i.e. REPSIL, requires estimation of the loading functions at the tip of the rod for the duration of penetration due to resistance of the target. The loading functions are applied in two stages as shown in Figure 5. The effect of surface and internal friction is ignored. Both the transverse load and bending moment are superposed on the axial load at the tip of the rod for the duration of Stage I from the initial impact until the hemispherical nose is fully embedded in the target. During the second stage from full embedding of the nose until complete perforation, both the bending moment and transverse loads are zeroed out.

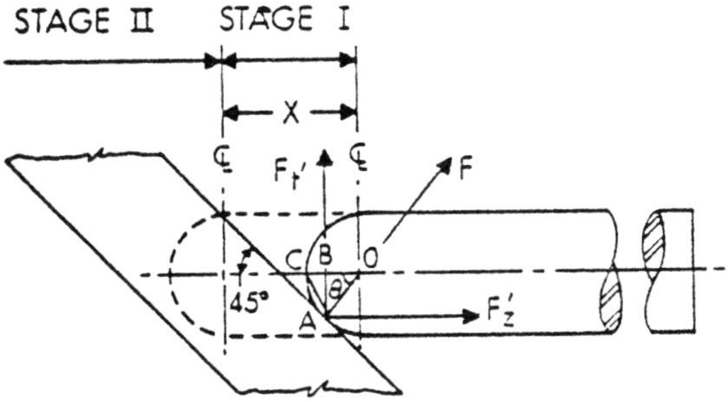

Figure 5. Initial projectile and target geometry upon impact.

Transferring the location of forces from the point of contact A to the tip of the rod C in Figure 5 results in an additional moment as shown in Figure 6.

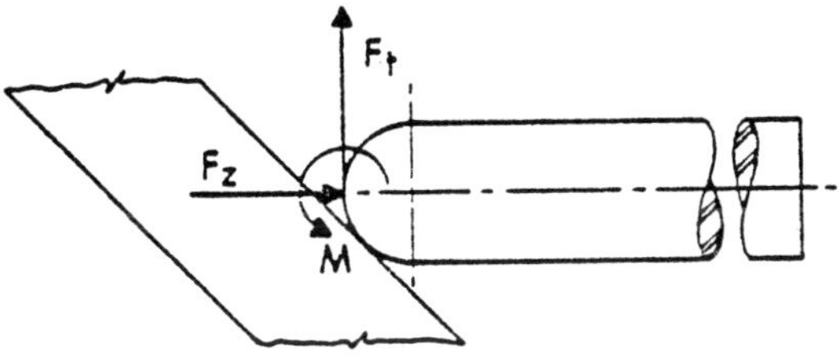

Figure 6. Equivalent axial force, transverse shear and moment at the tip of the rod.

The applied moment M could be calculated as

$$M = -r\{F_z \sin\theta + F_t (1-\cos\theta)\} \tag{1}$$

where F_t = transverse load at the tip of the rod
F_z = axial load at the tip of the rod
θ = angle of obliquity

The distance, X, penetrated by the rod until full embedding of the hemispherical nose could be found from the interface geometry in Figure 4 as

$$X = r\sec\theta + r\tan\theta \tag{2}$$

Also the transverse force F_t is estimated from the axial force from the relationship

$$\frac{F_t}{F_z} = \tan\theta \text{ where } 1<\tan\theta\leq 0$$

$$\text{or } \frac{\pi}{4} \leq \theta \leq 0 \qquad (3)$$

and θ is a function of the time of penetration.

The analysis is simplified by performing the calculations in three steps. Only the axial force at the tip estimated from normal penetration studies [17] is considered initially for the complete duration of penetration and the results are compared with the response at opposite gage locations B and D averaged to eliminate the response due to bending wave propagation alone.

Subsequently a transverse force and a bending moment computed from equations (1), (2), and (3) are successively superposed on the axial force at the tip for the duration of penetration from initial impact until full embedding of the hemispherical tip. By analyzing the problem in successive steps, the interactive coupling effect of each type of forcing function on the projectile response could be ascertained.

Once the rod experiences localized bending near the tip, the application of the resisting force in the axial direction of the rod could introduce additional spurious bending strains near the tip region. Hence, the code is modified to allow the resisting force to remain normal to the deformed tip at each time step.

Results and Discussions

The comparison between experimental and predicted results from REPSIL shows satisfactory agreement with all gage locations for a considerable time period as shown in Figures 7-10. Since a strain averaging method was applied at gage locations B and D to modify the estimated axial forcing function, good agreement was not entirely unexpected at these locations. However, application of the stepwise-loading procedure resulted in an improved model capable of predicting strain response reasonably well at other gage positions. The correlation is excellent during the elastic loading phase and deteriorates somewhat during the subsequent plastic phase of the deformation beyond the yield strain of 0.008.

Both computed and experimental strain readings show reversal of strain due to propagation of bending waves in addition to longitudinal wave propagation. However, when only axial and transverse forces are applied, the region near the tip of the rod bends upward resulting in compressive strain on the upper surface and tensile strain on the lower surface. Conversely, when the transverse force is deleted, the applied moment due to the eccentric location of the point of impact causes downward bending of the tip. Thus, for the oblique impact case, the relative magnitude of the transverse force and bending moment appears to dictate the direction in which the tip would bend. Due to severe bending at late times in our assumed model, agreement between numerical and experimental predictions suffers beyond approximately 35 microseconds from impact.

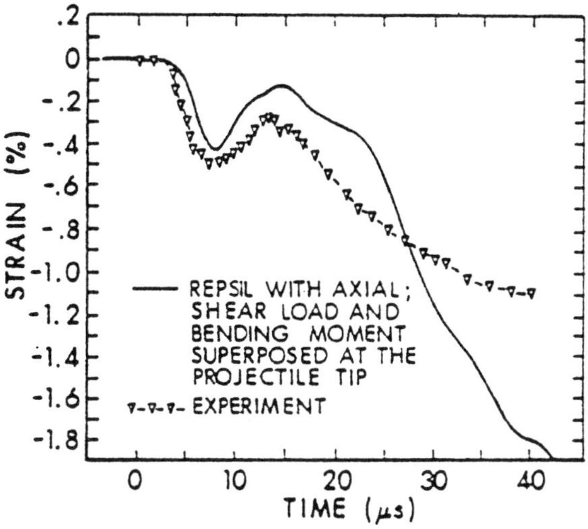

Figure 7. Comparison of experimental response with REPSIL prediction at gage location A on the projectile.

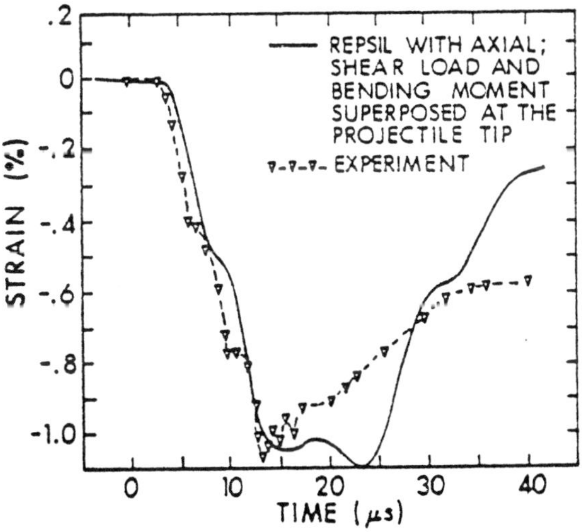

Figure 8. Comparison of experimental response with REPSIL prediction at gage location C on the projectile.

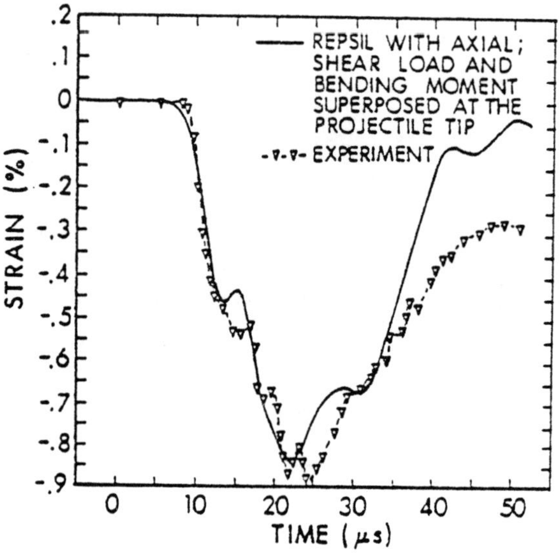

Figure 9. Comparison of experimental response with REPSIL prediction at gage location B on the projectile.

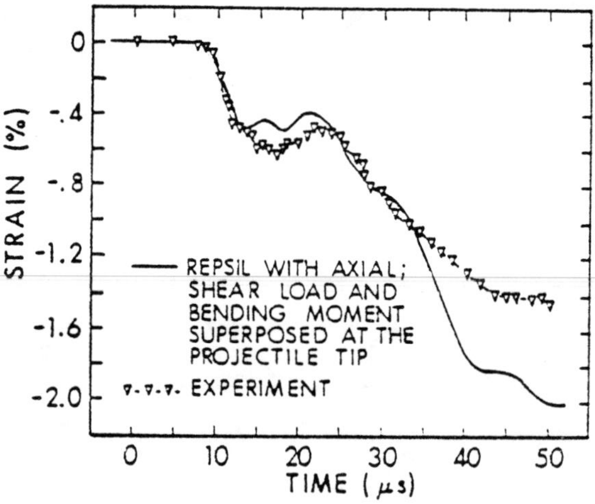

Figure 10. Comparison of experimental response with REPSIL prediction at gage location D on the projectile.

In the hard target impact problem the strains, computed and experimental, are true values. The strains at the 20 mm position are shown in Figure 11. At the onset of penetration compressive strains are induced in both top and bottom gages. Shortly thereafter the tip of the rod is forced to bend causing a rapid increase in compressive strains along the upper surface until the strain gage fails. In the meantime the lower surface undergoes stress release and a change in strain to that of tension before the bottom gage fails. The agreement between the computed and experimental results is excellent with only a difference of less than 1% strain. Due to problems in calibrating the test equipment data at the 40 mm position was not considered. At the 60 mm position shown in Figure 12 the experimental strains are consistently higher than the computed ones. Some bending of the rod is indicated by a increase in the difference of the strain measurements. However the computed strains reveal a pronounced bending of the rod occuring at 28 microseconds and going through a reversal at 38 microseconds.

Conclusions

Numerical techniques exist for calculating the response of projectiles during the penetration and perforation of inclined targets of finite thickness for oblique impacts in the ordnance impact regime. The analysis of ballistic impact at various angles of obliquity and related phenomena can now be conducted through simulation techniques such as demonstrated by the REPSIL and the EPIC codes.

The computer codes such as those discussed here have advanced to the point where they can be used in conjunction with experimental procedures to advance the state-of-the-art in penetrator and armor design and effectiveness studies involving realistic target-projectile configurations at various angles of obliquity. For the specific problems considered here the following conclusions can be made:

a. The Lagrangian method used in EPIC is well suited for strain hardening and history dependent failure as occuring in hard target penetration involving failure of penetrator material at the rod-target interface.

b. Surface strains in the EPIC code are most strongly affected by penetrator erosion and only marginally by failure of target material.

c. REPSIL code appears to be well suited for relatively soft target penetration problem involving minimal projectile erosion and is more economical to run particularly when several projectile target configurations are involved.

d. Both codes predict results that are in good agreement during the elastic loading phase deteriorating to some extent during the unloading phase of the deformation due to severe plastic bending near the tip region.

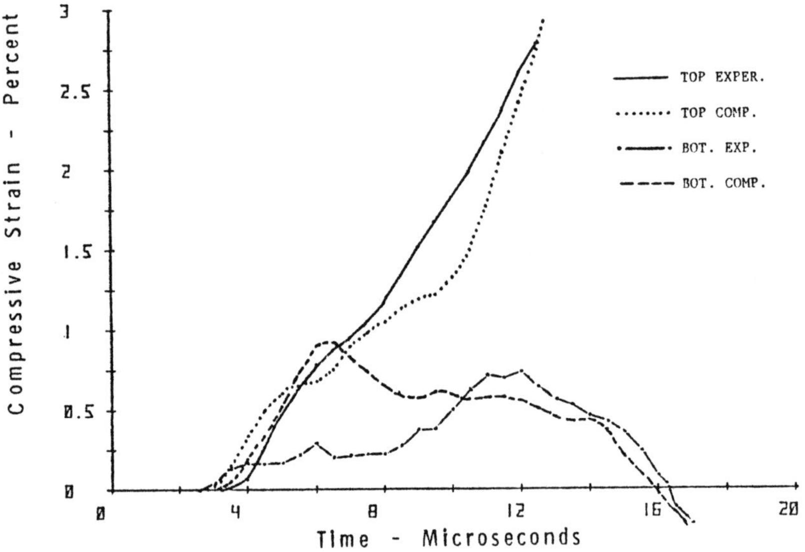

Figure 11. Strain measurements at the 20 mm position.

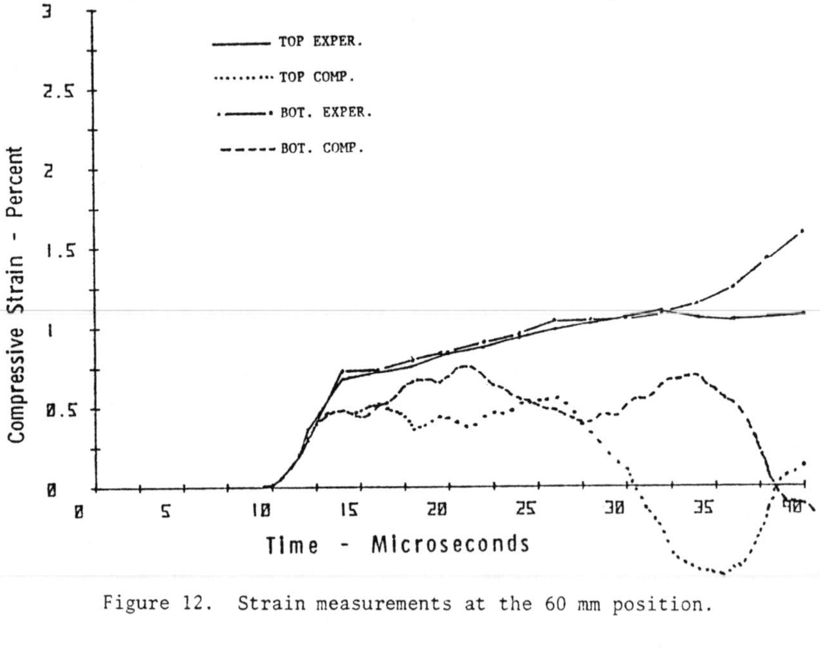

Figure 12. Strain measurements at the 60 mm position.

e. Although radial inertia and friction effects are ignored the REPSIL code is capable of rendering valuable insight into the effect of various interacting forces on the response of the rod during penetration, provided accurate estimates of the forcing functions are available. The inclusion of suitable failure criteria in the eroding rod model in the REPSIL could conceivably extend its capability to hard target penetration problems.

f. A complimentary use of a structural code, i.e. REPSIL, with a finite element Lagrangian code, i.e. EPIC, capable of accurate estimation of rod erosion and loading functions during the initial loading phase could be profitable. Such a technique could be extended to study the penetration problem of long rod penetrators involving improved armor and ballistic materials.

References

1. C. Zener and R. E. Peterson, "Mechanics of Armor Penetration," Watertown Arsenal Report No. 710/492, 1942.

2. R. Recht and T. W. Ipson, "Ballistic Perforation Dynamics," Journal of Applied Mechanics, Vol. 30, pp 384-390, 1963.

3. J. L. Summers, "Investigation of High Speed Impact; Regions of Impact at Oblique Angles," NASA TND-94, 1959.

4. W. Goldsmith and D. H. Cunningham, "Kinematic Phenomena Observed During the Oblique Impact of a Sphere on a Beam," Journal of Applied Mechanics, Vol. 23, pp 612-616, 1956.

5. J. Awerbuch and S. R. Bodner, "An Investigation of Oblique Perforation of Metallic Plates by Projectiles," Experimental Mechanics, pp 147-153, April 1977.

6. L. D. Bertholf, M. E. Kipp, and W. T. Brown, "Two-Dimensional Calculations for the Oblique Impact of Kinetic Energy Projectiles with a Multi-Layered Target," BRL-CR 333, March 1977.

7. G. E. Hauver, "Penetration with Instrumented Rods," International Journal of Engineering Science, Vol. 16, pp 871-877, 1978.

8. J. J. Misey, A. D. Gupta, and J. D. Wortman, "Simulation Methods for the Evaluation of Structural Response to High Strain Rates," in Emerging Technologies in Aerospace Structures, Structural Dynamics and Materials, J. R. Vison, ed, ASME 1980.

9. J. M. Santiago, H. L. Wisniewski, and N. J. Huffington, Jr., "A User's Manual for the REPSIL Code," BRL-R 1744, October 1974.

10. J. M. Santiago, "Formulation of the Large Deflection Shell Equations for Use in Finite Difference Structural Response Computer Codes," BRL-R 1571, February 1972.

11. G. R. Johnson, "Analysis of Elastic-Plastic Impact Involving Severe Distortions," Journal of Applied Mechanics, Vol. 98, No. 3, September 1976.

12. A. D. Gupta and J. D. Wortman, "An Eroding Long Rod Penetrator Model for Hard Target Penetration," Proceedings of the Third ASCE/EMD Specialty Conference, September 17-19, 1979, The University of Texas, Austin, Texas, pp 714-717.

13. A. D. Gupta, G. E. Hauver, and J. D. Wortman, "The Response of a Steel KE Penetrator During the Oblique Penetration of an Aluminum Target," Proceedings of the Fourth International Symposium on Ballistics, 17-19 October 1978, Monterey, California.

14. J. F. Bell, "Theoretical and Experimental Studies of Shock Waves in Solids," Progress Report, Contract DAAD05-76-C-0722, BRL, APG, MD, May 1977.

15. E. A. Murray and J. H. Suckling, "Quasi-Static Compression Stress-Strain Curves-1, Data Gathering and Reduction Procedures; Results for 1066 Steel," BRL-MR 2399, July 1974.

16. E. W. Bloore, private communication.

17. A. D. Gupta and J. D. Wortman, "Effect of Material Property Variation on the Response of KE Penetrators," presented at the 98th Winter Annual Meeting, ASME, Atlanta, Georgia, November 1977.

VIBRATION AND DYNAMICS

FORCED OSCILLATIONS OF SYSTEMS HAVING QUADRATIC NONLINEARITIES

Ali H. Nayfeh

University Distinguished Professor
Engineering Science and Mechanics Department
Virginia Polytechnic Institute and State University
Blacksburg, Virginia 24061
and
Yarmouk University, Irbid, Jordan

The method of multiple scales is used to determine an approximate solution to coupled oscillators having quadratic nonlinearities; such oscillators arise in the motions of pendulums, ships, beams, arches, composite plates, and shells. For a system with two degrees of freedom in which $\omega_2 \approx 2\omega_1$, where the ω_n are the linear natural frequencies, the analysis shows that a saturation phenomenon exists when the external excitation Ω is approximately equal to the frequency of the second mode ω_2. When the amplitude F of the external excitation is small, only the second mode is excited. As F reaches a critical value F_c, which depends on the damping coefficients and detunings of the two modes, the second mode saturates and the first mode begins to grow. As F increases further, all the additional input energy goes into the first mode due to the internal resonance (i.e., $\omega_2 \approx 2\omega_1$). When $\Omega \approx \omega_1$, the analysis shows that under some conditions there exists no steady-state solution in spite of the damping. In this case, the energy is continuously exchanged between these interacting modes without being attenuated.

1. Introduction

In analyzing the forced response of many elastic and dynamical systems such as elastic pendulums, ships, beams, arches, composite plates, and shells, one encounters systems of coupled, inhomogeneous ordinary-differential equations with quadratic nonlinearities. Steady-state solutions of such systems exhibit particularly complicated behavior when their linear undamped natural frequencies are commensurable [1]; that is, when these systems possess internal resonances. Mettler and Weidenhammer [2] and Sethna [3] used the method of averaging to analyze primary resonances of systems governed by equations with quadratic nonlinearities when one linear natural frequency is twice another. Nayfeh, Mook, and Marshall [4] and Mook, Marshall, and Nayfeh [5] used the method of multiple scales [6,7] to analyze a simple system of two coupled oscillators with quadratic nonlinearities as a model for the coupling of pitch and yaw motions of ships. They investigated both primary and secondary resonances. When $\omega_2 \cong 2\omega_1$ and $\Omega \cong \omega_2$, where the ω_n are the linear natural frequencies and Ω is the excitation frequency, they demonstrated the existence of a saturation phenomenon. Moreover, when $\omega_2 \cong 2\omega_1$ and $\Omega \cong \omega_1$, they showed that there are conditions for which stable periodic steady-state motions do not exist. Instead, there exist amplitude-modulated motions in which the energy is continuously exchanged between the two modes. Both phenomena are discussed later in this paper. Later, Yamamoto and Yasuda [8] observed amplitude modulated steady-state motions in the forced response of systems with quadratic and cubic nonlinearities when one frequency is twice the other. Nayfeh and Mook [1] used the method of multiple scales to analyze the response of a beam to a harmonic excitation. They accounted for the interaction of lateral and longitudinal vibrations. Nayfeh and Mook [9] demonstrated the existence of the saturation phenomenon for three-degree-of-freedom systems with quadratic nonlinearities when $\omega_3 \cong \omega_1 + \omega_2$ and $\Omega \cong \omega_3$. Mook, Plaut, and HaQuang [10] used the method of multiple scales [6,7] to analyze primary and secondary resonances for systems with quadratic and cubic nonlinearities and applied their results to arches.

We consider the forced response of a system with quadratic nonlinearities to a harmonic excitation. To simplify the algebra, we consider the problem of the forced response of ships that are constrained to pitch and roll only. Nayfeh, Mook, and Marshall [10] used an energy-perturbation approach to develop the following equations describing the forced response of a ship that is constrained to pitch and roll only:

$$I_{xx}\ddot{\phi} - I_{xz}\dot{\phi}\dot{\theta} = -V_1\phi - d_1\dot{\phi} - I_1\ddot{\phi} - V_2\phi\theta - I_7\ddot{\phi}\theta - I_2\theta\ddot{\phi} + (I_8 - I_2)\dot{\phi}\dot{\theta} + R(t) \tag{1}$$

$$I_{yy}\ddot{\theta} + I_{xz}\dot{\phi}^2 = -V_3\phi - d_2\dot{\theta} - I_3\ddot{\theta} - \frac{1}{2}V_2\phi^2 - \frac{3}{2}V_4\theta^2 - I_7\phi\ddot{\phi} - I_4\theta\ddot{\theta}$$
$$+ \frac{1}{2}(I_2 - 2I_7 - 2I_8)\dot{\phi}^2 - \frac{1}{2}I_4\dot{\theta}^2 + M(t) \tag{2}$$

where ϕ is the roll angle, θ is the pitch angle, the I's are the moments and added moments of inertia, the V's are the coefficients of the potential energy, the d's are the linear damping coefficients, and M and R are the pitching and rolling moments exerted by the water waves.

To make explicit the assumption that we are considering small motions, we introduce a small positive dimensionless parameter ε that is the order of the amplitudes and let

$$\phi = \varepsilon u_1, \quad \theta = \varepsilon u_2 \tag{3}$$

Moreover, we consider only single frequency harmonic excitations; that is,

$$M = F_2 \cos(\Omega t + \tau_2), \qquad R = F_1 \cos(\Omega t + \tau_1) \tag{4}$$

To the first approximation, any resonant excitations must be balanced by the nonlinear and damping terms. Using Eqs. (3) and (4) and the above assumptions, we rewrite Eqs. (1) and (2) as

$$\ddot{u}_1 + \omega_1^2 u_1 = -2\varepsilon\mu_1 \dot{u}_1 - \varepsilon[\Gamma_{11} u_1 u_2 + \Gamma_{12} u_1 \ddot{u}_2 + \Gamma_{13} \dot{u}_1 \dot{u}_2 + \Gamma_{14} \ddot{u}_1 u_2 - f_1 \cos(\Omega t + \tau_1)] \tag{5}$$

$$\ddot{u}_2 + \omega_2^2 u_2 = -2\varepsilon\mu_2 \dot{u}_2 - \varepsilon[\Gamma_{21} u_1^2 + \Gamma_{22} u_2^2 + \Gamma_{23} u_1 \ddot{u}_1 + \Gamma_{24} \dot{u}_1^2 + \Gamma_{25} \dot{u}_2^2 + \Gamma_{25} u_2 \ddot{u}_2 - f_2 \cos(\Omega t + \tau_2)] \tag{6}$$

where

$$\omega_1^2 = \frac{V_1}{I_{xx} + I_1}, \qquad \omega_2^2 = \frac{V_3}{I_{yy} + I_3} \tag{7a}$$

$$2\varepsilon\mu_1 = \frac{d_1}{I_{xx} + I_1}, \qquad 2\varepsilon\mu_2 = \frac{d_2}{I_{xx} + I_3} \tag{7b}$$

$$[\Gamma_{11}, \Gamma_{12}, \Gamma_{13}, \Gamma_{14}] = (I_{xx} + I_1)^{-1} [V_2, I_7, -(I_{xz} + I_8 - I_2), I_2] \tag{8}$$

$$[\Gamma_{21}, \Gamma_{22}, \Gamma_{23}, \Gamma_{24}, \Gamma_{25}, \Gamma_{26}] = (I_{yy} + I_3)^{-1} [\tfrac{1}{2} V_2, \tfrac{3}{2} V_4, I_7, -(I_{xz} + \tfrac{1}{2} I_2 - I_7 - I_8), \tfrac{1}{2} I_4, I_4] \tag{9}$$

$$\varepsilon^2 f_1 = \frac{F_1}{I_{xx} + I_1}, \qquad \varepsilon^2 f_2 = \frac{F_2}{I_{yy} + I_3} \tag{10}$$

2. Analysis

We use the method of multiple scales to analyze the solutions of Eqs. (5) and (6) for the case of primary resonance. We consider the important case of internal resonance in which $\omega_2 \simeq 2\omega_1$ and introduce the detuning parameter σ defined according to

$$\omega_2 = 2\omega_1 - \varepsilon\sigma_2 \tag{11}$$

According to the method of multiple scales [6,7], we seek a first-order uniform expansion in the form

$$u_n(t;\varepsilon) = u_{n_0}(T_0,T_1) + \varepsilon u_{n_1}(T_0,T_1) + \ldots \tag{12}$$

where n = 1 and 2 and

$$T_0 = t \quad \text{and} \quad T_1 = \varepsilon t \tag{13}$$

Then, the time derivative transforms according to

$$\frac{d}{dt} = D_0 + \varepsilon D_1 + \ldots \tag{14}$$

where $D_n = \partial/\partial T_n$. Substituting Eq. (12) into Eqs. (5) and (6), using Eq. (14), and equating coefficients of like powers of ε, we obtain

Order ε^0

$$D_0^2 u_{n_0} + \omega_n^2 u_{n_0} = 0, \quad n = 1 \text{ and } 2 \tag{15}$$

Order ε

$$\begin{aligned} D_0^2 u_{11} + \omega_1^2 u_{11} = {} & -2D_0 D_1 u_{10} - 2\mu_1 D_0 u_{10} - \Gamma_{11} u_{10} u_{20} - \Gamma_{12} u_{10} D_0^2 u_{20} \\ & - \Gamma_{13}(D_0 u_{10})(D_0 u_{20}) - \Gamma_{14} u_{20} D_0^2 u_{10} + f_1 \cos(\Omega T_0 + \tau_1) \end{aligned} \tag{16}$$

$$\begin{aligned} D_0^2 u_{21} + \omega_2^2 u_{21} = {} & -2D_0 D_1 u_{20} - 2\mu_2 D_0 u_{20} - \Gamma_{21} u_{10}^2 - \Gamma_{22} u_{20}^2 - \Gamma_{23} u_{10} D_0^2 u_{10} \\ & - \Gamma_{24}(D_0 u_{10})^2 - \Gamma_{25}(D_0 u_{20})^2 - \Gamma_{26} u_{20} D_0^2 u_{20} + f_2 \cos(\Omega T_0 + \tau_2) \end{aligned} \tag{17}$$

where $\cos(\Omega t + \tau_n)$ was expressed in terms of the fast scale T_0.

The solution of Eqs. (15) can be expressed as

$$u_{n_0} = a_n(T_1)\cos(\omega_n T_0 + \theta_n) \tag{18}$$

or as

$$u_{n_0} = A_n(T_1)e^{i\omega_n T_0} + \bar{A}_n(T_1)e^{-i\omega_n T_0} \tag{19}$$

where

$$A_n = \frac{1}{2} a_n e^{i\theta_n} \tag{20}$$

The A_n are unknown at this level of approximation; they are determined by imposing the solvability condition at the next level of approximation. Substituting Eq. (19) into Eqs. (16) and (17), we have

$$D_0^2 u_{11} + \omega_1^2 u_{11} = -2i\omega_1(A_1' + \mu_1 A_1)e^{i\omega_1 T_0} + \Lambda_1 A_2 \bar{A}_1 e^{i(\omega_2-\omega_1)T_0} + \frac{1}{2} f_1 e^{i(\Omega T_0+\tau_1)} + cc + NST \tag{21}$$

$$D_0^2 u_{21} + \omega_2^2 u_{21} = -2i\omega_2(A_2' + \mu_2 A_2) + \Lambda_2 A_1^2 e^{2i\omega_1 T_0} + \frac{1}{2} f_2 e^{i(\Omega T_0+\tau_2)} + cc + NST \tag{22}$$

where

$$\begin{aligned} \Lambda_1 &= \omega_1^2\Gamma_{14} + \omega_2^2\Gamma_{12} - \omega_1\omega_2\Gamma_{13} - \Gamma_{11} \\ \Lambda_2 &= \omega_1^2\Gamma_{23} + \omega_1^2\Gamma_{24} - \Gamma_{21} \end{aligned} \tag{23}$$

Here, cc stands for the complex conjugate of the preceding terms and NST stands for terms that do not produce secular terms. To eliminate the secular terms from the particular solutions of u_{11} and u_{21}, we need to distinguish between the case in which the lower mode is resonantly excited (i.e., $\Omega \tilde{=} \omega_1$) and the case in which the higher mode is resonantly excited (i.e., $\Omega \tilde{=} \omega_2$).

2.1. The Case of Ω Near ω_2

To express the nearness of Ω to ω_2, we introduce the detuning parameter σ_1 defined according to

$$\Omega = \omega_2 + \epsilon\sigma_1 \tag{24}$$

and write

$$\Omega T_0 = \omega_2 T_0 + \sigma_1 T_1 \tag{25}$$

Moreover, it follows from Eq. (11) that

$$\omega_2 T_0 = 2\omega_1 T_0 - \sigma_2 T_1 \tag{26}$$

Substituting Eqs. (25) and (26) into Eqs. (21) and (22) and eliminating the secular terms from the resulting equations, we obtain

$$2i\omega_1(A_1' + \mu_1 A_1) = \Lambda_1 A_2 \bar{A}_1 e^{-i\sigma_2 T_1} \tag{27}$$

$$2i\omega_2(A_2' + \mu_2 A_2) = \Lambda_2 A_1^2 e^{i\sigma_2 T_1} + \frac{1}{2} f_2 e^{i\sigma_1 T_1} \tag{28}$$

Substituting Eq. (20) into Eqs. (27) and (28) and separating the result into real and imaginary parts, we obtain

$$a_1' = -\mu_1 a_1 + \frac{1}{4}\omega_1^{-1}\Lambda_1 a_1 a_2 \sin\gamma_2 \tag{29}$$

$$a_2' = -\mu_2 a_2 - \frac{1}{4}\omega_2^{-1}\Lambda_2 a_1^2 \sin\gamma_2 + \frac{1}{2}\omega_2^{-1} f_2 \sin\gamma_1 \tag{30}$$

$$a_1\theta_1' = -\frac{1}{4}\omega_1^{-1}\Lambda_1 a_1 a_2 \cos\gamma_2 \tag{31}$$

$$a_2\theta_2' = -\frac{1}{4}\omega_2^{-1}\Lambda_2 a_1^2 \cos\gamma_2 - \frac{1}{2}\omega_2^{-1} f_2 \cos\gamma_1 \tag{32}$$

where

$$\gamma_1 = \sigma_1 T_1 - \beta_2 + \tau_2 \tag{33}$$

$$\gamma_2 = \beta_2 - 2\beta_1 - \sigma_2 T_1 \tag{34}$$

We note that under the change of variables

$$a_1 = \frac{\hat{a}_1}{\sqrt{\Lambda_1\Lambda_2}}, \qquad a_2 = \frac{\hat{a}_2}{\Lambda_1}, \qquad f_2 = \frac{\hat{f}_2}{\Lambda_1}$$

Eqs. (29)-(32) transform into Eqs. (6.5.17)-(6.5.20) of Ref. [1], where a detailed discussion of their solutions could be found. In this paper, we summarize some of the results and conclusions.

For the steady-state response, $a_n' = 0$ and $\gamma_n' = 0$, and there are two possibilities. The first is essentially the solution of the linear problem; that is,

$$a_1 = 0 \quad \text{and} \quad a_2 = \frac{f_2}{2\omega_2\sqrt{\sigma_2^2 + \mu_2^2}} \tag{35}$$

The second is

$$\hat{a}_2 = \hat{a}_2^* = 2\omega_1[4\mu_1^2 + (\sigma_1 - \sigma_2)^2]^{1/2} \tag{36}$$

$$\hat{a}_1 = 2\Big\{-2\omega_1\omega_2[\sigma_1(\sigma_2 - \sigma_1) + 2\mu_1\mu_2] \mp \left[\frac{1}{4}\hat{f}_2^2 - 4\omega_1^2\omega_2^2\Big(2\sigma_1\mu_1 - \mu_2(\sigma_2 - \sigma_1)\Big)^2\right]^{1/2}\Big\}^{1/2} \tag{37}$$

We note a very interesting feature of the second solution, namely, the amplitude $\hat{a}_2$ of the directly excited mode is independent of the excitation amplitude $\hat{f}_2$, whereas, the amplitude $\hat{a}_1$ of the mode that is not directly excited is a function of the excitation amplitude. Thus, the second solution is a result of the nonlinearity. Figures 1 and 2 show the two possible steady-state solutions for a given set of parameters [1]. To determine which of these solutions is physically realizable, one usually performs a linear stability analysis by examining the variational equations. The unstable solutions are represented by the broken lines. When more than one stable solution exists as in Figure 1, the initial conditions determine which of them gives the response. The analytical results were verified in Ref. [4] by numerically integrating the original governing equations.

Figure 1 exhibits a saturation phenomenon. As f_2 increases from zero, a_2 increases linearly with f_2 until it reaches the critical value a_2^* in accordance with the linear solution, whereas $a_1 = 0$. At this point, a_2 has its maximum value, and further increases in f_2 do not affect a_2 because the linear solution is unstable. In other words, the second mode is saturated. In contrast, the amplitude of the first mode increases with f_2.

2.2 The Case of Ω Near ω_1

To express the nearness of Ω to ω_1, we introduce a detuning parameter σ_1 defined by

$$\Omega = \omega_1 + \varepsilon\sigma_1 \tag{38}$$

and write

$$\Omega T_0 = \omega_1 T_0 + \sigma_1 T_1 \tag{39}$$

Substituting Eqs. (26) and (39) into Eqs. (21) and (22) and eliminating the secular terms from the resulting equations, we obtain

$$2i\omega_1(A_1' + \mu_1 A_1) = \Lambda_1 A_2 \bar{A}_1 e^{-i\sigma_1 T_1} + \frac{1}{2} f_1 e^{i(\sigma_1 T_1 + \tau_1)} \tag{40}$$

$$2i\omega_2(A_2' + \mu_2 A_2) = \Lambda_2 A_1^2 e^{i\sigma_2 T_1} \tag{41}$$

Substituting the polar form (20) into Eqs. (40) and (41) and separating real and imaginary parts, we have

$$a_1' = -\mu_1 a_1 + \frac{1}{4}\,\omega_1^{-1}\Lambda_1 a_1 a_2 \sin\gamma_2 + \frac{1}{2}\,\omega_1^{-1} f_1 \sin\gamma_1 \tag{42}$$

$$a_2' = -\mu_2 a_2 - \frac{1}{4}\,\omega_2^{-1}\Lambda_2 a_1^2 \sin\gamma_2 \tag{43}$$

$$a_1\theta_1' = -\frac{1}{4}\,\omega_1^{-1}\Lambda_1 a_1 a_2 \cos\gamma_2 - \frac{1}{2}\,\omega_1^{-1} f_1 \cos\gamma_1 \tag{44}$$

$$a_2\theta_2' = -\frac{1}{4}\,\omega_2^{-1}\Lambda_2 a_1^2 \cos\gamma_2 \tag{45}$$

where γ_2 is defined by Eq. (34) and

$$\gamma_1 = \sigma_1 T_1 - \beta_1 + \tau_1 \tag{46}$$

Under the change of variables

$$a_1 = \frac{\hat{a}_1}{\sqrt{\Lambda_1\Lambda_2}}, \qquad a_2 = \frac{\hat{a}_2}{\Lambda_1}, \qquad f_1 = \frac{\hat{f}_1}{\sqrt{\Lambda_1\Lambda_2}} \tag{47}$$

Eqs. (42)-(45) transform into Eqs. (6.5.34)-(6.5.37) of Ref. [1], where a detailed discussion of their solutions can be found. A summary is presented here.

For a steady-state solution, $a_n' = 0$ and $\gamma_n' = 0$. Then, Eqs. (42)-(46) and (34) can be manipulated to give [1]

$$\hat{a}_2^3 + 8\omega_1(\sigma_1 \cos\gamma_2 - \mu_1 \sin\gamma_2)\hat{a}_2^2 + 16\omega_1^2(\mu_1^2 + \sigma_1^2)\hat{a}_2 - \Gamma\omega_2^{-1}\hat{f}_1^2 = 0 \tag{48}$$

$$\hat{a}_1 = 2\sqrt{\frac{\omega_2\hat{a}_2}{\Gamma}} \tag{49}$$

$$\sin\gamma_2 = -\mu_2\Gamma, \qquad \cos\gamma_2 = -(\sigma_2 + 2\sigma_1)\Gamma \tag{50}$$

$$\Gamma = [\mu_2^2 + (\sigma_2 + 2\sigma_1)^2]^{-1/2} \tag{51}$$

Figure 2 shows the variation of $\hat{a}_1$ and $\hat{a}_2$ with σ_1 for a given set of parameters. Again, the unstable solutions are represented by broken lines. In addition to the jump phenomenon, Figure 2 shows that there is a range of σ_1 near 0 for which there is no stable steady-state solution. Integration of the original equations in this range for long time yields records such as those in Figure 3. It clearly shows that the response consists of amplitude modulated oscillations with the energy being continuously exchanged between the two modes. Recently, Sethna and Bajaj [12] used a theorem by Hopf to show that the amplitude of modulation is determined by the motion of a point on a torus.

This work was supported by the Office of Naval Research under Contract No. N00014-79-C-0103.

References

1. Nayfeh, A. H. and Mook, D. T., Nonlinear Oscillations, Wiley-Interscience, 1979.

2. Mettler, E. and Weidenhammer, F., "Zum Problem des kinetischen Durschschlangens Schwack gekrümmter Stäbe", Ing.-Arch., Vol. 31, 1962, pp. 421-432.

3. Sethna, P. R., "Vibrations of Dynamical Systems with Quadratic Nonlinearities", J. Appl. Mech., Vol. 32, 1965, pp. 576-582.

4. Nayfeh, A. H., Mook, D. T., and Marshall, L. R., "Nonlinear Coupling of Pitch and Roll Modes in Ship Motions", J. Hydro., Vol. 7, 1973, pp. 145-152.

5. Mook, D. T., Marshall, L. R., and Nayfeh, A. H., "Subharmonic and Superharmonic Resonances in the Pitch and Roll Modes of Ship Motions", J. Hydro., Vol. 8, 1974, pp. 32-40.

6. Nayfeh, A. H., Perturbation Methods, Wiley-Interscience, 1973.

7. Nayfeh, A. H., Introduction to Perturbation Techniques, Wiley-Interscience, 1981.

8. Yamamoto, T. and Yasuda, K., "On the Internal Resonance in a Nonlinear Two-Degree-of-Freedom System", Bull. JSME, Vol. 20, 1977, pp. 168-175.

9. Nayfeh, A. H. and Mook, D. T., "A Saturation Phenomenon in the Forced Response of Systems with Quadratic Nonlinearities", Proceedings of the VIIIth International Conference on Nonlinear Oscillations, Prague, 1978, pp. 511-516.

10. Nayfeh, A. H., Mook, D. T., and Marshall, L. R., "Perturbation-Energy Approach for the Development of the Nonlinear Equations of Ship Motion", J. Hydro., Vol. 8, 1974, pp. 130-136.

11. Mook, D. T., Plaut, R. H., and HaQuang, N., "The Response of Multidegree-of-Freedom Systems with Quadratic and Cubic Nonlinearities to Harmonic Excitation, with Application to a Shallow Arch", Submitted for publication, SIAM Review.

12. Sethna, P. R. and Bajaj, A. K., "Bifurcations in Dynamical Systems with Internal Resonance", J. Appl. Mech., Vol. 45, 1978, pp. 895-902.

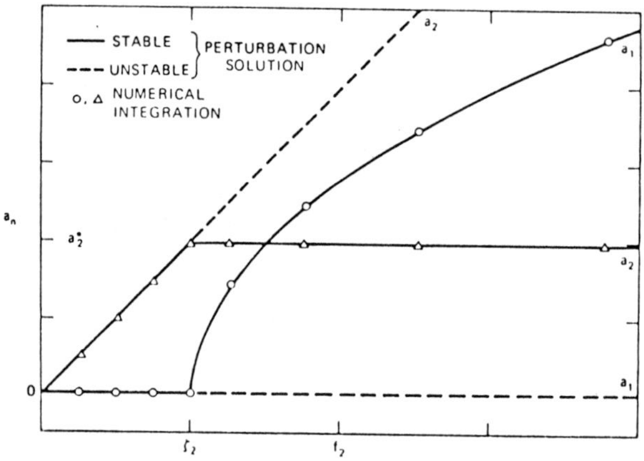

Figure 1. Amplitudes of the response as functions of the amplitude of the excitation; $\sigma_1 = \sigma_2 = 0$; $\Omega \simeq \omega_2$.

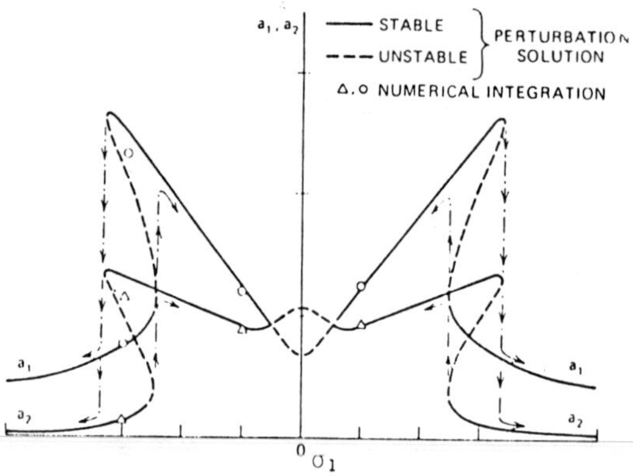

Figure 2. Frequency-response curves $\sigma_2 = 0$, $\Omega \simeq \omega_1$.

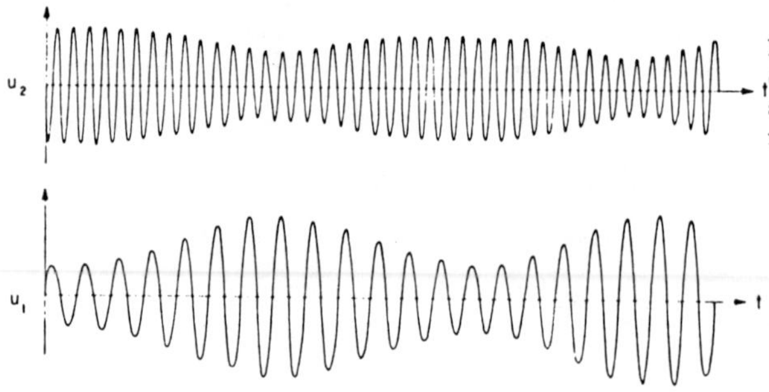

Figure 3. Nonexistence of periodic motions in a system with quadratic nonlinearities.

RESPONSE OF A TWO DEGREE OF FREEDOM NON-LINEAR SYSTEM
BY THE EQUIVALENT LINEARIZATION METHOD

Joseph T. Kayani

Research Fellow, Polytechnic Institute of N.Y.

John R. Curreri

Professor of M.E., Polytechnic Institute of N.Y.

Abstract: The system response is developed for a two degree of freedom dynamic system with hardening type non-linearity in the restoring elements and with white excitation having a normal distribution acting on the primary mass. The method involved the evaluation of the equivalent linear stiffness by minimizing the mean-squared error. The joint probability density of the displacements required in the evaluation is assumed to be that of a linear system. The joint moment integrals were evaluated by using Price's theorem. The mean square responses were obtained by the transfer function procedure. Numerical results were evaluated for certain system parameters. It is found that attaching a secondary system having a hardening spring to an existing primary system can result is significant reduction in primary mass response for low values of the uncoupled frequency ratio. However, there are frequency ratio ranges for which the primary mass response for the non-linear system is greater than for the linear system.

NOMENCLATURE

m_1, m_2	Primary and Secondary masses
μ	Mass ratio (m_2/m_1)
ξ_1, ξ_2	Viscous damping factors
k_1, k_2	Spring constants
x,y	System displacement variables
z	(y-x)
$\dot{x}$, $\dot{y}$	Velocity variables

$F_1(x)$, $F_2(z)$	Restoring spring forces
$F(t)$	Force excitation
$f(t)$	Force excitation per unit mass
ε_1^2, ε_2^2	Non-linearity factors
ω_1, ω_2	Uncoupled natural frequencies
α	Uncoupled natural frequency ratio (ω_2/ω_1)
So	Constant value of ensemble spectral density
$E\{x^2\}$, $E\{y^2\}$	Mean square values
$E\{\ \}$	Expected value of quantities in parenthesis
σ_1^2, σ_2^2	Constants for a system as defined in eq. (10)
r	Coefficient of correlation as defined in eq. (10)
$\omega_1{}^2{}_{eq}$, $\omega_2{}^2{}_{eq}$	Parameters
$p(x,y)$	Jointly normal density distribution

INTRODUCTION

Multi-degree of freedom non-linear systems are treated in a general way using statistical linearization techniques by Caughey [1], Foster, Jr. [2], Iwan and Yang [3]. However, few numerical results are given and no specific problem is solved.

Using the equivalent linearization method, the system response is developed for a two degree of freedom dynamic system with a hardening type non-linearity in the restoring elements and with white excitation having a normal distribution acting on the primary mass. The method involved the evaluation of the equivalent linear stiffness for the spring elements. This was achieved by minimizing the mean squared error term with respect to the equivalent stiffness. This required the evaluation of some joint moments whose distribution was not available. The joint probability density of the displacements needed in the evaluation of joint moments is assumed to be that of a linear system [4,5]. The required moments were completely evaluated using Price's theorem in probability [5]. Having obtained the equivalent linear stiffness, the system mean square response were evaluated by the transfer function procedure. Numerical results were evaluated for certain system parameters.

SYSTEM DESCRIPTION

A two degree of freedom mechanical model is considered, as shown in Fig. 1. It consists of two masses which are coupled by hardening type restoring elements. The mass m_1 is called the primary mass and m_2 the secondary mass. The primary mass is attached to a rigid foundation through a hardening type spring and a dashpot. The system excitation is assumed to be white noise having normal distribution and acting on the primary mass. The springs and dampers are considered as massless.

SOLUTION METHOD

Let x(t) and y(t) be the two independent coordinates corresponding to the primary and secondary masses to describe the complete configuration. The non-linear restoring spring elements are considered to be cubic hardening of the form:

$$F_1\,(x) = k_1\,(x + \varepsilon_1^2\,x^3)$$

$$F_2\,(z) = k_2\,(z + \varepsilon_2^2\,z^3) \tag{1}$$

Using Newton's second law, the governing equations of motion for the masses are as follows:

$$m_1\ddot{x} + c_1\dot{x} - c_2\dot{z} + F_1(x) - F_2(z) = F(t)$$

$$m_2\ddot{y} + c_2\dot{z} + F_2(z) = 0 \tag{2}$$

Dividing the first of eq. (2) by m_1, the second by m_2 and defining the following quatities

$$\frac{k_1}{m_1} = \omega_1^2 \qquad \frac{k_2}{m_2} = \omega_2^2$$

$$\xi_1 = \frac{C_1}{C_{C_1}} \qquad \xi_2 = \frac{C_2}{C_{C_2}} \qquad C_{C_1} = 2\sqrt{k_1 m_1}$$

$$C_{C_2} = 2\sqrt{k_2 m_2} \qquad \frac{C_1}{m_1} = 2\xi_1\omega_1 \qquad \frac{C_2}{m_2} = 2\xi_2\omega_2$$

$$\frac{m_2}{m_1} = \mu \qquad \frac{k_2}{m_1} = \mu\,\omega_2^2 \qquad \frac{F(t)}{m_1} = f(t) \tag{3}$$

the equations of motion may be rewritten as

$$\ddot{x} + 2\xi_1\omega_1\dot{x} - 2\mu\xi_2\omega_2(\dot{y}-\dot{x}) + \omega_1^2[x + \varepsilon_1^2x^3]$$

$$- \mu\omega_2^2[(y-x) + \varepsilon_2^2(y-x)^3] = f(t)$$

$$\ddot{y} + 2\xi_2\omega_2(\dot{y} - \dot{x}) + \omega_2^2[(y-x) + \varepsilon_2^2(y-x)^3] = 0 \quad (4)$$

The system is assumed to be lightly damped and weakly non-linear. Equation (4) may be rewritten as follows:

$$\ddot{x} + 2\xi_1\omega_1\dot{x} - 2\mu\xi_2\omega_2(\dot{y}-\dot{x}) + {\omega_1}^2_{eq}x - \mu{\omega_2}^2_{eq}(y-x) + e_1(x,y) = f(t)$$

$$\ddot{y} + 2\xi_2\omega_2(\dot{y}-\dot{x}) + {\omega_2}^2_{eq}(y-x) + e_2(x,y) = 0 \quad (5)$$

where

$$e_1(x,y) = (\omega_1^2 - {\omega_1}^2_{eq})x - (\mu\omega_2^2 - \mu{\omega_2}^2_{eq})(y-x)$$

$$+ \omega_1\varepsilon_1^2x^3 - \mu\omega_2^2\varepsilon_2^2(y-x)^3$$

$$e_2(x,y) = (\omega_2^2 - {\omega_2}^2_{eq})(y-x) + \omega_2^2\varepsilon_2^2(y-x)^3 \quad (6)$$

The parameters ${\omega_1}^2_{eq}$ and ${\omega_2}^2_{eq}$ are initially unknown but will subsequently be chosen so as to optimize the equivalent linearization. If the non-linearities were zero, then ${\omega_1}^2_{eq} = \omega_1^2$ and ${\omega_2}^2_{eq} = \omega_2^2$ and hence $e_1 = e_2 = 0$. Then eq. (5) reduces to a set of complex equations which can be readily solved. When the non-linearities are present, it is not possible to select the parameters $\omega^2_{1_{eq}}$ and $\omega^2_{2_{eq}}$ so that error terms are identically zero. However, the logical choice of $\omega^2_{1_{eq}}$ and $\omega^2_{2_{eq}}$ are those values that make e_1 and e_2 a minimum. The choice of the minimization procedures is somewhat arbitrary. A simple statistical measure of the magnitude of the error is its mean square. Therefore, a natural optimization is to choose $\omega^2_{1_{eq}}$ and $\omega^2_{2_{eq}}$ so as to minimize the mean squared error. This requires that

$$\frac{\partial E\{e_1^2\}}{\partial {\omega_1}^2_{eq}} = 0 \quad , \quad \frac{\partial E\{e_2^2\}}{\partial {\omega_2}^2_{eq}} = 0 \quad (7)$$

Taking the expectation on both sides and setting $\frac{\partial E\{e_i^2\}}{\partial \omega_{i\,eq}^2} = 0$, we have

$$0 = -2\,(\omega_1^2 - \omega_{1\,eq}^2)\,E\{x^2\} - 2\,\mu\,(-\omega_2^2 + \omega_{2\,eq}^2)\,E\{xy-x^2\}$$

$$-2\,\omega_1^2\,\varepsilon_1^2\,E\{x^4\} + 2\,\mu\,\omega_2^2\,\varepsilon_2^2\,E\{x(y-x)^3\}$$

which gives

$$\omega_{1\,eq}^2 = \omega_1^2 + \mu\,(\omega_2^2 - \omega_{2\,eq}^2)\,[1 - \frac{E\{xy\}}{E\{x^2\}}]$$

$$+ \omega_1^2\varepsilon_1^2\,\frac{E\{x^4\}}{E\{x^2\}} - \mu\,\omega_2^2\varepsilon_2^2\,\frac{E\{x(y-x)^3\}}{E\{x^2\}} \qquad (8)$$

$$\omega_{2\,eq}^2 = \omega_2^2 + \frac{\omega_2^2\varepsilon_2^2\,E\{y^4-4y^3x+6y^2x^2-4yx^3+x^4\}}{E\{y^2-2xy+x^2\}} \qquad (9)$$

The various expectations required in eq. (8) and (9) must now be evaluated but it is not known exactly how the joint probability density of the displacements is distributed. Now assume that x and y have a joint distribution as in a linear system in which the following Gaussian distribution may be used [4,5].

$$p(x,y) = \frac{1}{2\pi\sigma_1\sigma_2\sqrt{1-r^2}}\;e^{-\frac{1}{2(1-r^2)}\left[\frac{x^2}{\sigma_1^2} - \frac{2rxy}{\sigma_1\sigma_2} + \frac{y^2}{\sigma_2^2}\right]} \qquad (10)$$

where

$$\sigma_1^2 = \frac{\pi\,So\omega_1}{2\xi_1} \qquad \sigma_2^2 = \frac{\sigma_1^2}{r^2} \qquad r = \sqrt{\frac{\mu\,\alpha^2}{1+\mu\,\alpha^2}}$$

It should be noted here that the unit of S_o is assumed as in^2/rad. per sec. If S_o is given in terms of g^2/cycle per sec., then σ_1^2 should be defined as $\sigma_1^2 = \frac{\pi So}{2\xi_1\omega_1^3}$.

Using the probability distribution given by eq. (10) the various expectations required in equations (8) and (9) can be developed. It is found that the use of Price's theorem in Probability [4,5] simplified the evaluation of the different expectations in equations (8) and (9).

Making use of eq. (10), eq. (8) is evaluated as

$$\omega_{1\,eq}^{2} = \omega_1^2 + 3\,\omega_1^2\,\varepsilon_1^2\,\sigma_1^2 \tag{11}$$

Similarly

$$\omega_{2\,eq}^{2} = \omega_2^2 + 3\,\frac{\omega_2^2\,\varepsilon_2^2\,\sigma_1^2}{\mu\,\alpha^2} \tag{12}$$

Knowing $\omega_{1\,eq}^{2}$ and $\omega_{2\,eq}^{2}$ the mean square responses can be found from the transfer function as in a linear system.

From linear systems, it can be shown that

$$H_X(\omega) = \frac{X}{X_s} = \frac{\omega_1^2}{\Delta}\,(\omega_{2\,eq}^{2} - \omega^2 + 2i\,\xi_2\omega_2\omega)$$

$$H_Y(\omega) = \frac{Y}{X_s} = \frac{\omega_1^2}{\Delta}\,(2\,i\,\xi_2\omega_2\omega + \omega_{2\,eq}^{2}) \tag{13}$$

and after simplification

$$\begin{aligned} \Delta = \omega^4 &- i\,\omega^3\,[2\xi_2\omega_2(1+\mu) + 2\xi_1\omega_1] \\ &- \omega^2\,[\omega_{1\,eq}^{2} + \omega_{2\,eq}^{2}\,(1+\mu) + 4\,\xi_1\xi_2\omega_2\omega_1] \\ &+ i\omega\,[2\xi\,\omega\,\omega^{2} + 2\,\xi\,\omega^{2}\,\omega] + \omega^{2}\,\omega_{2\,eq}^{2} \end{aligned} \tag{14}$$

The mean square responses will be evaluated with the standard form in reference [6].

If $H(\omega) = \dfrac{-i\omega^3 B_3 - \omega^2 B_2 + i\omega B_1 + B_0}{\omega^4 A_4 - i\omega^3 A_3 - \omega^2 A_2 + i\omega A_1 + A_o}$, then for this case

$B_3 = 0$ $\quad$ $B_2 = \omega_1^2$ $\quad$ $B_1 = 2\xi_2\omega_2\omega_1^2$

$B_o = \omega_1^2\omega_{2\,eq}^{2}$ $\quad$ $A_4 = 1$ $\quad$ $A_3 = [2\xi_2\omega_2(1+\mu) + 2\,\xi_1\omega_1]$

$A_2 = [\omega_{1\,eq}^{2} + \omega_{2\,eq}^{2}\,(1+\mu) + 4\,\xi_1\xi_2\,\omega_2\omega_1]$

$A_1 = [2\xi_1\omega_1\omega_{2\,eq}^{2} + 2\xi_2\,\omega_{1\,eq}^{2}\,\omega_2]$ $\quad$ $A_o = \omega_{1\,eq}^{2}\,\omega_{2\,eq}^{2}$

Therefore the mean square response of the primary mass is obtained as

$$\frac{E\{x^2\}}{\pi S_o \omega_1^4} = \frac{\begin{aligned}&(2\omega_1\omega_{2\,eq}^{2})\ [\ \frac{\xi_2\omega_2\omega_{2\,eq}^{2}}{\omega_1\omega_{1\,eq}^{2}}(1+\mu)^2 + \frac{\omega_{2\,eq}^{2}}{\omega_{1\,eq}^{2}}\xi_1\mu + 4\xi_1\xi_2\frac{\omega_2^2}{\omega_{1\,eq}^{2}}(1+\mu) \\ &+ 4\xi_1^2\xi_2\frac{\omega_1\omega_2}{\omega_{1\,eq}^{2}} + 4\xi_2^3\frac{\omega_2^3}{\omega_1\omega_{2\,eq}^{2}}(1+\mu) - 2\xi_2\frac{\omega_2}{\omega_1} \\ &-\xi_2\frac{\omega_2}{\omega_1}\mu + 4\xi_1\xi_2^2\frac{\omega_2^2}{\omega_{2\,eq}^{2}} + \xi_2\frac{\omega_2}{\omega_1}\frac{\omega_{1\,eq}^{2}}{\omega_{2\,eq}^{2}}\ ]\end{aligned}}{\begin{aligned}&(4\xi_1\omega_1^3\omega_2\omega_{2\,eq}^{2})\ [\xi_2\alpha^2\frac{\omega_{2\,eq}^{2}}{\omega_2^2}(1+\mu)^2 + \xi_1\alpha\frac{\omega_{2\,eq}^{2}}{\omega_2^2}\mu \\ &+ 4\xi_1\xi_2^2\alpha(1+\mu) + 4\xi_1^2\xi_2 + \frac{\xi_2^2}{\xi_1}\alpha\frac{\omega_{1\,eq}^{2}}{\omega_{2\,eq}^{2}}\frac{\omega_{1\,eq}^{2}}{\omega_1^2}\mu \\ &+ \xi_2\frac{\omega_{1\,eq}^{2}}{\omega_1^2}\frac{\omega_{1\,eq}^{2}}{\omega_{2\,eq}^{2}} + 4\xi_2^3\alpha^2\frac{\omega_{1\,eq}^{2}}{\omega_{2\,eq}^{2}}(1+\mu) \\ &+ 4\xi_1\xi_2^2\alpha\frac{\omega_{1\,eq}^{2}}{\omega_{2\,eq}^{2}} - 2\xi_2\frac{\omega_{1\,eq}^{2}}{\omega_1^2}\ ]\end{aligned}} \tag{15}$$

In a similar manner, the mean square displacement of the secondary mass is obtained as

$$\frac{E\{Y^2\}}{\pi S_o \omega_1^4} = \frac{\begin{aligned}&(2\,\omega_{2\,eq}^{2}\,\omega_2)\ [\mu\xi_2 + \xi_1\frac{\omega_1}{\omega_2} + \xi_2(1+\mu)^2\frac{\omega_{2\,eq}^{2}}{\omega_{1\,eq}^{2}} \\ &+ \xi_1\mu\frac{\omega_1}{\omega_2}\frac{\omega_{2\,eq}^{2}}{\omega_{1\,eq}^{2}} + 4\,\xi_1\xi_2^2\frac{\omega_2}{\omega_1}\frac{\omega_1^2}{\omega_{1\,eq}^{2}}(1+\mu) + 4\,\xi_1^2\xi_2\frac{\omega_1^2}{\omega_{1\,eq}^{2}}\ ]\end{aligned}}{\begin{aligned}&(4\,\xi_1\omega_1^3\,\omega_2\omega_{2\,eq}^{2})[(1+\mu)^2\xi_2\frac{\omega_2^2}{\omega_1^2}\frac{\omega_{2\,eq}^{2}}{\omega_1^2} + \mu\xi_1\frac{\omega_2}{\omega_1}\frac{\omega_{2\,eq}^{2}}{\omega_2^2} \\ &+ 4\,(1+\mu)\frac{\omega_2}{\omega_1}\xi_1\xi_2^2 + 4\xi_1^2\xi_2 + \mu\frac{\xi_2^2}{\xi_1}\frac{\omega_2}{\omega_1}\frac{\omega_{1\,eq}^{2}}{\omega_1^2}\frac{\omega_{1\,eq}^{2}}{\omega_{2\,eq}^{2}}\end{aligned}}$$

$$+ \xi_2 \frac{\omega_1{}^2_{eq}}{\omega_1^2} \frac{\omega_1{}^2_{eq}}{\omega_2{}^2_{eq}} + 4\xi_2^3 (1+\mu) \frac{\omega_2^2}{\omega_1^2} \frac{\omega_1{}^2_{eq}}{\omega_2{}^2_{eq}}$$

$$+ 4\xi_1\xi_2^3 \frac{\omega_2}{\omega_1} \frac{\omega_1{}^2_{eq}}{\omega_2{}^2_{eq}} - 2\xi_2 \frac{\omega_1{}^2_{eq}}{\omega_1^2}] \tag{16}$$

Since $\alpha = \frac{\omega_2}{\omega_1}$, from equations (11) and (12)

$$\frac{\omega_2{}^2_{eq}}{\omega_1{}^2_{eq}} = \alpha^2 \left[\frac{1 + 3 \frac{\varepsilon_2^2\sigma_1^2}{\mu\alpha^2}}{1 + 3 \varepsilon_1^2 \sigma_1^2} \right] \tag{17}$$

Substitution of (17) in (15) and (16) leads to the following expressions for the mean square responses.

$$\frac{E\{x^2\}}{\sigma_1^2} = \frac{\frac{1}{\alpha} [\xi_2 (1+\mu)^2\alpha^3 \left(\frac{1 + 3 \frac{\varepsilon_2^2\sigma_1^2}{\mu\alpha^2}}{1 + 3 \varepsilon_1^2\sigma_1^2}\right) + \xi_1 \mu\alpha^2 \frac{(1 + 3 \frac{\varepsilon_2^2\sigma_1^2}{\mu\alpha^2})}{(1 + 3 \varepsilon_1^2\sigma_1^2)} + 4 \xi_1\xi_2^2 (1+\mu) \alpha^2 \left(\frac{1}{1 + 3 \varepsilon_1^2\sigma_1^2}\right) + 4 \xi_1^2\xi_2\alpha \left(\frac{1}{1 + 3 \varepsilon_1^2\sigma_1^2}\right) + 4\xi_2^3 (1+\mu) \alpha \left(\frac{1}{1 + 3 \frac{\varepsilon_2^2\sigma_1^2}{\mu \alpha^2}}\right) - 2 \xi_2\alpha - \xi_2 \mu \alpha + 4 \xi_1\xi_2^2 \left(\frac{1}{1 + 3 \frac{\varepsilon_2^2\sigma_1^2}{\mu \alpha^2}}\right) + \frac{\xi_2}{\alpha} \frac{(1 + 3 \varepsilon_1^2\sigma_1^2)}{(1 + 3 \frac{\varepsilon_2^2\sigma_1^2}{\mu \alpha^2})}]}{[\xi_2(1+\mu)^2 \alpha^2(1 + 3 \frac{\varepsilon_2^2\sigma_1^2}{\mu \alpha^2}) + \xi_1 \mu \alpha (1 + 3 \frac{\varepsilon_2^2\sigma_1^2}{\mu \alpha^2}) + 4 \xi_1\xi_2^2 (1+\mu) \alpha + 4 \xi_1^2\xi_2 + \frac{\mu}{\alpha} \frac{\xi_2^2}{\xi_1} \frac{(1 + 3 \varepsilon_1^2\sigma_1^2)^2}{(1 + 3 \frac{\varepsilon_2^2\sigma_1^2}{\mu \alpha^2})} \ldots} \tag{18}$$

$$+ \frac{\xi_2}{\alpha^2} \frac{(1 + 3\,\varepsilon_1^2\sigma_1^2)^2}{(1 + 3\,\frac{\varepsilon_2^2\sigma_1^2}{\mu\,\alpha^2})} + 4\,\xi_2^3\,(1+\mu)\,\frac{(1 + 3\,\varepsilon_1^2\sigma_1^2)}{(1 + 3\,\frac{\varepsilon_2^2\sigma_1^2}{\mu\,\alpha^2})}$$

$$+ \frac{4\,\xi_1\xi_2^3}{\alpha} \frac{(1 + 3\,\varepsilon_1^2\sigma_1^2)}{(1 + 3\,\frac{\varepsilon_2^2\sigma_1^2}{\mu\,\alpha^2})} - 2\,\xi_2\,(1 + 3\,\varepsilon_1^2\sigma_1^2)]$$

Similarly

$$\frac{E\{Y^2\}}{\sigma_1^2} = \frac{\begin{array}{l}[\mu\xi_2 + \frac{\xi_1}{\alpha} + \xi_2(1+\mu)^2\alpha^2\,\frac{(1+3\,\frac{\varepsilon_2^2\sigma_1^2}{\mu\alpha^2})}{(1+3\,\varepsilon_1^2\sigma_1^2)} \\ + \xi_1\mu\alpha\,\frac{(1+3\,\frac{\varepsilon_2^2\sigma_1^2}{\mu\alpha^2})}{(1+3\,\varepsilon_1^2\sigma_1^2)} + \frac{4\xi_1\xi_2^2(1+\mu)\alpha}{(1+3\,\varepsilon_1^2\sigma_1^2)} + \frac{4\xi_1^2\xi_2}{(1+3\,\varepsilon_1^2\sigma_1^2)}\,]\end{array}}{\begin{array}{l}[\xi_2(1+\mu)^2\alpha^2(1+3\,\frac{\varepsilon_2^2\sigma_1^2}{\mu\alpha^2}) + \mu\xi_1+(1+3\,\frac{\varepsilon_2^2\sigma_1^2}{\mu\alpha^2}) \\ + 4(1+\mu)\xi_1\xi_2^2\alpha + 4\xi_1^2\xi_2 + \frac{\mu\xi_2^2(1+3\varepsilon_1^2\sigma_1^2)^2}{\alpha\xi_1(1+3\,\frac{\varepsilon_2^2\sigma_1^2}{\mu\alpha^2})} \\ + \frac{\xi_2}{\alpha^2}\,\frac{(1+3\,\varepsilon_1^2\sigma_1^2)^2}{(1+3\,\frac{\varepsilon_2^2\sigma_1^2}{\mu\alpha^2})} + 4(1+\mu)\,\xi_2^3\,\frac{(1+3\,\varepsilon_1^2\sigma_1^2)}{(1+3\,\frac{\varepsilon_2^2\sigma_1^2}{\mu\alpha^2})} \\ + \frac{4\xi_1\xi_2^3}{\alpha}\,\frac{(1+3\,\varepsilon_1^2\sigma_1^2)}{(1+3\,\frac{\varepsilon_2^2\sigma_1^2}{\mu\alpha^2})} - 2\xi_2(1+3\varepsilon_1^2\sigma_1^2)]\end{array}} \qquad (19)$$

In Figures 2 through 7 the primary mass mean square response (eq. (18)) is plotted against the uncoupled frequency ratio for different values and combinations of non-linearities. Similarly Figures 8 through 10 show the variation of the secondary mass mean square response (eq. (19)) as a function of α. In all these cases the critical damping ratio is taken as $\xi_1 = \xi_2 = 0.05$ and σ_1^2 as unity.

The root mean square response may be obtained by taking the square root of equations (18) and (19).

DISCUSSION

In Fig. 2 equation (18) is plotted against the uncoupled frequency ratio and for unity mass ratio with damping $\xi_1 = \xi_2 = 0.05$. Equal non-linearities of zero, 0.005, 0.010 and 0.015 are assumed for the primary and secondary springs. The results indicate that for low values of α (approximately less than 0.50, i.e. weak secondary spring) the mean square response is lower than the linear response. For α between 0.5 and 1.0 the response is less than linear, but the reduction is not significant. When α is increased the response becomes less than the linear case.

Similar parameters as in Fig. 2 are used in Fig. 3 except that the secondary spring is made linear. Primary spring non-linearities of 0.005, 0.010 and 0.015 are used. The response is generally lower than the linear response. The reduction in response for α less than 0.5 is significantly less than when both springs are non-linear. For α between 0.5 and 1.0 the response reduction is slightly more than when both springs are non-linear. As α becomes greater than 1.0 the response is more or less the same as in the case when both springs are equally non-linear.

In Fig. 4 the same parameters as in Fig. 2 are used except that the primary spring is made linear. The results indicate that for α less than 0.5 the response is less than the linear case. However the response is not as low as when both the springs are equally non-linear. As α increases the response is found to be larger than the linear response. Beyond $\alpha = 1.1$ the response essentially becomes linear.

Figure 5 shows the variation of the primary mass mean square response against the uncoupled frequency ratio for $\mu = 0.1$. Equal non-linearities of zero, 0.005, 0.010 and 0.015 are assumed for both springs. The response for a non-linearity of $\varepsilon_1^2 = \varepsilon_2^2 = 0.005$ is found to be smaller than the linear response for α less than 0.9. For α between 0.9 and 1.45 the response is found to be higher than the linear case. As α is increased further, the response drops below linear level. When the non-linearity is made 0.010, the response is found to be less than for α below 0.85. For α between 0.85 and 1.46 the response is larger than linear. The response drops below the linear level for further increase of α. In general, when the non-linearities are increased, the magnitude of the minimum response becomes higher occurring a at lower value of the uncoupled frequency ratio.

Similarly in Fig. 6, the same parameters as in Fig. 5 are used except that the secondary spring is made linear. Primary spring non-linearities of 0.010 and 0.015 are considered. The response is lower than the linear case for α less than 0.65. However, the response is not as low as when both the springs are equally linear. For α between 0.65 and 0.85 the response is essentially linear. When α is increased further the response is below linear level.

Figure 7 has the same parameters as in Fig. 5 with the exception that the primary spring is made linear. When the secondary spring non-linearity is 0.010, the response is less than linear for α less than 0.8, afterwards the response increases more than linear until α reaches 2.2 beyond which the response essentially becomes linear. Generally it is seen that the dynamic absorber effect is very significant in the low α region. All other trends remain the same as in Figure 5.

In Fig. 8 the secondary mass mean square response as given by equation (19) is plotted against the uncoupled frequency ratio for unity mass ratio. Equal non-linearities of zero, 0.005, 0.010 and 0.015 are considered for the primary and secondary springs. When the non-linearities are 0.005 on each spring, the response is found to be more than linear for α less than 0.5. For α between 0.5 to 0.7 the response is nearly linear. As α is increased beyond 0.7, the response is below linear. The trend remains almost the same for higher non-linearities even though the response magnitude and cross-over values of α are slightly different.

Figure 9 has identical parameters as in Fig. 8 except that the secondary spring is assumed as linear. In this case the response is always less than the linear response. In general, the larger the non-linearity in the primary spring the lower is the response. In the high α region the response in Fig. 8 and Fig. 9 are practically identical.

In Fig. 10 the same parameters as in Fig. 8 are used with the exception that the primary spring is made linear. It is seen that for α less than 1.0, the secondary mass mean square response is larger than the linear response. For α greater than 1.0, the response nearly becomes linear.

CONCLUSION

The mean square response of a two degree of freedom dynamic system with hardening type non-linearity in the restoring elements and with white excitation having normal distribution acting on the primary mass has been developed using the equivalent linearization method.

A parametric study of the solution indicated that the primary mass response is significantly reduced for lower values of the uncoupled frequency ratio (see for example Fig. 7) when the absorber system is of the hardening type as opposed to a linear absorber. However, there are frequency ratio ranges for which the primary mass response for the non-linear system is greater than for the linear system (see for example Figures 5 and 7).

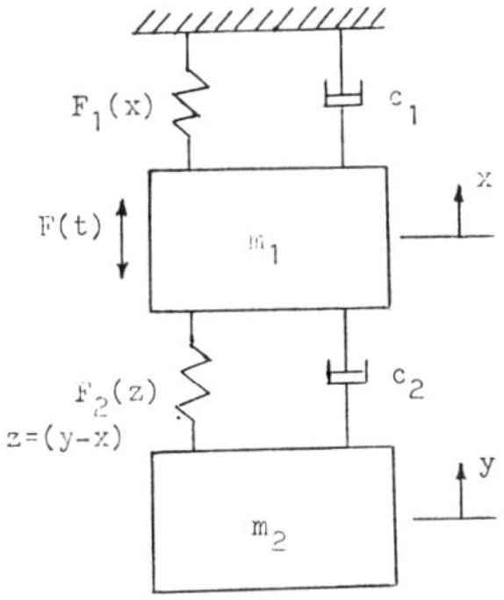

FIG. 1. TWO DEGREE OF FREEDOM SYSTEM

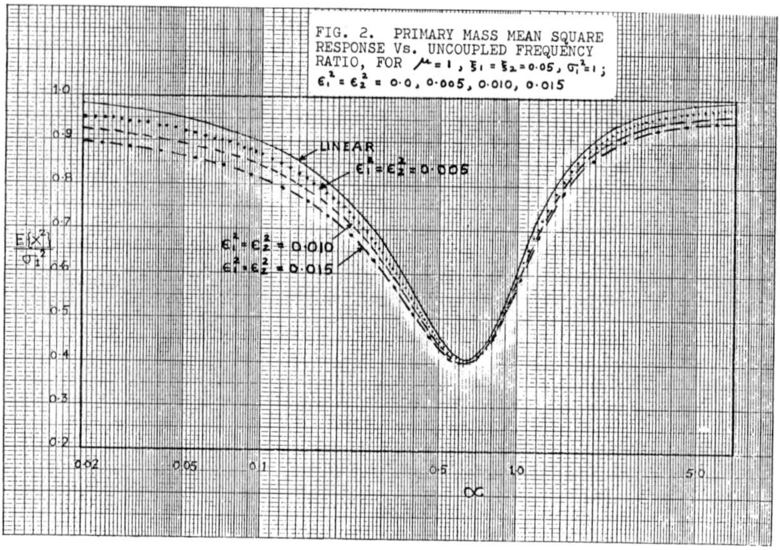

FIG. 2. PRIMARY MASS MEAN SQUARE RESPONSE Vs. UNCOUPLED FREQUENCY RATIO, FOR $\mu = 1$, $\xi_1 = \xi_2 = 0.05$, $\sigma_1^2 = 1$; $\epsilon_1^2 = \epsilon_2^2 = 0.0, 0.005, 0.010, 0.015$

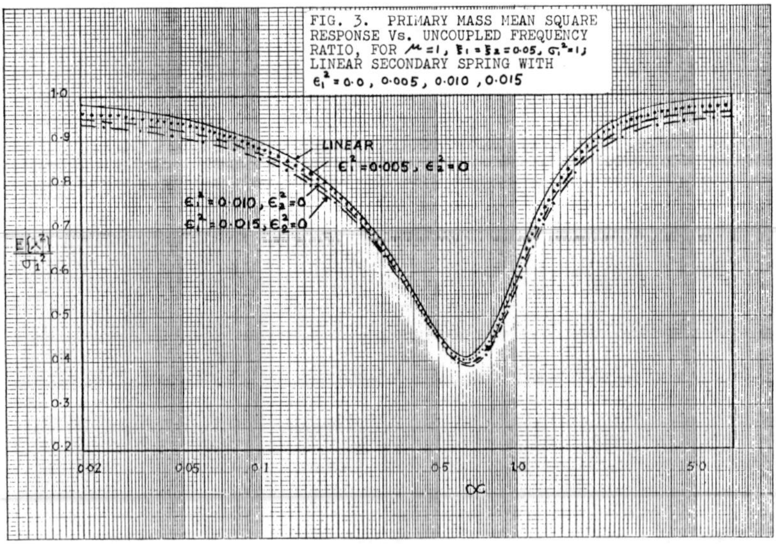

FIG. 3. PRIMARY MASS MEAN SQUARE RESPONSE Vs. UNCOUPLED FREQUENCY RATIO, FOR $\mu = 1$, $\xi_1 = \xi_2 = 0.05$, $\sigma_1^2 = 1$; LINEAR SECONDARY SPRING WITH $\epsilon_1^2 = 0.0, 0.005, 0.010, 0.015$

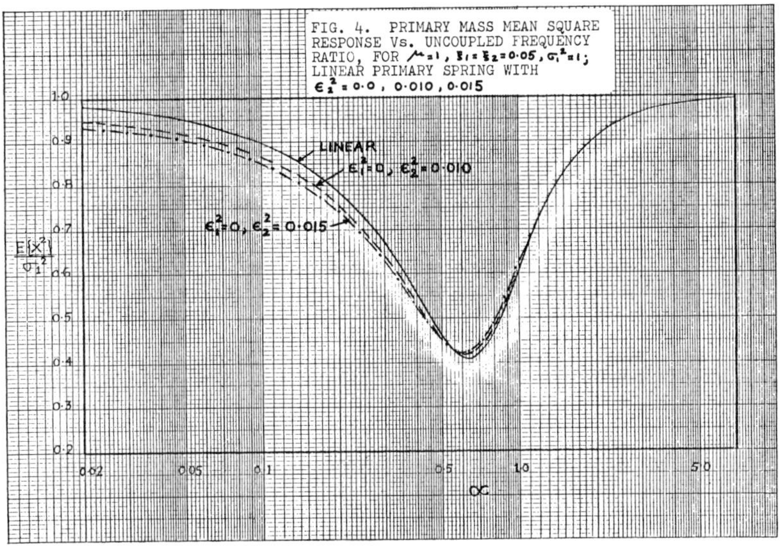

FIG. 4. PRIMARY MASS MEAN SQUARE RESPONSE Vs. UNCOUPLED FREQUENCY RATIO, FOR $\mu=1$, $\xi_1=\xi_2=0.05$, $\sigma_1^2=1$; LINEAR PRIMARY SPRING WITH $\epsilon_2^2 = 0.0,\ 0.010,\ 0.015$

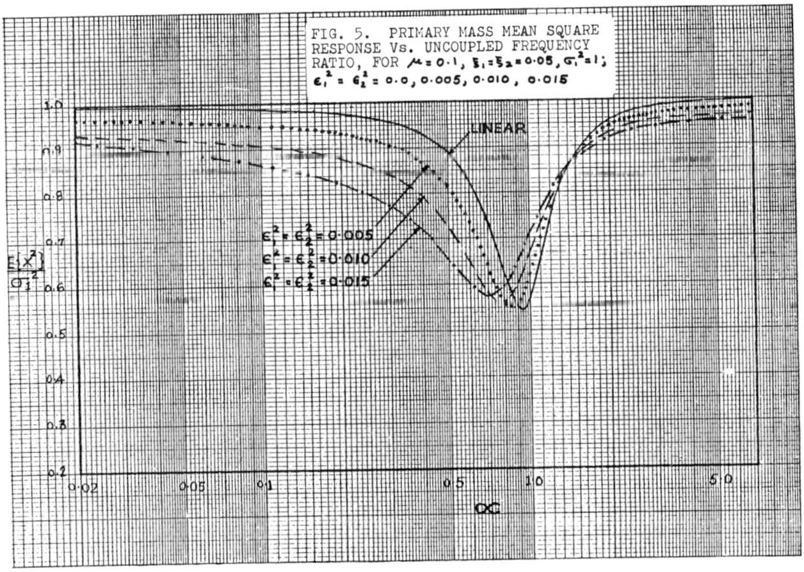

FIG. 5. PRIMARY MASS MEAN SQUARE RESPONSE Vs. UNCOUPLED FREQUENCY RATIO, FOR $\mu=0.1$, $\xi_1=\xi_2=0.05$, $\sigma_1^2=1$; $\epsilon_1^2=\epsilon_2^2=0.0,\ 0.005,\ 0.010,\ 0.015$

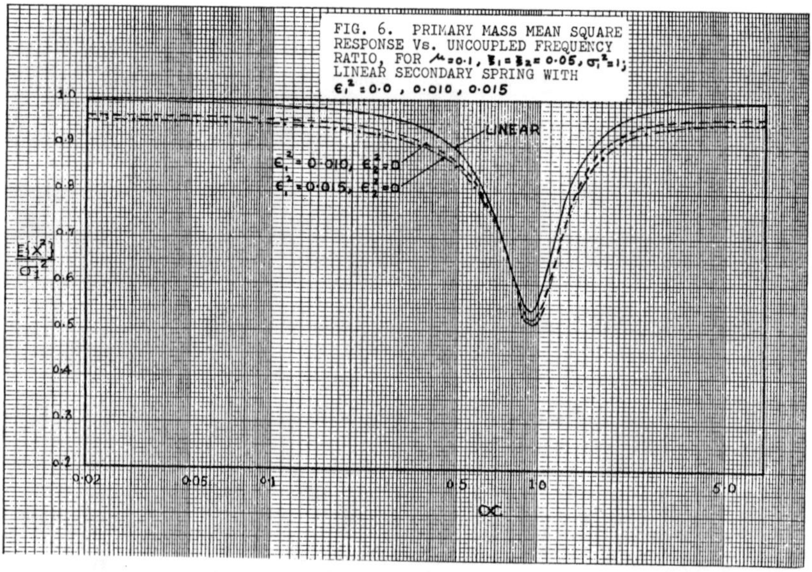

FIG. 6. PRIMARY MASS MEAN SQUARE RESPONSE Vs. UNCOUPLED FREQUENCY RATIO, FOR $\mu = 0.1$, $\xi_1 = \xi_2 = 0.05$, $\sigma_1^2 = 1$; LINEAR SECONDARY SPRING WITH $\epsilon_1^2 = 0.0$, 0.010, 0.015

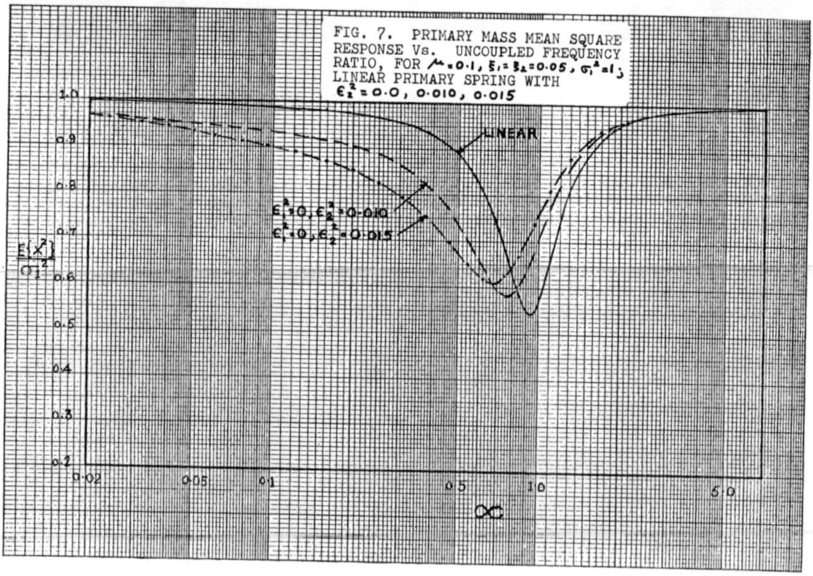

FIG. 7. PRIMARY MASS MEAN SQUARE RESPONSE Vs. UNCOUPLED FREQUENCY RATIO, FOR $\mu = 0.1$, $\xi_1 = \xi_2 = 0.05$, $\sigma_1^2 = 1$; LINEAR PRIMARY SPRING WITH $\epsilon_2^2 = 0.0$, 0.010, 0.015

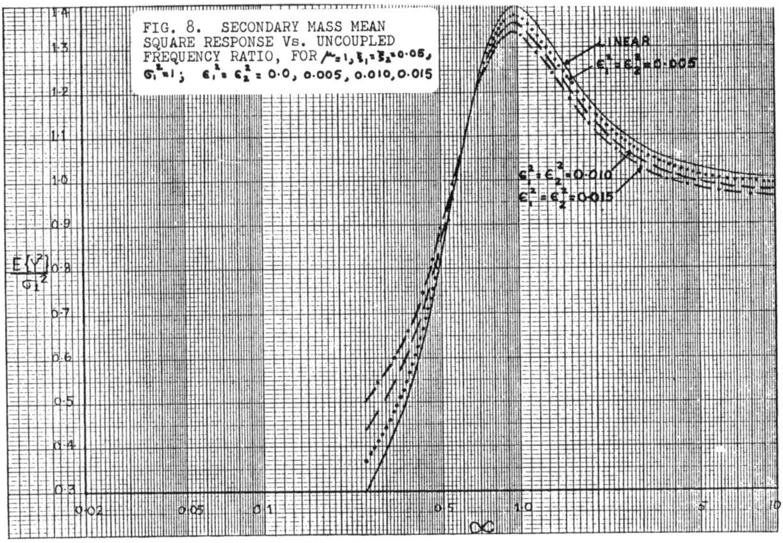

FIG. 8. SECONDARY MASS MEAN SQUARE RESPONSE Vs. UNCOUPLED FREQUENCY RATIO, FOR $\mu=1, \xi_1=\xi_2=0.05$, $\sigma_1^2=1$; $\epsilon_1^2=\epsilon_2^2=0.0, 0.005, 0.010, 0.015$

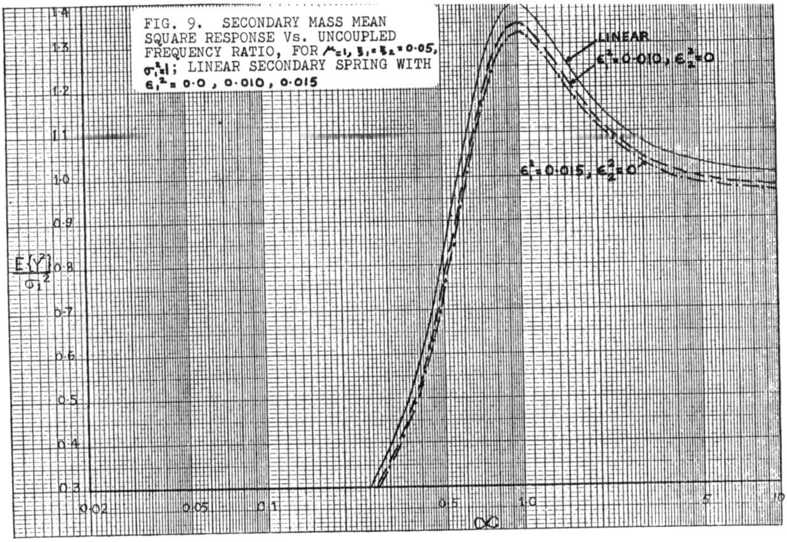

FIG. 9. SECONDARY MASS MEAN SQUARE RESPONSE Vs. UNCOUPLED FREQUENCY RATIO, FOR $\mu=1, \xi_1=\xi_2=0.05$, $\sigma_1^2=1$; LINEAR SECONDARY SPRING WITH $\epsilon_1^2=0.0, 0.010, 0.015$

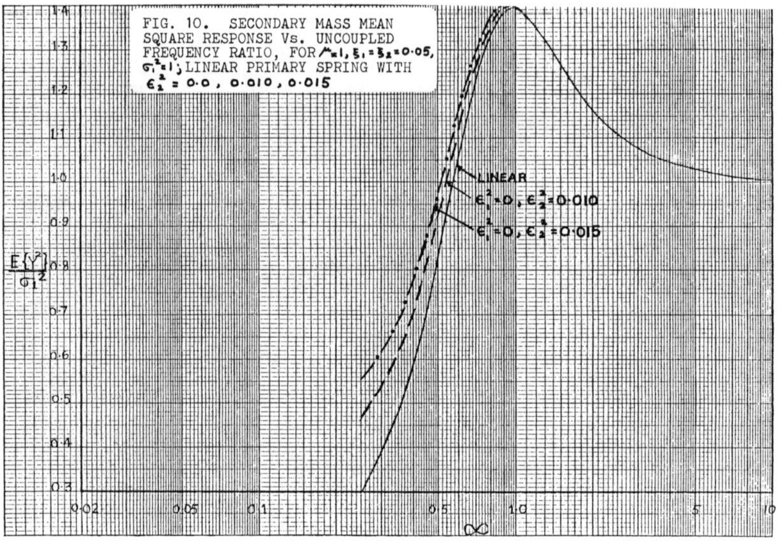

FIG. 10. SECONDARY MASS MEAN SQUARE RESPONSE Vs. UNCOUPLED FREQUENCY RATIO, FOR $\mu=1, \xi_1=\xi_2=0.05, \sigma_1^2=1$; LINEAR PRIMARY SPRING WITH $\epsilon_2^2 = 0.0,\ 0.010,\ 0.015$

REFERENCES

1. Caughey, T.K., "Equivalent Linearization Technique," Jn. of Acoust. Soc. of Am., Vol. 35, N11, 1963.

2. Foster, Jr., F.T., "Semilinear Random Vibrations in Discrete Systems," ASME paper 68-APM-U 1968 5P.

3. Iwan, W.D., and Yang, I.M., "Application of Statistical Linearization Techniques to Non-Linear Multi-Degree of Freedom Systems," Jn. of App. Mech., Trans. ASME Series E, 39, 2, pg. 545-550, 1972.

4. Kayanickupurathu, J.T., "Random Vibrations of Discrete Non-Linear Mechanical Systems," (Ph.D. Thesis), Polytech. Inst. of New York, N.Y., 1975.

5. Papoulis, A., Probability, Random Variables and Stochastic Processes, McGraw Hill, 1965, pg. 226.

6. Crandall, S.H., and Mark, W.D., Random Vibration in Mechanicsl Systems, Academic Press, 1963.

VIBRATION PROBLEM OF VISCOELASTICALLY DAMPED STRUCTURES.

Marc VAN OVERMEIRE
Vrije Universiteit Brussel
Faculty of Applied Sciences
Pleinlaan, 2
B-1050 Brussels (Belgium).

ABSTRACT

For the solution of vibrational problems, involving viscoelastic damping layers, the application of a modal superposition technique requires the solution of an eigenproblem, $([K(\omega)] - \lambda[M])\{u\} = 0$, with frequency dependent stiffness matrix $[K(\omega)]$.
This paper presents a straightforward method, namely a slightly modified Sturm Sequence Method, for the solution of this eigenproblem.

INTRODUCTION

In recent years, there has been an increasing interest in the reduction of vibration levels in structures, to optimise performance regarding acoustical noise, human comfort, longevity, and others.
These reduction can be achieved in two ways :

1. The load frequencies may certainly not coincide, or even lay in a close neighbourhood of the eigenfrequencies.
2. If it is, for various reasons, impossible to separate loading and eigenfrequencies, it is necessary to increase the internal damping capacity of the structure.

In the first case, it is necessary to determine the eigenfrequencies and modes of the structure. Before the introduction of the finite element method, eigenfrequencies and modes could only be obtained during a vibration test. Nowadays, due to calculation capacities of modern computers and the efficiency of numerical methods, it is possible to obtain eigenfrequencies and modes of complex structures. By changing the mass distribution or elastic properties of the structure, it is then possible to shift the eigenfrequencies.

But often, it is quire impossible, for design reasons, to alter the structure, or if the excitations possess a broad band frequency range, there are always resonance effects. The only solution here is an increase of the energy dissipating capacity of the structure. This objective can be achieved by utilizing the damping properties of viscoelastic materials. These damping properties are fully exploited, if enclosed viscoelastic layers are used (sandwich-technique).

This paper deals with one aspect of the above problems, the calculation of the eigenfrequencies and modes of moderately damped structures, whose behaviour is approximated with a finite element method. The resulting eigenvalue problem is solved by a modified Sturm Sequence method.

Moreover, once the eigenvalue problem solved, the structural response in the frequency domain, can be found by applying a normal mode method.

FORMULATION OF THE PROBLEM

The dynamic behaviour of the discretised structure is described by equation (1)

$$[M]\{\ddot{u}\} + [C]\{\dot{u}\} + [K]\{u\} = \{F(t)\} \qquad (1)$$

For an harmonic excitation $F(t) = F_o\, e^{j\omega t}$, the response is, in the case of linear geometric and material behaviour, also harmonic and may be written as :

$$\{u(t)\} = \{u_o\}\, e^{j\omega t} \qquad (2)$$

The addition of damping causes the real equations for an undamped system to develop into complex equations for the damped system $\{u_o\}$ must be written as a complex variable :

$$\{u_o\} = \{u_r\} + j\{u_i\} \qquad (3)$$

Substitution of (2) and (3) in (1) leads to a 2N system of equations (4) ; (N : number of degree of freedom of the discretised structure)

$$\begin{vmatrix} [K]-\omega^2[M] & -\omega[C] \\ \omega\,[C] & [K]-\omega^2[M] \end{vmatrix} \begin{Bmatrix} \{u_r\} \\ \{u_i\} \end{Bmatrix} = \begin{Bmatrix} \{F_o\} \\ \{0\} \end{Bmatrix} \qquad (4)$$

This set of 2N equations must be solved for each frequency, but allows frequency dependent material properties, as is the case for viscoelastic materials. This approach is called the frequency response method. The doubling of the order of the equations results in a more increased computation time, especially for three-dimensional structures, or when the frequency ranges of interest (near the resonant frequencies) are not known. An accurate information about the eigenfrequencies and modes is nearly indispensable.

In most cases, only the lowest eigenfrequencies and modes are necessary, because higher modes are often critically damped and unimportant.
The natural eigenvalue problem for the structure is given by equation (5) :

$$[K]\{\phi_i\} = \omega_i^2 \, [M]\{\phi_i\} \tag{5}$$

If the structure contains viscoelastic layers or strips, the stiffness matrix $[K]$ becomes frequency dependent, because we describe the behaviour of a viscoelastic material under harmonic excitation by its complex modulus $\widetilde{E}$

$$\widetilde{E} = E_o(\omega)\ (1 + j\eta(\omega)\ ) \tag{6}$$

The undamped eigenfrequencies of such a composite structure are found by solving

$$[K(\omega_i)]\{\phi_i\} = \omega_i^2 \, [M]\{\phi_i\} \tag{7}$$

Possible solution methods will be given in next section.
Once eigenfrequencies ω_i and $\{\phi_i\}$ are found, a modal superposition technique (also called normal mode method) can be applied : displacement $\{u\}$ is expressed as a linear combination of the eigenmodes $\{\phi_i\}$

$$\{u\} = \sum_{i=1}^{m} x_i \, \{\phi_i\} \tag{8}$$

Equation (1) becomes, after premultiplication with ${}^{T}[\phi]$

$$[m]\{\ddot{x}\} + [c]\{\dot{x}\} + [k]\{x\} = \{f\}$$

with

$$\begin{aligned} [m] &= {}^{T}[\phi][M][\phi] \\ [c] &= {}^{T}[\phi][C][\phi] \\ [k] &= {}^{T}[\phi][K][\phi] \\ \{f\} &= {}^{T}[\phi]\{F\} \end{aligned} \tag{9}$$

For harmonic excitation, it is obvious that equation (9) becomes

$$(-\omega^2[m] + j\omega[c] + [k])\{x_o\} = \{f_o\} \tag{10}$$

For composite structures, the most important contribution to the structure's energy dissipation possibility is caused by the viscoelastic layers. Compared with the damping caused by this viscoelastic material, the viscous damping and also the material damping in the elastic members (normally very small for steel, aluminum) can be neglected.
As a result, only the damping by the viscoelastic material is considered. The stiffness matrix of the viscoelastic material may be written as

$$[\widetilde{K}_v] = [K_v] + j\eta[K_v] \tag{11}$$

using equation (6).

The global stiffness matrix becomes then

$$[K] = [K_r] + j\eta\,[K_i] \qquad (12)$$

The only non-zero terms in $[K_i]$ are those of $[K_v]$, because we only consider the damping, caused by the viscoelastic material.
In normal coordinates, $[k]$ becomes then

$$[k] = [k_r] + j\eta\,[k_i] \qquad (13)$$

Substitution of (13) in (10)

$$(-\omega^2[m] + j\eta[k_i] + [k_p])\{x_o\} = \{f_o\} \qquad (14)$$

$\{x_o\}$ is again a complex variable, due to the damping term $j\eta[k_i]$.
If the dynamic response of the structure is described by the m lowest modes, the solution of only 2m simultaneous equations gives the structural frequency response. This normal mode method is much more efficient than the frequency response method, because it makes repeated use of natural frequency and mode shape solution, which is obtained just once.

However, the proposed method has some limitations. It is assumed that every undamped frequency has a corresponding damped frequency. This is certainly true for moderately damped structures, but in case of heavy damping, clustering of the eigenfrequencies occurs and the resonant peaks are suppressed. In this case, the frequency response method must be used. Another possibility is to use, in the normal mode method, complex natural frequencies and modes, instead of undamped real frequencies and modes. The solution of this complex eigenvalue problem is rather cumbersome, and will not be considered here.

SOLUTION OF THE EIGENVALUE PROBLEM

Because stiffness properties of a composite damped structure are frequency dependent, the solution of this eigenvalue problem (equation 7), is usually found in two steps, as showed in [1].
The characteristic $E_o(\omega)$ and $\eta(\omega)$ are known (or obtained by experiment). The modal response in the neighbourhood of each frequency can be obtained once the values of E_o and η are known at that frequency. We suppose here again, that the natural frequencies of the structure are sufficiently separated, and start with two appropriate values of E_o, say E_o' and E_o''. With these two values, a set of natural frequencies ω_1', ω_2', ω_3' and ω_1'', ω_2'', ω_3'' is calculated. Then it is possible to calculate the variation of each natural frequency as function of E_o. Therefore a perturbation formula, based on the Rayleigh quotient, and often used in the design stage in order to predict effects of proposed inertia and mass changes, is used (equation 5), [2] :

$$\Delta\omega_i = \frac{1}{2\omega_i} \frac{\Delta E_o}{E_o} \frac{^r\{\phi_i\}[K_i]\{\phi_i\}}{^r\{\phi_i\}[M]\{\phi_i\}} \tag{15}$$

with $\Delta\omega_i$ variation of the i-th natural frequency and ΔE_o the variation of the E-modulus. The variation of ω_i is nearly linear. The material characteristics at the natural frequencies are found as an intersection of $E_o(\omega)$ characteristic and the lines representing equation (15). The eigenvalue problem can be solved, and structural response near the eigenfrequencies can be found (figure 1).

The eigenfrequencies can however be found in a straightforward manner too. For this purpose, use is made of a Sturm Sequence Solution Method [3],[4]. At this point, it is interesting to recall the underlying theorems and basic steps of this algorithm. It is primarily based on two theorems.

The first one states, that consecutive leading minors $p_s(\lambda) = \det([K]_s - \lambda[M]_s)$, ($\lambda = \omega^2$ and s = 1,N) possess the Sturm Sequence property, namely that the number of changes in sign (called sign count, (sc)) of consecutive terms of the sequence $p_r(\lambda)$, [with $p_o(\lambda) = 1$] equals the number of eigenvalues, smaller than λ. The number of eigenvalues in an interval (a,b) equals then sc(b) - sc(a).

A second theorem proves, that the number of eigenvalues smaller than λ_o, equals the number of negative pivots, when a triangularisation on $[K]-\lambda_o[M]$ is performed.

Applying these theorems, following solution procedure is derived :

1. Determination of the number of eigenvalues in (λ_ℓ,λ_u) by calculating $sc(\lambda_u) - sc(\lambda_\ell)$, which are obtained by a factorisation of $[K]-\lambda[M]$. λ_ℓ and λ_u are respectively lower and upper bounds, which are initially given as input data.
2. Repeated bisection on (λ_ℓ,λ_u) in order to isolate the lowest eigenvalue.
3. Once the lowest eigenvalue is isolated, it is calculated by a faster converging technique (like a regula falsi).
4. The next interval is $(\lambda_2^\ell,\lambda_2^u)$, on which steps 1, 2, 3 are executed.

In each bisection step, $[K]-\omega^2[M]$ is evaluated, or for a viscoelastic material behaviour, $[K(\omega)] - \omega^2[M]$. This is the only modification of the original Sturm sequence.

For viscoelastic materials, convergence of this algorithm is assured. For viscoelastic materials, the problem can be reduced to an elastic problem. Suppose that the linear viscoelastic behaviour can be formulated as

$$E = E(\omega) = E_o\, f(\omega) \tag{16}$$

with $f(\omega)$ a monotone growing function of ω.

The stiffness matrix can be written as (supposing ν (= Poisson ratio) frequency independent)

$$[K] = [K_o]\ f(\omega) \tag{17}$$

or the eigenvalue problem becomes

$$([K_o]\ f(\omega) - \omega^2[M])\{\phi\} = \{0\} \tag{18}$$

Supposing $f(\omega) \neq 0$, which is always the case for real material behaviour, one gets

$$([K_o] - \frac{\omega^2}{f(\omega)}[M])\{\phi\} = \{0\} \tag{19}$$

Setting now

$$\lambda' = \frac{\omega^2}{f(\omega)} \tag{20}$$

equation (21) is obtained

$$([K_o] - \lambda'[M])\{u\} = 0 \tag{21}$$

which is the classical eigenvalue problem, and is certainly convergent. Again for real materials, a single value of λ corresponds with obtained value for λ'.
For composite materials, where the frequency dependency is less, due to the presence of elastic material parts, there is evidently convergence.

NUMERICAL EXAMPLE

Sandwich beam [5]

A sandwich beam, composed of a viscoelastic layer, constrained with two elastic layers, is presented in figure 2.
The elastic members are made of machined steel, while the viscoelastic layer is made of an acrylic base polymer.
The complex modulus G of the viscoelastic layer is approximated by

$$G = G_o\ (1 + j\eta)$$

where

$$G_o = 21.47\ f^{0.579}\ \text{kPa}$$

$$\eta = 1.008\ f^{0.022}$$

The viscoelastic material is assumed to be incompressible. Loss factors in shear and direct deformation are equal, and Poisson's ratio is taken as 0.48 to avoid numerical difficulties.
For the modelling of the beam, 10 isoparametric 9-node elements are used for each layer, as shown in figure 3.
Numerical results are obtained with the program EIGEN80 [6].

The eigenfrequencies are presented in table 1 and a mechanical driving point impedance is presented for an harmonic excitation at the midsection of the beam (figure 4). The values were compared with the analytical and experimental values, presented in reference [5].

NR	Natural frequency (Hz)	Mode type
1	116	S
2	272	A
3	507	S
4	821	A
5	1237	S
6	1774	A

Table 1.

Only the symmetric bending modes are excited by a force, at the middle of the beam, because this midpoint is a nodal point for the asymmetric modes.

In the neighbourhood of the resonance frequencies, the results of the normal mode method were also checked against the values of the frequency response method, and no significant differences were observed.

CONCLUSION

A Sturm Sequence Solution Method, together with a normal mode approach, is able to predict the frequency response of moderately viscoelastically damped structures.

ACKNOWLEDGEMENT

Author acknowledges the "Belgian National Foundation for Scientific Research" for their financial support.

REFERENCES

[1] LALANNE, M. ; PAULARD, M. ; TROMPETTE, P., Response of thick structures damped by viscoelastic material with application to layered beam and plates, The Shock and Vibration Bulletin, N°45/5, p. 65-72, 1972.

[2] CRANDALL, S.H. ; CALLEY, R.B., The Shock and Vibration Handbook, vol. 2, Ch. 28, Mc Graw-Hill, 1961.

[3] VAN OVERMEIRE, M. ; DECONINCK, J., A Sturm Sequence Solution Method for Medium Band Matrices, Proc. CanCam 1977, vol. 2, pp. 1033-1034, 1977.

[4] VAN OVERMEIRE, M., A Sturm Sequence Solution Method for General Eigenvalue Problems, in Euromech 112 on Bracketing of Eigenfrequencies of Continuous Structures", Ed. Hungarian Academy of Sciences, pp. 357-370, 1980.

[5] LU, Y.P. ; EVERSTINE, C.C., More on Finite Element Modelling of Damped Composite Systems, Journal of Sound and Vibration, 69(2), pp. 199-205, 1980.

[6] VAN OVERMEIRE, M., User's Manual of EIGEN80, a modular finite element program for structural dynamics, Report N°1980/04, Department of Civil Construction,Vrije Universiteit Brussel, 70 pages, 1980.

LIST OF USED SYMBOLS

$[C]$ damping matrix

$[F]$ force vector

$[K]$ stiffness matrix

$[K_i]$ imaginary part stiffness matrix, when using viscoelastic materials

$[K_r]$ real part stiffness matrix

$[K_v]$ stiffness matrix for viscoelastic material

$[M]$ mass matrix

$[c]$ modal damping matrix

$[f]$ modal force vector

$[k]$ modal stiffness matrix

$[k_i]$ imaginary part modal stiffness matrix

$[k_r]$ real part modal stiffness matrix

$[m]$ modal mass matrix

$\{u\}$ displacement vector

$\{x\}$ modal participation vector

$\{\phi_i\}$ i-th eigenmode

$[\phi]$ matrix of m eigenmodes

$[\{\phi_1\} \ \{\phi_2\} \qquad \{\phi_m\}]$

$\tilde{E}$ complex elasticity modulus

E real part elasticity modulus

j $\sqrt{-1}$

$p_s(\lambda)$ s-th term of the Sturm Sequence, equals the s-th leading minor

$$\left[\begin{array}{cccc:c} a_{11} & a_{22} & & a_{1s} & a_{1n} \\ a_{21} & & & & \\ & & & & \\ a_{s1} & & & a_{ss} & \\ \hdashline a_{n1} & & & & a_{nn} \end{array}\right]$$

λ eigenvalue

λ' transformed eigenvalue

η loss factor of the viscoelastic material

ν Poisson ratio

Δ increment operator

. time derivative

r transposed of a vector or matrix

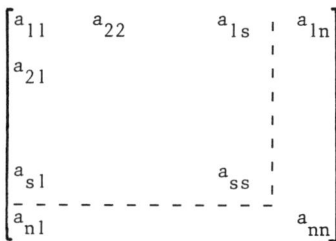

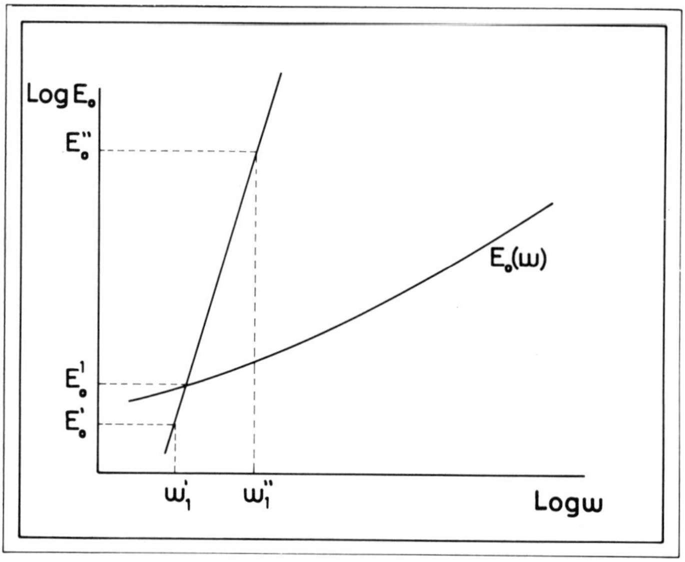

Figure 1 : Determination of the Young modulus.

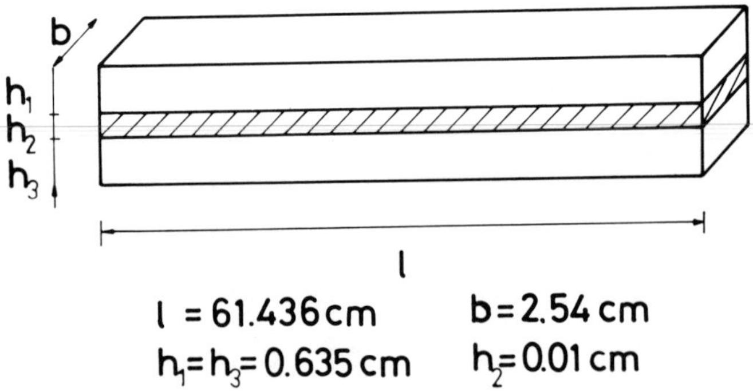

Figure 2 : Composite beam.

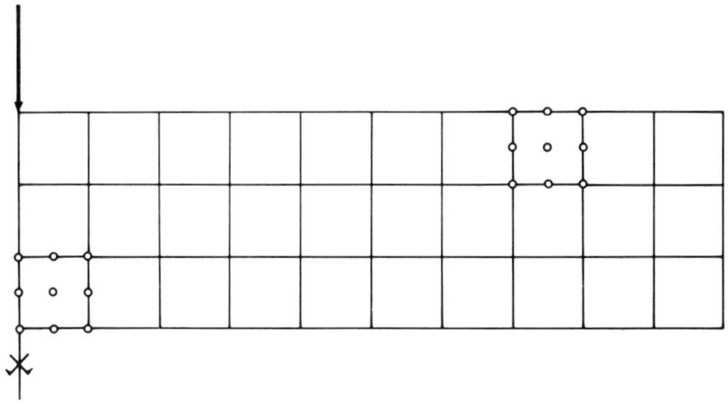

Figure 3 : Modelling of the composite beam by (3x10) element grid.

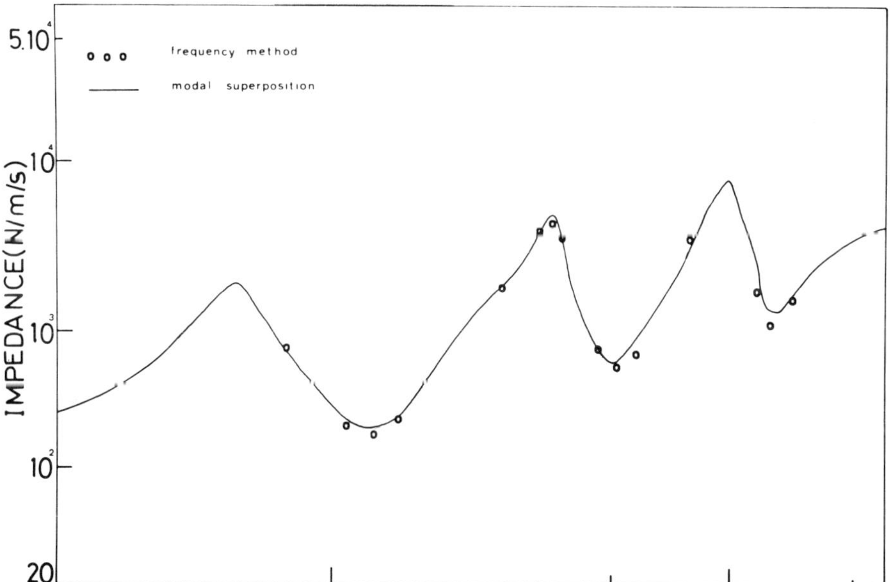

Figure 4 : Mechanical driving point impedance of composite beam.

TWO-DIMENSIONAL DYNAMIC MODELING OF HUMAN JOINTS

Ali Erkan Engin
Professor of Biomechanics & Mechanics

and

Manssour H. Moeinzadeh
Graduate Research Associate

Department of Engineering Mechanics
The Ohio State University
Columbus, Ohio 43210

ABSTRACT

This paper is concerned with a two-dimensional mathematical modeling of an articulating joint defined by the contact surfaces of two body segments which execute a dynamic motion within the constraints of the joint ligaments. The joint ligaments are modeled as nonlinear elastic springs of realistic stiffness properties. Nonlinear equations of motion coupled with nonlinear constraint conditions are solved numerically. Time derivatives are approximated by Newmark difference formulas and the resulting nonlinear algebraic equations are solved employing the Newton-Raphson iteration scheme. Several dynamic loads are applied to the center of mass of the moving body segment and the ensuing motion is investigated. By considering the human knee joint as an example some numerical results are presented on the magnitudes of the ligament and contact forces at the joint.

INTRODUCTION

Biodynamic simulation of the response of the total human body to external forces provide essential input for the establishment of the injury prediction criteria and subsequent design and development of crash protection systems. The most sophisticated versions of the total-human-body models are articulated and multi-segmented to simulate all the major articulating joints and segments of the human body. A brief review of the mathematical models of the human body was provided by the first author [1,2] and an extensive treatment of the same subject can be found in [3].

Effectiveness of the multi-segmented models to predict accurately live human response depends heavily on the proper biodynamic description and simulation of the articulating joints. Because of the extreme complexity of incorporating articulating joint structures possessing realistic geometries and ligament behavior to the multi-segmented models of the total human body, these models, thus far, have been employing simple geometric shapes for their joints. Although in the literature [4], there are mathematical joint models that consider both the geometry of the joint surfaces and behavior of the joint ligaments, these models are quasi-static in nature, and employ the so-called inverse method. In the inverse method the ligament forces caused by a specified set of translations and rotations along the specified directions are determined by comparing the geometries of the initial and displaced configurations of the joint. Furthermore, in the inverse method utilized in [4] it is necessary to specify the external force required for the preferred equilibrium configuration. Such an approach is applicable only in a quasi-static analysis. In a dynamic analysis, on the other hand, the equilibrium configuration preferred by the joint is the unknown and the mathematical analysis itself is to provide that dynamic equilibrium configuration.

The purpose of this paper is to present development of a two-dimensional mathematical model of an articulating joint by including both geometric and material nonlinearities. The most general and realistic dynamic model of an articulating joint should obviously be three-dimensional; however, it is believed that a simpler two-dimensional model can be helpful and rewarding in an understanding of the problems that may rise from the three-dimensional dynamic model. The highly nonlinear set of model equations are solved numerically and some representative results are presented by taking the human knee joint as an example.

FORMULATION OF AN ARTICULATING JOINT MODEL

An articulating joint connects two body segments which are designated as segments 1 and 2 as shown in Fig. 1. An inertial coordinate system (x,y) with unit vectors $\hat{i}$ and $\hat{j}$ is connected to the fixed body segment, while the coordinate system (x',y') with unit vectors $\hat{i}'$ and $\hat{j}'$ is attached to the center of mass of the moving body segment 1. The motion of the moving (x',y') system relative to the fixed (x,y) system may be characterized by three quantities: the translational movement of the origin of the (x',y') system in the x and y-directions, and its rotation, α, with respect to the fixed (x,y) system.

In two-dimensional formulation the articulating contact surfaces can be represented by smooth mathematical functions of the following form:

$$y = f_1(x) \quad \text{and } y' = f_2(x'). \tag{1}$$

Denoting the position vectors of the contact point, C, by $\bar{r}_c$ and $\bar{\rho}_c'$ in the bases $(\hat{i},\hat{j})$ and $(\hat{i}',\hat{j}')$, respectively, we can write

$$[r_c] = [r_o] + [T][\rho_c'] \quad , \tag{2}$$

where $\bar{r}_o$ is the position vector of the origin of the (x',y') system in the fixed (x,y) system and $[T]$ is the transformation matrix between $(\hat{i},\hat{j})$ and $(\hat{i}',\hat{j}')$ bases. Note that the relationship expressed with Eq. (2) also holds for any arbitrary point belonging to the moving segment 1. Eq. (2) is a part of the geometric compatibility condition for the two contacting surfaces. In addition, the unit normals, $\hat{n}_1$ and $\hat{n}_2$ to the surfaces of the fixed and moving body segments, respectively, must be colinear. The colinearity of the normals at the contact point C requires that:

$$\hat{n}_1 \times \hat{n}_2 = 0 \quad \text{at} \quad x = x_c \quad \text{and} \quad x' = x_c' \quad . \tag{3}$$

It can be shown [5] that the colinearity of the normals takes the following form:

$$\sin\alpha \left[1 + \left(\frac{df_1}{dx}\right)_{x=x_c} \left(\frac{df_2}{dx'}\right)_{x'=x'_c}\right] - \cos\alpha \left[\left(\frac{df_1}{dx}\right)_{x=x_c} - \left(\frac{df_2}{dx'}\right)_{x'=x'_c}\right] = 0 \quad . \tag{4}$$

During its motion the segment 1 is subjected to the ligament forces, contact

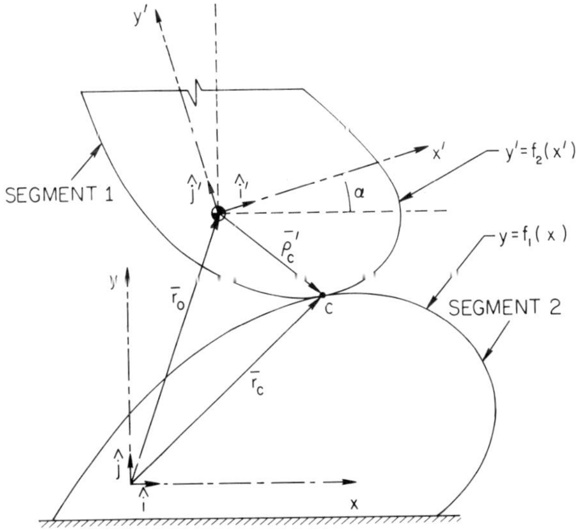

Fig. 1. A two-body segmented joint is illustrated.

forces and the externally applied forces and moments, Fig. 2. The contact force and the ligament forces are the unknowns of the problem and the external forces and moments will be specified. These forces are discussed in some detail in the following paragraphs.

Ligament Forces

The joint ligaments and a portion of the joint capsule are modeled as nonlinear elastic springs. More specifically, the following force-elongation relationship is assumed for a particular ligament, j:

$$|\bar{F}_j| = k_j(L_j - \ell_j)^2 \qquad \text{for} \quad L_j > \ell_j \tag{5}$$

in which k_j is the spring constant, L_j and ℓ_j are, respectively the current and initial lengths of the ligament, j. It is assumed that the ligaments cannot carry any compressive force; accordingly:

$$|\bar{F}_j| = 0 \qquad \text{for} \quad L_j < \ell_j \tag{6}$$

The direction of the force, F_j, exerted by a ligament on the articulating body segment coincides with the direction of the line segment through the origin

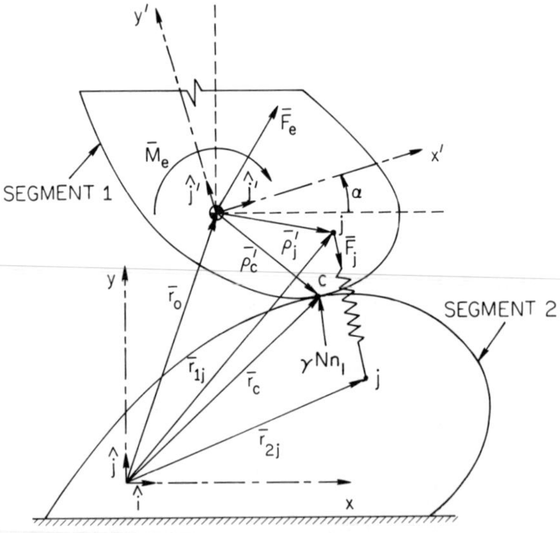

Fig. 2. Forces acting on the moving body segment of a two-body segmented joint and relevant vectors for the contact point, C, and ligament, j, are illustrated.

and insertion points of that ligament. The length, ℓ_j, of the ligament, j is taken to be equal to the distance between its insertion and origin points.

Let $(\bar{\rho}'_j)_m$ be the position vector in the base $(\hat{i}',\hat{j}')$ of the insertion point of the ligament, j, in the moving body segment. The position vector of the origin point of the same ligament, j, in the fixed body segment is denoted by $(\bar{r}_{2j})_f$ in the base $(\hat{i},\hat{j})$. Here the subscripts m and f outside the parenthesis denote "moving" and "fixed", respectively. The current length of the ligament is given by:

$$L_j = \sqrt{[(\bar{r}_{2j})_f - \bar{r}_o - T(\bar{\rho}'_j)_m] \cdot [(\bar{r}_{2j})_f - \bar{r}_o - T(\bar{\rho}'_j)_m]} \tag{7}$$

The unit vector, $\hat{\lambda}_j$, along the ligament, j, directed from the moving to the fixed body segment is:

$$\hat{\lambda}_j = \frac{1}{L_j}[(\bar{r}_{2j})_f - \bar{r}_o - T(\bar{\rho}'_j)_m] \tag{8}$$

Thus, the axial force in the ligament, j, in its vectorial form, can be written as $\bar{F}_j = F_j\hat{\lambda}_j$.

Contact Forces

Due to presence of synovial fluid, with excellent lubricating characteristics [6], on the articulating surfaces, the friction force between the moving and fixed body segments can be neglected. Thus, the contact force will be in the direction of the normal to the surface at the point of contact. The contact force, $\bar{N}$, acting on the moving body segment is given by:

$$\bar{N} = \gamma\, N\, \hat{n}_1 \quad \text{at} \quad x=x_c \tag{9}$$

where N is the magnitude of the contact force and

$$\gamma = (d^2f_1/dx^2)/|d^2f_1/dx^2| \quad \text{at} \quad x = x_c \tag{10}$$

γ is either +1 or -1 and it ensures the correct direction of the contact force acting on the moving body segment.

Equations of Motion

The dynamic equations of motion of the moving body relative to the fixed body segment are as follows:

$$(F_e)_x + \gamma N(n_1)_x + \sum_{j=1}^{p} F_j(\lambda_j)_x = M\ddot{x}_o \tag{11a}$$

$$(F_e)_y + \gamma N(n_1)_y + \sum_{j=1}^{p} F_j(\lambda_j)_y = M\ddot{y}_o \tag{11b}$$

$$M_e + (T\bar{\rho}_c') \times (\gamma N\bar{n}_1) + \sum_{j=1}^{p} (T\bar{\rho}_j') \times (F_j\bar{\lambda}_j) = I_z\ddot{\alpha} \tag{11c}$$

where p is the number of ligaments and the subscripts, x and y, denote the components of the related quantities in the x and y-directions. The mass of the moving body segment is denoted by M and the dots denote derivatives with respect to time, t. The mass moment of inertia of the moving body segment about the z-axis is I_z and $\ddot{\alpha}$ designates its angular acceleration. $(F_e)_x$, $(F_e)_y$ and M_e are the components in the base $(\hat{i},\hat{j})$ of the external force and moment, respectively, which are applied to the center of mass of the moving body segment. The problem description is completed by assigning the initial conditions, which are:

$$\dot{x}_o = \dot{y}_o = \dot{\alpha} = 0 \tag{12}$$

along with the specified values for x_o, y_o and α at t = 0.

Three nonlinear second order differential equations, Eqs. 11 a, b, and c, along with the geometric compatibility and contact conditions, given by Eqs. (2) and (4), provide the necessary relationships to determine the following unknowns:

a) x_o and y_o; the components of vector $\bar{r}_o$,

b) x_c and x_c'; the x and x'-coordinates of the contact point, C, in the base $(\hat{i},\hat{j})$ and $(\hat{i}',\hat{j}')$, respectively,

c) α; the orientation angle of the moving (x',y') system relative to the fixed (x,y) system, and

d) N; the magnitude of the contact force.

The numerical procedure employed in the solution of the governing equations is described in the following section.

NUMERICAL METHOD OF SOLUTION

The first step in obtaining a numerical solution of the model equations is the replacement of the time derivatives with a temporal operator; in the present work, the Newmark operators [7] are chosen for this purpose. For instance, $\ddot{x}_o$ is expressed in the following form:

$$\ddot{x}_o^t = \frac{4}{(\Delta t)^2}(x_o^t - x_o^{t-\Delta t}) - \frac{4}{\Delta t}\dot{x}^{t-\Delta t} - \ddot{x}_o^{t-\Delta t} , \tag{13a}$$

$$\dot{x}_o^t = \dot{x}_o^{t-\Delta t} + \frac{\Delta t}{2}\ddot{x}_o^{t-\Delta t} + \frac{\Delta t}{2}\ddot{x}_o^t \quad , \tag{13b}$$

in which Δt is the time increment and the superscripts refer to the time stations. Similar expressions are used for $\ddot{y}_o$ and $\ddot{\alpha}$. In the application of Eqs. 13a and b, the conditions at the previous time station ($t-\Delta t$) are, of course, assumed to be known.

After the time derivatives in Eqs. 11 a, b, and c are replaced with the temporal operators defined, the governing equations take the form of a set of nonlinear algebraic equations. The solution of these equations is accomplished by an iteration method. In this work, the Newton-Raphson [8] iteration process is used for the solution. To linearize the resulting set of simultaneous algebraic equations, we assume:

$$_k x_o^t = {}_{k-1}x_o^t + \Delta x_o \tag{14}$$

and similar expressions for the other variables are written. Here, the right subscripts denote the time station under consideration and the left subscripts denote the iteration number. At each iteration, k, the values of the variables at the previous (k-1) iteration are assumed to be known. The delta quantities denote incremental values. Equation (14) and the corresponding ones for the other variables are substituted into the governing nonlinear algebraic equations and the higher order terms in the delta quantities are dropped. The set of n simultaneous algebraic (now linearized) equations can be put into matrix form:

$$[K]\ \{\Delta\} = \{D\} \tag{15}$$

where [K] is an nxn coefficient matrix, $\{\Delta\}$ is a vector of incremental quantities and {D} is a vector of known values.

The iteration process at a fixed time station continues until the delta quantities of all the variables become negligibly small. A solution is accepted and the iteration process is terminated when the delta quantities become less than or equal to a prescribed percentage of the previous values of the corresponding variables. The converged solutions of each variable is then used as the initial value for the next time step and the process is repeated for consecutive time steps. The only problem that the Newton-Raphson process may present in the solution of dynamic problems is due to the fact that the period of the forced motion of the system may turn out to be quite short. In this case it becomes necessary to use very small time steps; otherwise a significantly large number of iterations is required for convergence.

RESULTS AND DISCUSSION

The human knee joint is chosen for the application of the mathematical formulation presented in this paper. Detailed discussions of various anatomical and functional aspects of the human knee joint can be found in [9,10]. The first task in obtaining numerical results is determination of the functions $f_1(x)$ and $f_2(x')$ from an X-ray of a human knee joint. A number of points on the two-dimensional profiles of the femoral and tibial articulating surfaces were utilized to obtain quartic and quadratic polynomials, respectively.

Two types of external forces which pass through the center of mass of the tibia and perpendicular to the long axis of the tibia are considered. The first one is a rectangular pulse of duration t_o, and the second one is an exponentially decaying sinusoidal pulse of the same duration. A dynamic loading in the form of a rectangular pulse is extremely difficult to simulate experimentally. Exponentially decaying sinusoidal pulse has been previously used [11] as a typical dynamic load in head impact analysis. The effect of pulse duration on the dynamic response of the knee joint is examined by taking t_o = 0.05, 0.10, and 0.15 seconds for both rectangular and exponentially decaying sinusoidal pulses. The effect of pulse amplitude, A, is also investigated by taking A = 20, 60, 100, 140, and 180 Newtons for both types of pulses. Because of the space limitations, only some representative results are presented in Figs. 3-5. In Figs. 4 & 5 the values in parentheses indicate the flexion angles of the corresponding times. In conclusion, several remarks can be made on the ligament and contact forces. When the knee joint is extended dynamically all major ligaments with the exception of the posterior ligament are elongated. The magnitudes of the anterior cruciate ligament forces and the corresponding contact forces in response to a particular forcing function are comparable.

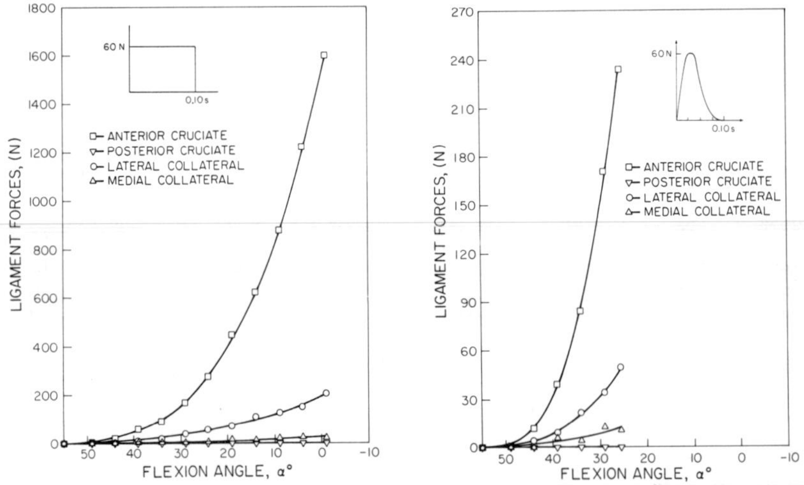

Fig. 3. Ligament forces versus flexion angle, α, are displayed for a rectangular and exponentially decaying sinusoidal pulses of 0.1 second duration.

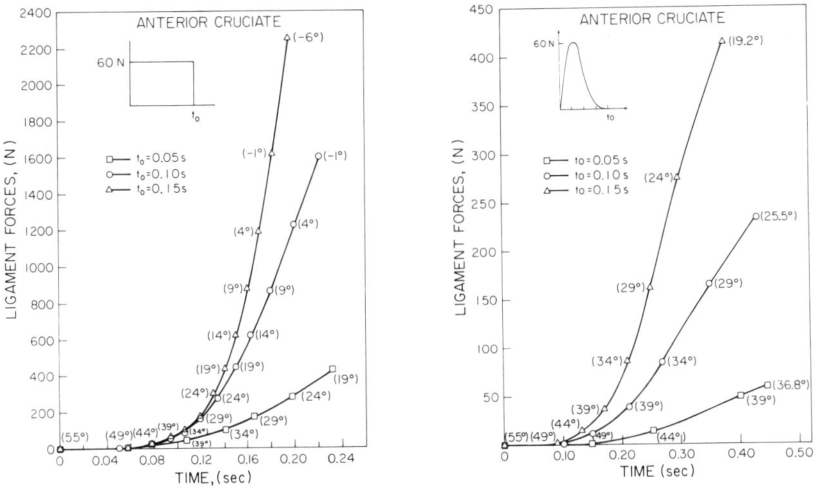

Fig. 4. Anterior cruciate ligament forces versus time are plotted for a rectangular and exponentially decaying sinusoidal pulses of various durations.

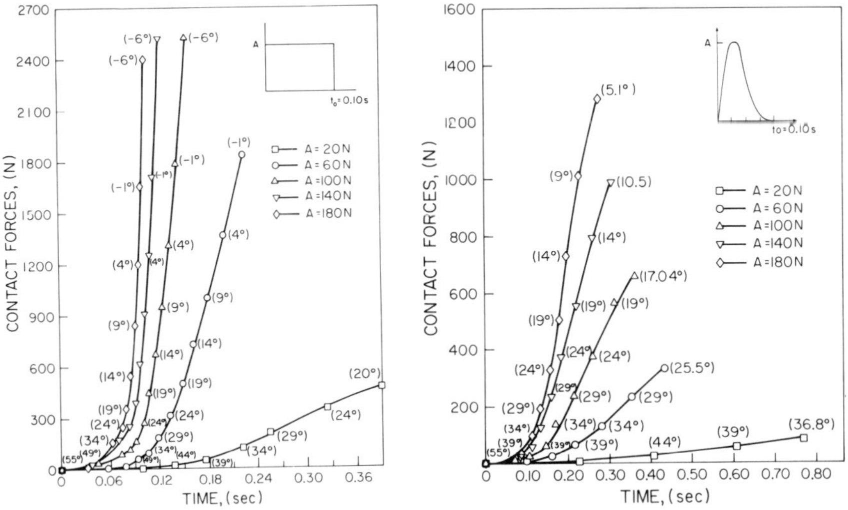

Fig. 5. Joint contact forces versus time are shown for a rectangular and exponentially decaying sinusoidal pulses of various amplitudes.

Acknowledgements-This work has been supported by the Mathematics and Analysis Branch of the Aerospace Medical Research Laboratory of the U.S. Air Force and a NATO Grant (No. 012.80) awarded to the senior author. The assistance of Dr. N. Akkaş of the Middle East Technical University, Ankara, Turkey on the numerical solution phase of the research program is also acknowledged.

REFERENCES

1. Engin, A.E., "Passive Resistive Torques about Long Bone Axes of Major Human Joints," Aviation, Space, and Environmental Medicine, Vol. 50, No. 10, pp. 1052-1057, 1979.

2. Engin, A.E., "Resistive Force and Moments in Major Human Joints," Proceedings of the Eighth Canadian Congress of Applied Mechanics, pp. 181-200, 1981.

3. King, A.I., and C.C. Chou, "Mathematical Modeling, Simulation and Experimental Testing of Biomechanical System Crash Response," Journal of Biomechanics, Vol. 9, pp. 301-317, 1976.

4. Wismans, J., F. Veldpaus, J. Janssen, A. Huson, and P. Struben, "A Three-Dimensional Mathematical Model of the Knee-Joint," Journal of Biomechanics, Vol. 13, pp. 677-686, 1980.

5. Engin, A.E., and M.H. Moeinzadeh, Modeling of Human Joint Structures, AF AMRL-TR-81-117, 1981.

6. Radin, E.L., and I.L. Paul, "A Consolidated Concept of Joint Lubrication," Journal of Bone and Joint Surgery, Vol. 54-A, pp. 607-616, 1972.

7. Bathe, K.J., and E.L. Wilson, Numerical Methods in Finite Element Analysis, Prentice-Hall, Englewood Cliffs, New Jersey, 1976.

8. Kao, R., "A Comparison of Newton-Raphson Methods and Incremental Procedures for Geometrically Nonlinear Analysis," Computers and Structures, Vol. 4, pp. 1091-1097, 1974.

9. Engin, A.E., and M.S. Korde, "Biomechanics of Normal and Abnormal Knee Joint," Journal of Biomechanics, Vol. 7, pp. 325-334, 1974.

10. Engin, A.E., "Mechanics of the Knee Joint: Guidelines for Osteotomy in Osteoarthritis," in Orthopaedic Mechanics: Procedures and Devices, edited by D.N. Ghista and R. Roaf, Academic Press, London, England, pp. 55-98, 1978.

11. Engin, A.E., and N. Akkas, "Application of a Fluid-Filled Spherical Sandwich Shell as a Biodynamic Head Injury Model for Primates," Aviation, Space and Environmental Medicine, Vol. 49, No. 1, pp. 120-124, 1978.

ON JUMP PHENOMENON IN A NONLINEAR VISCOELASTIC OSCILLATOR

I. C. Jong
Professor of Mechanical Engineering
and Engineering Science
University of Arkansas at Fayetteville

and

C. C. W. Tu
Graduate Student
University of Arkansas at Fayetteville

ABSTRACT

A nonlinear viscoelastic oscillator under sinusoidal excitations is examined. It is found that the phenomenon of sudden jumps in the amplitude-versus-forcing-frequency response curves of the oscillator, containing either a softening or hardening spring, can always be suppressed by choosing a small enough value of the forcing level. However, the phenomenon of sudden jumps in the amplitude-versus-forcing-level response curves of the oscillator can always be suppressed by choosing a large enough value of the forcing frequency if it contains a softening spring, or by choosing a small enough value of the forcing frequency if it contains a hardening spring.

INTRODUCTION

The existence of the phenomenon of sudden jumps in amplitude of a nonlinear viscoelastic oscillator, as either the forcing frequency or the forcing level is gradually varied, is well known in the theory of nonlinear vibrations [1-3]. However, complete answers to questions concerning when and how the jump phenomenon may cease to exist in the nonlinear viscoelastic oscillator have not been found in the literature. The present study is, therefore, aimed at determining the bifurcation of the presence and absence of the phenomenon of sudden jumps in amplitude of a nonlinear viscoelastic oscillator as either the forcing frequency or the forcing level is gradually varied. Results of the study reveal features different from those of a hysteretic oscillator [4,5].

FORMULATION

The system for study may be represented by a one-degree-of-freedom system where a mass m moving on a frictionless horizontal surface is connected to a fixed wall by a nonlinear viscoelastic restoring mechanism and is acted on by a horizontal rectilinear pulsating forcing function $P\cos\omega t$. The restoring mechanism is composed of a nonlinear spring in parallel with a linear dashpot.

When the mechanism is stretched or compressed by an amount of x, the nonlinear spring will develop a cubic restoring force of $kx + bkx^3$, while the dashpot will develop a restoring force of $c\dot{x}$, where k, b, and c are constants, $k > 0$, $0 < |b| << 1$, and $\dot{x}$ is the first derivative of x with respect to the time variable t. Thus, the equation of motion of the mass may be written as

$$m\ddot{x} + c\dot{x} + kx + bkx^3 = P\cos\omega t \tag{1}$$

where $\ddot{x}$ indicates the second derivative of x with respect to t. This equation is of the form of Duffing's equation with viscous damping. Letting the natural length of the spring be x_o, the amplitude of displacement of the mass be A, and using the notations

$$\xi = x/x_o \tag{2}$$

$$R = A/x_o \tag{3}$$

$$\omega_o^2 = k/m \tag{4}$$

$$\tau = \omega_o t \tag{5}$$

$$\eta = \omega/\omega_o \tag{6}$$

$$\zeta = c/(2m\omega_o) \tag{7}$$

$$B = bx_o^2 \tag{8}$$

$$G = P/(kx_o) \tag{9}$$

we can write the dimensionless form of Eq. (1) as

$$\xi'' + 2\zeta\xi' + \xi + B\xi^3 = G\cos\eta\tau \tag{10}$$

where the prime is used to denote the derivative with respect to the dimensionless time variable τ.

STEADY-STATE EQUATIONS

Since b, and hence B, has been assumed to be small, we may write the steady-state solution of Eq. (10) as

$$\xi = R\cos\theta \tag{11}$$

$$\theta = \eta\tau - \phi \tag{12}$$

where R and ϕ are slowly varying functions of τ. Neglecting higher-order terms, we write

$$\xi' = -\eta R\sin\theta \tag{13}$$

$$\xi'' = -\eta^2 R\cos\theta \tag{14}$$

The equation of motion, as given by Eq. (10), may be written as

$$\xi'' + \lambda\xi' + \kappa\xi + \varepsilon = G\cos\eta\tau \tag{15}$$

where

$$\varepsilon = 2\zeta\xi' + \xi + B\xi^3 - \lambda\xi' - \kappa\xi \tag{16}$$

By minimizing the mean square of ε with respect to λ and κ over one cycle of θ, we find that

$$\lambda = 2\zeta \tag{17}$$

$$\kappa = 1 + \frac{3}{4}BR^2 \tag{18}$$

Having minimized the mean square of ε, we may discard ε in Eq. (15) and use Eqs. (17) and (18) to obtain the equivalent linearized equation

$$\xi'' + 2\zeta\xi' + (1 + \frac{3}{4}BR^2)\xi = G\cos\eta\tau \tag{19}$$

Using Eqs. (11) through (14), we can deduce from Eq. (19) the following governing steady-state equations:

$$\tan\phi = 8\zeta\eta/(4 + 3BR^2 - 4\eta^2) \tag{20}$$

$$R^2(4 + 3BR^2 - 4\eta^2)^2 + 64\zeta^2\eta^2R^2 - 16G^2 = 0 \tag{21}$$

CONDITION OF RESONANCE

Equation (21) may be manipulated to yield

$$\eta = \{(1 + \frac{3}{4}BR^2 - 2\zeta^2) \pm [(\frac{G}{R})^2 - 4\zeta^2(1 + \frac{3}{4}BR^2 - \zeta^2)]^{\frac{1}{2}}\}^{\frac{1}{2}} \tag{22}$$

Thus, in each steady-state R-versus-η response curve, there are two values of η corresponding to each value of R. Two representative families of such steady-state response curves are drawn and shown in Figs. 1 and 2, where the dotted line in each figure indicates the backbone curve, which is the locus of the resonance peaks.

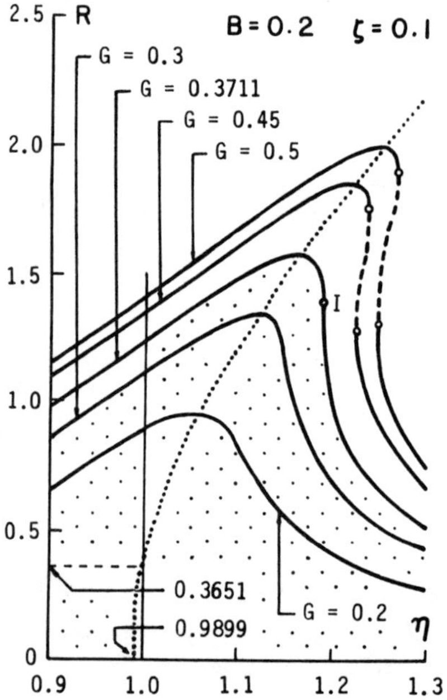

Fig. 1 Steady-state response curve of R versus η for B = 0.2 and ζ = 0.1

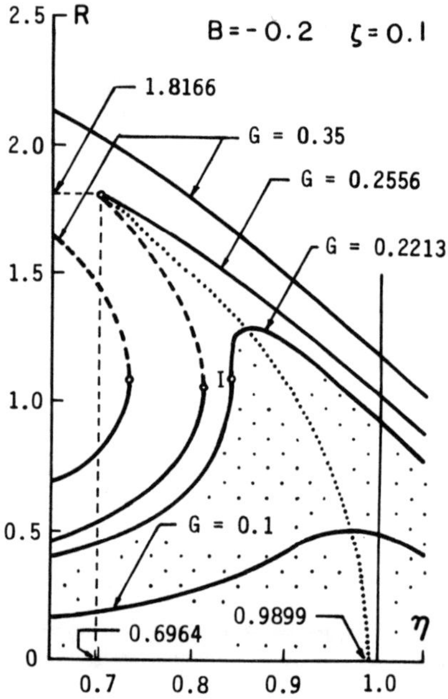

Fig. 2 Steady-state response curve of R versus η for B = -0.2 and ζ = 0.1

At the resonance peak, we have

$$\left(\frac{G}{R_r}\right)^2 - 4\zeta^2\left(1 + \frac{3}{4}BR_r^2 - \zeta^2\right) = 0 \tag{23}$$

$$\eta_r = \left(1 + \frac{3}{4}BR_r^2 - 2\zeta^2\right)^{\frac{1}{2}} \tag{24}$$

where the subscript r indicates the condition of resonance. We find from Eq. (23) that the resonance amplitude is given by

$$R_r = \left\{\frac{-2\zeta(1 - \zeta^2) + [4\zeta^2(1 - \zeta^2)^2 + 3BG^2]^{\frac{1}{2}}}{3B\zeta}\right\}^{\frac{1}{2}} \tag{25}$$

We note from Eq. (25) that, with a hardening spring ($B > 0$), a resonance peak always exists for any given positive values of ζ and G at a certain value of η. Since $\eta_r \geq 0$, we can deduce from Eqs. (24) and (25) that, with a softening spring ($B < 0$), a resonance peak exists at a certain value of η if

$$0 < \zeta \leq \frac{\sqrt{3}}{3} \quad \text{and} \quad G < \frac{2\zeta(1 - \zeta^2)}{\sqrt{-3B}} \tag{26}$$

or

$$\frac{\sqrt{3}}{3} < \zeta \leq \frac{\sqrt{2}}{2} \quad \text{and} \quad G < 4\zeta^2\sqrt{\frac{1 - 2\zeta^2}{-3B}} \tag{27}$$

If the above conditions are not true, no resonance peak exists for the oscillator containing a softening spring, but its amplitude is still bounded at the frequency $\eta = 0$, which corresponds to the static loading.

It is to be noted from Eq. (24) and the backbone curves in Figs. 1 and 2 that when the resonance amplitude parameter R_r as well as the forcing level parameter G approaches a vanishing value, the resonance forcing frequency parameter η_r will approach the value of $(1 - 2\zeta^2)^{1/2}$, which is less than 1. In other words, when the forcing level is vanishingly small, the resonance forcing frequency ω_r is always smaller than the natural frequency ω_0 of the otherwise undamped linear oscillator.

PRESENCE AND ABSENCE OF SUDDEN JUMPS IN STEADY-STATE RESPONSE CURVES

Equation (21) may also be manipulated to yield

$$G = \frac{R}{4}[(4 + 3BR^2 - 4\eta^2)^2 + 64\zeta^2\eta^2]^{\frac{1}{2}} \tag{28}$$

This equation is used to plot two representative families of steady-state R-versus-G response curves as shown in Figs. 3 and 4. The phenomenon of sudden jumps in steady-state amplitude R as either the forcing frequency η or the forcing level G is gradually varied is known to exist in most, but not all, cases for a nonlinear viscoelastic oscillator containing either a hardening spring or a softening spring. This is readily observed in Figs. 1 through 4, where the dashed segment in each response curve is, through a stability analysis, found to be unstable and cannot be realized in practice. Note that, except the inflection point I, a small circle in the figures is used to mark the point where a vertical tangent exists and whence a jump may occur. No jumps exist for curves lying in the region shaded by dots in Figs. 1 and 2.

In the $R\eta$ plane, a steady-state response curve will not have a sudden jump in amplitude if it possesses no more than one vertical tangent, where $dR/d\eta = \infty$ or $d\eta/dR = 0$. It will have a sudden jump if the response curve contains an unstable segment and possesses two vertical tangents, one at each end of the unstable segment. Thus, the critical case that separates the cases of no jump from the cases of sudden jumps is one where the response curve contains an inflection point at which the tangent is vertical. In other words, the unstable segment of the response curve in the critical case shrinks to a point and the two vertical tangents at the ends of that segment merge into one at this point.

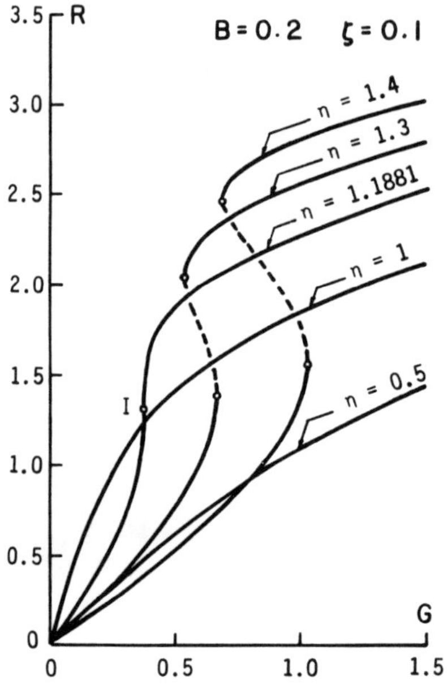

Fig. 3 Steady-state response curve of R versus G for $B = 0.2$ and $\zeta = 0.1$

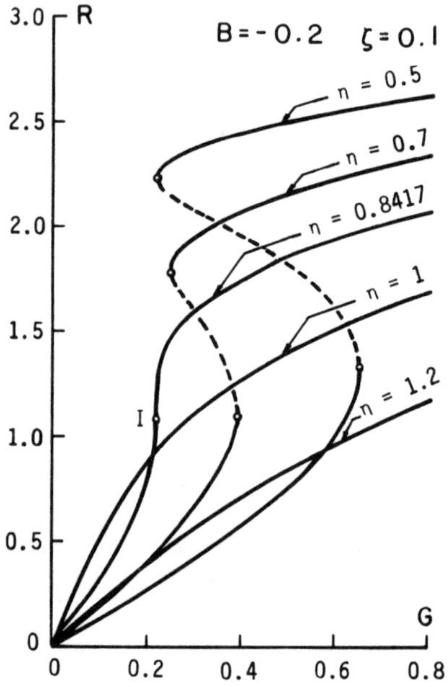

Fig. 4 Steady-state response curve of R versus G for $B = -0.2$ and $\zeta = 0.1$

To investigate the bifurcation of the presence and absence of the jump phenomenon in the $R\eta$ plane, we accordingly hold G as constant and take the first and second derivatives of both sides of the equal sign in Eq. (21) with respect to R and then set $d\eta/dR = d^2\eta/dR^2 = 0$ to obtain two additional equations which may be written as

$$E^2 + 6BR^2E + 64\zeta^2\eta^2 = 0 \tag{29}$$

$$E^2 + 30BR^2E + 36B^2R^4 + 64\zeta^2\eta^2 = 0 \tag{30}$$

where

$$E = 4 + 3BR^2 - 4\eta^2 \tag{31}$$

Equations (21) and (29) through (31) yield the following relations which exist in the critical case separating the cases of no jump from the cases of sudden jumps:

$$R = \frac{4}{3}\left[\frac{\zeta(3\zeta \pm \sqrt{9\zeta^2 + 3})}{B}\right]^{\frac{1}{2}} \tag{32}$$

$$G = \frac{16}{9\sqrt{\pm B}}[\zeta(\pm 3\zeta + \sqrt{9\zeta^2 + 3})]^{3/2} \tag{33}$$

$$\eta = [1 + 2\zeta(3\zeta \pm \sqrt{9\zeta^2 + 3})]^{\frac{1}{2}} \tag{34}$$

where the upper sign in ± applies if $B > 0$ and the lower sign in ± applies if $B < 0$.

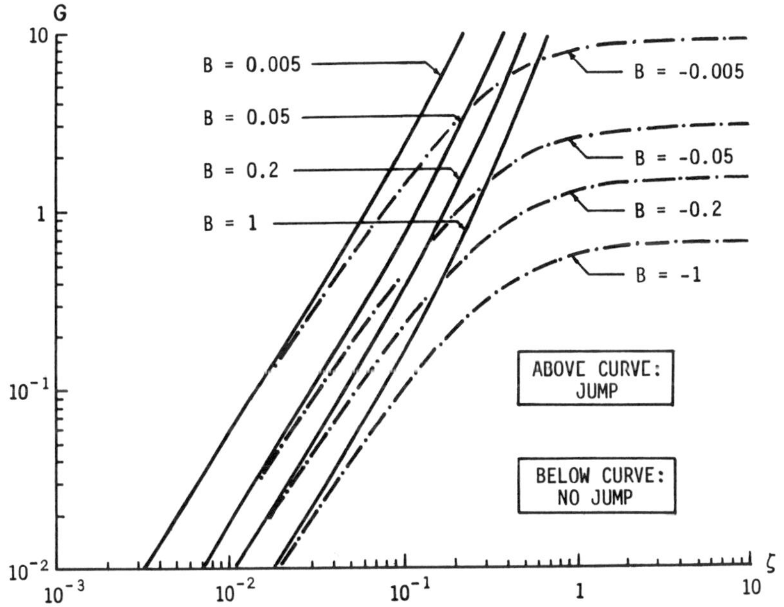

Fig. 5 Bifurcation of jump and no jump in R-versus-η response curves

It is examined and found that in the Rη plane the jump phenomenon ceases to exist if

$$G \le \frac{16}{9\sqrt{B}}[\zeta(3\zeta + \sqrt{9\zeta^2 + 3})]^{3/2} \quad \text{and} \quad B > 0 \tag{35}$$

or

$$G \le \frac{16}{9\sqrt{-B}}[\zeta(-3\zeta + \sqrt{9\zeta^2 + 3})]^{3/2} \quad \text{and} \quad B < 0 \tag{36}$$

These two results are illustrated by the curves in Fig. 5 where the solid curves are for $B > 0$ and the dash-dot curves are for $B < 0$.

Following a similar reasoning and procedure, we may investigate the bifurcation of the presence and absence of the jump phenomenon in the RG plane by holding η as constant and taking the first and second derivatives of both sides of the equal sign in Eq. (21) with respect to R and then setting $dG/dR = d^2G/dR^2 = 0$ to obtain two additional equations, which happen to be the same as Eqs. (29) and (30). Thus, Eqs. (32) through (34) are also the relations which exist in the critical case separating the cases of no jump from the cases of sudden jumps in the RG plane. It is examined and found that in the RG plane the jump phenomenon ceases to exist if

$$\eta \le [1 + 2\zeta(3\zeta + \sqrt{9\zeta^2 + 3})]^{\frac{1}{2}} \quad \text{and} \quad B > 0 \tag{37}$$

or

$$\eta \ge [1 + 2\zeta(3\zeta - \sqrt{9\zeta^2 + 3})]^{\frac{1}{2}} \quad \text{and} \quad B < 0 \tag{38}$$

These two results are illustrated by the curves in Fig. 6, where the solid curve is for $B > 0$ and the dash-dot curve is for $B < 0$.

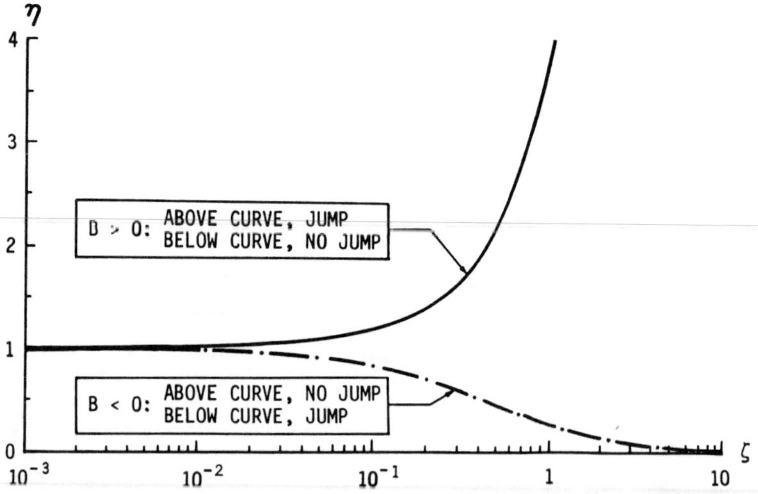

Fig. 6 Bifurcation of jump and no jump in R-versus-G response curves

CONCLUDING REMARKS

The present study investigates the bifurcation of the presence and absence of the phenomenon of sudden jumps in the steady-state amplitude of a nonlinear viscoelastic oscillator as either the forcing frequency or the forcing level is gradually varied. Results of the study are analytically expressed in Eqs. (35) through (38), which are illustrated in Figs. 5 and 6. Thses results provide complete answers to questions concerning when and how the jump phenomenon may cease to exist in the steady-state response curves of the nonlinear viscoelastic oscillator.

It has been shown in an earlier study [5] that sudden jumps in the amplitude-versus-forcing-frequency response curves of a binonlinear hysteretic oscillator can be suppressed only if the oscillator contains a hardening spring. The feasibility of suppressing sudden jumps in such response curves of a nonlinear viscoelastic oscillator containing either a softening spring or a hardening spring, as manifested in the results of the present study, is thus a contrasting feature of the nonlinear viscoelastic oscillator when it is compared with that of a corresponding binonlinear hysteretic oscillator.

REFERENCES

1. Cunningham, W. J., Introduction to Nonlinear Analysis, McGraw-Hill Book Company, Inc., New York, 1958.

2. Minorsky, N., Nonlinear Oscillations, D. Van Nostrand Company, Inc., Princeton, New Jersey, 1962.

3. Dinca, F., and Teodosiu, C., Nonlinear and Random Vibrations, Academic Press, Inc., New York, 1973.

4. Jong, I. C., "On Stability of a Circulatory System with Bilinear Hysteresis," Journal of Applied Mechanics, Vol. 32, Trans. ASME, Vol. 87, Series E, pp. 76-82, 1969.

5. Jong, I. C., Boatright, K. E., and Tu, C. O. W., "Jump Phenomenon and Its Absence in a Binonlinear Hysteretic Oscillator," Developments in Theoretical and Applied Mechanics, Vol. 10, University of Tennessee, Knoxville, TN, pp. 729-740, 1980.

COMPOSITES

TRANSIENT THERMAL STRESS ANALYSIS IN GRAPHITE/EPOXY COMPOSITE LAMINATES

C.T. Sun and J.K. Chen
School of Aeronautics and Astronautics
Purdue University
West Lafayette, Indiana 47907

Abstract

Transient responses of AS/3501-6 graphite/epoxy laminates subjected to time-dependent thermal loading are investigated by using a 9-node isoparametric finite element. The temperature field is obtained using a one-dimensional finite difference code. The effects of ablation, degradation of material properties at elevated temperatures, and radiation and convective heat losses are included in the formulation. Numerical solutions for both quasi-static and dynamic conditions are obtained and it is found that the dynamic effect is insignificant. The transient heat conduction coupled with the variation of material properties as a function of the temperature makes thermal stress responses less tractable intuitively.

Introduction

Most of the previous work in the thermal stress analysis of laminated composites dealt with static problems [1-5]. The response of an orthotropic material due to a thermal shock was first investigated by Woo et al [6-7]. However, in [6-7], the temperature through the thickness of the plate was assumed to be linear. This assumption is not adequate especially when transient heat conduction is considered. Further, the mechanical properties of the material were assumed unchanged with respect to temperature. In general, this is only reasonable in the small range of change of temperature.

When subjected to intense heat, the matrix material in the epoxy-based fiber composites softens and evaporates. As a result, the ply stiffness as well as the ply strength may decrease significantly. In the extreme situation, a burnthrough may take place and create a stress concentration for the remaining laminate.

In this paper, transient responses of laminated composites subjected to thermal loadings are presented. The thermomechanical coupling was neglected. The structural response of the laminate was analyzed using an isoparametric plate finite element developed based on the Mindlin type of plate theory. The heat source was a laser irradiation for which the temperature distribution was calculated following a finite difference method used by Griffis et al [8]. The temperature-dependent mechanical properties of the composites have been taken into account. In the analysis of the dynamic response, Newmark integration scheme was employed.

Temperature

The thermal loading was assumed to be a uniform high-intensity energy flux applied over an area whose characteristic dimension is significantly larger than the thickness of the laminate and the time duration is short. Since graphite/epoxy composites are poor heat conductors, the heat conduction in the lateral directions in the laminate was neglected and the heat conduction was assumed to take place only in the thickness direction. In other words, the heat conduction was assumed to be one-dimensional.

Due to the high temperature induced, material properties vary over the thickness of the laminate. In addition, the material was assumed to undergo a sublimation reaction when the surface temperature reached a critical value (the ablation temperature). Mathematically, this surface ablation may be considered equivalent to a classical melting problem in which the melt is instantaneously removed.

With respect to the moving coordinate as shown in Fig. 1, the one-dimensional heat conduction equation is given by

$$\frac{\partial}{\partial \bar{z}'}\left[\kappa(T)\frac{\partial T}{\partial \bar{z}'}\right] + \rho(T)\,C_p(T)\,V\frac{\partial T}{\partial \bar{z}'} = \rho(T)\,C_p(T)\frac{\partial T}{\partial t} \tag{1}$$

where $T(\bar{z}', t)$ indicates the temperature, κ, ρ, and C_p are the thermal conductivity, the mass density, and the heat capacity, respectively; $\bar{z}' = \bar{z} - \bar{z}_b$ where $\bar{z}'$ and $\bar{z}$ are the convected and global coordinates in the thickness direction, respectively; $\bar{z}_b$ denotes the intantaneous position of the receding surface; and $V \equiv d\bar{z}_b/dt$ is the surface recession rate which is regarded as an unknown function of time as the material ablates.

Let t_0 and T_S denote the time of onset of ablation and ablation temperature, respectively. At the irradiation surface ($\bar{z}' = 0$), the boundary conditions include

$$T(0, t) = \begin{cases} \text{variable} & t < t_o \\ T_S = \text{constant} & t_o < t \end{cases} \tag{2}$$

and a straightforward energy balance given by

$$\rho H_S \dot{V} = \alpha I_0 + \kappa \frac{\partial T(0, t)}{\partial \bar{z}'} + I_r + I_c \tag{3}$$

in which H_S is an effective heat of ablation (assumed constant) and I_0 represents the incident heat flux; I_r and I_c are the surface energy losses due to radiation and convection, respectively; and α denotes the fraction of the incident energy which is absorbed by the composite. In this study, α was assumed to be 0.92 [8].

The radiation loss cited in Eq. (3) is computed according to

$$I_r = \bar{\sigma}\, \bar{\varepsilon}\, [T_0^4 - T^4 (0, t)] \tag{4}$$

where $\bar{\sigma}$ is the Stefan-Boltzman constant; $\bar{\varepsilon}$ is the surface emissity; and T_0 is the surrounding air temperature. The value of $\bar{\varepsilon}$ was taken to be 0.92 same as that used in [8].

The convective heat loss, I_c, resulting from the flow of air parallel to the irradiated surface is given by

$$I_c = h_c\, [T_r - T(0, t)] \tag{5}$$

where h_c and T_r are the convection coefficient and air recovering temperature, respectively. The convection coefficient is evaluated from the semi-empirical expression given in [9] as

$$h_c = 1.14252 \times 10^{-4} \left(\frac{T_e^{3/2}}{T_e + 201.6}\right) \left(\frac{T_o + 198.6}{T_e^{5/2}}\right)^{1/2} \tag{6}$$

$$\times (1. - 8.889 \times 10^{-6}\, T_e + 3.58 \times 10^{-8}\, T_e^2)$$

in which

$$T_e = 0.5\, T(0, t) + 266.7$$

The unit of h_c is BUT/in^2 - sec - ^{o}R.

The rear surface is assumed to be insulated, i.e.,

$$\frac{\partial T}{\partial \bar{z}'} = 0 \quad \text{at} \quad \bar{z}' = h - \bar{z}_b \tag{7}$$

where h is the initial laminate thickness.

Griffis et al [8] used a modified Crank-Nicholson finite difference method to solve the nonlinear boundary value problem defined by Eqs. (1-7) for a twenty-ply AS/350 1-6 graphite/epoxy composite with a symmetric layup subjected to different intensities of laser beam heating. The thermo-physical properties for this composite are presented in Table 1. Good agreement between their numerical solutions and experimental data were obtained.

Table 1 Thermo-physical properties of the graphite/epoxy composite

Temp. (°F)	Conductivity (BTU/sec-in-°F)	Heat Capacity (BTU/lbm-°F)	Density (lbm/in^3)
50	0.185185×10^{-4}	0.30	0.057685
625	0.871101×10^{-5}	0.491667	
675	0.785818×10^{-5}	0.197592	
925	0.395405×10^{-5}	1.172508	
975	0.305347×10^{-5}	0.381046	0.041534
1050	0.275×10^{-5}	0.384314	
1150	0.250×10^{-5}	0.388671	
1300	0.240×10^{-5}	0.395206	
1500	0.231482×10^{-5}	0.403921	
6000	0.231482×10^{-5}	0.60	

In the present study, two laminate systems were considered. One system is a twenty-ply $[+45, -45, 90, 0, 0, 0, -45, +45, 0, 0]_s$ laminate and the other is a four-ply $[0, 90]_s$ laminate. The ply thickness is 0.005 inch. The intensity of the incident laser beam is given by

$$I_0 = \begin{cases} 17.1 \text{ BTU/sec-in}^2 & t < 0.2 \text{ sec.} \\ 0 & t > 0.2 \text{ sec.} \end{cases} \tag{8}$$

The same finite difference scheme used in [8] was employed. In view of the expected severe temperature gradient near the heat source, the finite difference grid was finer near the heat recipient (or the front) surface and coarser with increasing distance from it. However, in each ply, the grid spacing was uniform. The temperature time-history in the four-ply laminate is presented in Table 2.

Table 2 Temperatures at the interfaces in the 4-ply laminate; location 1 is the front surface

Locat. / Temp(°F) / Time(sec)	1	2	3	4	5	Total Thickness (in.)
0.01	2517.	256.6	95.56	79.04	77.56	0.02
0.02	6001.	385.3	144.0	88.78	81.18	0.02
0.03	6001.	486.0	194.9	108.3	90.99	0.01988
0.04	6001.	581.6	240.7	133.1	107.9	0.01983
0.05	6001.	649.2	284.1	160.1	129.3	0.01973
0.06	6001.	723.4	323.1	188.1	153.2	0.01966
0.07	6001.	791.9	359.0	216.2	178.1	0.01956
0.10	6001.	1204.	454.4	296.7	253.2	0.01930
0.11	6001.	1398.	483.5	321.8	277.2	0.01921
0.12	6001.	1466.	510.4	346.1	300.5	0.01912
0.15	6001.	1792.	575.8	412.1	365.8	0.01883
0.16	6001.	1874.	594.1	431.8	385.9	0.01873
0.17	6001.	1979.	610.3	450.9	405.0	0.01864
0.20	6001.	2490.	664.7	502.9	457.4	0.01835
0.25	2643.	2258.	758.0	577.3	531.2	0.01835
0.30	2024.	1976.	838.3	635.3	593.0	0.01835

Laminate Deformation and Stress

A nine-node isoparametric plate finite element (see Fig. 2) has been developed for the analysis of the thermoelastic response of the laminated composites. Mindlin type of plate theory was used. The displacement components in the laminate are given by

$$u = u_0(x, y, t) - z\,\psi_x(x, y, t)$$
$$v = v_0(x, y, t) - z\,\psi_y(x, y, t) \tag{9}$$
$$w = w_0(x, y, t)$$

where u_0, v_0, and w_0 are the mid-plane displacement components in the x-, y-, and z-axis, respectively; and ψ_x and ψ_y are the rotations of the cross-sections perpendicular to the x- and y-axis, respectively.

Each lamina is considered as an orthotropic solid. For each lamina, the stress-strain relations are given by

$$\begin{Bmatrix} \sigma_{xx} \\ \sigma_{yy} \\ \sigma_{xy} \\ \sigma_{yz} \\ \sigma_{xz} \end{Bmatrix} = \begin{bmatrix} \bar{Q}_{11} & \bar{Q}_{12} & \bar{Q}_{16} & 0 & 0 \\ \bar{Q}_{12} & \bar{Q}_{22} & \bar{Q}_{26} & 0 & 0 \\ \bar{Q}_{16} & \bar{Q}_{26} & \bar{Q}_{66} & 0 & 0 \\ 0 & 0 & 0 & \bar{Q}_{44} & \bar{Q}_{45} \\ 0 & 0 & 0 & \bar{Q}_{45} & \bar{Q}_{55} \end{bmatrix} \begin{Bmatrix} \varepsilon_{xx} - \alpha_x \Delta T \\ \varepsilon_{yy} - \alpha_y \Delta T \\ \gamma_{xy} - \alpha_{xy} \Delta T \\ \gamma_{yz} \\ \gamma_{xz} \end{Bmatrix} \tag{10}$$

where $\bar{Q}_{ij}$ (i, j = 1, 2, 6) are the reduced in-plane stiffnesses for a state of generalized plane stress [10], T is the temperature rise from the reference temperature, and α_x, α_y, and α_{xy} are the transformed thermal expansion coefficients of the lamina which are related to the thermal expansion coefficients in the fiber direction α_1 and the transverse direction α_2 as [10]

$$\alpha_x = \cos^2\theta\,\alpha_1 + \sin^2\theta\,\alpha_2$$
$$\alpha_y = \sin^2\theta\,\alpha_1 + \cos^2\theta\,\alpha_2 \tag{11}$$
$$\alpha_{xy} = 2(\alpha_1 - \alpha_2)\cos\theta\sin\theta$$

in which θ is the angle between the x-axis and the fiber orientation. In Eq. (10), the strain components are related to the mid-plane displacements of the laminated plate as

$$\{\varepsilon\} = \begin{Bmatrix} \varepsilon_{xx} \\ \varepsilon_{yy} \\ \gamma_{xy} \\ \gamma_{yz} \\ \gamma_{xz} \end{Bmatrix} = \begin{Bmatrix} \varepsilon^{\circ} \\ 0 \end{Bmatrix} + \begin{Bmatrix} z\ \kappa \\ \gamma_z^{\circ} \end{Bmatrix} \tag{12}$$

where

$$\{\varepsilon^{\circ}\} = \begin{Bmatrix} \dfrac{\partial u_0}{\partial x} \\ \dfrac{\partial v_0}{\partial y} \\ \dfrac{\partial u_0}{\partial y} + \dfrac{\partial v_0}{\partial x} \end{Bmatrix} \tag{13}$$

$$\{\kappa\} = \begin{Bmatrix} -\dfrac{\partial \psi_x}{\partial x} \\ -\dfrac{\partial \psi_y}{y} \\ -\dfrac{\partial \psi_x}{\partial y} - \dfrac{\partial \psi_y}{\partial x} \end{Bmatrix} \tag{14}$$

$$\{\gamma_z^{\circ}\} = \begin{Bmatrix} \dfrac{\partial w}{\partial y} - \psi_y \\ \dfrac{\partial w}{\partial x} - \psi_x \end{Bmatrix} \tag{15}$$

are the in-plane strains, the slopes of rotations and the transverse shear strains, respectively.

Integration of the stresses and stresses multiplied with z over the thickness of the laminate, h, yields the in-plane resulting forces, moments and transverse shear forces as

$$\{N\} = \begin{Bmatrix} N_x \\ N_y \\ N_{xy} \end{Bmatrix} = [A]\{\varepsilon^\circ\} + [B]\{\kappa\} - \{N^T\} \tag{16}$$

$$\{M\} = \begin{Bmatrix} M_x \\ M_y \\ M_{xy} \end{Bmatrix} = [B]\{\varepsilon^\circ\} + [D]\{\kappa\} - \{M^T\} \tag{17}$$

$$\{Q\} = \begin{Bmatrix} Q_y \\ Q_x \end{Bmatrix} = [H]\{\gamma_z^\circ\} \tag{18}$$

In Eqs. (16-18),

$$(A_{ij}, B_{ij}, D_{ij}) = \int_{-h/2}^{h'} \bar{Q}_{ij}\,(1, z, z^2)\,dz, \quad i, j = 1, 2, 6$$

$$H_{ij} = \int_{-h/2}^{h'} \bar{Q}_{ij}\,dz \qquad i, j = 4, 5$$

$$(N_i^T, M_i^T) = \int_{-h/2}^{h'} \bar{Q}_{ij}\,(1, z)\,\bar{\alpha}_j\,\Delta T\,dz \qquad i, j = 1, 2, 6$$

where $h' = \frac{h}{2} - \bar{z}_b(t)$, $\bar{\alpha}_1 = \alpha_x$, $\bar{\alpha}_2 = \alpha_y$, $\bar{\alpha}_6 = \alpha_{xy}$, $N_1^T = N_x^T$, $N_2^T = N_y^T$, $N_6^T = N_{xy}^T$, $M_1^T = M_x^T$, $M_2^T = M_y^T$, and $M_6^T = M_{xy}^T$. In this study, we assume that the original mid plane coincides with the plane $z = 0$. The front (top) surface may recede if the surface temperature reaches the temperature of ablatron.

Finite Element Formulation

The finite element used in this study is a 9-node isoparametric element, see Fig. 2. Following the standard isoparametric finite element formulation, the plate displacements are expressed in the form

$$\begin{Bmatrix} u_0 \\ v_0 \\ w_0 \\ \psi_x \\ \psi_y \end{Bmatrix} = \sum_{i=1}^{9} [\phi]_i \{\delta\}_i \tag{19}$$

where $\{\delta\}_i = [u_{oi}, v_{oi}, w_{oi}, \psi_{xi}, \psi_{yi}]^T$ are the nodal displacements at node i, and

$$[\phi]_i = \phi_i[I]$$

in which $[I]$ is the 5x5 identity matrix and ϕ_i are the shape functions given by

$$\phi_1 = \frac{1}{4}(1-\xi)(1-\eta) - \frac{1}{2}\phi_5 - \frac{1}{2}\phi_8 - \frac{1}{4}\phi_9$$

$$\phi_2 = \frac{1}{4}(1+\xi)(1-\eta) - \frac{1}{2}\phi_5 - \frac{1}{2}\phi_6 - \frac{1}{4}\phi_9$$

$$\psi_3 = \frac{1}{4}(1+\zeta)(1-\eta) - \frac{1}{2}\phi_6 \;\; \frac{1}{2}\phi_7 \;\; \frac{1}{4}\phi_9$$

$$\phi_4 = \frac{1}{4}(1+\xi)(1-\eta) - \frac{1}{2}\phi_7 - \frac{1}{2}\phi_8 - \frac{1}{4}\phi_9$$

$$\phi_5 = \frac{1}{2}(1-\xi^2)(1-\eta) - \frac{1}{2}\phi_9$$

$$\phi_6 = \frac{1}{2}(1-\xi)(1-\eta^2) - \frac{1}{2}\phi_9$$

$$\phi_7 = \frac{1}{2}(1-\xi^2)(1+\eta) - \frac{1}{2}\phi_9$$

$$\phi_8 = \frac{1}{2}(1-\xi)(1-\eta^2) - \frac{1}{2}\phi_9$$

$$\phi_9 = (1-\xi^2)(1-\eta^2)$$

In this study, the mechanical loading was not included. Following the standard procedure for finite element formulations, the equations of motion for the discretized system are obtained as

$$[M]\{\ddot{\delta}\} + [K]\{\delta\} = \{F^T\}$$

where $\{\delta\}$ is the nodal displacement vector, $[M]$, $[K]$ and $\{F^T\}$ are the system mass matrix, stiffness matrix, and the generalized thermal loading vector, respectively. These system matrices are obtained by assembling the element mass matrix $[m]$, stiffness matrix $[k]$, and the generalized thermal loading vector $\{f^T\}$, which are given as

$$[m] = \int_{A_e} [\phi]^T \begin{bmatrix} \bar{\rho} & & & & \\ & \bar{\rho} & & 0 & \\ & & \bar{\rho} & & \\ & 0 & & I & \\ & & & & I \end{bmatrix} [\phi] dA \tag{20}$$

$$[k] = \int_{A_e} [B_0]^T \begin{bmatrix} [A] & [B] & 0 \\ [B] & [D] & 0 \\ 0 & 0 & [H] \end{bmatrix} [B_0] dA \tag{21}$$

$$\{f^T\} = \int_{A_e} [B_0]^T \begin{Bmatrix} \{N^T\} \\ \{M^T\} \\ 0 \\ 0 \end{Bmatrix} dA \tag{22}$$

in which A_e is the area of the finite element and

$$\bar{\rho} = \int_{-h/2}^{h'} \rho \, dz \tag{23}$$

$$I = \int_{-h/2}^{h'} \rho z^2 \, dz \tag{24}$$

$$[B_o] = [B_{01} \; B_{02} \; \cdots \; B_{09}] \tag{25}$$

where

$$
[B_{oi}] = \left[\begin{array}{cc|ccc}
\frac{\partial \phi_i}{\partial x} & 0 & & & \\
0 & \frac{\partial \phi_i}{\partial y} & & 0 & \\
\frac{\partial \phi_i}{\partial y} & \frac{\partial \phi_i}{\partial x} & & & \\
\hline
 & & 0 & -\frac{\partial \phi_i}{\partial x} & \\
 & & 0 & 0 & -\frac{\partial \phi_i}{\partial y} \\
 & 0 & 0 & -\frac{\partial \phi_i}{\partial y} & -\frac{\partial \phi_i}{\partial x} \\
 & & \frac{\partial \phi_i}{\partial x} & -\phi_i & 0 \\
 & & \frac{\partial \phi_i}{\partial y} & 0 & -\phi_i
\end{array}\right] \tag{26}
$$

Rectangular elements with 9 nodes were used in the present study. The resulting mass and stiffness matrices are 45 x 45 for each element.

In using the finite element developed based upon Mindlin type of plate theory, numerical inaccuracy often arises due to the small transverse shear deformation in thin plates. The so-called reduced integration technique has been employed to achieve better results [11]. In the present work, the 3x3 Gaussian rule was used to evaluate the stiffness coefficients for the in-plane and bending deformations, and the 2x2 Gaussian rule for the stiffness coefficients for the transverse shear deformation.

Numerical Results

AS/3501-6 graphite/epoxy laminates were considered. The layups are $[0, 90]_s$ and $[+45, -45, 90, 0, 0, 0, -45, +45, 0, 0]_s$. The temperature-dependent properties of the composite are presented in Table 3. The properties at other temperatures not listed in Table 3 were assumed to vary linearly between those at two nearest temperatures given in the table.

Table 3 Temperature-dependent mechanical properties of the graphite/epoxy composite, $G_{13} = G_{23} = G_{12}$

Temp(°F) properties	75	400	500	1100	6000
$E_1(10^6$ psi)	20.5	20.5	20.5	20.5	20.5
$E_2(10^6$ psi)	1.9	1.5	0.02	0.01	0.01
$G_{12}(10^6$ psi)	1.35	1.08	0.01	0.0005	0.0005
ν_{12}	0.28	0.28	0.28	0.28	0.28
$\alpha_1(10^{-6}/°F)$	0.01	0.03	0.03	0.03	0.03
$\alpha_2(10^{-6}/°F)$	12.0	21.0	21.0	21.0	21.0

Three different thermal loading and boundary conditions as shown in Fig. 3 were considered. Case 1 is a clamped 4-ply laminate, case 2 is a simply-supported 4-ply laminate, and case 3 involves the 20-ply laminate with simply-supported boundary conditions. These laminates are square and 10" in each direction. For the simply-supported boundary conditions we set $u_0 = v_0 = w_0 = \psi_y$ (or ψ_x) = 0 along the edges in the x-direction (or y-direction), and for the clamped edges we set $u_0 = v_0 = w_0 = \psi_x = \psi_y = 0$.

In the heated region, the temperature was assumed to be uniform in the x-y plane. The temperature variation in the z-direction was calculated using the finite difference code described in Section 2. The temperature variation through the thickness in each lamina was approximated by a linear distribution. For the region outside of the directly heated area, the temperature was assumed to stay at the reference temperature.

For the $[0, 90]_s$ laminate under the loadings considered, x- and y-axes are axes of symmetry for both in-plane and flexural deformations. Thus a quadrant of the plate was taken for analysis. For the 20-ply laminate, there is no such symmetry, and the whole plate was analyzed. The finite element mesh for each case is shown in Fig. 3.

Figures 4-5 show the thermal forces $\{N^T\}$ and moments $\{M^T\}$ in the heated region of the 4-ply laminate subjected to a heat source given by Eq. (8). It is interesting to note that the thermal forces and moments do not attain maximum values at the end of the heating as one might have expected. It is also noted that there are several abrupt turns in the curves. These phenomena occur as a result of the ablation as well as the material degradation at higher temperatures.

The time-histories of the thermal loadings were then incorporated into the finite element program for both dynamic and quasi-static analyses of the laminates. The Newmark integration method was used to integrate the time variable in the equations of motion. The time step used in the numerical integration was chosen to be 0.01 sec.

The results for case 1 are presented in Figs. 6-10. Figure 6 shows the deflections at the center of the plate according to the dynamic and quasi-static analyses. It is seen that the plate oscillates first and then approaches to zero. The deflection as well as the in-plane forces and moments reduce to zero toward the end of the heating period as the thermal loading decreases. In Figs. 6 and 9, dynamic solutions are also shown for comparison. It is evident that the dynamic effect is quite insignificant. Although not shown in the figures, the dynamic solutions for M_{yy} and the in-plane forces are almost identical to the quasi-static solutions.

Results for case 2 are shown in Figs. 11-15, which bear close resemblance to those for case 1. However, the dynamic effect seems to be more pronounced in the deflection and M_{xx}, probably due to a larger heated area. The dynamic effect on M_{yy}, M_{xy} and the in-plane forces is negligible.

Figures 16-19 present results for the 20-ply laminate. It is noted that after the heating ceases, the deflection and the thermal stresses do not drop to zero as in the 4-ply laminate. It is interesting to note that M_{xx} undergoes a complete reversal in direction. In this case, both M_{xx} and M_{yy} show some dynamic effects.

References

1. W.H. Pell, "Thermal Deflections of Anisotropic Thin Plate", Q. Appl. Math., Vol. 4, pp. 27-44, 1946.

2. Y. Stavsky, "Thermoelasticity of Heterogeneous Aeolotropic Plates", J. Eng. Mech. Div., Proc. ASCE, Vol. 89, pp. 89-105, 1963.

3. C.H. Wu and T.R. Tauchert, "Thermoelastic of Laminated Plates 1: Symmetric specially Orthotropic Laminates", J. Therm. Stress, Vol. 3, pp. 247-259, 1980.

4. C.H. Wu and T.R. Tauchert, "Thermoelastic Analysis of Laminated Plates 2: Antisymmetric cross-ply and angle-ply Laminates", J. Therm. Stress, Vol. 3, pp. 365-378, 1980.

5. J.N. Reddy and Y.S. Hsu, "Effects of Shear Deformation and Anisotropy on the Thermal Bending of Layered Composite Plates", J. Therm. Stress, Vol. 3, pp. 475-493, 1980.

6. C.L.D. Huang and H.K. Woo, "The Fundamental Heat Conduction Solution for a Thermal Shock on a Finite Orthotropic Cylindrical Thin Shell", Nucl. Eng. Des., Vol. 45, pp. 67-79, 1978.

7. C.L. D. Huang and H.K. Woo, "Thermal Stresses and Displacements Induced in a Finite, Orthotropic, Cylindrical, Thin Shell by an Instantaneous Shock", J. Therm. Stress, Vol. 3, pp. 277-293, 1980.

8. C.A. Griffis, R.A. Masumura, and C.I. Chang, "Thermal Response of Graphite Epoxy Composite Subjected to Rapid Heating", NRL Memorandom Report, 4479, March 31, 1981.

9. N.P. Hobbs, T.A. Dalton, and R.F. Smiley, "TRAP2-A Digital Computer Program for Calculating the Response of Mechanically Loaded Structures to Laser Irradiation", KATR-143 Kamon Avidyne (Burlington, Mass.), Oct. 1977.

10. R.M. Jones, Mechanics of Composite Materials, Scripta Book Company, Washington, D.C., 1975.

11. O.C. Zienkiewicz, R.L. Taylor, and J.M. Too, "Reduced Integration Technique in General Analysis of Plate and Shell", Int. J. Numer. Meth. Eng., Vol. 3, pp. 575-586, 1971.

Acknowledgments

This work was sponsored by Naval Research Laboratory under Contract No. N00014-81-2021 with Purdue University. The authors are grateful to Drs. C.I. Chang and C.A. Griffis for their valuable discussions.

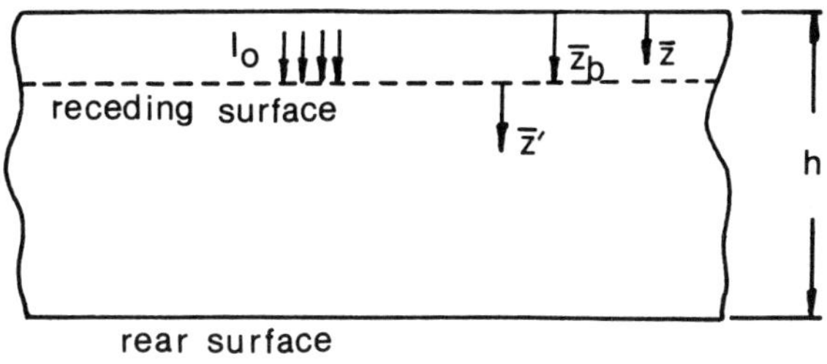

Fig. 1 Moving coordinate for one-dimension heat transfer model.

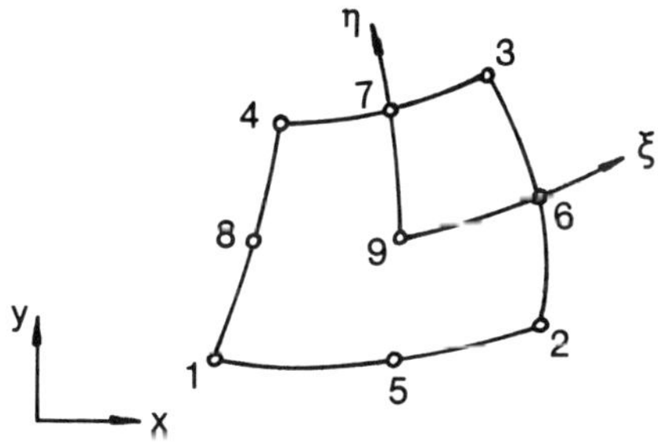

Fig. 2 The 9-node isoparametrid finite element.

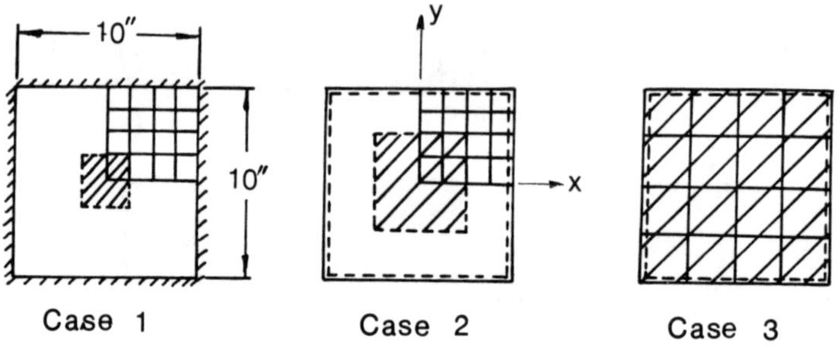

Fig. 3 Three cases of thermal loadings and boundary conditions; shaded areas indicate the heated regions.

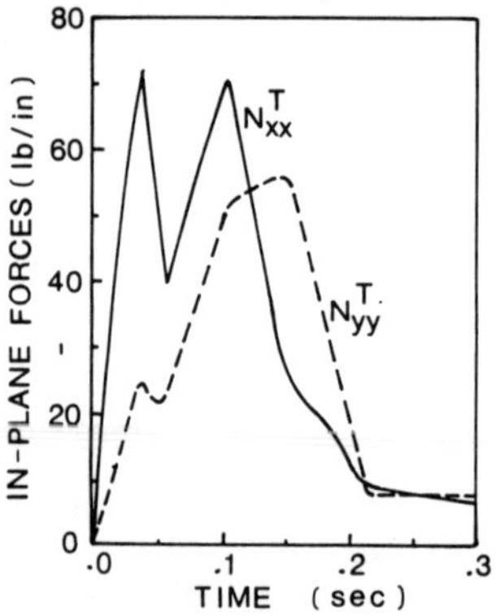

Fig. 4 In-plane thermal forces N^T_{xx}, N^T_{yy} (case 1 and case 2)

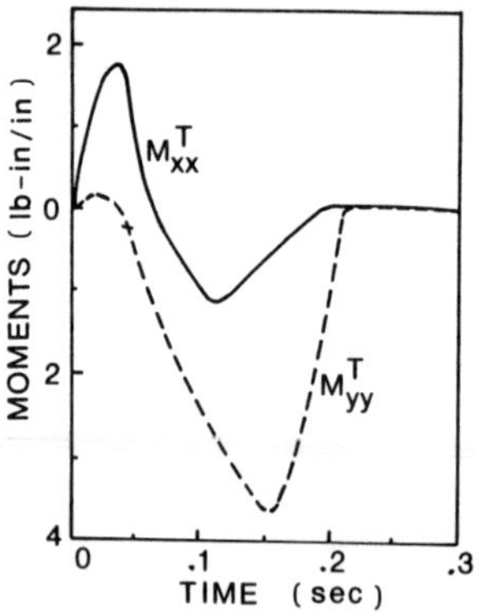

Fig. 5 Thermal moments M^T_{xx} and M^T_{yy} (case 1 and case 2)

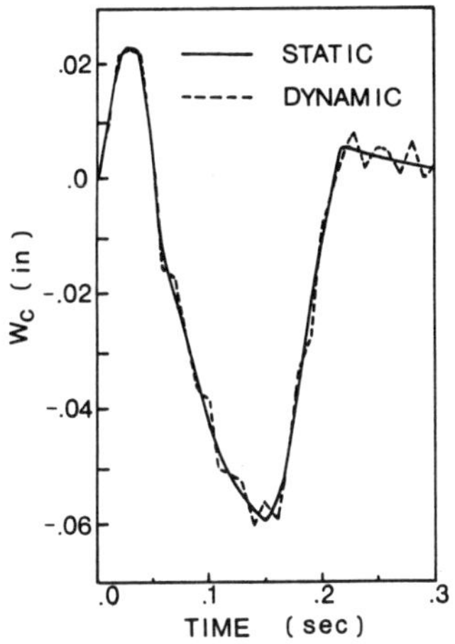

Fig. 6 Center deflection time-history (case 1).

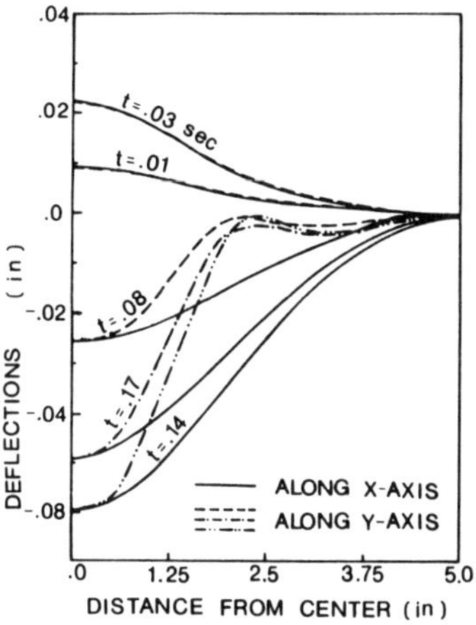

Fig. 7 Profiles of deflection at different times according to the dynamic analysis (case 1).

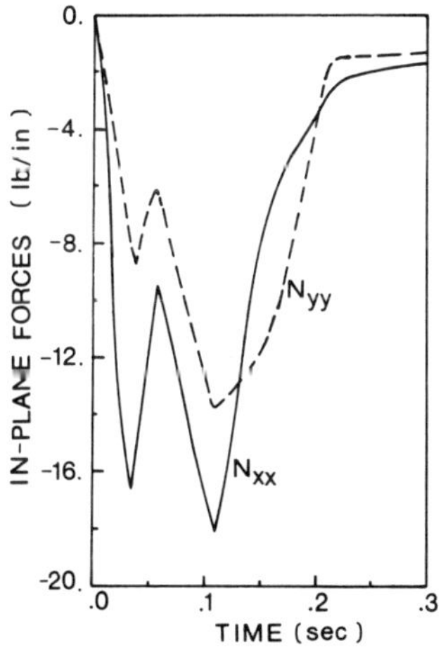

Fig. 8 In-plane forces N_{xx} and N_{yy} at the center of plate (case 1).

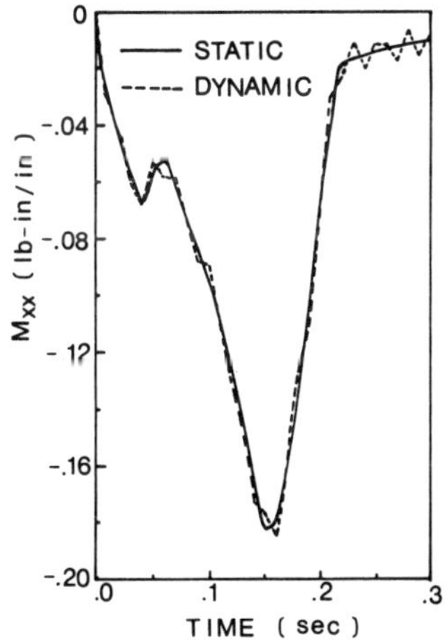

Fig. 9 Bending moment M_{xx} at the center of plate (case 1).

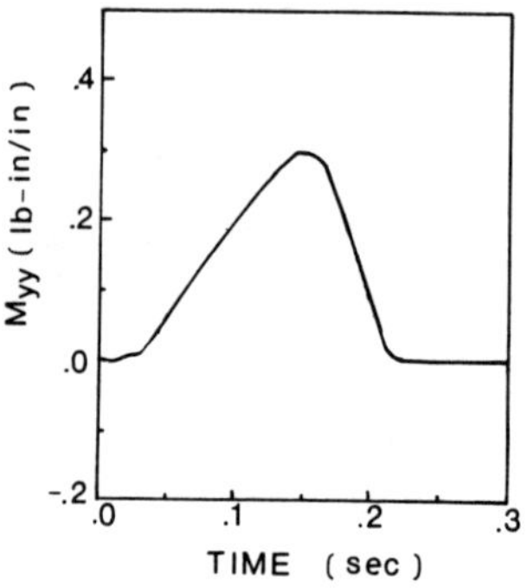

Fig. 10 Bending moment M_{yy} at the center of plate (case 1).

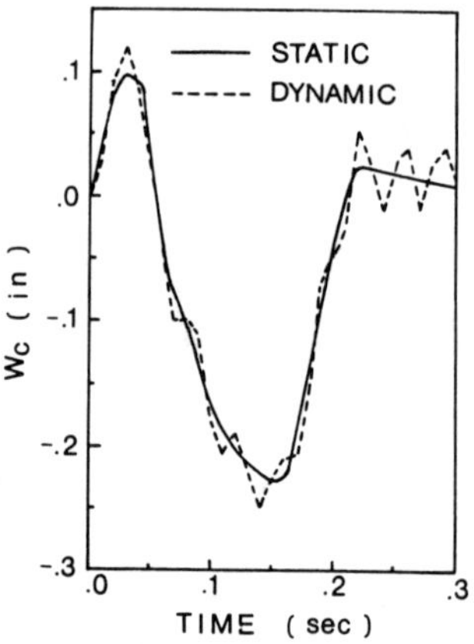

Fig. 11 Center deflection time-history (case 2).

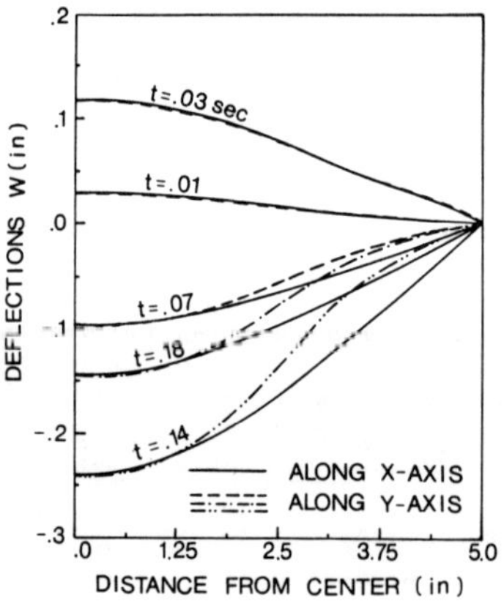

Fig. 12 Profiles of deflection at different times according to the dynamic analysis (case 2).

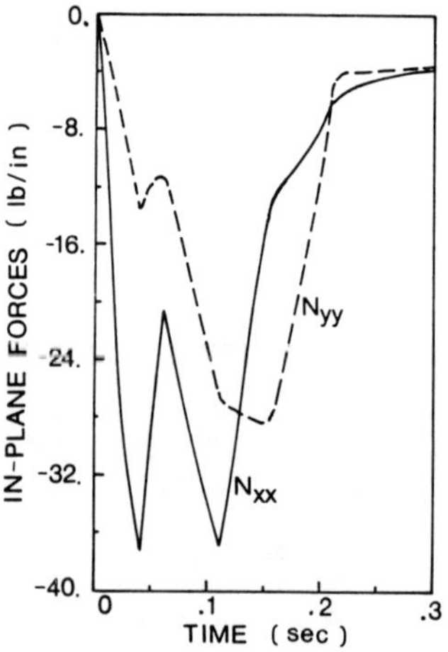

Fig. 13 In-plane forces at the center of plate (case 2)

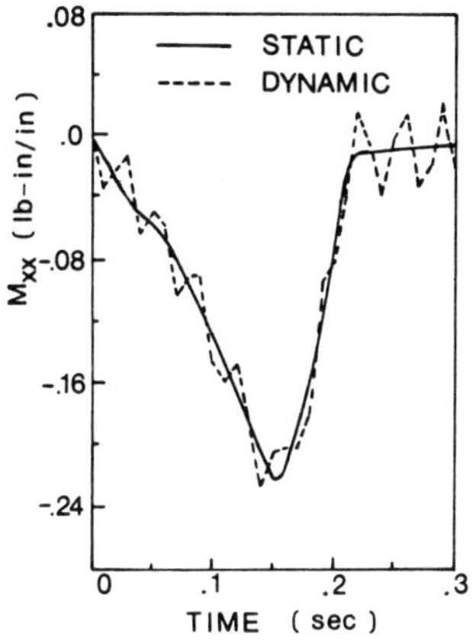

Fig. 14 Bending moment M_{xx} at the center of plate (case 2)

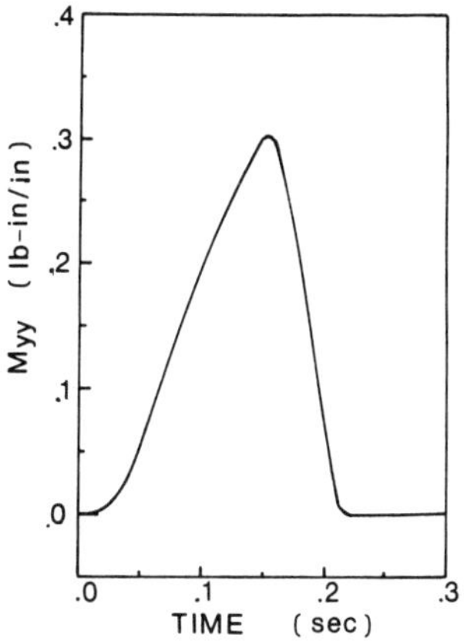

Fig. 15 Bending moment M_{yy} at the center of plate (case 2).

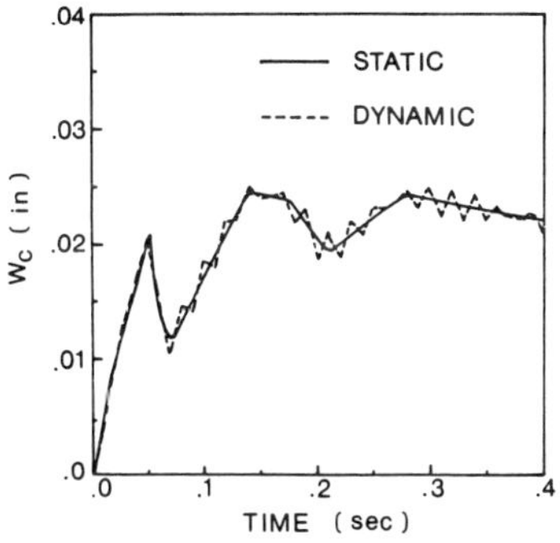

Fig. 16 Deflection at the center of plate (case 3).

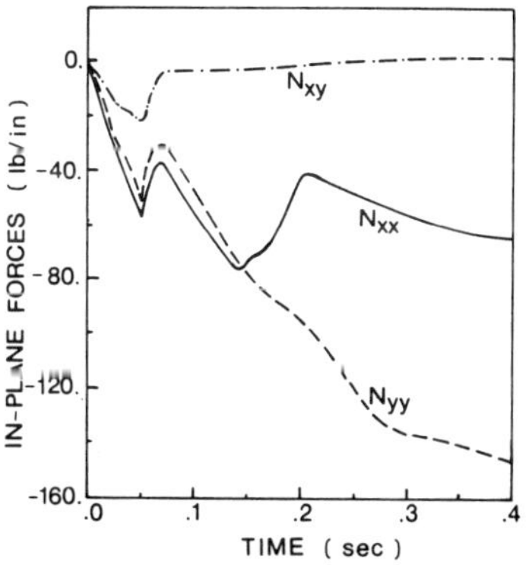

Fig. 17 In-plane forces at the center of plate (case 3).

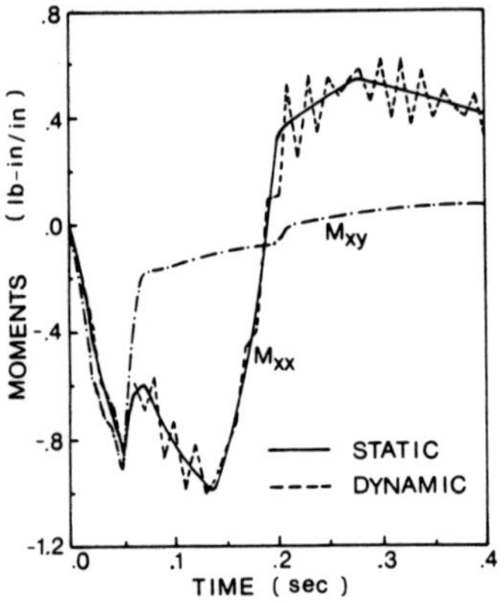

Fig. 18 Bending moment M_{xx} and twisting moment M_{xy} at the center of plate (case 3).

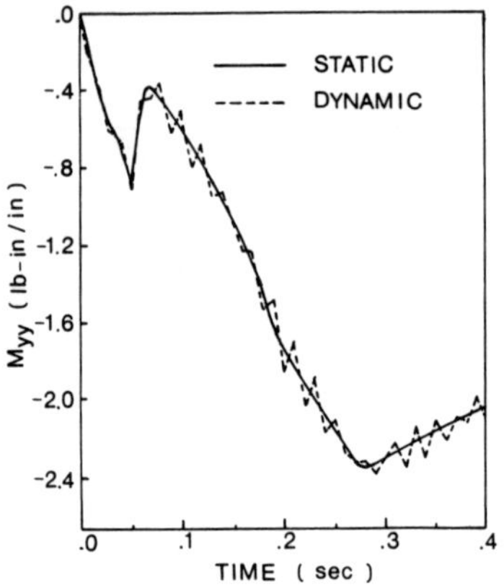

Fig. 19 Bending moment M_{yy} at the center of plate (case 3).

NON-LINEAR VIBRATION OF RECTANGULAR COMPOSITE PLATES WITH RECTANGULAR CUTOUTS

J. N. Reddy
Department of Engineering Science and Mechanics
Virginia Polytechnic Institute and State University
Blacksburg, Virginia 24061

ABSTRACT

Layered composite plates of rectangular planform with rectangular cutouts are analyzed for nonlinear (in the von Karman sense) vibration using the shear-flexible element developed previously by the author. Numerical results are presented showing the effects of side-to-thickness ratio, aspect ratio, plate side-to-cutout ratio, and lamination scheme on linear and nonlinear natural frequencies. Present results are in good agreement with those available for linear frequencies of isotropic plates.

INTRODUCTION

The dynamic behavior of plates with cutouts has been a topic of design interest for many years. Cutouts not only lighten the structure but also provide the designer an added flexibility to alter the resonant frequency.

Most of the previous investigations have been confined to isotropic plates, and plates with circular holes [1-6]. Moreover, these investigations neglected the transverse shear strains in calculating the natural frequencies. The natural frequencies calculated using the classical thin plate theory are higher than those obtained by including the transverse shear strains in the analysis. Due to the low transverse shear modulus relative to the in-plane Young's moduli, the transverse shear deformation effects are even more pronouncned in composite plates.

The present analysis is concerned with the natural vibration of rectangular composite plates with rectangular cutouts. The shear-flexible element developed by the present author [7-9] is employed in the present analysis. Only a limited number of numerical results are included in the present paper, and for additional results the reader is referred to the recent work of the author [10].

THEORY AND FORMULATION

Consider a plate of uniform thickness h, laminated of anisotropic layers with the material axes of each layer being arbitrarily oriented with respect to the midplane (R) of the plate. The midplane of the plate forms the x-y plane of the coordinate system, with the z-axis perpendicular to the midplane of the plate. In order to account for the midplane stretching due to large deflections, the shear deformable theory of Whitney and Pagano [11] is modified to include large rotations (in von Karman sense). The displacement field is assumed to be of the form,

$$u_1(x,y,z,t) = u(x,y,y) + z\ \psi_x(x,y,t),$$
$$u_2(x,y,z,t) = v(x,y,t) + z\ \psi_y(x,y,t), \tag{1}$$
$$u_3(x,y,z,t) = w(x,y,t).$$

Here t is the time; u_1, u_2, u_3 are the displacements in x, y, z directions, respectively; u, v, w are the associated midplane displacements; and ψ_x abd ψ_y are the slopes in the xz and yz planes due to bending only. Assuming that the plate is moderately thick and strains are much smaller than the rotations, we employ the nonlinear strain-displacement equations of the von-Karman theory (see Ref. 10):

$$\varepsilon_1 = \frac{\partial u}{\partial x} + \frac{1}{2}\left(\frac{\partial w}{\partial x}\right)^2 + z\frac{\partial \psi_x}{\partial x} \equiv \varepsilon_1^0 + z\kappa_1$$

$$\varepsilon_2 = \frac{\partial v}{\partial y} + \frac{1}{2}\left(\frac{\partial w}{\partial y}\right)^2 + z\frac{\partial \psi_y}{\partial y} \equiv \varepsilon_2^0 + z\kappa_2$$

$$\varepsilon_6 = \frac{\partial u}{\partial y} + \frac{\partial v}{\partial x} + \frac{\partial w}{\partial x}\frac{\partial w}{\partial y} + z\left(\frac{\partial \psi_x}{\partial y} + \frac{\partial \psi_y}{\partial x}\right) \equiv \varepsilon_6^0 + z\kappa_6 \tag{2}$$

$$\varepsilon_5 = \psi_x + \frac{\partial w}{\partial x}\ , \quad \varepsilon_4 = \psi_y + \frac{\partial w}{\partial y}$$

wherein the products of ψ_x, ψ_y, $\partial u_1/\partial x$ and $\partial u_2/\partial y$ are neglected. Since the constitutive relations, to be given shortly, are based on the plane-stress assumption, strain ε_3 does not come into the equations.

Neglecting the body moments and surface shearing forces, we write the equations of motion (in the absence of surface and body forces) as,

$$N_{1,x} + N_{6,y} = Pu_{,tt} + R\psi_{x,tt}$$
$$N_{6,x} + N_{2,y} = Pv_{,tt} + R\psi_{y,tt}$$
$$Q_{1,x} + Q_{2,y} + N(N_i, w) = Pw_{,tt} \tag{3}$$
$$M_{1,x} + M_{6,y} - Q_1 = I\psi_{x,tt} + Ru_{,tt}$$

$$M_{6,x} + M_{2,y} - Q_2 = I\psi_{y,tt} + Rv_{,tt}$$

where P, R and I are the normal, coupled normal-rotary, and rotary inertia coefficients respectively; N_i, Q_i, and M_i are the stress and moment resultants; $N(\cdot)$ is the nonlinear differential operator:

$$(P,R,I) = \int_{-h/2}^{h/2} (1,z,z^2)\rho dz = \sum_m \int_{z_m}^{z_{m+1}} (1,z,z^2)\rho^{(m)} dz \tag{4}$$

$$(N_i, M_i) = \int_{-h/2}^{h/2} (1,z)\sigma_i \, dz \quad , \quad (Q_1, Q_2) = \int_{-h/2}^{h/2} (\sigma_5, \sigma_4) \, dz, \tag{5}$$

$$N(w, N_i) = \frac{\partial}{\partial x}\left(N_1 \frac{\partial}{\partial x}\right) + \frac{\partial}{\partial y}\left(N_6 \frac{\partial w}{\partial x}\right) + \frac{\partial}{\partial x}\left(N_6 \frac{\partial w}{\partial y}\right) + \frac{\partial}{\partial y}\left(N_2 \frac{\partial w}{\partial y}\right) \tag{6}$$

Here $\rho^{(m)}$ denote the material density of the m-th layer and σ_i $(i = 1,2,\ldots,6)$ denote the stress components ($\sigma_1 = \sigma_x$, $\sigma_2 = \sigma_y$, $\sigma_3 = \sigma_z$, $\sigma_4 = \sigma_{zy}$ $\sigma_5 = \sigma_{xz}$, and $\sigma_6 = \sigma_{xy}$).

Assuming the existence of one plane of elastic symmetry parallel to the plane of the layer, for each layer, the constitutive equations for the m-th layer (in the plate coordinates) are given by

$$\begin{Bmatrix} \sigma_1 \\ \sigma_2 \\ \sigma_6 \end{Bmatrix} = [Q_{ij}^{(m)}] \begin{Bmatrix} \varepsilon_1 \\ \varepsilon_2 \\ \varepsilon_6 \end{Bmatrix}, \quad \begin{Bmatrix} \sigma_4 \\ \sigma_5 \end{Bmatrix} = \begin{bmatrix} Q_{44}^{(m)} & Q_{45}^{(m)} \\ Q_{45}^{(m)} & Q_{55}^{(m)} \end{bmatrix} \begin{Bmatrix} \varepsilon_4 \\ \varepsilon_5 \end{Bmatrix} \tag{7}$$

where $Q_{ij}^{(m)}$ are the stiffness coefficients of the m-th layer in the plate coordinates. Combining eqns. (5) and (7), we obtain the plate constitutive equations,

$$\begin{Bmatrix} N_i \\ M_i \end{Bmatrix} = \begin{bmatrix} A_{ij} & B_{ij} \\ B_{ji} & D_{ij} \end{bmatrix} \begin{Bmatrix} \varepsilon_j^0 \\ \kappa_j \end{Bmatrix}, \quad \begin{Bmatrix} Q_2 \\ Q_1 \end{Bmatrix} = \begin{bmatrix} A_{44} & A_{45} \\ A_{45} & A_{55} \end{bmatrix} \begin{Bmatrix} \varepsilon_4 \\ \varepsilon_5 \end{Bmatrix} \tag{8}$$

The A_{ij}, B_{ij}, D_{ij} $(i,j = 1,2,6)$, and A_{ij} $(i,j = 4,5)$ are the respective inplane, bending-inplane coupling, bending or twisting, and thickness-shear stiffnesses, respectively:

$$(A_{ij},B_{ij},D_{ij}) = \sum_m \int_{z_m}^{z_{m+1}} Q_{ij}^{(m)} (1,z,z^2)dz \ , \ A_{ij} = \sum_m \int_{z_m}^{z_{m+1}} k_i k_j Q_{ij}^{(m)} dz \quad (9)$$

Here z_m denotes the distance from the mid-plane to the lower surface of the m-th layer, and k_i are the shear correction coefficients.

Toward constructing a finite element model of the equations, we present a variational formulation of the eqns. (3) and (8), with u, v, w, ϕ_x, and ϕ_y as the primary dependent variables. Multiplying the five equations in eqn. (3) with variations δu, δv, δw, $\delta\phi_x$, and $\delta\phi_y$ respectively, and integrating by parts once, we arrive at the expression,

$$\begin{aligned}
0 = \int_R \{ & \delta u(Pu_{,tt} + R\phi_{x,tt}) + \delta u_{,x}N_1 + \delta u_{,y}N_6 + \delta v(Pv_{,tt} + P\psi_{y,tt}) \\
& + \delta v_{,x}N_6 + \delta v_{,y}N_2 + \delta w(Pw_{,tt}) + \delta w_{,x}Q_1 + \delta w_{,y}Q_2 \\
& + \frac{\partial \delta w}{\partial x}\frac{\partial w}{\partial x}N_1 + \frac{\partial \delta w}{\partial y}\frac{\partial w}{\partial x}N_6 + \frac{\partial \delta w}{\partial x}\frac{\partial w}{\partial y}N_6 + \frac{\partial \delta w}{\partial y}\frac{\partial w}{\partial y}N_2 \\
& + \delta\phi_x(I\phi_{x,tt} + Ru_{,tt}) + \delta\phi_{x,x}M_1 + \delta\phi_{x,y}M_6 + \delta\phi_x Q_1 \\
& + \delta\phi_y(I\phi_{y,tt} + Rv_{,tt}) + \delta\phi_{y,x}M_6 + \delta\phi_{y,y}M_2 + \delta\phi_y Q_2\}\, dxdy \\
& + \int_{C_n}(\delta u_n N_n + \delta u_s N_{ns})ds + \int_{C_q}\delta wq ds + \int_{C_m}(\delta\phi_n M_n + \delta\phi_s M_{ns})ds \quad (10)
\end{aligned}$$

wherein N_i, M_i and Q_i are given in terms of the generalized displacements by eqn. (8). The boundary terms N_n, N_{ns}, q, M_n and M_{ns} (notation used is very standard) in eqn. (10) get cancelled at interelement boundaries, and equal to those specified at the plate boundary. It is clear from the variational formulation that we have the following essential and natural boundary conditions:

essential: specify, u_n, u_s, w, ψ_n, ψ_s .

natural: specify, N_n, N_{ns}, q, M_n, M_{ns}. (11)

wherein u_n and u_s, for example, denote the normal and tangential components of the in-plane displacement vector, $\underset{\sim}{u} = (u,v)$.

The finite-element formulation of (10) is straightforward, and it will not be repeated here (see Refs. 8 and 9).

RESULTS AND DISCUSSION

The finite-element formulation of eqn. (10) leads to the standard equation of structural dynamics.

$$[K]\{\Delta\} + [M]\{\ddot{\Delta}\} = \{F\} \tag{12}$$

Here $\{\Delta\}$ is the column vector of the nodal values of the generalized displacements, $[K]$ is the matrix of stiffness coefficients (which depends on the generalized displaements), $[M]$ is the matrix of mass coefficients, and $\{F\}$ is the column vector containing the boundary contributions. The elements of $[K]$ and $[M]$ are given in Appendix A of Reference 8. In the present study we assume that

$$u = U\cos^2\omega t,\ v = V\cos^2\omega t,\ \phi_x = X\cos^2\omega t,\ \phi_y = Y\cos^2\omega t$$

$$w = W\cos\omega t,\ \text{and } U = \Sigma U_i\phi_i,\ \text{etc.} \tag{13}$$

where ω denotes the frequency of natural vibration. To further simplify the analysis, we neglect the in-plane and rotatory inertia terms (not a limitation of the present formulation, but computationally quite inexpensive). As a result, we obtain the standard eigenvalue problem for the composite plate:

$$([K] - \omega^2[M])\{\Delta\} = \{0\}. \tag{14}$$

In the following, numerical results are presented for rectangular plates of planeform dimensions a and b in x and y directions, respectively, and rectangular cutouts of dimensions c and d in x and y directions, respectively (see Fig. 1). In the present study the nine-node isoparametric rectangular element (with five degrees of freedom per node) was employed and only quarter-plate model was used whenever the biaxial symmetry existed. A shear correction factor of $k_1 = k_2 = k = 5/6$ was used.

In order to validate the present results, linear fundamental frequencies obtained in the present study are compared with those available in the literature. Table 1 contains nondimensionalized natural frequencies of a simply supported isotropic ($\nu = 0.3$) square plate with a square cutout for various ratios of side-to-thickness ratio, and plate side-to-cutout side. The following boundary conditions were used (see Fig. 1 for the finite element

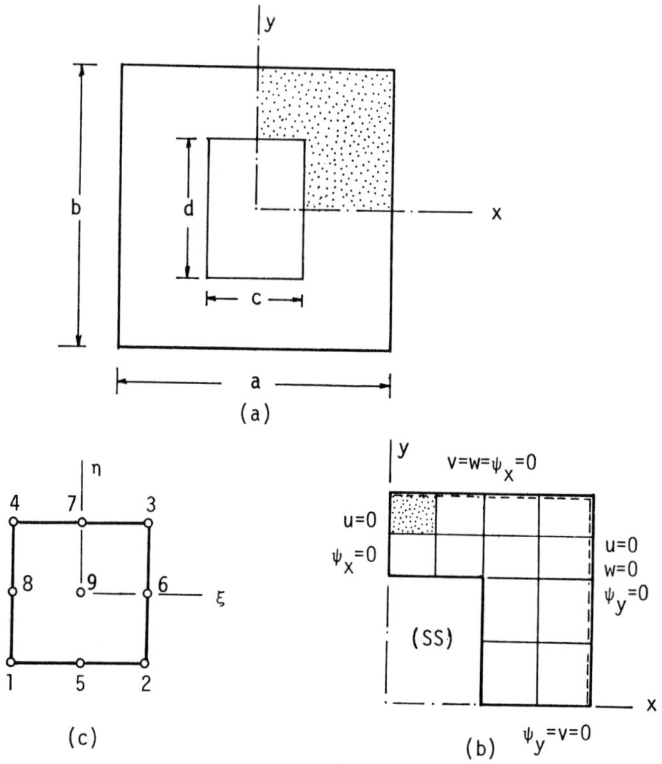

Fig. 1. Finite element discretization of rectangular plates with rectangular cutouts (a) plate geometry, (b) finite element mesh of a quadrant, and (c) the nine-node isoparametric rectangular finite element.

mesh and boundary conditions):

SS: in-plane displacement normal to the side is zero, tangential slope is zero, and the transverse deflection is zero on all sides of the plate.

Clearly, the present thin-plate (a/h = 1000) results are in good agreement with those of the other investigators. The results indicate that the inclusion of transverse shear strains reduces the natural frequencies for thick plates ($a/h \leq 10$).

TABLE 1
COMPARISON OF NONDIMENSIONALIZED FREQUENCY (λ) FOR
A SIMPLY SUPPORTED ISOTROPIC (ν = 0.3) SQUARE PLATE

($\lambda = \omega_{11} a^2 \sqrt{\rho h/D}$, a = b; c = d)

a/h \ c/a	Present Results			Thin-Plate Results			
	5	10	1000	FEM[5]	R-Ritz [5]	Ref. [3]	Ref. [6]
0.0	17.458	19.077	19.752	19.816	19.739	19.630	19.739
0.2	17.452	18.679	19.200	18.446	19.274	-	-
0.4	19.163	20.246	20.807	20.650	20.705	-	-
0.5	21.554	22.804	23.515	23.241	23.364	25.45	24.88
0.6	25.688	27.379	28.453	28.445	27.857	-	-
0.8	44.069	51.465	57.512	57.706	58.885	-	-

Table 2 presents a comparison of nondimensionalizes fundamental frequency ($\lambda = \omega_{11} a^2 \sqrt{\rho h/D}$) for a simply supported isotropic (ν = 0.3) square plate with rectangular cutouts (d/c = 2). The present results for a/h = 1000 are bounded by the finite element and Rayleigh-Ritz solutions of Ali and Atwal [5]. Similar results are presented for orthotropic ($E_1/E_2 = 40$, $G_{12}/E_2 = 0.5$, $\nu_{12} = 0.25$) square plate in Table 3.

Figure 2 shows plots of the nondimensionalized linear fundamental frequency (λ_{11}) and the second symmetric mode (λ_{13}) versus the side-to-thickness ratio for square plates with cutouts. The orthotropic materials properties are: $E_1/E_2 = 40$, $G_{12}/E_2 = 0.5$, $G_{13} = G_{23} = G_{12}$, $\nu_{12} = 0.25$. Several observations can be made from these plots: (i) The effect of shear deformation is more pronounced in the case of clamped plates (all of the degrees of freedom are set to zero on the edges) than in simply supported plates; the effect is more in the higher modes than in the fundamental mode. Also, the shear deformation effect is more pronounced for c/a = 0.8 compared to c/a < 0.8. (ii) The shear deformation effect is more pronounced in the case of 2-layer angle-ply (45°/-45°) plates (B_{16}, B_{26} and B_{66} are zero).

Plots of the ratio of linear to nonlinear frequency versus the ratio of amplitude to thickness are shown in Fig. 3 for isotropic square plates with

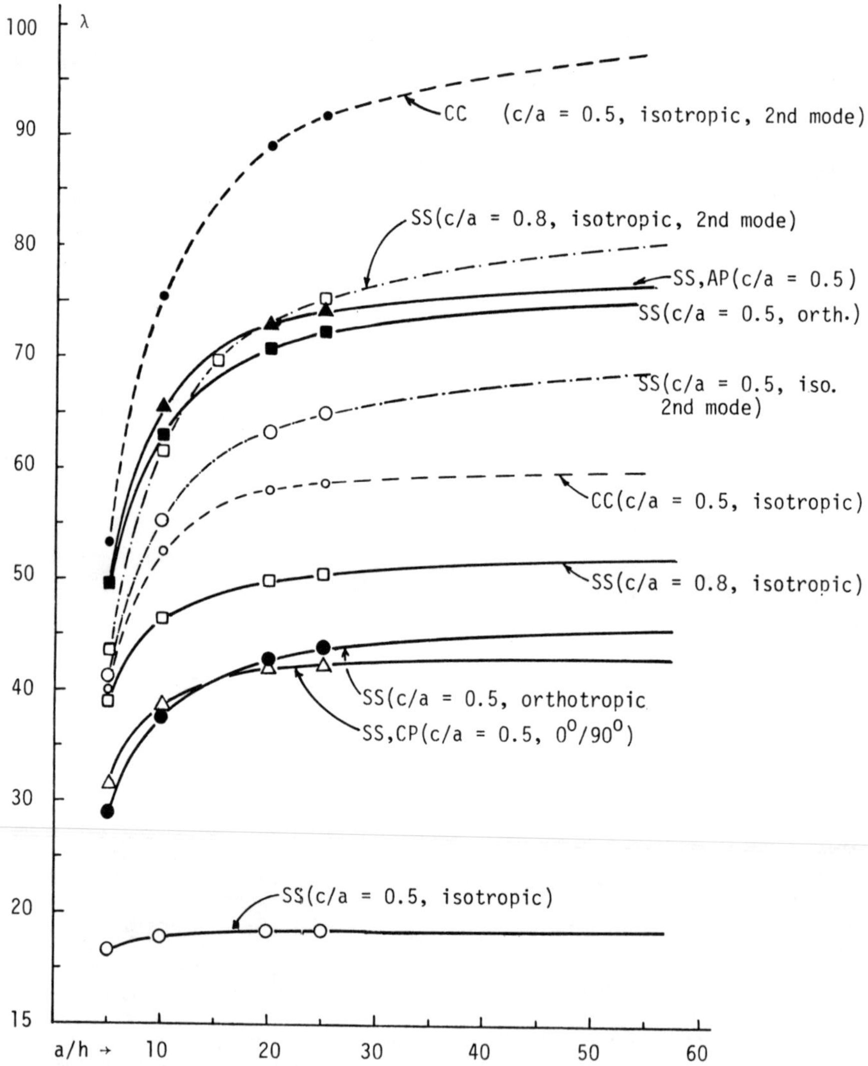

Fig. 2 Effect of the side-to-thickness ratio on the nondimensionalized linear frequencies of square plates with cutouts (SS = simply supported, CC = clamped, AP = angle-ply, CP = cross-ply).

various cutout sizes. Two sets of results are presented: (i) three degrees of freedom (w, ϕ_x, ϕ_y), and (ii), five degrees of freedom (u, v, w, ϕ_x, ϕ_y) results. From these plots one observes that the five degrees-of-freedom model (c/a = 0.2, a/h = 1000) exhibits relatively less nonlinearity compared to the three degrees-of-freedom model (c/a = 0.2, a/h = 1000). Also, the nonlinear effect is less pronounced in the higher modes. Plates with cutouts exhibit lower ratios of linear to nonlinear frequencies than those without cutouts.

TABLE 2

COMPARISON OF NONDIMENSIONALIZED FREQUENCY (λ) FOR A SIMPLY SUPPORTED ISOTROPIC (ν = 0.3) SQUARE PLATE WITH RECTANGULAR CUTOUTS (d/c = 2)

c/a	Present results		Ali & Atwal [5]	
	a/h=10	a/h=1000	FEM	R-R
0.05	19.972	20.556	18.595	19.809
0.20	19.470	20.035	18.740	21.029
0.25	19.977	20.588	19.449	21.803
0.30	20.882	21.608	20.520	22.953
0.35	22.103	23.051	-	-
0.40	-	-	23.607	25.983

TABLE 3

COMPARISON OF NONDIMENSIONALIZED FREQUENCY (λ) FOR A SIMPLY SUPPORTED ORTHOTROPIC SQUARE PLATE WITH RECTANGULAR CUTOUTS (d/c = 2,3).

c/d	a/h \ c/a	0.0	0.05	0.2	0.25	0.3	0.35	0.4
2	1000	122.18	62.84	44.50	39.80	35.87	32.76	30.82
	10	99.68	47.16	36.31	33.20	30.66	28.86	28.03
3	1000	-	60.06	32.69	27.28	24.79	-	-
	10	-	45.28	27.95	24.51	23.32	-	-

Finally, Table 4 presents ratios of linear to nonlinear frequencies for simply supported square plates of two-layer cross-ply and angle-ply (45°/-45°) construction. Results are presented for two cutouts (c/a = 0.2, 0.5). These results should serve as bench mark results for future investigations.

CONCLUSIONS

Large-amplitude free vibration of rectangular plates with rectangular cutouts is investigated using a shear deformable finite element that accounts for the transverse shear strains and large rotations. The element yields

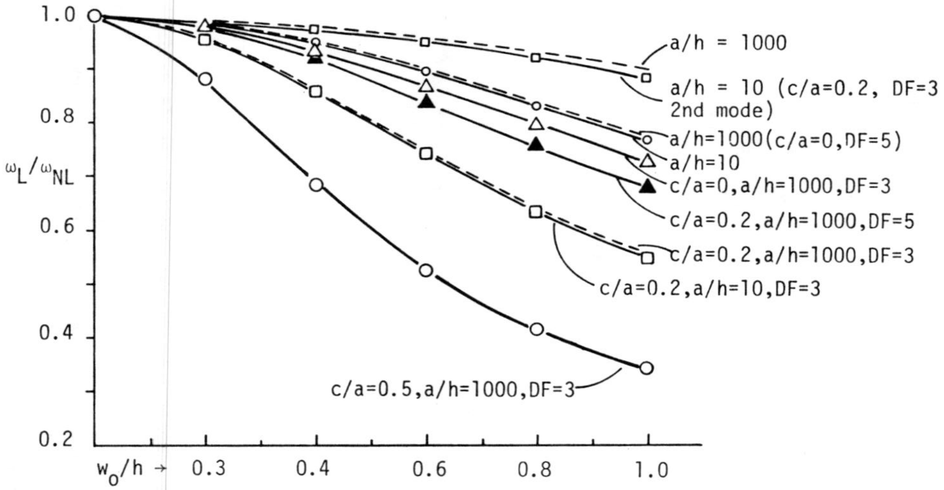

Fig. 3. Plots of linear to nonlinear frequencies of isotropic, simply-supported square plates with cutouts (DF = degree of freedom)

results which compare very well with the other small-deflection results available in the literature. New finite element results for ratios of linear to nonlinear natural frequencies are presented for orthotropic plates and two-layer composite plates. These should serve as check cases for any future investigations.

TABLE 4

RATIO OF LINEAR TO NONLINEAR FUNDAMENTAL FREQUENCIES OF TWO-LAYER SIMPLY SUPPORTED SQUARE PLATES WITH SQUARE CUTOUTS

c/a	cross-ply (0°/90°)		angle-ply (45°/-45°)	
	a/h=10	a/h=1000	a/h=10	a/h=1000
0.2	-	0.9625(0.2)†	-	0.8594(0.2)
0.5	0.9766(0.2)	-	0.9772(0.2)	-
	0.9762(0.4)	-	0.9763(0.4)	-
	0.573 (1.0)	0.573 (1.0)	0.6598(1.0)	0.5997(1.0)

†value in the parenthesis indicates the amplitude-to-thickness ratio.

ACKNOWLEDGMENT

The support of this work by the Structural Mechanics Section, the Air Force Office of Scientific Research, through Grant No. AFOSR-81-0142 is gratefully acknowledged.

REFERENCES

1. C. V. JOGA RAO and G. PICKETT, Vibrations of plates of irregular shapes and plates with holes, Journal of Aeronautical Society of India, 13, (1961), pp. 83-88.

2. G. PICKETT, Bending, buckling and vibrations of plates with holes, Proceedings of the Second Southeastern Conference on Theoretical and Applied Mechanics, Atlanta, Georgia, 1964.

3. P. PARAMASIVAM, Free vibration of square plates with square openings, Journal of Sound and Vibration, 30(2), (1973), pp. 173-178.

4. G. AKSU and R. ALI, Determination of dynamic characteristics of rectangular plates with cutouts using a finite difference formulation, Journal of Sound and Vibration, 44(1), (1976), pp. 147-158.

5. R. ALI and S. J. ATWAL, Prediction of natural frequencies of vibration of rectangular plates with rectangular cutouts, Computers and Structures, 12 (6), (1980), pp. 819-823.

6. N. L. BASDEKAS and M. CHI, Dynamic response of plates with cutouts, The Shock and Vibration Bulletin, 41(7), (1979), pp. 29-35.

7. J. N. REDDY and W. C. CHAO, Large-deflection and large-amplitude free vibrations of laminated composite-material plates, Computers and Structures, 13, (1981), pp. 341-347.

8. J. N. REDDY, Nonlinear vibration of layered composite plates including transverse shear and rotatory inertia, 1981 ASME Vibration Conference, Sept. 20-23, Hartford, Connecticut, ASME Paper No. 81-DET-144.

9. J. N. REDDY and W. C. CHAO, Nonlinear oscillations of laminated, anisotropic, thick rectangular plates, 1981 ASME Winter Annual Meeting, November 15-20, Washington, D.C.

10. J. N. REDDY, Large-amplitude flexural vibration of layered composite plates with cutouts, in review.

11. J. M. WHITNEY and N. J. PAGANO, Shear deformation in heterogeneous anisotropic plates, Journal of Applied Mechanics, 37, (1970), pp. 1031-1036.

A THEORY OF TWO COMPONENT LINEAR VISCOELASTIC COMPOSITES

Matthew F. McCarthy
Department of Engineering Science and Mechanics
Virginia Polytechnic Institute and State University
Blacksburg, Virginia 24061

Abstract

The general theory of two component heat conducting linear viscoelastic composites is developed. The thermodynamic restrictions placed on the relaxation functions of the composite are derived. The implications of invariance under time reversal on the symmetry properties of the relaxation functions are investigated.

1. Introduction

In this work we study linear viscoelastic composites which have two identifiable constituents and following Tiersten and Jahanmir [1] we model the composite as a mixture of interpenetrating solid continua in which each of the component continua has its own motion but interacts with the other component of the composite. Mixture theories of elastic composites have been developed by a number of authors [2-5] while mixture theories of viscoelastic composites have also been developed [6-9].

The purpose of the present work is to present a theory of linear viscoelastic two component composites based on the approaches taken in [1] and [9]. The kinematics of the motion as well as the balance laws and thermodynamic principles which govern it are discussed in Section 3. The implications of the second law of thermodynamics for the forms which constitutive equations may assume are explored in Section 3 and some of the resulting symmetry properties of the relaxation functions are established. The dissipation inequality appropriate to the description of two component linear viscoelastic composites is also derived in Section 3. In Section 4 the restrictions placed upon the relaxation functions are established and the analogues of many of the results given earlier [12-14] for single component media are presented for composites. The implications of invariance under time reversal for the symmetry properties of the relaxation functions of the composite are also explored in Section 4 and many powerful new symmetry relations for viscoelastic composites are derived. The foregoing results are illustrated in Section 5 where isotropic composites are treated.

2. Basic Equations and Formulae

We consider a body which is macroscopically homogeneous and is made up of two distinct solid interpenetrating continua. Initially, at time $t = 0$, say, both continua occupy the same region of space and, consequently, have the same material Cartesian coordinates x_j. The subsequent motions of each of the particles of the two interpenetrating continua which are simultaneously located at the point $\underset{\sim}{x}$ at time $t = 0$ are given by

$$\underset{\sim}{y}^{(i)} = \underset{\sim}{x} + \underset{\sim}{u}^{(i)}(\underset{\sim}{x},t) = \underset{\sim}{y}(\underset{\sim}{x},t) + w^{(i)}(\underset{\sim}{x},t), \quad i = 1,2 \tag{2.1}$$

where $\underset{\sim}{u}^{(i)}(\underset{\sim}{x},t)$, $i = 1,2$, are the infinitesimal displacements at time t of the particles of the interpenetrating continua whose positions coincide at the point $\underset{\sim}{x}$ at time $t = 0$. The position of the centre of mass, at time t, of the particles which coincide at time $t = 0$ is

$$\underset{\sim}{y}(\underset{\sim}{x},t) = \underset{\sim}{x} + \underset{\sim}{u}(\underset{\sim}{x},t), \quad \underset{\sim}{u}(\underset{\sim}{x},t) = (r\underset{\sim}{u}^{(1)}(\underset{\sim}{x},t) + \underset{\sim}{u}^{(2)}(\underset{\sim}{x},t)/(1+r) \tag{2.2}$$

and here $r = \rho^{(1)}/\rho^{(2)}$ denotes the ratio of the mass densities of the interpenetrating continua. The functions $\underset{\sim}{w}^{(1)}(\underset{\sim}{x},t)$, $\underset{\sim}{w}^{(2)}(\underset{\sim}{x},t)$ describe the displacements of the particles of the interpenetrating continua, whose common material coordinates are x_j, relative to the centre of mass of these particles. Equations (2.1) - (2.2) together imply that

$$\underset{\sim}{w}^{(1)} = \underset{\sim}{w}^{(1)}(\underset{\sim}{x},t) = (\underset{\sim}{u}^{(1)} - \underset{\sim}{u}^{(2)})/(1+r) = -\underset{\sim}{w}^{(2)}/r \tag{2.3}$$

so that

$$\rho^{(1)}\underset{\sim}{w}^{(1)} + \rho^{(2)}\underset{\sim}{w}^{(2)} = \underset{\sim}{0}. \tag{2.4}$$

We define the deformation gradient at the centre of mass and at the material points $\underset{\sim}{x}$ of the interpenetrating continua by the formulae

$$F_{ij} = y_{i,j} = \delta_{ij} + u_{i,j}, F_{ij}^{(k)} = u_{i,j}^{(k)} = \delta_{ij} + u_{i,j}^{(k)} = F_{ij} + w_{i,j}^{(k)}, \tag{2.5}$$

where $k = 1,2$ and a comma denotes partial differentiation with respect to x_j. We call the quantities

$$\varepsilon_{ij} = \varepsilon_{ij}(\underset{\sim}{x},t) = \frac{1}{2}(u_{i,j} + u_{j,i}), \quad p_{ij} = w_{i,j}^{(1)} \tag{2.6}$$

the **infinitesimal strain tensor** and the **relative deformation gradient,** respectively. Let $\underset{\sim}{u}(\underset{\sim}{x},\tau)$ and $\underset{\sim}{w}^{(1)} = \underset{\sim}{W}(\underset{\sim}{x},\tau)$ be continuous and have continuous partial derivatives of any desired degree in the interval $-\infty < \tau \leqslant t$ and tend to zero as $\tau \to -\infty$. Let us put

$$\varepsilon_1 = \sup_{\tau} |u_{i,j}(\underset{\sim}{x}, \tau)|, \quad \varepsilon_2 = \sup_{\tau} |w_{i,j}(\underset{\sim}{x}, \tau)|, \tag{2.7}$$

where $|\cdot|$ denotes magnitude and sup denotes least upper bound. We say that the deformation of the composite is **infinitesimal** at all times τ if $\varepsilon = \max (\varepsilon_i) \ll 1$.

As in [1], the interpenetrating continua are assumed to interact with each other by means of defined local equal and opposite force fields $\mathscr{F} = {}^{L}F^{12} = -{}^{L}F^{12}$ which are located at the position $\underset{\sim}{y}$, where the first superscript denotes the continuum being acted on and the second, the continuum producing the action and defined equal and opposite local material couples $C = -C$. Each continuum interacts with neighbouring elements of the same continuum by means of a traction force per unit area $t = n\cdot\tau$, where $\underset{\sim}{n}$ is a unit vector and $\underset{\sim}{\tau}$ is the stress tensor of the ith continuum.

A straightforward application of the laws of balance of linear momentum, angular momentum and conservation of energy leads to the equations, [1],

$$\tau_{ji,j} + \rho f_i = \rho \dot{v}_i, \quad D_{ji,j} + \rho \tilde{f}_i + \mathscr{F}_i = \rho^{(1)} \ddot{\eta}_i, \tag{2.8), (2.9}$$

$$\tau^{A}_{ij} = \frac{1}{2}\{D_{ki}w_{j,k} - D_{kj}w_{i,k} - \mathscr{F}_i w_j + \mathscr{F}_j w_i\}, \tag{2.10}$$

$$\rho \dot{e} = \tau_{ij}\dot{u}_{j,i} + D_{ij}\dot{w}_{j,i} - \mathscr{F}_i \dot{w}_i - q_{i,i}, \tag{2.11}$$

and in these equations the following definitions have been used:

$$\tau_{ij} = \tau^{(1)}_{ij} + \tau^{(2)}_{ij}, \quad \rho = \rho^{(1)} + \rho^{(2)}, \quad \rho f_i = \rho^{(1)} f^{(1)}_i + \rho^{(2)} f^{(2)}_i, \quad v_i = \dot{y}_i,$$

$$D_{ij} = \tau^{(1)}_{ij} - r\tau^{(2)}_{ij}, \quad \tilde{f}_i = f^{(1)}_i - f^{(2)}_i, \quad \underset{\sim}{\eta} = (1+r)\underset{\sim}{w}^{(1)}, \quad \tau^{A}_{ij} = \frac{1}{2}(\tau_{ij} - \tau_{ji}). \tag{2.12}$$

In (2.12) $f^{(\alpha)}_i$ is the body force per unit mass acting on the αth constituent, in (2.11) e is the specific internal energy per unit mass while $\underset{\sim}{q}$ is the heat flux vector. In the foregoing equations, and in what follows, a superposed dot denotes differentiation with respect to t.

Notice that the right hand side of (2.10) is $O(\varepsilon^2)$ while τ_{ij} is $O(\varepsilon)$ and consequently in the theory of linear viscoelastic composites developed here τ_{ij} is a symmetric tensor.

The **entropy production rate** per unit volume γ is defined through the relation

$$\rho\gamma = \rho\dot{\eta} + (q_i/\theta)_{,i} \tag{2.13}$$

where η is the specific entropy and θ (> 0) is the temperature. When the energy equation (2.14) is used, we can write

$$\rho\gamma = \rho\dot{\eta} + \frac{1}{\theta}\{\tau_{ij}v_{j,i} + D_{ij}\dot{w}_{j,i} - \mathscr{F}_i\dot{w}_i - \rho\dot{e}\} - q_i\theta_{,i}/\theta^2. \tag{2.14}$$

The second law of thermodynamics requires that the Clausius-Duhem inequality be satisfied i.e.

$$\gamma \geqslant 0. \tag{2.15}$$

3. Consequences of the Second Law of Thermodynamics

Let us define the free energy per unit mass by the formula

$$\phi = e - \theta\eta \tag{3.1}$$

so that the Clausius-Duhem inequality may be recast in the form

$$\rho\theta\gamma = -\rho\dot{\phi} + \tau_{ij}\dot{\varepsilon}_{ji} + D_{ij}\dot{p}_{ji} - \mathscr{F}_i\dot{w}_i - \rho\eta\dot{\theta} - q_i\theta_{,i}/\theta \geqslant 0, \tag{3.2}$$

where $p_{ij} = w_{i,j}$.

The inequality (3.2) may be written in more revealing form and the subsequent algebra is considerably simplified by introducing a shorthand notation at this point. Let

$$\underset{\sim}{\Lambda} = (\underset{\sim}{a}, \underset{\sim}{b}, \underset{\sim}{c}, \alpha) \tag{3.3}$$

be an ordered quadruple of dimension 22, where a is a second order symmetric tensor, b is a second order tensor, c is a vector and α is a scalar. Then, under the usual conditions, the space of all such elements form a vector space $\mathscr{V}$ of dimension 22, with natural inner product

$$\underset{\sim}{\Lambda}^{(1)} \cdot \underset{\sim}{\Lambda}^{(2)} = a_{ij}^{(1)} a_{ji}^{(2)} + b_{ij}^{(1)} b_{ji}^{(2)} + c_i^{(1)} c_i^{(2)} + \alpha^{(1)} \alpha^{(2)} \tag{3.4}$$

and norm $|\underset{\sim}{\Lambda}| = (\underset{\sim}{\Lambda} \cdot \underset{\sim}{\Lambda})^{1/2}$.

Let $\mathscr{L}(\mathscr{V},\mathscr{V})$ denote the space of all linear transformations $\underset{\sim}{A}$ from $\mathscr{V}$ into $\mathscr{V}$. The norm of $\underset{\sim}{A}$ is defined to be

$$\|\underset{\sim}{A}\| = \sup_{|\underset{\sim}{\Lambda}|=1} |\underset{\sim}{A}\underset{\sim}{\Lambda}| \tag{3.5}$$

while the transpose of $\underset{\sim}{A}$ is denoted by $\underset{\sim}{A}^T$ and is defined by the relation

$$\underset{\sim}{\Lambda}^{(1)} \cdot \underset{\sim}{A}\underset{\sim}{\Lambda}^{(2)} = \underset{\sim}{\Lambda}^{(2)} \cdot \underset{\sim}{A}^T \underset{\sim}{\Lambda}^{(1)}. \tag{3.6}$$

If $\underset{\sim}{A}^T = \underset{\sim}{A}$, we say that $\underset{\sim}{A}$ is symmetric.

Any function $\underset{\sim}{M}(\cdot):[0,\infty) \to \mathscr{L}(\mathscr{V},\mathscr{V})$ which is continuous and bounded is called a **relaxation function.**

We call

$$\underset{\sim}{\Omega} = (\underset{\sim}{\varepsilon}, \underset{\sim}{p}, \underset{\sim}{w}, \theta) \text{ and } \underset{\sim}{\Sigma} = (\underset{\sim}{\tau}, \underset{\sim}{D}^T, -\mathscr{F} - \rho\eta) \tag{3.7}$$

the **generalized infinitesimal strain** and the **generalized infinitesimal stress**, respectively. Both Σ and M are elements of $\mathscr{V}$ and when (3.4) is used the inequality (3.2) may be recast in the form

$$\rho\theta\gamma = -\varepsilon\dot{\phi} + \underset{\sim}{\Sigma} \cdot \dot{\underset{\sim}{\Omega}} + \underset{\sim}{q} \cdot \underset{\sim}{g}/\theta \geq 0. \tag{3.8}$$

By a **generalized stress path** in we mean a function $\Omega(\cdot)$, defined on $-\infty < t < \infty$, which is continuously differentiable. This means that there are times t_1, t_2 and unique vectors $\underset{\sim}{\Omega}(-\infty)$, $\underset{\sim}{\Omega}(\infty)$ such that $\underset{\sim}{\Omega}(t) = \underset{\sim}{\Omega}(-\infty)$ for every $t \le t_1$ and $\underset{\sim}{\Omega}(t) = \underset{\sim}{\Omega}(\infty)$ for every $t \ge t_2$. If $\underset{\sim}{\Omega}(-\infty) = \underset{\sim}{\Omega}(\infty)$ the path is **closed** and if $\underset{\sim}{\Omega}(-\infty) = (\underset{\sim}{0}, \underset{\sim}{0}, \underset{\sim}{0}, \alpha)$, where $\alpha > 0$, the path starts from a **virgin state.**

Let $\underset{\sim}{\Omega}^t = \underset{\sim}{\Omega}(t-s)$, $s \in [0,\infty)$ denote the history of the function $\Omega(\cdot)$ up to time t. We assume that the composite is characterized by the constitutive equations

$$\underset{\sim}{\Sigma} = \hat{\underset{\sim}{\Sigma}}(\underset{\sim}{\Omega}^t), \quad \underset{\sim}{q} = -\underset{\sim}{K}\underset{\sim}{g}, \tag{3.9}$$

where $\underset{\sim}{K}$ is a second order symmetric tensor, while the functional $\hat{\phi}$ will be assumed to be of the form

$$\phi = \hat{\phi}(\underset{\sim}{\Omega}^t) = \frac{1}{2}\,\underset{\sim}{\Omega}\cdot\underset{\sim}{\Gamma}\underset{\sim}{\Omega} + \frac{1}{2}\int_0^\infty\int_0^\infty \underset{\sim}{\Omega}_d^t(s)\cdot\frac{\partial^2\underset{\sim}{R}(s,u)}{\partial s\,\partial u}\,\underset{\sim}{\Omega}_d^t(u)\,ds\,du$$

$$= \frac{1}{2}\,\underset{\sim}{\Omega}\cdot\underset{\sim}{\Gamma}\underset{\sim}{\Omega} + \frac{1}{2}\int_{-\infty}^{t}\int_{-\infty}^{t} \dot{\underset{\sim}{\Omega}}(\xi)\cdot\underset{\sim}{R}(t-\xi,t-\eta)\dot{\underset{\sim}{\Omega}}(\eta)\,d\xi\,d\eta \tag{3.10}$$

where $\underset{\sim}{\Gamma} = \underset{\sim}{\Gamma}^T$, and $\underset{\sim}{\Omega}_d^t(\cdot) = \underset{\sim}{\Omega}^t(\cdot) - \underset{\sim}{\Omega}$ is the **difference history** of the function $\underset{\sim}{\Omega}$. The relaxation function $\underset{\sim}{R}(\cdot,\cdot)$ is $C^{(1)}$ function of its arguments and has the properties

$$\underset{\sim}{R}^T(u,s) = \underset{\sim}{R}(s,u);\ \underset{\sim}{R}(u,\infty) = \underset{\sim}{R}(\infty,u) = \underset{\sim}{0}. \tag{3.11}$$

$\underset{\sim}{R}(\cdot,\cdot)$ has the following "component" form

$$\underset{\sim}{R}(u,s) = \begin{bmatrix} \underset{\sim}{G}(u,s) & \underset{\sim}{H}(u,s) & \underset{\sim}{A}(u,s) & \underset{\sim}{I}(u,s) \\ \underset{\sim}{H}^T(s,u) & \underset{\sim}{J}(u,s) & \underset{\sim}{B}(u,s) & \underset{\sim}{L}(u,s) \\ \underset{\sim}{A}^T(s,u) & \underset{\sim}{B}^T(s,u) & \underset{\sim}{N}(u,s) & \underset{\sim}{h}(u,s) \\ \underset{\sim}{I}^T(s,u) & \underset{\sim}{L}^T(s,u) & \underset{\sim}{h}(s,u) & g(u,s) \end{bmatrix} \tag{3.12}$$

where $\underset{\sim}{G}$, $\underset{\sim}{H}$, $\underset{\sim}{J}$ are fourth order tensors; $\underset{\sim}{A}$, $\underset{\sim}{B}$ are third order tensors; $\underset{\sim}{I}$, $\underset{\sim}{L}$ and $\underset{\sim}{N}$ are tensors of order two, while $\underset{\sim}{h}$ is a vector and g is a scalar. In (3.12), the notations

$$H^T_{k\ell mn} = H_{mnk\ell},\ A^T_{k\ell m} = A_{\ell mk},\ B^T_{k\ell m} = B_{\ell mk} \tag{3.13}$$

are employed. Of course, it follows from the definition of $\underset{\sim}{\Omega}$ that the components of $\underset{\sim}{R}(u,s)$ are endowed with the symmetry properties:

$$G_{k\ell mn}(u,s) = G_{\ell kmn}(u,s) = G_{\ell knm}(u,s) = G_{mnk\ell}(s,u),\ J_{k\ell mn}(u,s) = J_{mnk\ell}(s,u),$$

$$H_{k\ell mn}(u,s) = H_{\ell kmn}(u,s),\ H_{k\ell mn}(u,s) = H_{k\ell nm}(u,s),\ N_{k\ell}(u,s) = N_{\ell k}(u,s),$$

$$A_{k\ell m}(u,s) = a_{\ell km}(u,s),\ I_{k\ell}(u,s) = I_{\ell k}(u,s),\ h(u,s) = h(s,u). \tag{3.14}$$

$\underset{\sim}{\Gamma}$ has the "component" form

$$\underset{\sim}{\Gamma} = \begin{bmatrix} \overline{\underset{\sim}{G}} & \overline{\underset{\sim}{H}} & \overline{\underset{\sim}{A}} & \overline{\underset{\sim}{I}} \\ \overline{\underset{\sim}{H}}^T & \overline{\underset{\sim}{J}}^T & \overline{\underset{\sim}{B}} & \overline{\underset{\sim}{L}} \\ \overline{\underset{\sim}{A}}^T & \overline{\underset{\sim}{B}}^T & \overline{\underset{\sim}{N}} & \overline{\underset{\sim}{h}} \\ \overline{\underset{\sim}{I}}^T & \overline{\underset{\sim}{L}}^T & \overline{\underset{\sim}{h}} & \overline{\underset{\sim}{g}} \end{bmatrix} \tag{3.15}$$

and the symmetry properties of the components of $\underset{\sim}{\Gamma}$ follow from (3.14) on setting u = s = 0 and replacing $\underset{\sim}{G}$ by $\overline{\underset{\sim}{G}}$ etc.

When (3.10) is used in the inequality (3.8) and a standard argument of the type employed by Coleman [17] is employed it follows that

$$\underset{\sim}{\Sigma} = (\underset{\sim}{\Gamma} + \underset{\sim}{R}(0,0))\underset{\sim}{\Omega} + \int_0^\infty \partial\underset{\sim}{R}(0,s)/\partial s\ \underset{\sim}{\Omega}(t-s)ds = \int_{-\infty}^{t} \underset{\sim}{M}(t-u)\dot{\underset{\sim}{\Omega}}(u)du \tag{3.16}$$

where $\underset{\sim}{M}(u) = \underset{\sim}{R}(0,u) + \underset{\sim}{\Gamma}$. We note that in general $\underset{\sim}{M}(u)$ is not symmetric. If $\underset{\sim}{M}(u)$ is expressed in component form as

$$\underset{\sim}{M} = \begin{bmatrix} \underset{\sim}{\mathscr{G}} & \underset{\sim}{\mathscr{H}} & \underset{\sim}{\mathscr{A}} & \underset{\sim}{\mathscr{I}} \\ \overset{*}{\underset{\sim}{\mathscr{H}}} & \underset{\sim}{\mathscr{J}} & \underset{\sim}{\mathscr{B}} & \underset{\sim}{\mathscr{L}} \\ \overset{*}{\underset{\sim}{\mathscr{A}}} & \overset{*}{\underset{\sim}{\mathscr{B}}} & \underset{\sim}{\mathscr{N}} & \overset{*}{\underset{\sim}{h}} \\ \overset{*}{\underset{\sim}{\mathscr{I}}} & \overset{*}{\underset{\sim}{\mathscr{L}}} & \overset{*}{h} & \overset{*}{g} \end{bmatrix} \tag{3.17}$$

then (3.16) is equivalent to the relations

$$\begin{aligned}
\underset{\sim}{\tau}(t) &= \int_{-\infty}^{t} \{\underset{\sim}{\mathscr{G}}(t-\beta)\dot{\underset{\sim}{\varepsilon}}(\beta) + \underset{\sim}{\mathscr{H}}(t-\beta)\dot{\underset{\sim}{p}}(\beta) + \underset{\sim}{\mathscr{A}}(t-\beta)\dot{\underset{\sim}{w}}(\beta) + \underset{\sim}{\mathscr{I}}(t-\beta)\dot{\theta}(\beta)\}d\beta, \\
\underset{\sim}{D}^T(t) &= \int_{-\infty}^{t} \{\overset{*}{\underset{\sim}{\mathscr{H}}}(t-\beta)\dot{\underset{\sim}{\varepsilon}}(\beta) + \underset{\sim}{\mathscr{J}}(t-\beta)\dot{\underset{\sim}{p}}(\beta) + \underset{\sim}{\mathscr{B}}(t-\beta)\dot{\underset{\sim}{w}}(\beta) + \underset{\sim}{\mathscr{L}}(t-\beta)\dot{\theta}(\beta)\}d\beta, \\
\underset{\sim}{\mathscr{Y}}(t) &= -\int_{-\infty}^{t} \{\overset{*}{\underset{\sim}{\mathscr{A}}}(t-\beta)\dot{\underset{\sim}{\varepsilon}}(\beta) + \overset{*}{\underset{\sim}{\mathscr{B}}}(t-\beta)\dot{\underset{\sim}{p}}(\beta) + \underset{\sim}{\mathscr{N}}(t-\beta)\dot{\underset{\sim}{w}}(\beta) + \overset{*}{\underset{\sim}{h}}(t-\beta)\dot{\theta}(\beta)\}d\beta, \\
\eta(t) &= -\frac{1}{\rho}\int_{-\infty}^{t} \{\overset{*}{\underset{\sim}{\mathscr{I}}}(t-\beta)\dot{\underset{\sim}{\varepsilon}}(\beta) + \overset{*}{\underset{\sim}{\mathscr{L}}}(t-\beta)\dot{\underset{\sim}{p}}(\beta) + \overset{*}{\underset{\sim}{h}}(t-\beta)\dot{\underset{\sim}{w}}(\beta) + \overset{*}{g}(t-\beta)\dot{\theta}(\beta)\}d\beta,
\end{aligned} \tag{3.18}$$

and these are the constitutive equations which describe the response of a two component linear viscoelastic composite.

The symmetry properties of the relaxation functions occuring in (3.18) follow from the definitions of $\underset{\sim}{M}(u)$ and $\underset{\sim}{\Gamma}$ as well as the relations (3.14). Thus we have

$$\begin{aligned}
&\mathscr{G}_{ijk\ell}(s) = \mathscr{G}_{jik\ell}(s) = \mathscr{G}_{ji\ell k}(s) = \mathscr{G}_{ij\ell k}(s), \mathscr{H}_{ijk\ell}(s) = \mathscr{H}_{jik\ell}(s), \\
&\overset{*}{\mathscr{H}}_{ijk\ell}(s) = \overset{*}{\mathscr{H}}_{ij\ell k}(s), \mathscr{A}_{ijk}(s) = \mathscr{A}_{jik}(s), \mathscr{I}_{k\ell}(s) = \mathscr{I}_{\ell k}(s),
\end{aligned} \tag{3.19}$$

$$\overset{*}{\mathscr{A}}_{ijk}(s) = \overset{*}{\mathscr{A}}_{ikj}(s), \mathscr{N}_{ij}(s) = \mathscr{N}_{ji}(s), \overset{*}{\mathscr{I}}_{ij}(s) = \overset{*}{\mathscr{I}}_{ji}(s)$$

for all $s \geqslant 0$, while additionally we have

$$\begin{aligned}
&\mathscr{G}_{ijk\ell}(0) = \mathscr{G}_{jik\ell}(0) = \mathscr{G}_{ij\ell k}(0) = \mathscr{G}_{ij\ell k}(0) = \mathscr{G}_{k\ell ij}(0), \\
&\mathscr{J}_{ijk\ell}(0) = \mathscr{J}_{k\ell ij}(0), \mathscr{H}_{ijk\ell}(0) = \mathscr{H}_{jik\ell}(0) = \overset{*}{\mathscr{H}}_{k\ell ij}(0), \\
&\mathscr{A}_{ijk}(0) = \mathscr{A}_{jik}(0) = \overset{*}{\mathscr{A}}_{kij}(0), \mathscr{B}_{ijk}(0) = \overset{*}{\mathscr{B}}_{kij}(0), \\
&\mathscr{I}_{k\ell}(0) = \overset{*}{\mathscr{I}}_{\ell k}(0), \mathscr{L}_{ij}(0) = \overset{*}{\mathscr{L}}_{ji}(0)
\end{aligned} \tag{3.20}$$

with identical symmetry relations holding when 0 is replaced by ∞ in (3.20).

Of course, the forms of the relaxation functions will be further restricted by the symmetry properties of the composite. In particular, in a centrosymmetric composite the relaxation functions $\underset{\sim}{\mathscr{A}}, \underset{\sim}{\overset{*}{\mathscr{A}}}, \underset{\sim}{\mathscr{B}}, \underset{\sim}{\overset{*}{\mathscr{B}}}$ and $\underset{\sim}{\overset{*}{h}}$ vanish.

In view of (3.8) and (3.16) we have the residual inequality

$$\rho\theta\gamma = \rho\theta\sigma - \underset{\sim}{q}\cdot\underset{\sim}{g}/\theta \geqslant 0 \tag{3.21}$$

where

$$\begin{aligned}
\rho\sigma\theta &= \rho\delta\hat{\phi}\left(\underset{\sim}{\Omega}^t \,\Big|\, \frac{d\underset{\sim}{\Omega}^t(s)}{ds}\right) = -\int_0^\infty\int_0^\infty \underset{\sim}{\Omega}_d^t(s)\cdot\frac{\partial^2\underset{\sim}{R}(s,u)}{\partial s\,\partial u}\,\frac{d\underset{\sim}{\Omega}^t(u)}{du}\,ds\,du \\
&= -\int_0^\infty\int_0^\infty \frac{d\underset{\sim}{\Omega}^t(s)}{ds}\cdot\frac{\partial\underset{\sim}{R}(s,u)}{\partial u}\,\frac{d\underset{\sim}{\Omega}^t(u)}{du}\,ds\,du \\
&= -\frac{1}{2}\int_{-\infty}^{t}\int_{-\infty}^{t} \frac{d\underset{\sim}{\Omega}(\xi)}{d\xi}\,\frac{\partial\underset{\sim}{R}(t-\xi, t-\eta)}{\partial t}\,\frac{d\underset{\sim}{\Omega}(\eta)}{d\eta}\,d\xi\,d\eta,
\end{aligned} \tag{3.22}$$

σ being the **internal dissipation** per unit mass of the composite. It follows from (3.21) that the thermal conductivity tensor is positive definite and consequently we have the dissipation inequality

$$\sigma \geqslant 0. \tag{3.21}$$

4. Restrictions on Relaxation Functions

Suppose that the composite is in a state of equilibrium with $\underset{\sim}{\Omega}(\tau) = \underset{\sim}{\Omega}^{(o)}$, a constant, at all times $-\infty < \tau \leqslant t$. It follows from (3.10) that the specific free energy is given by

$$\psi^{(\infty)}(\underset{\sim}{\Omega}^{(o)}) = \frac{1}{2} \underset{\sim}{\Omega}^{(o)} \cdot \underset{\sim}{\Gamma}\underset{\sim}{\Omega}^{(o)}. \tag{4.1}$$

Furthermore, we have shown in [9] that the free energy is a minimum in an equilibrium state and consequently if we consider the particular equilibrium state described by the virgin state $\underset{\sim}{\Omega}^{(o)} = (0,0,0,\alpha)$, where $\alpha > 0$, then it follows that the **equilibrium response modulus** $\underset{\sim}{M}(\infty) = \underset{\sim}{\Gamma}$ of the composite is positive semi definite i.e.

$$\underset{\sim}{\Lambda} \cdot \underset{\sim}{M}(\infty) \underset{\sim}{\Lambda} = \underset{\sim}{\Lambda} \cdot \underset{\sim}{\Gamma}\underset{\sim}{\Lambda} \geqslant 0. \tag{4.2}$$

A further consequence of the fact that ψ is a minimum in an equilibrium state is, c.f. inequality (6.17) of [9], that

$$\underset{\sim}{\Lambda} \cdot (\underset{\sim}{M}(0) - \underset{\sim}{M}(\infty))\underset{\sim}{\Lambda} = \underset{\sim}{\Lambda} \cdot \underset{\sim}{R}(0,0) \underset{\sim}{\Lambda} \geqslant 0. \tag{4.3}$$

The relaxation function $M(\cdot)$ is further restricted by the dissipation inequality (3.32) and when an argument due to Kolpashchikov and Shnipp [10] is employed it follows, c.f. [9], on writing

$$\phi(u) = \underset{\sim}{\Lambda} \cdot (\underset{\sim}{M}(u) - \underset{\sim}{M}(\infty))\underset{\sim}{\Lambda} = \underset{\sim}{\Lambda} \cdot \underset{\sim}{R}(0,u) \underset{\sim}{\Lambda}, \tag{4.4}$$

that the Fourier cosine transform of $\phi(s)$ is positive semi definite i.e.

$$\bar{\phi}^c(\omega) = \int_0^\infty \phi(s) \cos(\omega s) ds \geqslant 0. \tag{4.5}$$

Let us define the rate of working of the generalized stress in any process to be

$$\underset{\sim}{\Sigma} \cdot \underset{\sim}{\dot{\Omega}} = \tau_{ij} \dot{\varepsilon}_{ji} + D_{ij}\dot{P}_{ji} - \mathscr{F}_i w_i = \rho\eta\dot{\theta}. \tag{4.6}$$

The **generalized work*** done on the path $\underset{\sim}{\Omega}(\cdot)$ is

$$w(\underset{\sim}{\Omega}(\cdot)) = \int_{-\infty}^\infty \underset{\sim}{\Sigma}(t) \cdot \underset{\sim}{\dot{\Omega}}(t) dt, \tag{4.7}$$

---

*In the case of one component media, the generalized work is called "total work-heat" by Truesdell and Noll [11].

and, in view of $(3.16)_2$, this may be rewritten in the form

$$w(\underset{\sim}{\Omega}(\cdot)) = \int_{-\infty}^{\infty}\int_{-\infty}^{t} \dot{\underset{\sim}{\Omega}}(t)\cdot\underset{\sim}{M}(t-j)\dot{\underset{\sim}{\Omega}}(u)dudt. \tag{4.8}$$

In [9] it was shown that the Clausius-Duhem inequality implies that the generalized work is non negative for every closed generalized strain path i.e.

$$w(\underset{\sim}{\Omega}(\cdot)) \geqslant 0 \tag{4.9}$$

for every closed generalized strain path $\underset{\sim}{\Omega}(\cdot)$ and when (4.9) holds we say that the relaxation function $\underset{\sim}{\Omega}(\cdot)$ is **compatible with thermodynamics.** The symmetry properties of the components of $\underset{\sim}{M}(0)$ and $\underset{\sim}{M}(\infty)$ as expressed in the relations (3.20), as well as the inequalities (4.2), (4.3) and (4.5), are also consequences of compatibility with thermodynamics.

The relaxation function $\underset{\sim}{M}(\cdot)$ is said to be **dissipative,** if and only if

$$w(\underset{\sim}{\Omega}(\cdot)) \geqslant 0 \tag{4.10}$$

for every generalized strain path starting from the virgin state. The consequences of dissipativity in the context of linear viscoelasticity have been discussed by number of authors [12,13,14]. The results obtained in the context of single component linear viscoelastic media by Gurtin and Herrera [13] may be readily generalized to show that in order that $M(\cdot)$ should be dissipative it is necessary and sufficient that the inequality (4.5) holds for $\omega \geqslant 0$ and it follows that in order that the relaxation function be compatible with thermodynamics it is necessary that it be dissipative. An immediate consequence is that as well as satisfying the inequality (4.3), $\underset{\sim}{M}(u) - \underset{\sim}{M}(\infty) = \underset{\sim}{R}(o,u)$ must also satisfy the inequalities

$$\underset{\sim}{\Lambda}\cdot(\underset{\sim}{M}(0) - \underset{\sim}{M}(\infty))\underset{\sim}{\Lambda} \geqslant + \underset{\sim}{\Lambda}(M(u) - M(\infty))\underset{\sim}{\Lambda},\ \underset{\sim}{\Lambda}\cdot\underset{\sim}{M}'(0)\underset{\sim}{\Lambda} \leqslant 0. \tag{4.11}$$

The symmetry relations (3.20) are properties of the components of $\underset{\sim}{M}(0)$ and $\underset{\sim}{M}(\infty)$. We now seek to establish the conditions which must be satisfied by the relaxation function $\underset{\sim}{M}(\cdot)$ if the relation

$$\underset{\sim}{M}(u) = \underset{\sim}{M}^T(u) \tag{4.12}$$

is to hold for all u. There is no a priori reason why (4.12) should hold and as has been demonstrated in the case of single component materials by Shu and Onat [12] and Day [14], neither compatibility with thermodynamics or dissipativity implies the relations (4.12). The condition which must be satisfied in order that (4.12) should hold in a single phase material was originally determined by Day [15] for the purely mechanical relaxation function and Day's results were subsequently extended to include thermal and thermomechanical relaxation functions by Gurtin [16].

Let $\Omega(\cdot)$ be a closed generalized strain path starting from a virgin state. The **time reversal** of $\underset{\sim}{\Omega}(\cdot)$ is defined by

$$\tilde{\underset{\sim}{\Omega}}(t) = \underset{\sim}{\Omega}(-t). \tag{4.13}$$

The arguments given by Day [15] and Gurtin [16] generalize immediately to show that the relaxation function M(u) is symmemtric if and only if the generalized work done on every closed path $\underset{\sim}{\Omega}(\cdot)$ starting from the virgin state is invariant under time reversal i.e.

$$w(\underset{\sim}{\Omega}(\cdot)) = w(\tilde{\underset{\sim}{\Omega}}(\cdot)). \tag{4.14}$$

Thus, if we assume that (4.14) holds for every closed process which starts from a virgin state, it follows from (3.17) and (4.12) that for every $u \geq 0$:

$$\underset{\sim}{M}(u) = \begin{bmatrix} \underset{\sim}{\mathscr{G}}(u) & \underset{\sim}{\mathscr{H}}(u) & \underset{\approx}{\mathscr{A}}(u) & \underset{\sim}{\mathscr{I}}(u) \\ \underset{\sim}{\mathscr{H}}^T(u) & \underset{\sim}{\mathscr{J}}(u) & \underset{\approx}{\mathscr{B}}(u) & \underset{\sim}{\mathscr{L}}(u) \\ \underset{\sim}{\mathscr{A}}^T(u) & \underset{\approx}{\mathscr{B}}^T(u) & \mathscr{M}(u) & \underset{\sim}{\overset{*}{h}}(u) \\ \underset{\sim}{\bar{g}}(u) & \underset{\sim}{\mathscr{L}}^T(u) & \underset{\sim}{\overset{*}{h}}(u) & \overset{*}{g}(u) \end{bmatrix} \tag{4.15}$$

where $\underset{\sim}{\mathscr{G}}(\cdot)$, $\underset{\sim}{\mathscr{J}}(\cdot)$, $\underset{\sim}{\mathscr{M}}(\cdot)$ and $\underset{\sim}{\mathscr{I}}(\cdot)$ are symmetric so that

$$\mathscr{G}_{ijk\ell}(s) = \mathscr{G}_{klij}(s), \; \mathscr{J}_{ijkl}(s) = \mathscr{J}_{klij}(s), \mathscr{N}_{ij}(s) = \mathscr{N}_{ji}(s), \mathscr{I}_{kl}(s) = \mathscr{I}_{lk}(s). \tag{4.16}$$

5. Constitutive Equations for Isotropic Composites

In this section we obtain the constitutive equations for an isotropic centrosymmetric composite material with two constituents. We ignore thermal effects. In view of the assumed symmetry of the material the only nonvanishing components of the relaxation function $M(\cdot)$ are $\mathscr{G}(\cdot)$, $\mathscr{J}(\cdot)$, $\mathscr{H}(\cdot)$ and $\mathscr{M}(s)$. When the isotropy of the composite and the symmetry properties (3.26), (3.27) and (4.12) are taken into account it follows that the only nonvanishing components of $\underset{\sim}{M}(\cdot)$ may be represented as follows:

$$\mathscr{G}_{ijkl}(s) = \hat{\lambda}(s)\delta_{ij}\delta_{kl} + \hat{\mu}(s)(\delta_{ik}\delta_{jl} + \delta_{il}\delta_{jk}),$$

$$\mathscr{J}_{ijkl}(s) = \hat{b}_1(s)\delta_{ij}\delta_{kl} + \hat{b}_2(s)(\delta_{ik}\delta_{jl} + \delta_{il}\delta_{jk}) + \hat{b}_3(s)(\delta_{ik}\delta_{ji} - \delta_{il}\delta_{jk}), \tag{5.1}$$

$$\mathscr{H}_{ijkl}(s) = \hat{\beta}_1(s)\delta_{ij}\delta_{kl} + \frac{1}{2}\hat{\beta}_2(s)(\delta_{ik}\delta_{jl} + \delta_{il}\delta_{jk}), \mathscr{N}_{ij}(s) = -\hat{a}_1(s)\delta_{ij}.$$

The scalar functions, e.g. $\hat{\lambda}(s)$, occuring on the right hand sides of equations (5.1) are continuous and bounded.

It follows from equations (3.18) and (5.1) that the response of an isotropic two component linear viscoelastic composite is described by the constitutive relations

$$\tau_{ij} = \int_{-\infty}^{t} \{\hat{\lambda}(s)\dot{\varepsilon}_{kk}(s)\delta_{ij} + 2\hat{\mu}(s)\dot{\varepsilon}_{ij}(s) + \hat{\beta}_1(s)\dot{p}_{kk}(s)\delta_{ij} + \frac{1}{2}\hat{\beta}_2(s)(\dot{p}_{ij}(s) + \dot{p}_{ji}(s)\}ds,$$

$$D_{ji} = \int_{-\infty}^{t} \{\hat{\beta}_1(s)\dot{\varepsilon}_{kk}(s)\delta_{ij} + \hat{\beta}_2(s)\dot{\varepsilon}_{ij}(s) + \hat{b}_1(s)\dot{p}_{kk}(s)\delta_{ij} + \hat{b}_2(s)(\dot{p}_{ij}(s) + \dot{p}_{ji}(s)) + \hat{b}_3(s)(\dot{p}_{ij}(s) - \dot{p}_{ji}(s))\}ds, \tag{5.2}$$

$$\mathscr{F}_i = -\int_{-\infty}^{t} \hat{a}_1(s)\dot{w}_i(s)ds.$$

The behavior of the coefficients $\hat{\lambda}(s)$, $\hat{\mu}(s)$, etc., occuring in (5.1) is restricted by the inequalities (4.2), (4.3), (4.5) and (4.11). Let f(s) denote any of the functions occuring on the right hand side of (5.1) and define the function $\tilde{f}(s)$ by the relation

$$\tilde{f}(s) = \hat{f}(s) - \overline{f}, \lim_{s\to\infty} \hat{f}(s) = \overline{f}. \tag{5.3}$$

In order that (4.2) be satisfied for the composite we must have

$$\overline{\mu} > 0, 3\overline{\lambda} + 2\overline{\mu} > 0, \overline{b}_2 > 0, 3\overline{b}_1 + 2\overline{b}_2 > 0, \overline{b}_3 < 0, \overline{a}_1 < 0, \tag{5.4}$$

as well as

$$(\overline{\beta}_1 + 2\overline{\beta}_2)^2 \leqslant (\overline{\lambda} + 2\overline{\mu})(\overline{b}_1 + 2\overline{b}_2), \overline{\beta}_2^2 \leqslant \overline{\mu}\,\overline{b}_2. \tag{5.5}$$

Next, the inequality (4.3) implies that

$$\tilde{\mu}(0) > 0, 3\tilde{\lambda}(0) + 2\tilde{\mu}(0) > 0, \tilde{b}_2(0) > 0,$$

$$3\tilde{b}_1(0) + 2\tilde{b}_2(0) > 0, \tilde{b}_3(0) < 0, \tilde{a}_1(0) < 0, \tag{5.6}$$

$$(\tilde{\beta}_1(0) + 2\tilde{\beta}_2(0)^2 \leqslant (\tilde{\lambda}(0) + 2\tilde{\mu}(0))(\tilde{b}_1(0) + 2\tilde{b}_2(0)),$$

$$\tilde{\beta}_2(0)^2 \leqslant \tilde{\mu}(0)\tilde{b}_2(0),$$

and, of course, the inequalities (4.2) and (4.3) together imply that the inequalities (5.6) are also satisfied by the quantities $\mu(0)$, $\lambda(0)$, etc. Finally, the inequality $(4.11)_2$ implies that

$$\tilde{\mu}'(0) < 0,\ 3\tilde{\lambda}'(0) + 2\tilde{\mu}'(0) < 0,\ \tilde{b}_2'(0) < 0,$$

$$3\tilde{b}_1'(0) + 2\tilde{b}_2'(0) < 0,\ \tilde{b}_3'(0) > 0,\ \tilde{a}_1'(0) > 0,$$

$$(\tilde{\beta}_1'(0) + 2\tilde{\beta}_2'(0))^2 \leqslant (\tilde{\lambda}'(0) + 2\tilde{\mu}'(0))(\tilde{b}_1'(0) + 2\tilde{b}_2'(0)), \qquad (5.7)$$

$$\tilde{\beta}_2'(0)^2 < \tilde{\mu}'(0)\tilde{b}_2'(0).$$

The importance of the foregoing inequalities is illustrated in a forthcoming paper on wave propagation in linear viscoelastic composites.

References

1. Tiersten, H. F. and M. Jahanmir, Arch. Rational Mech. Anal., 65, 153-192, 1977.

2. Bedford, A. and M. Stern, Acta Mechanica, 14, 85-102, 1972.

3. Aboudi, J. and Y. Benveniste, J. Sound Vib., 41, 163-175, 1975.

4. Hegemier, G. A., G. A. Gurtman and A. H. Nayfeh, Int. J. Solids Structures, 9, 1973.

5. Khoroshun, L. P., Sov. Appl. Mech. 13, 124-132, 1978.

6. Marinov, P., Bull. Roy. Soc. de Liege, No. 1-2, 106-118, 1974.

7. Marinov, P., Int. J. Engrg. Sci. 16, 533-544, 1978.

8. Bedford, A. and M. Stern, J. Appl. Mech., 38, 8-14, 1971.

9. McCarthy, M. F. and H. F. Tiersten, A theory of viscoelastic composites modelled as interpenetrating continua with memory. (To appear, 1982.)

10. Kolpashchikov, V. L. and A. L. Shnipp, Int. J. Engrg. Sci. 16, 503-514, 1978.

11. Truesdell, C. and W. Noll, The Nonlinear Field Theories of Mechanics, Handbuch der Physik III/2, Springer, Berlin, 1965.

12. Shu, L. S., and E. T. Onst, Proc. Fourth Symp. on Naval Structural Mechanics, Pergamon, Oxford, 1966.

13. Gurtin, M. E. and I. Herrera, Quart. Appl. Math. 23, 235-245, 1965.

14. Day, W. A., Quart, J. Mech. App. Math. 24, 487-497, 1971.

15. Day, W. A. Arch. Rational Mech. Anal. 40, 155-159, 1971.

16. Gurtin, M. E. Arch. Rational. Mech. Anal. 44, 387-399, 1972.

17. Coleman, B. D., Arch. Rational Mech. Anal. 17, 1-46, 1964.

SPECIAL TOPICS IN FLUIDS

STABILITY REGION OF CELLULAR CONVECTION WITH NEARLY INSULATING BOUNDARIES

N. Riahi

Department of Theoretical and Applied Mechanics
University of Illinois at Urbana-Champaign
Urbana, Illinois 61801, U.S.A.

Abstract

The problem of stability region for the preferred convection pattern in a horizontal layer with nearly insulating boundaries is investigated. Stability criteria for $\alpha \neq \alpha_c$(α_c being the critical wave number α which minimizes the Rayleigh number R) are found which determine the region of the preferred steady solution in the form of square cells in the R, α - plane.

1. Introduction

In a recent study of finite amplitude cellular convection in a horizontal layer with nearly insulating boundaries, Busse and Riahi [3] found that square pattern convection represents the preferred stable convection with respect to disturbances which have the same wave numbers $\dot{\alpha}$ and $\tilde{\alpha} = \alpha_c$, where α_c is the value of the wave number α of the steady squares which minimizes the Rayleigh number R. The analysis in [3] has carried out only for the case where $\alpha = \alpha_c$, since the main concern was the preferred flow pattern rather than the range of wave numbers of the realized convection flow. In the case $\alpha \neq \alpha_c$, however, the most strongly growing disturbances have wave numbers $\tilde{\alpha}$ which do not equal α. The present study is concerned with this latter case and includes a detailed stability analysis of the problem. In an earlier study of a related problem with isothermal boundaries, Busse [1] determined the stability region for the steady two-dimensional cells which is the preferred flow pattern for that case via a stability criterion. In the present problem, we find a more complicated stability criterion for the stability region of square cells. Although Busse's stability region depends on the prandtl number P and decreases slightly with increasing P, ours is independent of P, provided P does not tend to zero as was pointed out in [3]. Furthermore, in contrast to the case with isothermal boundaries, the stability region in the present problem is located to the left of the line $\alpha = \alpha_c$ in the (R, α) space.

The problem of the realizable range of wavelengths of three dimensional convection cells in the form of squares studied in this paper compliments that in [3] for convection in a layer with nearly insulating boundaries. Although these studies are on the basis of the weakly non-linear theory, the results will be useful for the interpretation of instabilities occurring in the more complex forms of convection at higher Rayleigh numbers.

2. Governing equations

We consider an infinite horizontal layer of fluid of depth d bounded by two infinite half spaces with the thermal conductivity λ^e which is assumed to be small compared to the thermal conductivity λ of the fluid. In the steady state, a constant heat flux transverses the system such that the temperatures T_1 and T_2 are attained at the upper and lower boundaries of the fluid. Under the usual Boussinesq approximation, the non-dimensional forms of the equations for momentum, heat and conservation of mass can be simplified by using the general representation for the velocity vector as the sum of the poloidal $\nabla \times \nabla \times \underset{\sim}{\lambda} v$ and toroidal $\nabla \times \underset{\sim}{\lambda} \psi$ vectors where $\underset{\sim}{\lambda}$ represents a unit vector in vertical direction. The detailed analysis, though omitted here, indicates that the terms containing ψ in the basic equations are essentially insignificant, since the toroidal component of the velocity vector is of the order of the mth power of the amplitude ε $(m \geq 3)$ and thus cannot enter the small amplitude analysis discussed in this paper. Therefore, we simply set $\psi = 0$ in the basic equations for the subsequent analysis to be discussed in the next section. Vertical component of the curl of the curl of the momentum equation together with the heat equation then yield the following governing equations

$$\Delta_2(\nabla^4 v - \theta - P^{-1} \frac{\partial}{\partial t} \nabla^2 v) = P^{-1} \underset{\sim}{\delta} \cdot (\underset{\sim}{\delta} \cdot \nabla \underset{\sim}{\delta} v), \qquad (1a)$$

$$\nabla^2 \theta - R\Delta_2 v - \frac{\partial \theta}{\partial t} = \underset{\sim}{\delta} v \cdot \nabla \theta , \qquad (1b)$$

where $\underset{\sim}{\delta} = \nabla \times \nabla \times \underset{\sim}{\lambda}$, θ is the deviation of the temperature from its static value, $R = a(T_2 - T_1)gd^3 \rho_o c/(\nu\lambda)$ is the Rayleigh number, $P = \nu\rho_o c\lambda^{-1}$ is the Prandtl number, ρ_o is the reference density, c is the specific heat at constant pressure, a is the coefficient of thermal expansion, ν is the kinematic viscosity, g is the acceleration due to gravity and Δ_2 is the horizontal Laplacian.

Following [3], the conditions for v and θ at rigid boundaries are

$$\left.\begin{aligned} & v = \frac{\partial v}{\partial z} = 0 , \\ & \theta = \theta^e, \ \frac{\partial \theta}{\partial z} = \beta \frac{\partial \theta^e}{\partial z} , \end{aligned}\right\} \text{at} \quad z = \pm \frac{1}{2} \qquad (2)$$

where $\beta = \lambda^e/\lambda$ and θ^e describes the deviation from the static temperature distribution in the space $|z| \geq \frac{1}{2}$. The linear planeform function w(x,y) has the representation

$$w(x,y) = \sum_{n=-N}^{N} c_n w_n \equiv \sum_{n=-N}^{N} c_n \exp(I \underset{\sim}{K}_n \cdot \underset{\sim}{r}) , \tag{3}$$

and satisfies the relation

$$\Delta_2 w = - \alpha^2 w, < ww > = 1 . \tag{4}$$

Here an angular bracket indicates an average over the fluid layer, α is the horizontal wave number of the linear planeform function $w(x,y)$, $\underset{\sim}{r}$ is the position vector and the horizontal wave number vectors $\underset{\sim}{K}_n$ satisfy the properties

$$\underset{\sim}{K}_n \cdot \underset{\sim}{\lambda} = 0, \quad |\underset{\sim}{K}_n| = \alpha, \quad \underset{\sim}{K}_{-n} = - \underset{\sim}{K}_n . \tag{5}$$

The coefficients c_n satisfy the conditions

$$\sum_{n=-N}^{N} c_n c_n^* = 1, \quad c_n^* = c_{-n} , \tag{6}$$

where C_n^* denotes the complex conjugate of c_n. The function w and (4)-(6) are now standard in thermal convection theory. The reader is referred to the recent review of the problem of thermal convection, [2], for details on the subject.

3. Stability analysis

The analysis discussed in this section is based on the simplified assumption that $P = \infty$. The equation (1a) then reduces to

$$\Delta_2 (\nabla^4 v - \theta) = 0 . \tag{7}$$

However, an analogous discussion to that in [3] concludes that the results derived in this paper do indeed hold for arbitrary P, provided P does not tend to zero.

Since the finite amplitude steady problem of the present study is identical to that discussed in [3], it is found again that the value α_c which minimizes R is proportional to $\beta^{1/3}$. Thus it is assumed that $\alpha^2 = \eta^2 \gamma$, where $\gamma \equiv \beta^{2/3}$ and η is of the order unity for the convection modes of physical interest. We then seek solutions in terms of a double series in powers of γ and ε :

$$\begin{pmatrix} v \\ \theta \end{pmatrix} = \sum_{n=0,\ m=1} \varepsilon^m \gamma^n \begin{pmatrix} v_m^{(n)} \\ \theta_m^{(n)} \end{pmatrix}, \quad R = \sum_{m,n=0} \varepsilon^m \gamma^n R_m^{(n)} . \tag{8}$$

As we mentioned above, the linear and nonlinear analysis has already been discussed in [3]. For the linear problem, it was found in [3] that

$$\theta_1^{(0)} = w(x,y), \quad v_1^{(0)} = w(x,y)\,(z^2 - \tfrac{1}{4})^2/4! ,$$

$$\eta_c = (462/17)^{1/3}, \quad R_o^{(0)} = 720, \quad R_o^{(1)} = 720(2\beta\alpha^{-1} + \eta_c^{-3}\alpha^2)\gamma^{-1} . \tag{9}$$

We now define

$$R_o = R_o^{(0)} + \gamma R_o^{(1)} . \tag{10}$$

Minimizing R_o with respect to α yields

$$R_c = 720\,(1 + 3\gamma\eta_c^{-1}), \quad R_c \equiv \min_\alpha R_o . \tag{11}$$

Using (9)-(10), the approximate expression of R_o for small values of $|\alpha - \alpha_c|$ can be written as

$$R_o \simeq R_c + \mu(\alpha - \alpha_c)^2 , \quad \text{where} \quad \mu = \frac{1}{2} \frac{\partial^2 R_o}{\partial \alpha^2}\Big|_{\alpha=\alpha_c} = 3R_o^{(0)} \eta_c^{-3} \tag{12a}$$

(12b)

For the nonlinear problem, it is found in [3] that the expression $\varepsilon^2 \gamma R_2^{(1)}$ is turned out to be the largest non-zero term in the nonlinear domain $(m > 1)$, in the expansion (8) for R. The expression for $R_2^{(1)}$ is found after forming the soluability condition for the heat equation in the order $\varepsilon^3 \gamma^2$. After some considerable algebra, this condition yields the following system of equations,

$$R_2^{(1)} c_n = \alpha^2 R_o^{(0)} [-10 \sum_{\ell \neq -p} \phi_{\ell p} \phi_{mn} c_m c_\ell c_p \langle w_n^* w_m w_\ell w_p \rangle + 3 c_n]/10! ,$$

where
$$\phi_{mn} = (\underset{\sim}{K}_m \cdot \underset{\sim}{K}_n)\alpha^{-2} . \tag{13}$$

It was shown in [3] that the preferred flow pattern convection is that of squares for sufficiently small ε. In order to determine the region of stability of squares in (α, R) space, we consider the system of equations for the time dependent infinitesimal disturbances $\tilde{v}$, $\tilde{\theta}$

$$\Delta_2 (\nabla^4 \tilde{v} - \tilde{\theta}) = 0 , \tag{14a}$$

$$(\nabla^2 - \sigma)\tilde{\theta} - R \Delta_2 \tilde{v} = \underset{\sim}{\delta} \tilde{v} \cdot \nabla\theta + \underset{\sim}{\delta} v \cdot \nabla\tilde{\theta} , \tag{14b}$$

$$\tilde{v} = \frac{\partial \tilde{v}}{\partial z} = \tilde{\theta} - \tilde{\theta}^e = \frac{\partial}{\partial z} (\tilde{\theta} - \tilde{\theta}^e) = 0 \quad \text{at} \quad z = \pm \frac{1}{2} \tag{14c}$$

where $\tilde{\theta}^e$ describes the temperature perturbation for the disturbances in the space $|z| \geq \frac{1}{2}$, and we have introduced a growth rate σ by $\frac{\partial \tilde{v}}{\partial t} \equiv \sigma\tilde{v}$. The equations (14) are solved by using (8) and the following expansion

$$\begin{pmatrix} v \\ \theta \end{pmatrix} = \sum_{n=1,m=0} \varepsilon^{n-1} \gamma^m \begin{pmatrix} v_n^{(m)} \\ \theta_n^{(m)} \end{pmatrix} , \quad \sigma = \sum_{n,m=0} \varepsilon^n \gamma^m \sigma_n^{(m)} . \tag{15}$$

In the lowest order of ε and γ equations (14) became

$$\frac{\partial^4 \tilde{v}_1^{(0)}}{\partial z^4} = \tilde{\theta}_1^{(0)} , \tag{16a}$$

$$\left(\frac{\partial^2}{\partial z^2} - \sigma_o^{(0)}\right)\tilde{\theta}_1^{(0)} = 0 , \tag{16b}$$

$$\tilde{v}_1^{(0)} = \frac{\partial \tilde{v}_1^{(0)}}{\partial z} = \frac{\partial \tilde{\theta}_1^{(0)}}{\partial z} = 0 \qquad \text{at} \quad z = \pm \frac{1}{2} . \tag{16c}$$

The solution to (16a)-(16b) subject to the boundary conditions (16c) are

$$\tilde{\theta}_1^{(0)} = \tilde{w}(x,y), \quad \sigma_o^{(0)} = 0 , \tag{17a,b}$$

$$\tilde{v}_1^{(0)} = \tilde{w}(x,y) \left(z^2 - \frac{1}{4}\right)^2 /4! , \tag{17c}$$

where $$\tilde{w}(x,y) = \sum_n \tilde{c}_n \tilde{w}_n \equiv \sum_n \tilde{c}_n \exp(I \underset{\sim}{\tilde{K}}_n \cdot \underset{\sim}{r}) . \tag{18}$$

The horizontal wave number vectors $\underset{\sim}{\tilde{K}}_n$ for the disturbances are assumed to be in the direction of $\underset{\sim}{K}_n$, have the same magnitude $\tilde{\alpha}$ and satisfy the properties

$$\tilde{K}_n \cdot \lambda = 0, \quad \tilde{K}_{-n} = -\tilde{K}_n . \tag{19}$$

The coefficients $\tilde{c}_n$ satisfy the conditions

$$\sum_n \tilde{c}_n \tilde{c}^*_n = 1, \quad \tilde{c}^*_n = \tilde{c}_{-n} . \tag{20}$$

In deriving (17), a normalization condition of the form

$$\langle \tilde{\theta}_n^{(m)} \tilde{\theta}_1^{(0)} \rangle = \delta_{n1} \delta_{m0} \tag{21}$$

has been assumed. This condition is used throughout the present study to determine the solutions $\tilde{\theta}_n^{(m)}$.

We shall assume that the horizontal wave numbers of the steady solution and the disturbances differ from α_c only by a small amount of the order $\varepsilon\gamma^{1/2}$ and that $\tilde{\alpha} \approx \alpha + O(\varepsilon\gamma^{1/2})$.

In the order γ equations (14) become

$$\frac{\partial^4 \tilde{v}_1^{(1)}}{\partial z^4} - \tilde{\theta}_1^{(1)} = 2\tilde{\eta}^2 \frac{\partial^2 \tilde{v}_1^{(0)}}{\partial z^2} , \tag{22a}$$

$$-\sigma_o^{(1)} \tilde{\theta}_1^{(0)} + \frac{\partial^2 \tilde{\theta}_1^{(1)}}{\partial z^2} = \tilde{\eta}^2 (\tilde{\theta}_1^{(0)} - R_o^{(0)} \tilde{v}_1^{(0)}) , \tag{22b}$$

$$\tilde{v}_1^{(1)} = \frac{\partial \tilde{v}_1^{(1)}}{\partial z} = \frac{\partial \tilde{\theta}_1^{(1)}}{\partial z} = 0 \qquad \text{at} \qquad z = \pm \frac{1}{2} , \tag{22c}$$

where $\tilde{\eta} = \tilde{\alpha}\gamma^{-1/2}$.

Multiplying the equation (22b) by $\tilde{\theta}_1^{(0)}$, averaging it over the fluid layer, using (9) and (17)-(20), we find that

$$\sigma_o^{(1)} = 0 . \tag{23}$$

The solution to (22) then becomes

$$\tilde{\theta}_1^{(1)} = \tilde{\eta}^2 (-z^6 + \frac{5}{4} z^4 - \frac{7}{16} z^2 + \frac{31}{1344}) \tilde{w}(x,y) , \tag{24a}$$

$$\tilde{v}_1^{(1)} = \tilde{\eta}^2 \tilde{w}(x,y) [-(2z)^{10} + 15(2z)^8 + 126(2z)^6 - 810(2z)^4 + 1187(2z)^2 - 517]/(315.2^{14}) \tag{24b}$$

In the order γ^2 equation (14b) becomes

$$-\sigma_o^{(2)}\,\tilde{\theta}_1^{(0)} + \frac{\partial^2\tilde{\theta}_1^{(2)}}{\partial z^2} + \tilde{\eta}^2(R_o^{(0)}\,\tilde{v}_1^{(1)} + R_o^{(1)}\,\tilde{v}_1^{(0)} - \tilde{\theta}_1^{(1)}) = 0\ . \tag{25a}$$

The corresponding boundary condition is

$$\frac{\partial^2\tilde{\theta}_1^{(2)}}{\partial z^2} \pm \tilde{\eta}\,\tilde{\theta}_1^{(0)} = 0 \qquad \text{at} \qquad z = \pm\frac{1}{2}\ . \tag{25b}$$

The solvability condition for the inhomogeneous boundary-value problem is obtained in the same way as condition (23) for $\sigma_o^{(1)}$ and yields

$$\gamma^2\,\sigma_o^{(2)} = -2\,\tilde{\alpha}\,\gamma^{3/2} - \tilde{\alpha}^4\,\eta_c^{-3} + \tilde{\alpha}^2\,R_o^{(1)}/R_o^{(0)}\ . \tag{26}$$

After some manipulations, (26) can be written in the following form

$$\gamma^2\,\sigma_o^{(2)} = 3\tilde{\alpha}^2\,\eta_c^{-3}\,[(\alpha - \alpha_c)^2 - (3\tilde{\alpha})^{-1}\,(2\alpha_c + \tilde{\alpha})(\tilde{\alpha} - \alpha_c)^2]. \tag{27}$$

It can be seen from (27) that in the limit $\tilde{\alpha} \to \alpha$,

$$\sigma_o^{(2)} \quad 0 \quad \text{for} \quad \alpha > \alpha_c\ . \tag{28}$$

In the order $\epsilon\gamma$ equations (14) become

$$\frac{\partial^4\tilde{v}_2^{(1)}}{\partial z^4} = \tilde{\theta}_2^{(1)} \tag{29a}$$

$$\gamma\,\frac{\partial^2\tilde{\theta}_2^{(1)}}{\partial z^2} - \gamma\sigma_1^{(1)}\,\tilde{\theta}_1^{(0)} = \delta\tilde{\underset{\sim}{v}}_1^{(0)}\cdot\nabla\theta_1^{(0)} + \delta\underset{\sim}{v}_1^{(0)}\cdot\nabla\tilde{\theta}_1^{(0)}\ , \tag{29b}$$

$$\tilde{v}_2^{(1)} = \frac{\partial}{\partial z}\,\tilde{v}_2^{(1)} = [\![\,\tilde{\theta}_2^{(1)}\,]\!] = \frac{\partial}{\partial z}\,(\tilde{\theta}_2^{(1)} - [\![\,\tilde{\theta}_2^{(1)}\,]\!]) = 0 \quad \text{at} \quad z = \pm\frac{1}{2}, \tag{29c}$$

where open brackets denote a horizontal average. The reader is referred to [3] for discussion on the thermal boundary conditions. Multiplying (29b) by $\tilde{\theta}_1^{(0)}$ and averaging over the layer yields

$$\sigma_1^{(1)} = 0\ . \tag{30}$$

The solution to (29b) then becomes

$$\tilde{\theta}_2^{(1)} = \gamma^{-1}\left(-\frac{z^5}{5} + \frac{z^3}{6} - \frac{z}{16}\right) \sum_{\ell \neq -p} [c_\ell \tilde{c}_p (\underset{\sim}{K}_\ell \cdot \tilde{\underset{\sim}{K}}_p) w_\ell \tilde{w}_p + \tilde{c}_\ell c_p (\tilde{\underset{\sim}{K}}_\ell \cdot \underset{\sim}{K}_p) \cdot$$

$$\tilde{w}_\ell w_p]/4! + \alpha \tilde{\alpha} \gamma^{-1}\left(\frac{z^5}{5} - \frac{z^3}{6} + \frac{7z}{240}\right) \sum_{\ell=-N}^{N} (c_\ell^* \tilde{c}_\ell w_\ell^* \tilde{w}_\ell + \tilde{c}_\ell^* c_\ell \tilde{w}_\ell^* w_\ell)/4! . \quad (31)$$

The expression for $\tilde{v}_2^{(1)}$ is not given here, since it is not needed in the present study.

In the order $\varepsilon^2 \gamma^2$ equation (14b) for $\tilde{\theta}_3^{(2)}$ becomes

$$\gamma \frac{\partial^2}{\partial z^2} \tilde{\theta}_3^2 = \gamma \sigma_2^{(2)} \tilde{\theta}_1^{(0)} + R_2^{(1)} \Delta_2 \tilde{v}_1^{(0)} + \delta \tilde{\underset{\sim}{v}}_1^{(0)} \cdot \nabla \theta_2^{(1)} + \delta \underset{\sim}{v}_1^{(0)} \cdot \nabla \tilde{\theta}_2^{(1)} + \delta \tilde{\underset{\sim}{v}}_2^{(1)} \cdot \nabla \theta_1^{(0)} +$$

$$+ \delta \underset{\sim}{v}_2^{(1)} \cdot \nabla \tilde{\theta}_1^{(0)} . \quad (32a)$$

The boundary conditions are

$$[\![\tilde{\theta}_3^{(2)}]\!] = \frac{\partial}{\partial z} (\tilde{\theta}_3^{(2)} - [\![\tilde{\theta}_3^{(2)}]\!]) = 0 \qquad \text{at} \qquad z = \pm \frac{1}{2} . \quad (32b)$$

Multiplying (32a) by $\tilde{w}_n^*$ and averaging over the layer yields

$$[-\sigma_2^{(2)} + \tilde{\eta}^2 R_2^{(1)}/720]\gamma \tilde{c}_n = \langle \tilde{w}_n^* (\delta \tilde{v}_1^{(0)} \cdot \nabla \theta_2^{(1)} + \delta v_1^{(0)} \cdot \nabla \tilde{\theta}_2^{(1)}) \rangle . \quad (33)$$

After some algebra, (33) can be simplified to the form

$$[-\sigma_2^{(2)} + \tilde{\eta}^2 R_2^{(1)}/720]\gamma^2 \tilde{c}_n = 3\alpha^4 [\tilde{c}_n + c_n \sum_{\ell=-N}^{N} (c_\ell \tilde{c}_\ell^* + c_\ell^* \tilde{c}_\ell)]/10!$$

$$- \sum_{\ell \neq -p} [(\underset{\sim}{K}_\ell \cdot \underset{\sim}{K}_p)(\tilde{\underset{\sim}{K}}_m \cdot \tilde{K}_n) \tilde{c}_m c_\ell c_p \langle \tilde{w}_n^* \tilde{w}_m w_\ell w_p \rangle + (\tilde{\underset{\sim}{K}}_\ell \cdot \underset{\sim}{K}_p)(\underset{\sim}{K}_m \cdot \tilde{\underset{\sim}{K}}_n) c_m \cdot$$

$$c_\ell c_p \langle \tilde{w}_n^* w_m \tilde{w}_\ell w_p \rangle + (\underset{\sim}{K}_\ell \cdot \tilde{\underset{\sim}{K}}_p)(\underset{\sim}{K}_m \cdot \tilde{\underset{\sim}{K}}_n) c_m c_\ell \tilde{c}_p \langle \tilde{w}_n^* w_m w_\ell \tilde{w}_p \rangle]/9! . \quad (34)$$

If one now defines

$$\sigma^{(2)} = \sigma_o^{(2)} + \sigma_2^{(2)} \varepsilon^2 , \quad (35)$$

then (27) and (34) can be combined into the following form

$$[-\varepsilon^{-2}\sigma^{(2)} + \tilde{\eta}^2 R_2^{(1)}/720]\gamma\tilde{c}_n + 3\varepsilon^{-2}\tilde{\alpha}^2\eta_c^{-3}[(\alpha-\alpha_c)^2 - (3\tilde{\alpha})^{-1}(2\alpha_c+\tilde{\alpha})$$

$$(\tilde{\alpha}-\alpha_c)^2]\tilde{c}_n - 3\alpha^4[\tilde{c}_n + c_n \sum_{\ell=-N} (c_\ell\tilde{c}^*_\ell + c^*_\ell\tilde{c}_\ell)]/10! = - \sum_{\ell\neq-p} [(\underset{\sim}{K}_\ell\cdot\underset{\sim}{K}_p)$$

$$(\tilde{\underset{\sim}{K}}_m\cdot\tilde{\underset{\sim}{K}}_n)\tilde{c}_m c_\ell c_p \, \langle\tilde{w}^*_n\tilde{w}_m w_\ell w_p\rangle + (\tilde{\underset{\sim}{K}}_\ell\cdot\underset{\sim}{K}_p)(\underset{\sim}{K}_m\cdot\tilde{\underset{\sim}{K}}_n)c_m\tilde{c}_\ell c_p \, \cdot$$

$$\langle\tilde{w}^*_n\tilde{w}_\ell w_m w_p\rangle + (\underset{\sim}{K}_\ell\cdot\tilde{\underset{\sim}{K}}_p)(\underset{\sim}{K}_m\cdot\tilde{\underset{\sim}{K}}_n)\, c_m c_\ell\tilde{c}_p \, \langle\tilde{w}^*_n w_m w_\ell\tilde{w}_p\rangle]/9! \tag{36}$$

Since the steady flow pattern is in the form of square cells, the expressions (6) and (13) for $|c_n|^2$ and $R_2^{(1)}$ can be simplified to the forms

$$|c_n|^2 = 1/4, \; R_2^{(1)} = 11\,\eta^2 R_o^{(0)}/(2\cdot 10!) \tag{37}$$

The right hand side of the expression (36) is non-zero only if the average products such as $\langle\tilde{w}^*_n w_m \tilde{w}_\ell w_p\rangle$ is non-zero. That is, if

$$\tilde{\underset{\sim}{K}}_n = \underset{\sim}{K}_m + \tilde{\underset{\sim}{K}}_\ell + \underset{\sim}{K}_p \; . \tag{38}$$

For the steady solution of the form (37), (38) is equivalent to one of the following relations

$$\begin{aligned}
&\tilde{\underset{\sim}{K}}_n = \tilde{\underset{\sim}{K}}_\ell \pm \underset{\sim}{K}_1 \pm \underset{\sim}{K}_1 \;, &&\tilde{\underset{\sim}{K}}_n = \tilde{\underset{\sim}{K}}_\ell \pm \underset{\sim}{K}_1 \mp \underset{\sim}{K}_1 \;,\\
&\tilde{\underset{\sim}{K}}_n = \tilde{\underset{\sim}{K}}_\ell \pm \underset{\sim}{K}_2 \pm \underset{\sim}{K}_2 \;, &&\tilde{\underset{\sim}{K}}_n = \tilde{\underset{\sim}{K}}_\ell \pm \underset{\sim}{K}_2 + \underset{\sim}{K}_2 \;,\\
&\tilde{\underset{\sim}{K}}_n = \tilde{\underset{\sim}{K}}_\ell \pm \underset{\sim}{K}_1 \pm \underset{\sim}{K}_2 \;, &&\tilde{\underset{\sim}{K}}_n = \tilde{\underset{\sim}{K}}_\ell \pm \underset{\sim}{K}_1 \mp \underset{\sim}{K}_2 \;.
\end{aligned} \tag{39}$$

An extensive search for the possible disturbances that satisfy (36) leads to two cases which yield qualitatively different results.

Case A) Disturbances that satisfy the relation

$$\tilde{\underset{\sim}{K}}_n = \tilde{\underset{\sim}{K}}_m \pm \underset{\sim}{K}_1 \mp \underset{\sim}{K}_1 \quad \text{or} \quad \tilde{\underset{\sim}{K}}_n = \tilde{\underset{\sim}{K}}_m \pm \underset{\sim}{K}_2 \mp \underset{\sim}{K}_2 \; . \tag{40}$$

Using (40) in (36), we find the following result

$$(-\varepsilon^{-2}\sigma^{(2)} + \tilde{\eta}^2 R_2^{(1)}/720)\gamma^2 \tilde{c}_n + 3\,\varepsilon^{-2}\tilde{\alpha}^2 \eta_c^{-3}[(\alpha-\alpha_c)^2 - (3\tilde{\alpha})^{-1}.$$

$$(2\alpha_c + \tilde{\alpha})(\tilde{\alpha}-\alpha_c)^2]\tilde{c}_n = 6\,\alpha^4(\frac{4}{3}\tilde{c}_n + c_n \sum_{\ell=-2}^{2} c_\ell \tilde{c}_\ell^*)/10! \ . \tag{41}$$

In deriving (41), the expression in the right-hand side of (36) is replaced by its value in the limit $\tilde{\alpha} = \alpha$. Using (37), the characteristic equation for the growth rate $\sigma^{(2)}$ in (41) leads to the result that the sign of $\sigma^{(2)}$ depends on the sign of

$$\mu[(\alpha-\alpha_c)^2 - (3\tilde{\alpha})^{-1}(2\alpha_c + \tilde{\alpha})(\tilde{\alpha}-\alpha_c)^2] - \frac{5}{11}\gamma\varepsilon^2 R_2^{(1)} \ . \tag{42}$$

The stability boundary is reached when the expression (42) vanishes with $\tilde{\alpha} \to \alpha$. Using the expression $R-R_c - \mu(\alpha-\alpha_c)^2$ in the place of $\gamma\varepsilon^2 R_2^{(1)}$, we can conclude that squares are unstable for

$$R-R_c - \mu(\alpha-\alpha_c)^2(37 - 22\,\alpha^{-1}\alpha_c)/15 < 0 \ . \tag{43}$$

Case B) Disturbances that satisfy the relation

$$\tilde{\underset{\sim}{K}}_n = \tilde{\underset{\sim}{K}}_m \pm \underset{\sim}{K}_1 \pm \underset{\sim}{K}_1 \quad \text{or} \quad \tilde{\underset{\sim}{K}}_n = \tilde{\underset{\sim}{K}}_m \pm \underset{\sim}{K}_2 \pm \underset{\sim}{K}_2 \ . \tag{44}$$

This case approaches the previous case as $\tilde{\alpha} \to \alpha$. Using (44) in (36) and following the same procedure as in the case A, we find that the sign of $\sigma^{(2)}$ depends on the sign of

$$\mu[(\alpha-\alpha_c)^2 - (3\tilde{\alpha})^{-1}(2\alpha_c + \tilde{\alpha})(\tilde{\alpha}-\alpha_c)^2] \ . \tag{45}$$

Comparing (45) with (42) leads to the conclusion that the disturbances of type B are indeed stronger than those of type A. In the limit of $\tilde{\alpha} \to \alpha$, (45) yields

$$\alpha \le \alpha_c \tag{46}$$

as necessary condition for stability. It can be seen from (45) that for disturbances of type B, the highest value of the growth rate $\sigma_2^{(2)}$ is zero. Hence, the growth rate $\sigma_o^{(2)}$ for the linear problem can itself lead to instability for $\alpha > \alpha_c$ as can be seen from (27) and (46) follows.

From the above discussions, it is clear that the stability criteria (43) and (46) determine the region of the stable steady solution. Stability boundary of the stable squares corresponding to the criteria (43) and (46) is shown graphically in Figure 1.

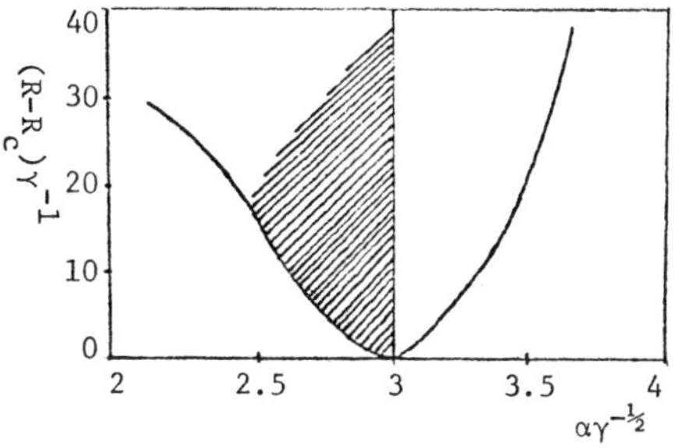

Figure 1. Stability boundary corresponding to the criteria (43) and (46). $R-R_c$ is plotted as a function of α. Squares are stable in the shaded region.

Cellular convection in the form of squares with the wave number α is stable in the shaded region. It is interesting to note that the stability region is to the left of the line $\alpha = \alpha_c$ and that the stability boundary is independent of the Prandtl number. These results are in contrast to the results in the corresponding problem with isothermal boundaries [1] where it was found that the stability region was located to the right of the line $\alpha = \alpha_c$ and the stability boundary did depend on the Prandtl number.

Although there has been much progress on the experimental problems of convection with isothermal boundaries [2], observational data for convection in a layer with nearly insulating boundaries are not yet available to test our results derived here and in [3].

References

[1] Busse, F. H. Stability regions of cellular fluid flow. In instability of continuous systems (ed. H. Leipholz), 41-47 (1971).

[2] Busse, F. H. Nonlinear properties of convection, Rep. Prog. Phys. 41, 1929-1967 (1978).

[3] Busse, F. H. and Riahi, N. Nonlinear convection in a layer with nearly insulating boundaries, J. Fluid Mech. 96, 243-256 (1980).

RUN-UP AND SPIN-UP IN A VISCOELASTIC FLUID CONTAINED BETWEEN PARALLEL PLATES

Jacob Y. Kazakia

Center for the Application of Mathematics
Lehigh University, Bethlehem, Pa. 18015

Abstract

The transfer of momentum from the solid walls of a liquid filled cavity to the fluid contained within is examined in a simplified geometry. The liquid filler is taken to be an incompressible viscoelastic fluid. Special emphasis is given to the case of a Maxwellian fluid. The following problems are considered:

i) The fluid is contained between parallel rigid plates which at some time are subjected to a translational velocity (parallel to the plates) which is then held constant.

ii) The rigid plates of problem i are subjected to an instanteneous increase of angular velocity about an axis perpendicular to the plates.

It is seen that for a Maxwellian fluid the transfer of momentum takes place as a result of reflections back and forth of a velocity discontinuity wave. The secondary flow generated behind the wave front in problem ii is examined in detail and it is found that it has a direction opposite to that generated in a Newtonian fluid.

1. Introduction

The study of run-up and spin-up in liquid filled cavities is essential in understanding the stability of motion of liquid filled bodies. Denn and Porteous [1] and Strauss [2] have studied problems similar to i . Kazakia and Rivlin [3] have studied in detail both problems when the fluid is contained in an infinitely long circular cylinder. This paper describes briefly work carried out in collaboration with Professor R.S. Rivlin. A full account based on a more general constitutive relation, will be published elsewhere.

2. Constitutive equations

We consider a particle, of an incompressible non-Newtonian fluid, which at some fixed time T has material contravariant coordinates X^A in some orthogonal curvilinear coordinate system. We denote by x^i the contravariant coordinates of the same particle at time t and by $\overset{*}{x}{}^i$ its contravariant coordinates at time τ $(-\infty<\tau<t)$. Accordingly,

$$x^i = x^i(X^A,t) \quad \text{and} \quad \overset{*}{x}{}^i = \overset{*}{x}{}^i(X^A,\tau) = g^i(x^j,t,\tau) . \tag{2.1}$$

The functions g^i, which give the flow history of the particle, can be determined by the initial value problem:

$$\frac{\partial g^i}{\partial t} + v^j \frac{\partial g^i}{\partial x^j} = 0 , \quad g^i = x^i \quad \text{when} \quad t = \tau . \tag{2.2}$$

Here $v^j(x^\ell,t)$ are the contravariant components of the particle velocity at time t.

We now consider a material element in the neighborhood of the particle. The covariant components $\overset{*}{C}_{ij}$ of the Cauchy strain in it at time τ with respect to its configuration at time t are given by

$$\overset{*}{C}_{ij} = \overset{*}{\gamma}_{pq} \frac{\partial \overset{*}{x}{}^p}{\partial x^i} \frac{\partial \overset{*}{x}{}^q}{\partial x^j} = \overset{*}{\gamma}_{pq} \frac{\partial g^p}{\partial x^i} \frac{\partial g^q}{\partial x^j} , \tag{2.3}$$

where $\overset{*}{\gamma}_{pq}$ is the covariant metric tensor at $\overset{*}{x}{}^i$.

We denote by σ_{ij} the covariant components of the Cauchy stress at time t and we assume a constitutive relation of the form

$$\sigma_{ij} = - \bar{p}\gamma_{ij} + \int_{-\infty}^{t} f(t-\tau) \frac{d\overset{*}{C}_{ij}}{d\tau} d\tau , \tag{2.4}$$

where $d/d\tau$ denotes derivative with respect to τ at constant t and x^i, $\bar{p}$ is an arbitrary hydrostatic pressure and $f(\ )$ is a constitutive function.

3. Longitudinal run-up between parallel plates

The fluid is contained between two infinite parallel rigid plates situated normally to the x_2-axis of a rectangular cartesian coordinate system x, at $x_2 = \pm h$. Prior to time $t = 0$, both plates are at rest and at time $t = 0$, their velocities are increased impulsively to V, in a direction parallel to x_1-axis, and thereafter held constent. The consequent velocity v of each particle of the fluid is parallel to x_1-axis and independent of x_1 and x_3. Accordingly we obtain from (2.2)

$$g^1 = x_1 + (\tau - t)v(x_2,t) , \quad g^2 = x_2 , \quad g^3 = x_3 . \tag{3.1}$$

From (2.3) and (2.4) we obtain

$$\sigma_{11} = \sigma_{22} = \sigma_{33} = -\bar{p} , \quad \sigma_{12} = \sigma_{21} = \sigma = \int_{-\infty}^{t} f(t-\tau)v'(x_2,\tau)d\tau , \tag{3.2}$$

where the prime denotes differentiation with respect to x_2. The equation of motion in the x_1 direction is

$$\sigma' = \rho\dot{v} , \tag{3.3}$$

where ρ is the density of the fluid and the dot denotes differentiation with respect to t. We wish to obtain the solution of (3.2)-(3.3) which satisfies the initial condition

$$v(x_2,\tau) = 0 \quad \text{for} \quad \tau \leq 0 , \quad -h < x_2 < +h , \tag{3.4}$$

and the boundary conditions

$$v(\pm h,\tau) = VH(\tau) , \tag{3.5}$$

where $H(\tau)$ is the Heaviside function defined by

$$H(\tau) = \begin{cases} 0 & \text{for} \quad \tau < 0 , \\ 1 & \text{for} \quad \tau \geq 0 . \end{cases} \tag{3.6}$$

Using Laplace transforms we obtain

$$v = \frac{V}{2\pi i}\int_{\gamma-i\infty}^{\gamma+i\infty} e^{st} \frac{\cosh\{[\rho s/\bar{f}(s)]^{\frac{1}{2}}x_2\}}{s\cosh\{[\rho s/\bar{f}(s)]^{\frac{1}{2}}h\}} ds , \tag{3.7}$$

where γ is a real constant such that $\bar{v}$ is analytic in the region $\mathrm{Re}\, s \geq \gamma$, and we have used a bar to denote the Laplace transform of a function.

3a. The Maxwellian fluid

In the particular case when the fluid is Maxwellian,

$$f(t-\tau) = \frac{\eta}{\lambda} e^{-(t-\tau)/\lambda} \tag{3.8}$$

where η and λ are positive constants. Using (3.8) in eq. (3.7) we obtain

$$\frac{v}{V} = 1 + \sum_{n=1}^{\infty} (R_n^+ + R_n^-) , \tag{3.9}$$

where

$$R_n^+ + R_n^- = \Phi_n(x_2)e^{-t/2\lambda}(\cosh\frac{\phi_n t}{2\lambda} + \frac{1}{\phi_n}\sinh\frac{\phi_n t}{2\lambda})$$

$$\text{for} \quad n = 1,2,\ldots m \,,$$

$$\Phi_n(x_2) = \frac{4(-1)^n}{\pi(2n-1)}\cos\{\frac{(2n-1)\pi x_2}{2h}\} \quad , \tag{3.10}$$

$$\phi_n = |1-\frac{\pi^2\lambda}{\nu^2}(2n-1)^2|^{\frac{1}{2}} \quad , \quad \nu^2 = \frac{h^2\rho}{\eta}$$

the integer m is defined by

$$m \leq \frac{1}{2}(\frac{\nu}{\pi\lambda^{\frac{1}{2}}} + 1) < m + 1 \quad , \tag{3.11}$$

and the hyperbolic functions in (3.10)a must be replaced by cos, sin for $n > m$. In the asymptotic case when $\lambda = 0$, i.e. the fluid is Newtonian we obtain

$$\frac{v}{V} = 1 + \sum_{n=1}^{\infty} \Phi_n(x_2)\exp\{-\frac{(2n-1)^2\pi^2 t}{4\nu^2}\} \, . \tag{3.12}$$

Figure 1 depicts v/V against $y = x_2/h$ for various values of the non-dimensional time t/ν^2, in the cases $\lambda/\nu^2 = 0$, .05, .2, 1.

It is seen that when the fluid is Newtonian, momentum is transmitted from the boundaries to the interior of the fluid by a purely diffusive mechanism. On the other hand, when the fluid is Maxwellian the transfer of momentum involves a wave mechanism. It can be shown [3] that (3.8) represents a superposition of right and left waves whose fronts leave the boundary $y = 1(-1)$ at times $t = 2n\nu\lambda^{\frac{1}{2}}$, and travel in the negative (positive) direction of y with speed c where

$$c^2 = \frac{h^2}{\lambda\nu^2} = \frac{\eta}{\lambda\rho} \, . \tag{3.13}$$

The time t_c which the discontinuities reach the mid-plane for the first time is $t_c = h/c$. For times prior to t_c the velocity immediately behind the fronts is given by

$$v_F = Ve^{-t/2\lambda} \, . \tag{3.14}$$

4. Flow between rotating disks

We consider the fluid to be contained again between two rigid parallel disks of infinite radius a distance $2h$ apart. At time $t = 0$ they are given an angular velocity $\varepsilon\Omega$ about a common axis which is thereafter held constant. We choose a cylindrical polar coordinate system r, θ, z with its z-axix coinciding with the axis of rotation and its origin mid-way between the discs. The continuity and momentum equations for an axisymmetric flow are given by

$$\frac{\partial}{\partial r}(rv_r) + \frac{\partial}{\partial z}(rv_z) = 0 ,$$

$$\frac{\partial \sigma_{rr}}{\partial r} + \frac{1}{r}(\sigma_{rr} - \sigma_{\theta\theta}) + \frac{\partial \sigma_{rz}}{\partial z} = \rho a_r ,$$

$$\frac{1}{r^2}\frac{\partial}{\partial r}(r^2 \sigma_{r\theta}) + \frac{\partial \sigma_{\theta z}}{\partial z} = \rho a_\theta , \tag{4.1}$$

$$\frac{1}{r}\frac{\partial}{\partial r}(r\sigma_{rz}) + \frac{\partial \sigma_{zz}}{\partial z} = \rho a_z ,$$

where $v_\alpha(r,z,t)(\alpha=r,\theta,z)$ are the physical components of the velocity, $\sigma_{\alpha\beta}(r,z,t)(\alpha,\beta=r,\theta,z)$ are the physical components of the stress and a_r , a_θ , a_z are the physical components of the acceleration given by

$$a_r = \frac{\partial v_r}{\partial t} + v_r \frac{\partial v_r}{\partial r} - \frac{v_\theta^2}{r} + v_z \frac{\partial v_r}{\partial z} ,$$

$$a_\theta = \frac{\partial v_\theta}{\partial t} + v_r \frac{\partial v_\theta}{\partial r} + \frac{v_r v_\theta}{r} + v_z \frac{\partial v_\theta}{\partial z} , \tag{4.2}$$

$$a_z = \frac{\partial v_z}{\partial t} + v_r \frac{\partial v_z}{\partial r} + v_z \frac{\partial v_z}{\partial t} .$$

Equations (4.1) and (4.2) are to be solved subject to the conditions

$$v_r = v_z = 0 , \quad v_\theta = \varepsilon r \Omega H(t) \quad \text{at} \quad z = \pm h , \tag{4.3}$$

$$v_r = v_z = v_\theta = 0 \quad \text{at} \quad t = 0 , \text{ for all } r \text{ and } -h < z < h .$$

We can introduce the stream function $\psi(r,z,t)$ by

$$rv_r = -\frac{\partial \psi}{\partial z} , \quad rv_z = \frac{\partial \psi}{\partial r} , \tag{4.4}$$

and expanding the variables v_θ , ψ and $\bar{p}$ in the form

$$v_\theta = \varepsilon r\omega(z,t) + O(\varepsilon^3) ,$$
$$\psi = \varepsilon^2 r^2\phi(z,t) + O(\varepsilon^4) , \qquad (4.5)$$
$$\bar{p} = \varepsilon p_o + \varepsilon^2 p + O(\varepsilon^3) ,$$

we obtain

$$v_r = -\varepsilon^2 r\phi' , \quad v_z = \varepsilon^2 2\phi . \qquad (4.6)$$

Substituting from (4.5)a and (4.6) into (2.2) and solving the resulting equations we obtain, neglecting terms of degree higher than the second in ε ,

$$g^1 = r + \varepsilon^2 r\int_\tau^t \phi'(z,s)ds, \; g^2 = \theta - \varepsilon\int_\tau^t \omega(z,s)ds ,$$
$$g^3 = z - 2\varepsilon^2\int_\tau^t \phi(z,s)ds . \qquad (4.7)$$

Using (4.7) in (2.3) and the resulting expressions in (2.4) we obtain the stresses (to order ε^2):

$$\sigma_{rr} = \sigma_{\theta\theta} = -\varepsilon p_o - \varepsilon^2[p+2F\{\phi'\}] ,$$
$$\sigma_{zz} = -\varepsilon p_o + \varepsilon^2[-p+4F\{\phi'\}-2r^2F\{\omega'(z,\tau)\int_\tau^t \omega'(z,s)ds\}] \qquad (4.8)$$
$$\sigma_{r\theta} = 0 , \quad \sigma_{rz} = -\varepsilon^2 rF\{\phi''\} , \quad \sigma_{\theta z} = \varepsilon rF\{\omega'\} .$$

Here the operator F is given by

$$F\{\;\} = \int_{-\infty}^t f(t-\tau)\{\;\}d\tau . \qquad (4.9)$$

4.1 The first order problem

Using eqs. (4.5), (4.6), (4.8) in (4.2) and (4.1) we obtain, to order ε , the initial value problem:

$$\rho\,\frac{\partial\omega}{\partial t} - F\{\omega''\} = 0 ,$$
$$\omega(\pm h,t) = \Omega H(t) , \quad \omega(z,0) = 0(-h<z<h) . \qquad (4.10)$$

This problem is identical to the one solved in section 3. In this section we will be interested in the secondary flow for times less than t_c. During these times two angular velocity discontinuity fronts defined by

$$t = a_o(z) = (h-z)/c = t_c(1-z/h)$$
$$t = b_o(z) = (h+z)/c = t_c(1+z/h) \qquad (4.11)$$

move from the upper and lower plate towards the mid plane. The fluid in the core (i.e. between the two fronts) remains undisturbed up to time t_c. Considering the stress field associated with the flow (4.10) we notice that p_o can be taken equal to zero and hence the only stress present in the approximation of order ε^1 is $\sigma_{\theta z}$. At a fixed plane z, the shearing stress $\sigma_{\theta z}$ is zero prior to the arrival of the front. At time $t=a_o(z)$, the angular velocity $\omega(z,t)$ and consequently $\sigma_{\theta z}$ change discontinuously. Using the jump conditions at the front (see for example [4])

$$[v_z] = 0 \ , \quad [\sigma_{rz}] = \rho[v_r(v_z+c)] \ ,$$
$$[\sigma_{\theta z}] = \rho[v_\theta(v_z+c)] \ , \quad [\sigma_{zz}] = \rho[v_z(v_z+c)] \qquad (4.12)$$

we obtain for a Maxwellian fluid an expression for the stress $\sigma_{\theta z}$ behind the front, (i.e. for times t such that $a_o(z)<t<t_c$)

$$\sigma_{\theta z} = \varepsilon r[\rho c\Omega e^{-(2t-a_o)/2\lambda} + F_1\{\omega'\}] \ . \qquad (4.13)$$

Here the operator F_1 is defined by

$$F_1\{\ \} = \int_{a_o^+}^{t} \frac{\eta}{\lambda}\, e^{-(t-\tau)/\lambda}\{\ \}d\tau \ . \qquad (4.14)$$

4.2 <u>The second order problem</u>

It can be shown that in order to satisfy eqs. (4.1) to order ε^2 we must have

$$\frac{1}{r}\frac{\partial p}{\partial r} - \rho\omega^2 = \frac{\partial K}{\partial z} \ , \quad \frac{\partial}{\partial z}[-p+r^2S] = 2K \ , \qquad (4.15)$$

where

$$K = \rho\,\frac{\partial \phi}{\partial t} - F_1\{\phi''\} \ , \quad S = -\ 2F_1\{\omega'(z,\tau)\int_\tau^t \omega'(z,s)ds\} \qquad (4.16)$$

Eqs. (4.15)-(4.16) can be satisfied by taking

$$p = \frac{1}{2} r^2 P(z,t) + Q(z,t) , \tag{4.17}$$

and determining P , Q and ϕ according to the eqs.

$$P = - 4F_1\{\omega'(z,\tau)\int_\tau^t \omega'(z,s)ds\} , \quad \frac{dQ}{dz} = - 2K , \tag{4.18}$$

$$\rho \frac{\partial\phi}{\partial t} - F_1\{\phi''\} = G_o(t) + \int_h^z G(s,t)ds , \tag{4.19}$$

where $G_o(t)$ is an arbitrary function to be determined from the boundary conditions and

$$G(z,t) = - \rho\omega^2 + P(z,t) . \tag{4.20}$$

As we have seen, in section 4.1, at a constant level z , ω suffers a discontinuity at time $t = a_o(z)$ due to the passage of the front. Consequently the operator $F_1\{ \}$ must be correctly interpreted. This is done by the use of the jump conditions (4.12) and eq. (4.13). As a result we obtain, for a Maxwellian fluid, the following initial-boundary value problem for the determination of the function $\phi(z,t)$ in the region $a_o(z)<t<t_c$, $0<z<h$:

$$\begin{aligned} &\rho[\frac{\partial\phi}{\partial t} + e^{-(t-a_o)/\lambda} \frac{\partial\phi}{\partial t}\Big|_{t=a_o^+}]-F_1\{\phi''\} = \\ &= G_o(t) + \int_h^z G(s,t)ds , \end{aligned} \tag{4.21}$$

$$\phi = \frac{\partial\phi}{\partial z} = 0 \quad \text{for} \quad z = h \quad \text{and all} \quad t ,$$

$$\phi(z,a_o) = 0 \quad \text{for all} \quad z .$$

Here $F_1\{ \}$ is defined by (4.14) and $G(z,t)$ is given by

$$\begin{aligned} G(z,t) = &- \rho\omega^2 - 2\rho\Omega^2 e^{-t/\lambda}[1+F_o(t,z)] \\ &- 4F_1\{\omega'(z,\tau)\int_\tau^t \omega'(z,s)ds\} , \end{aligned} \tag{4.22}$$

where

$$F_o(t,z) = \frac{2c}{\Omega} e^{a_o/2\lambda} \int_{a_o^+}^{t} \omega'(z,s)ds \ . \qquad (4.23)$$

The problem defined by (4.21) was solved by an implicit finite deference marching technique. Figure 2 depicts the resulting streamlines at two different times $t/t_c = .2 , .5$ for the value of the parameter $t_c = .1$.

Acknowledgement

The results in this paper are based on a joint research with Dr. R. S. Rivlin, which is sponsored by the U.S. Army Research Office under Contract No. DAAG 29-79-C-0058 with Lehigh University.

References

[1] Denn, M. M. and Portenous, K. C. Chem. Eng. J. 2, 280 (1971)

[2] Strauss, K., Rheologica Acta 16, 385 (1977)

[3] Kazakia, J. Y. and Rivlin, R. S., Rheologica Acta 20, 111 (1981)

[4] Eringen, A. C., Mechanics of Continua, John Wiley (1967).

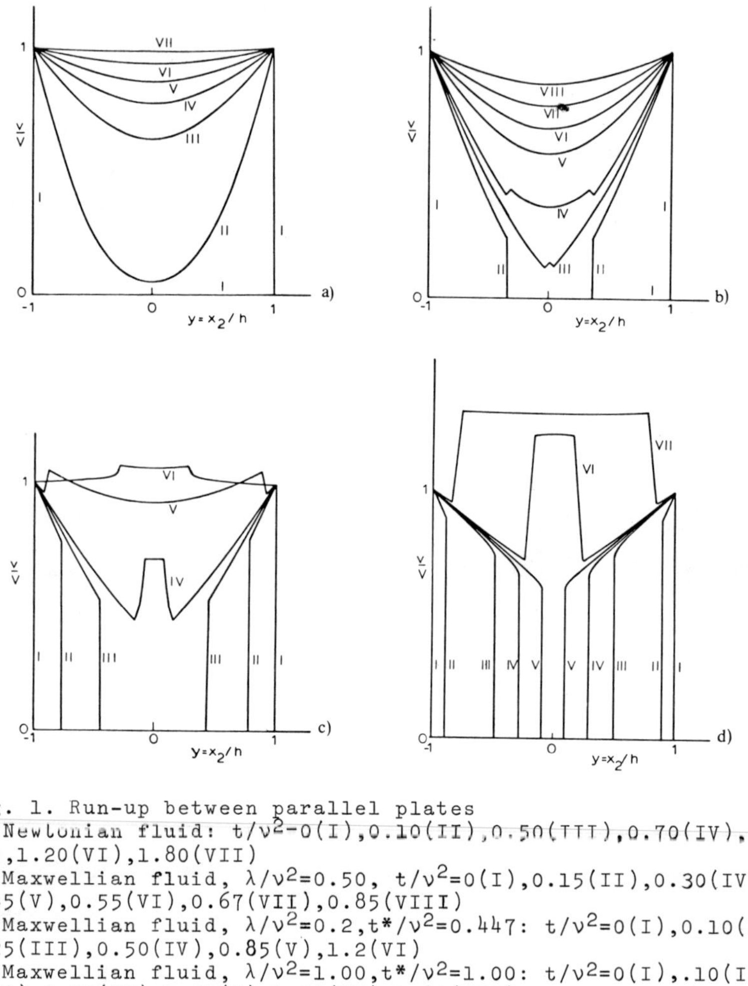

Fig. 1. Run-up between parallel plates
(a)Newtonian fluid: t/ν^2=0(I),0.10(II),0.50(III),0.70(IV),0.90 (V),1.20(VI),1.80(VII)
(b)Maxwellian fluid, λ/ν^2=0.50, t/ν^2=0(I),0.15(II),0.30(IV), 0.45(V),0.55(VI),0.67(VII),0.85(VIII)
(c)Maxwellian fluid, λ/ν^2=0.2,t^*/ν^2=0.447: t/ν^2=0(I),0.10(II), 0.25(III),0.50(IV),0.85(V),1.2(VI)
(d)Maxwellian fluid, λ/ν^2=1.00,t^*/ν^2=1.00: t/ν^2=0(I),.10(II),0.50 (III),0.70(IV),0.90(V),1.20(VI),1.80(VII)

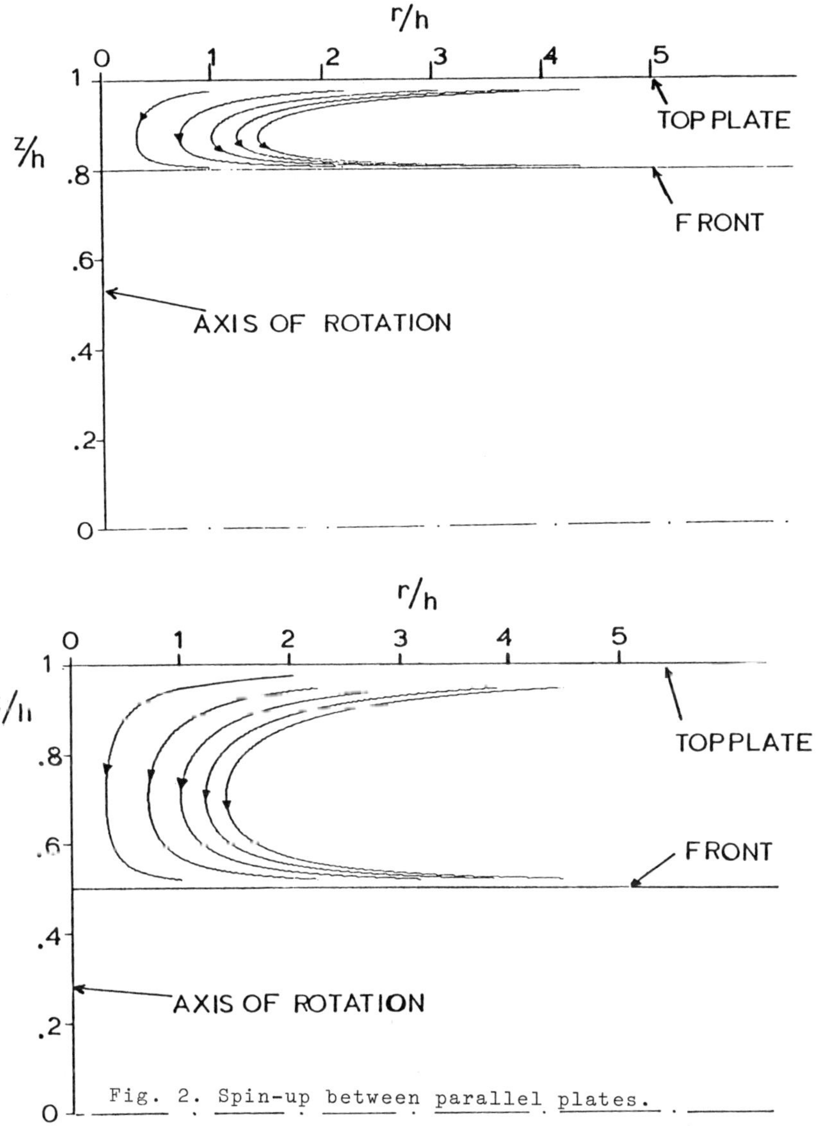

Fig. 2. Spin-up between parallel plates.

A GENERAL DERIVATION OF THE JUMP CONDITION
FOR THE VELOCITY POTENTIAL AND THE STREAM FUNCTION
ACROSS SHOCK WAVES

M. S. Cramer
Department of Engineering Science & Mechanics
Virginia Polytechnic Institute & State University
Blacksburg, Virginia 24061

Abstract

A general derivation of the jump conditions for the velocity potential and stream function across shock waves has been presented. It is shown that the jump in the velocity potential is an arbitrary function of time and that for the stream function is an arbitrary constant.

1. Introduction

In many areas of gasdynamics, aerodynamics and nonlinear acoustics a velocity potential is defined as follows

$$\phi(\underset{\sim}{x},\ t) = \int_C \underset{\sim}{v} \cdot d\underset{\sim}{x} + f(t),$$

where x is the velocity field and C is any curve connecting an arbitrary point $\underset{\sim}{x}_0$ to the point of interest $\underset{\sim}{x}$. When the flow is irrotational and the region of interest is simply connected, Stokes theorem may be used to show that $\phi(\underset{\sim}{x},t)$ is unique up to the additive function of time f(t); that is, ϕ is independent of the curve C. Different choices of f(t) have no effect on the physical quantities of interest, e.g., $\underset{\sim}{v} = \nabla\phi$, and we may therefore choose this to be any arbitrary function, e.g., $f(t) \equiv 0$. The velocity potential is frequently used even when shock waves are embedded in the flow. Because of the vorticity generated by the shock waves [1], the irrotationality is only valid in the limit of weak shocks; typically, both the linear theory and the first correction to it can be assumed irrotational.

In both analytical and numerical studies of such flows, it is necessary to know the jump condition for ϕ across shock waves; otherwise, the computations cannot be continued from one side of the discontinuity to the other. In many studies, the potential is taken to be continuous across the shock. However, I have been unable to find general proof of this and, in view of the importance of this result, the main objective of the present paper is to provide a derivation of this jump contition which is both general and simple. One of the advantages of the derivation presented here is that the condition is essentially local in nature; that is, it is not necessary to perform a global integration nor is it necessary that the shock strength vanish at some point. I have also found that the jump condition for the stream function across both wakes and shock waves can be obtained in a completely analogous manner; this proof is also outlined below.

2. Jump Condition for the Velocity Potential

Consider a surface σ over which the velocity $\underset{\sim}{v}$ suffers a finite jump; this could be either a vortex sheet or a shock. Quantities on one side of the surface will be denoted with a superscript + and quantities on the other side with a superscript −. The jump in a property across the discontinuity is denoted by double brackets, e.g., $[\![\phi]\!](\underset{\sim}{x},t) \equiv \phi^+ - \phi^-$ for points $\underset{\sim}{x}$ on σ. At any point, say $\underset{\sim}{x}^+$, on the + side of the surface σ, the potential can be written

$$\phi^+(\underset{\sim}{x}^+,t) = \int_{C^+} \underset{\sim}{v}^+ \cdot d\underset{\sim}{x} + f^+(t), \tag{1}$$

where C^+ is any curve which connects the two points $\underset{\sim}{x}_0^+$ and $\underset{\sim}{x}^+$. An analogous result can be written for any point $\underset{\sim}{x}^-$ on the other side of σ:

$$\phi^-(\underset{\sim}{x}^-,t) = \int_{C^-} \underset{\sim}{v}^- \cdot d\underset{\sim}{x} + f^-(t), \tag{2}$$

where C^- is any curve connecting $\underset{\sim}{x}_0^-$ and $\underset{\sim}{x}^-$. Now take the special case where C^- and C^+ both lie on the surface of discontinuity; that is, take

$$\underset{\sim}{x}_0^- = \underset{\sim}{x}_0^+ \equiv \underset{\sim}{x}_0^s, \quad \underset{\sim}{x}^- = \underset{\sim}{x}^+ \equiv \underset{\sim}{x}^s ,$$

and, in addition,

$$C^- = C^+ \equiv C^s ,$$

where the superscript s refers to points on the surface σ. When this is done and equations (1) and (2) are subtracted the following jump condition for ϕ is obtained

$$[\![\phi]\!]\,(\underset{\sim}{x}^s,t) = \int_{C^s} [\![\underset{\sim}{v}]\!] \cdot d\underset{\sim}{x} + F(t),$$

where

$$F(t) \equiv f^+(t) - f^-(t).$$

Once the physical quantity $[\![\underset{\sim}{v}]\!]$ is specified, $[\![\phi]\!]$ can be calculated explicitly; $[\![\underset{\sim}{v}]\!]$ is usually obtained by some jump condition. Here it is assumed that momentum is conserved across the surface of discontinuity; the jump in pressure p and velocity are therefore related as follows

$$\rho U [\![\underset{\sim}{v}]\!] + [\![p]\!] \underset{\sim}{n} = 0, \tag{3}$$

where ρ = fluid density, $U \equiv \underset{\sim}{v} \cdot \underset{\sim}{n} - u$, u = local speed of propagation of the surface, and $\underset{\sim}{n}$ is the unit normal to the surface at the point in question. Clearly, the fluid has also been assumed inviscid. Furthermore, the jump condition (3) can be applied to both wakes, i.e, vortex sheets, and shocks; see, for example, Coulson and Jeffrey [2]. The jump condition for shock waves alone can be found in any standard text on gasdynamics; see, e.g., Courant and Friedrichs [3].

Across a shock surface, there is a net mass flux and $u \equiv \underset{\sim}{v} \cdot \underset{\sim}{n} - u \neq 0$. Thus, (3) may be used to show

$$[\![\phi]\!] (\underset{\sim}{x}^s, t) = - \int_{C^s} \frac{[\![p]\!]}{\rho U} \underset{\sim}{n} \cdot d\underset{\sim}{x} + F(t).$$

Because $\underset{\sim}{n}$ is the unit normal to the surface and C^s lies in the surface, $\underset{\sim}{n} \cdot d\underset{\sim}{x} \equiv 0$ at every point; hence across a shock surface

$$[\![\phi]\!] (\underset{\sim}{x}^s, t) = F(t).$$

That is, the jump in ϕ is, at most, an arbitrary function of time. We are free to choose the functions f^+ and f^- at will; one choice which many authors have found to be convenient is to take $f^+ = f^-$, i.e., $[\![\phi]\!] = 0$. In general, we can expect a similar result whenever the components of velocity parallel to the shock surface remain continuous.

By way of comparison, consider a vortex sheet. There is no mass flux through a vortex sheet. Thus, $u \equiv \underset{\sim}{v} \cdot \underset{\sim}{n}$ at every point and (3) therefore only requires that $[\![p]\!] = 0$. The physical jump conditions therefore provide no constraint on $[\![\phi]\!]$ and the jump in ϕ across a wake must be determined by the solution to the complete flow problem.

3. Jump Condition for the Stream Function

In two-dimensional steady flow, the equation of mass conservation can be written

$$\frac{\partial}{\partial x} (\rho u) + \frac{\partial}{\partial y} (\rho v) = 0, \tag{4}$$

where the x and y components of the velocity vector are given by u and v. In flows such as these, a stream function $\psi = \psi(x,y)$ is usually defined

$$\psi(x,y) = \int_C \rho(u\,dy - v\,dx) + k \tag{5}$$

where C is any curve connecting point (x_0,y_0) to (x,y). The fact that ψ is unique up to the arbitrary additive constant k is easily seen by noting the ψ is just the potential of the vector function $\underset{\sim}{A}$. The x and y components of $\underset{\sim}{A}$ are given by

$$A_x \equiv -\rho v, \quad A_y \equiv \rho u.$$

Equations (4) and (5) then read

$$\nabla \times \underset{\sim}{A} = 0$$

$$\psi(x,y) = \int_C \underset{\sim}{A} \cdot d\underset{\sim}{x} + k$$

and the uniqueness of ψ then follows from Stokes theorem.

As in the case of the velocity potential, the jump in ψ over an arbitrary surface of discontinuity can be written

$$[\![\psi]\!](x^s,y^s) = \int_{C^s} [\![\underset{\sim}{A}]\!] \cdot d\underset{\sim}{x} + K,$$

where $K \equiv k^+ - k^-$ and the notation has the same meaning as in the previous section. If the equation of the surface of discontinuity is given by $y^s = f(x^s)$, then, the above integral can be written

$$\int_{C^s} [\![\rho u]\!]\,dy - [\![\rho v]\!]\,dx = -\int_{x_0^s}^{x^s} (1 + f'^2)^{1/2} [\![\rho v_n]\!]\,dx,$$

where $v_n \equiv \underset{\sim}{v}\cdot\underset{\sim}{n}$ is just the component of velocity normal to the surface to discontinuity, and $f' \equiv \frac{df}{dx}$.

Across any stationary surface of discontinuity, the jump condition corresponding to conservation of mass is

$$[\![\rho v_n]\!] = 0.$$

This, of course, can be found in any of the references mentioned above. Thus, the jump in the stream function across both shock waves and vortex sheets is, at most, an arbitrary constant.

It is always possible to define a unique stream function in flows which are steady and depend on no more than two independent variables; it is a straightforward exercise to extend the present proof to this general case.

4. Conclusion

The well-known shock jump conditions for the pressure, density and velocity have been used to show that the jump in ϕ and ψ across a shock is, at most, an arbitrary function of time and an arbitrary constant, respectively; the condition that ϕ and ψ are continuous across shock waves is clearly a particular choice of F(t) and K. Furthermore, the conditions derived are local, i.e., $\underset{\sim}{x}_0^s$ is arbitrary, and may therefore be applied in the same manner as a physical jump condition.

References

[1] Hayes, W. D., The Vorticity Jump Across a Gasdynamic Discontinuity, *J. Fl. Mech.*, Vol. 2, pp. 595-600, 1957.

[2] Coulson, C. A. and Jeffrey, A., *Waves; A Mathematical Approach to the Common Types of Wave Motion*, Longman Group Ltd., 1977.

[3] Courant, R. and Friedrichs, K. O., *Supersonic Flow and Shock Waves*, Interscience, 1948.

DISCHARGE MODULATED WATER JETS: THEORY AND EXPERIMENTS

Sedat Sami and David L. Eddingfield
Southern Illinois University, Carbondale, Illinois

Abstract

The results of an analytical and experimental investigation of the flow characteristics of a discharge modulated high velocity large diameter water jet are presented and discussed. A periodic velocity variation was impressed upon a continuous jet flow. The mathematical formulation of the problem involves the basic equations of motion and of continuity. The jet contours predicted by the linear model are compared with those obtained by the numerical solution of the nonlinear equations. Breakup lengths predicted by each method are compared against experimental data. An electrodynamic shaker programmed through a function generator was used to modulate the water jet. The amplitude and the frequency of the modulation were varied. High speed camera films and stroboscopic slides were used to estimate the breakup length. It appears that the jet contours are severely affected by the aerodynamic drag which bends and shears at their edges the discs formed by the radially moving fluid.

Introduction

It has been observed that, when a conventional water jet impacts upon a target material, the highest rate of failure would occur during the initial period of impact. This observation has led several investigators to study the initial impact characteristics of pulsating jets. The periodic interruption of a jet's discharge results in the firing of bursts of steady, unmodulated -constant diameter- discontinuous jet segments. Such a technique, however, offers some serious drawbacks. A more efficient method is to modulate the continuous jet and rely on the well known "klystron" effect to eventually cause the bunching and breakup of the jet. In such a scheme, the amplitude of modulation is relatively small -a few percent of the average flow rate- but since the faster part of each cycle eventually overtakes the slower portion of the flow, the jet stream becomes a train of separated fluid bunches.

The development of predictive models for the breakup of unmodulated water jets has been initiated by Lord Rayleigh [1,2], who in his linear stability analysis of an inviscid liquid jet has shown that an axisymmetric disturbance is stable or unstable depending on whether its wavelength is less or greater than the circumference of the jet. Capillary jet instability was further investigated by Tyler and Richardson [3], Schweitzer [4], and by

Merrington and Richardson [5]. More recently, Rayleigh's work has been extended to include the effects of viscosity [6] and nonlinearity [7, 8, 9, 10, 11]. However, work on modulated water jets has not received a similar degree of interest. The first account of a jet breakup due to an inertial type mechanism was reported by Crane et al. [12] and by McCormack et al. [13, 14]. They have reported some experimental evidence for flow modulation due to mechanical vibration of the injector. Their work has shown that even with high velocity turbulent jets, subjection of the injectors to high g frequency vibration will cause synchronized "bunching" in the liquid jet from each injector and thus produce a modulated flow of fuel into the combustion zone. The results of these studies have clearly shown that under high frequency vibrations inertial effects dominate the surface tension induced displacements. Fenwick and Bugler [15], while studying the combustion instability problem, have looked into the relationship between the combustion product release and the injection of the propellant at a location downstream of an injector. Recently Nebeker and Rodriguez [16] have developed a suitable flow modulator which operates by causing a cyclic variation of the discharge rate by the period variation of the flow resistance upstream of the nozzle.

In the present work, the fluid dynamics of a jet flow subject to a forced modulation rather than a capillary instability is formulated, a simple predictive model constructed and the analytical results, for both the linear and nonlinear models, compared with experimental data.

Dimensional Analysis and Mathematical Formulation

Consider a water jet (Fig. 1) issuing at time t from a nozzle of diameter D (and radius R) with a velocity given by

$$u(0,t) = U + \Delta U \cdot F(ft)$$

where U is a constant average jet exit velocity, ΔU and f are the respective amplitude and frequency of the waveform (with a wavelength λ) described by the dimensionless function F(ft). The jet contours will be described by the radius r_o, the value of which at any downstream station x will depend on the following physical quantities:

$$F(r_o, x, t, D, U, \Delta U, \lambda, \rho, \mu, \sigma) = 0$$

where x is the downstream distance from the nozzle; t, the time; D is the average jet diameter (nozzle diameter); U, the average jet velocity; ΔU and λ, the amplitude and wavelength of the modulation; ρ, μ, σ, the fluid density, viscosity, and surface tension respectively. The number of physical quantities involved is ten and they depend on three fundamental units. It is, therefore, to be expected that there will be seven dimensionless terms. These dimensionless groupings can be chosen by the application of the Buckingham-Pi Theorem:

$$\zeta = \frac{r_o}{R}, \quad \beta = \frac{x}{D}, \quad \tau = \frac{tU}{\lambda}, \quad M = \frac{\Delta U}{U}, \quad S = \frac{D}{\lambda}, \quad We = \frac{\rho U^2 D}{\sigma}, \quad Re = \frac{\rho U D}{\mu}, \qquad (1)$$

where R is the nozzle radius, and S, We, and Re represent the Stouhal, Weber and Reynolds numbers respectively. The final expression can thus be written in the form

$$\zeta = \phi(\beta, \tau, M, S, We, Re).$$

By neglecting the surface tension and viscous effects, the above relationship can be further reduced into the form

$$\zeta = \phi(\beta, \tau, M, S). \tag{2}$$

Considering an axisymmetrical jet flow and assuming uniform uniform distribution of axial velocity (i.e., flat profiles) the continuity equation in cylindrical coordinate system can be expressed as [17]

$$\frac{\partial (r_0^2)}{\partial t} + \frac{\partial (r_0^2 u)}{\partial x} = 0 \tag{3}$$

Under similar assumptions and neglecting the radial kinetic energy of the jet and the viscosity of the fluid, the governing equations of motion can be expressed as

$$\frac{\partial u}{\partial t} + u\frac{\partial u}{\partial x} = 0 \tag{4}$$

and

$$\frac{\partial v}{\partial t} \quad u\frac{\partial v}{\partial x} \quad v\frac{\partial v}{\partial r} = 0 \tag{5}$$

It is apparent that v, the radial component of the velocity, does not appear in Eqs. 3 and 4. So, in order to solve for $r_o(x,t)$ it is necessary to consider Eqs. 3 and 4 only. They can be rearranged and rewritten as

$$\frac{\partial r_o}{\partial t} + u\,\frac{\partial r_o}{\partial x} + \frac{r_o}{2}\,\frac{\partial u}{\partial x} = 0 \tag{6}$$

and

$$\frac{\partial u}{\partial t} + u\,\frac{\partial u}{\partial x} = 0 \tag{7}$$

An exact solution to this set of simultaneous, partial differential equations is not available. The two viable alternatives are either to reduce the equations to a set of first order linear partial differential equations and solve them by the method of Lagrange or to solve the set of hyperbolic nonlinear equations by a numerical technique.

Linear Analysis

The axial velocity u, when expressed in terms of a steady mean velocity U and a variation u' about the mean, can be given as

$$u(x,t) = U + u'(x,t).$$

If the above expression is substituted into Eqs. 6 and 7 and one considers the fact that $\partial U/\partial x = 0$ and $\partial U/\partial t = 0$, then one has

$$\frac{\partial r_o}{\partial t} + (U+u')\frac{\partial r_o}{\partial x} + \frac{r_o}{2}\cdot\frac{\partial u'}{\partial x} = 0$$

and

$$\frac{\partial u'}{\partial t} + (U+u')\frac{\partial u'}{\partial x} = 0$$

Assuming small amplitude modulations ($u' \ll U$), the above equations can be reduced to yield

$$\frac{\partial r_o}{\partial t} + U\frac{\partial r_o}{\partial x} + \frac{r_o}{2}\frac{\partial u'}{\partial x} = 0 \tag{8}$$

and

$$\frac{\partial u'}{\partial t} + U\frac{\partial u'}{\partial x} = 0 \tag{9}$$

By introducing the dimensionless parameters

$$\zeta = \frac{r_o}{R} \quad \upsilon = \frac{u'}{U} \quad \alpha = \frac{x}{\lambda} \quad \tau = \frac{tU}{\lambda} \tag{10}$$

one can render the above equations dimensionless. Furthermore, if a transformation given by $z = x - Ut$ or as $\eta = \alpha - \tau$ where $\eta = z/\lambda, \alpha = x/\lambda$, and $\tau = Ut/\lambda$, is applied, the governing equations become

$$\frac{\partial \zeta}{\partial \tau} = -\frac{\zeta}{2}\frac{\partial \upsilon}{\partial \eta} \tag{11}$$

$$\frac{\partial \upsilon}{\partial \tau} = 0 \tag{12}$$

The solution to the above system of simultaneous equations can be obtained by the method of Lagrange. The integration of the simultaneous system of subsidiary equations yields

$$\ln \zeta = -[F'(\eta) \cdot \tau]/2$$

where $F(\eta) = M \sin 2\pi\eta$ describes the periodic modulation . Hence, one has

$$\ln \zeta = -\pi M \tau \cos 2\pi\eta \tag{13}$$

or

$$\zeta = \exp[-\pi M \tau \cos 2\pi\eta] \tag{14}$$

Since the solution given by Eq. 14 is exponential, prediction of the breakup time or length can only be approximated by arbitrarily assuming that discontinuity will occur when the jet contour is reduced to a certain fraction of the initial jet diameter. Thus one has $\ln \zeta_b = -\pi M \tau_b$ where $\zeta_b = (r_o/R)_b$ represents such a limiting ratio. The dimensionless breakup length is then

$$\alpha_b = (x_b/\lambda) = -\ln \zeta_b/\pi M \tag{15}$$

Introducing the relationships $\eta = \alpha - \tau$ and $\alpha = S\beta$ where $S = fD/U = D/\lambda$ and $\beta = x/D$, Eq.14 can be rearranged to read

$$\zeta = \exp[-\pi M \tau \cos 2\pi(S\beta - \tau)] \tag{16}$$

This equation is in complete agreement with the relationship predicted by Eq.2 which was obtained from purely dimensional considerations. Figure 2 shows a plot of Eq. 16 for a ten percent modulation.

Nonlinear Analysis

The problem in its most general form requires the consideration of a set of two nonlinear simultaneous partial differential equations given by Eqs. 3 and 4:

$$\frac{\partial r_o}{\partial t} + \frac{\partial (r_o^2 u)}{\partial x} = 0 \tag{3}$$

and
$$\frac{\partial u}{\partial t} + u\frac{\partial u}{\partial x} = 0 \tag{4}$$

These equations can be modified by introducing a frame of reference translation defined by

$z = x - U\,t$

and a change in the expression for the axial velocity given by

$u(x,t) = U + u'(x,t)$

where u' represents the variation about the constant average jet velocity U. With the above transformations, Eqs. 3 and 4 become

$$\frac{\partial (r_o^2)}{\partial t} + \frac{\partial (r_o^2\, u)}{\partial z} = 0 \tag{3'}$$

and
$$\frac{\partial u'}{\partial t} + u'\frac{\partial u'}{\partial z} = 0 \tag{4'}$$

Eqs. 3' and 4' can, in turn, be normalized by introducing the dimensionless parameters

$\zeta = r_o/R$, $\upsilon = u'/U$, $\eta = z/\lambda$, and $\tau = t\,U/\lambda$.

Thus one has
$$\frac{\partial \zeta^2}{\partial \tau} + \frac{\partial (\zeta^2\upsilon)}{\partial \eta} = 0 \tag{17}$$

and
$$\frac{\partial \upsilon}{\partial \tau} + \frac{\partial (\upsilon^2/2)}{\partial \eta} = 0 \tag{18}$$

Eqs. 17 and 18 belong to a class of hyperbolic partial differential equations which can be transformed by introducing the following parameters:

$$\zeta^2 = s, \quad \zeta^2\upsilon = m \tag{19}$$

Direct substitution into Eq. 17 reduced it to

$$\frac{\partial s}{\partial \tau} + \frac{\partial m}{\partial \eta} = 0 \tag{20}$$

Eq. 18 can be manipulated to yield

$$\frac{\partial}{\partial \tau}\left(\frac{m}{s}\right) + \left(\frac{m}{s}\right)\frac{\partial}{\partial \eta}\left(\frac{m}{s}\right) = 0$$

After expanding terms and upon multiplication by s one gets

$$\frac{\partial m}{\partial \tau} - \frac{m\partial s}{s\partial \tau} + \frac{m}{s}\frac{\partial m}{\partial \eta} - \frac{m^2}{s^2}\frac{\partial s}{\partial \eta} = 0$$

Adding and subtracting $\frac{m\partial m}{s\partial \eta}$ and rearranging terms will yield

$$\frac{\partial m}{\partial \tau} - \frac{m}{s}\left[\frac{\partial s}{\partial \tau} + \frac{\partial m}{\partial \eta}\right] + \left[2\frac{m}{s}\frac{\partial m}{\partial \eta} - \frac{m^2}{s^2}\frac{\partial s}{\partial \eta}\right] = 0$$

The expression within the first bracket is zero (Eq. 20) and the second bracket is equal to

$$\frac{\partial}{\partial \eta}\left[\frac{m^2}{s}\right] .$$

Hence the equation is reduced to

$$\frac{\partial m}{\partial \tau} + \frac{\partial}{\partial \eta}\left[\frac{m^2}{s}\right] = 0 \tag{21}$$

A numerical solution to the set given by the Equations 20 and 21 can be obtained and used to determine the corresponding breakup length. The results are presented in Figure 3.

Experimental Studies

The experimental setup consisted of a water line equipped with a cavity connected to an electrodynamic shaker controlled via a function generator. The amplitude and waveform of the disturbance was monitored through an electronic counter and an oscilloscope. The jet modulation and the flow pattern was observed (i) by filming the emerging and bunching jet with a high speed camera capable of recording at 5000 frames a second (Fig 4a) and (ii) by taking slide pictures of the flow with open shutter cameras and an electronic (Fig 4b) stroboscope. The breakup length was estimated for each flow modulation condition by measuring on the screen the downstream distance at which discontinuity appears to occur for the first time. The experiments had a limited range insofar as the amplitudes were concerned. Three sets of runs were carried out, one at a relative modulation of one percent, the others at two and three percent. The modulation frequencies ranged from 100 to 2000 cycles per sec. The nozzles were made of brass and had a diameter of 1/8 inch. Some tests were also carried out with a 1/4 inch diameter nozzle. The jet velocities ranged between 40 and 100 fps.

Discussion of Results and Conclusion

The application of discharge modulation at the proper amplitude and frequency was shown to induce the "bunching" of the jet. The inertial effects clearly predominate over the capillary forces. Photographic evidence has shown the rapid build-up of jet contours, with the periodicity in bunching spacing (i.e., the wavelength) being in excellent agreement with the frequency of the programmed modulation. The jet contours, predicted by a simple analytical (linear) model and shown in Figure 2, confirms this rapid build-up.

However, a comparison between the linear and nonlinear models indicates that the variation of the stream contour, as time elapses, displays in the nonlinear model a non-symmetric wave form.

Experimental results given for the relative modulations of one, two, and three percent (Fig. 3) are in relatively good agreement with previously published data [14]. The discrepancy between the experimental results and the theoretical analysis stems, on the one hand, from the fact that the amplitude of the modulations was assumed to be very small when in fact it becomes gradually very large, and on the other hand, from the fact that the analysis has ignored the significant role played by aerodynamic drag. Indeed, a preponderance of photographic evidence suggests that the radial build-up of fluid leads initially to the formation of disc shaped fluid bunches which later are dissipated as the aerodynamic drag bends, shears, atomizes and eventually breaks them up at their edges.

Acknowledgments

This work was partly supported by the U.S. Department of Energy (Contract No. DE-AS22-79PC-20091) through a Special Research Agreement to the Coal Extraction and Utilization Research Center at Southern Illinois University, Carbondale, Illinois.

References

[1] Rayleigh, Lord, "On the Instability of Jets," Proc. London Math. Soc., V. X, 1879, p. 4.

[2] Rayleigh, Lord, "On the Capillary Phenomena of Jets," Proc. Royal Soc., V. XXIX, 1879, p. 71.

[3] Tyler, E. and Richardson, E. G., "The Characteristic Curves of Liquid Jets," Proc. Phys. Soc., V. 37, 1925, p. 297.

[4] Schweitzer, P. H., "Mechanism of Disintegration of Liquid Jets," J. Appl. Physics, V. 8, 1937, p. 513.

[5] Merrington, A. C. and Richardson, E. G., "The Breakup of Liquid Jets," The Proc. Phys. Soc., V. 59, p. 1, No. 331, 1947, p. 1.

[6] Weber, C., "Zum Zerfall eines Flussigkeitsstrables," ZAMM, V. 2, 1931, p. 136.

[7] Chaudhary, K. C. and Redekopp, L. G., "The Nonlinear Capillary Instability of a Liquid Jet. Part 1. Theory," J. Fluid Mech., V. 96, p. 2, 1980, p. 257.

[8] Chaudhary, K. C. and Maxworthy, T., "The Nonlinear Capillary Instability of a Liquid Jet. Part 2. Experiments on Jet Behavior Before Droplet Formation," J. Fluid Mech., V. 96, p. 2, 1980, p. 275.

[9] Lafrance, P., "Nonlinear Breakup of a Liquid Jet," The Physics of Fluids, V. 17, No. 10, 1974, p. 1913.

[10] Lee, H. C., "Drop Formation in a Liquid Jet," IBM J. Res. Develop., V. 18, 1974, p. 364.

[11] Nayfeh, A. H., "Nonlinear Stability of a Liquid Jet," The Physics of Fluids, V. 13, No. 4, 1970, p. 841.

[12] Crane, L., Birch, S. and McCormack, P. D., "The Effect of Mechanical Vibration on the Breakup of a Cylindrical Water Jet in Air," Brit. J. Appl. Phys., V. 15, 1964, p. 743.

[13] McCormack, P. D., "A Driving Mechanism for High Frequency Combustion Instability in Liquid Fuel Rocket Engines," J. Royal Aero. Soc., V. 68, 1964, p. 633.

[14] McCormack, P. D., Crane, L. and Birch, S., "An Experimental and Theoretical Analysis of Cylindrical Liquid Jets Subjected to Vibration," Brit. J. Appl. Phys., V. 16, 1965, p. 395.

[15] Fenwick, J. R. and Bugler, G. J., "Oscillating Flame Front Flowrate Amplification through Propellant Injection Ballistics (The Klystron Effect)," Chem. Propulsion Info. Agency Publication No. 138, V. 1, 1967, p. 417.

[16] Nebeker, E. B. and Rodriguez, S. E., "Development of Percussive Water Jets," Final Report, U.S. Dept. of Energy, 1979, 80 pages.

[17] Sami, S. and Ansari, H., "Governing Equations in a Modulated Liquid Jet," Proc. First U.S. Water Jet Symposium, Golden, Colorado, 1981.

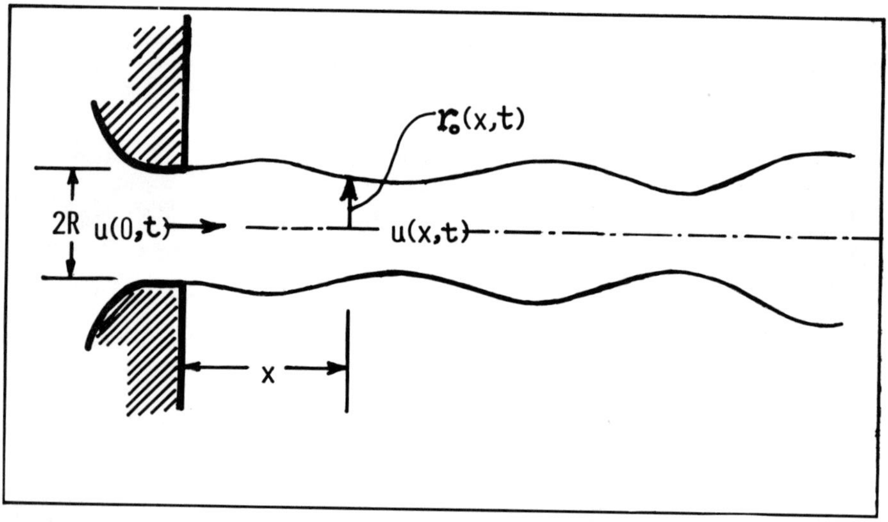

FIG. 1 - DEFINITION SKETCH

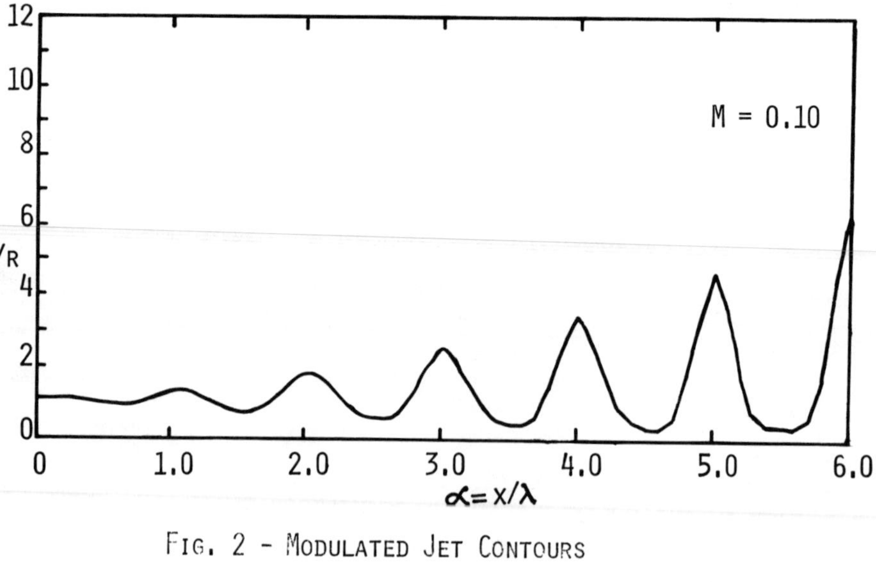

FIG. 2 - MODULATED JET CONTOURS

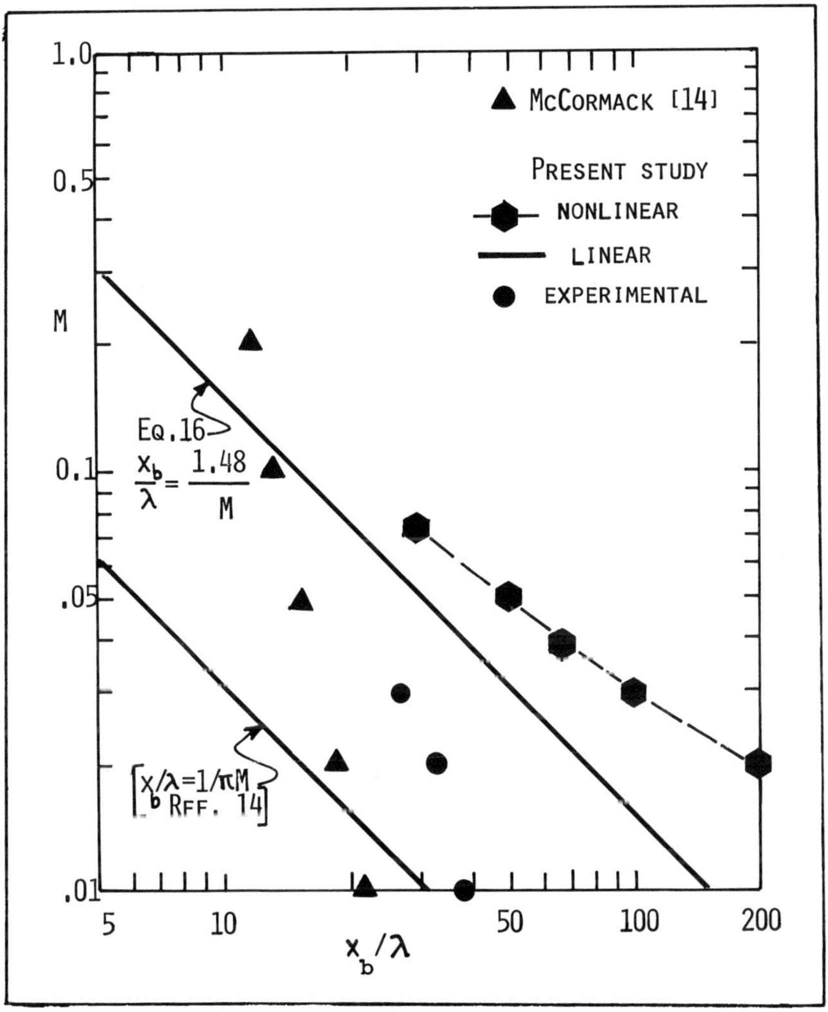

FIG. 3 - VARIATION OF BREAKUP LENGTH

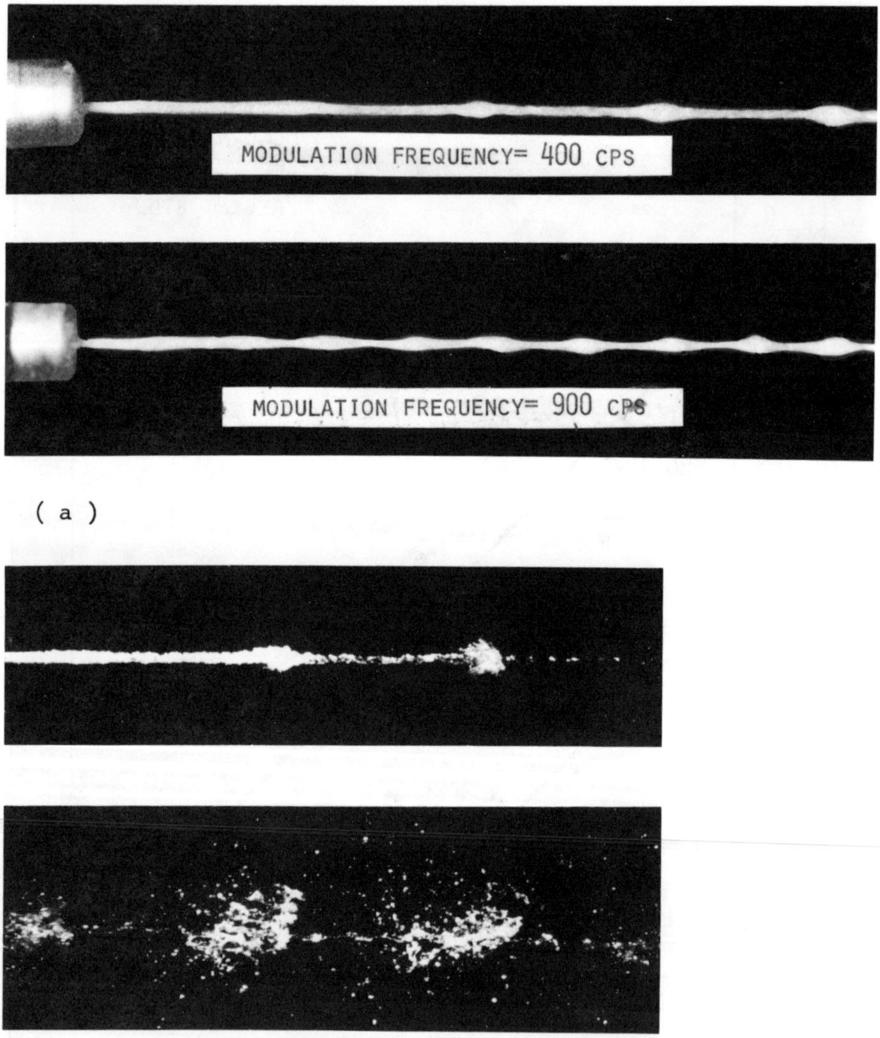

FIG. 4 - MODULATED JET : (a) INITIAL PHASE; (b) BREAKUP STAGE

MAGNETOHYDRODYNAMIC FLOW OF A SUSPENSION IN PIPES

Ampere A. Tseng, Martin Marietta Laboratories,
1450 South Rolling Road, Baltimore, Maryland 21227

and

Vireshwar Sahai, Tennessee Technological University,
Cookeville, Tennessee 38501

ABSTRACT

The effect of a transverse magnetic field on steady viscous flow of a solid-fluid suspension between parallel plates and in pipes of rectangular and elliptical cross sections is investigated. The fluid phase is electrically conducting and the solid phase is non-conducting. The governing equations are formulated and the one-dimensional problem is exactly solved. The two-dimensional problem is solved by using a finite difference scheme for pipes of both rectangular and elliptical cross sections. The results are checked against known solutions for the limiting cases of zero magnetic field strength and single-phase flow.

I. INTRODUCTION

This paper is concerned with magnetohydrodynamic (MHD) Poiseuille flow of a solid-fluid suspension in pipes. The flow is assumed to be steady, laminar, incompressible, viscous, and fully developed. The fluid phase is treated as electrically conducting and the particle phase as electrically nonconducting.

This study has been prompted by 1) a remark [1] that the performance of magnetohydrodynamic generators and accelerators may be affected by the presence of solid particles, e.g., combustion products like ash or soot, or controlled-size, deliberately introduced particles; and 2) magnetohydrodynamic generators involving two-phase flow [2]. However, in such bi-fluid generators both phases are electrically conducting, whereas in the present study only one phase is assumed to be conducting.

In the present analysis, both the fluid and the particle phases are treated as continua. The basic assumption in the theoretical analysis of such a suspension system is that the average properties of the particles can be described in terms of continuous variables. The suspension model used in the present study is the same as that used by Yang [3]. The fluid phase is Newtonian, and the solid phase consists of a large number of identical, minute, spherical particles. The partial pressure contributed by the solid phase is assumed to be negligible, but the viscosity of the particle cloud is not neglected. Because of the small size of the particles, the interactions between the phases are assumed to be given by Stoke's drag law for a sphere. The same assumption regarding the drag force on the particles in the present analysis was also made in a study of magnetohydrodynamic Stokes flow of a suspension [4].

Closed-form solutions are obtained for the case of flow between parallel plates. A successive over-relaxation finite difference scheme for both rectangular and elliptic cross sections solves for the corresponding two-dimensional case of flow in a pipe. These solutions yield Yang's results [3] when no magnetic field is present and yield the single-phase MHD solutions [6,8,10] when the particle density is zero.

II. GOVERNING EQUATIONS

The governing equations for the fluid phase of the flow situation described above may be obtained from the single-phase equations by adding an interphase force density $\vec{D}$. With the stated approximations, the particle phase is governed by a Navier-Stokes-type equation without the pressure term. The governing equations for the flow geometries considered here are easily shown as

$$\mu_f \nabla^2 u + \mu_e H_o \frac{\partial h_x}{\partial y} + (\rho_p/\tau)\,(u_p - u) = \frac{dP}{dx} \tag{1}$$

$$\mu_p \nabla^2 u_p + (\rho_p/\tau)(u - u_p) = 0 \tag{2}$$

$$\nabla^2 h_x + \sigma \mu_e H_o \frac{\partial u}{\partial y} = 0 \tag{3}$$

where $\nabla^2 = \partial^2/\partial y^2 + \partial^2/\partial z^2$. The x-components of the fluid velocity, the particle phase velocity, and the induced magnetic field are respectively denoted by u, u_p, and h_x. H_o is the strength of the applied magnetic field, μ is the viscosity, ρ is the density, σ is the electrical conductivity, and P is the pressure. The subscripts p and f represent the solid and fluid phases, respectively.

The term $(\rho_p/\tau)\,(u-u_p)$ represents the interphase drag-force density (see Marble [5]). The quantity τ, the relaxation time, is given by $\tau = m/6\pi c \mu_f$ where m is the mass and c is the radius of each identical spherical particle. It can be shown that the interphase drag term is negligibly affected by the magnetic induction.

Equations (1-3) can be normalized as:

$$\nabla^2 U + M\partial H/\partial Y + \gamma\alpha_1 (U_p - U) + G = 0 \tag{4}$$

$$\alpha_2\nabla^2 U_p + \alpha_1(U - U_p) = 0 \tag{5}$$

$$\nabla^2 H + M\partial U/\partial Y = 0. \tag{6}$$

where γ is the ratio of the densities of the two phases, G is the dimensionless pressure gradient, $M = \mu_e H_o a(\sigma/\mu_f)^{1/2}$ is the Hartmann number, $\alpha_1 = a^2/\tau\nu_f$ is the relaxation parameter for ordinary two-phase flow, and $\alpha_2 = \nu_p/\nu_f$ is the ratio of the kinematic viscosities of the two phases. The characteristic length, a, and velocity, U_1, have been used in the above transformation.

Boundary Conditions

It is assumed that the no-slip condition is applied to the fluid phase. Because the particle cloud may resemble a rarified continuum, a slip condition similar to that used in rarified gasdynamics is imposed on the particle phase. The slip boundary condition can then be written as

$$u_p - u_w = \alpha_3 u_p' \tag{7}$$

where u_w is the velocity of the boundary, α_3 is the slip coefficient, and u_p' is defined as du_p/dy for the one-dimensional case (the particle slip will be neglected for the two-dimensional case). In slip flow, α_3 is a function of the coefficient of viscosity. However, no attempt is made to relate α_3 to the internal properties of the suspension in the present work. The other boundary condition is for the induced magnetic field and arises from the fact that the walls are assumed to be nonconducting.

III. PLANE POISEUILLE FLOW

In this section, the flow between two fixed horizontal plates in the presence of a uniform, transverse magnetic field is considered. The distance between the two plates is a. Since in this case $\partial/\partial Z = 0$, Eqs. (4-6) reduce to the following ordinary differential equations:

$$U'' + MH' + \gamma\alpha_1(U_p - U) + G = 0 \tag{8}$$

$$\alpha_2 U''_p + \alpha_1 (U-U_p) = 0 \tag{9}$$

$$H'' + MU' = 0 . \tag{10}$$

Primes denote differentiation with respect to Y.

If the viscosity of the particle cloud is assumed to be zero, the viscosity ratio α_2 becomes zero. The flow then is in equilibrium with $U = U_p$ in Eq. (9), and the problem reduces to one-phase Hartmann Poiseuille flow [5]. If, on the other hand, α_2 is not negligible, Eqs. (8-10) must be solved simultaneously. Fortunately, because of their simplicity, these equations can be integrated in closed form,

$$U(Y) = \frac{C_1}{C_4}\cosh(R_1Y) + \frac{C_2}{C_4}\cosh(R_2Y) + \frac{C_3}{C_4} \tag{11}$$

$$U_p(Y) = \frac{C_1C_5}{C_4}\cosh(R_1Y) + \frac{C_2C_6}{C_4}\cosh(R_2Y) + \frac{C_3}{C_4} \tag{12}$$

$$H(Y) = -\frac{C_1M}{C_4R_1}\sinh(R_1Y) - \frac{C_2M}{C_4R_2}\sinh(R_2Y) - \frac{Y}{M} \tag{13}$$

where

$$C_i = \frac{(R_i^2 - M^2)}{\gamma\alpha_1}\cosh\frac{R_i}{2} - \alpha_3\left(1 - \frac{R_i^2}{\gamma\alpha_1} + \frac{M^2}{\gamma\alpha_1}\right)R_i\sinh\frac{R_i}{2}, \quad i = 1,2$$

$$C_3 = \left(\frac{R_2^2 - R_1^2}{\gamma\alpha_1}\right)\cosh\frac{R_1}{2}\cosh\frac{R_2}{2} + \alpha_3\left(1 - \frac{R_1^2}{\gamma\alpha_1} + \frac{M^2}{\gamma\alpha_1}\right)R_1\cosh\frac{R_2}{2}\sinh\frac{R_2}{2} - \\ - \left(1 - \frac{R_2^2}{\gamma\alpha_1} + \frac{M^2}{\gamma\alpha_1}\right)R_2\cosh\frac{R_1}{2}\sinh\frac{R_2}{2}$$

$$C_4 = 2M^2\left\{\left(\frac{M^2 - R_1^2}{\gamma\alpha_1R_2}\right)\sinh\frac{R_2}{2}\cosh\frac{R_1}{2} - \left(\frac{M^2 - R_2^2}{\gamma\alpha_1R_1}\right)\sinh\frac{R_1}{2}\cosh\frac{R_2}{2} + \right. \\ \left. + \left[\left(1 - \frac{R_1^2}{\gamma\alpha_1} + \frac{M^2}{\gamma\alpha_1}\right)\frac{R_1}{R_2} - \left(1 - \frac{R_2^2}{\gamma\alpha_1} + \frac{M^2}{\gamma\alpha_1}\right)\frac{R_2}{R_1}\right]\alpha_3\sinh\frac{R_1}{2}\sinh\frac{R_2}{2}\right\} \tag{14}$$

$$C_5 = \left(1 - \frac{R_1^2}{\gamma\alpha_1} + \frac{M^2}{\gamma\alpha_1}\right)$$

$$C_6 = \left(1 - \frac{R_2^2}{\gamma\alpha_1} + \frac{M^2}{\gamma\alpha_1}\right)$$

$$R_{1,2}^2 = \frac{1}{2}\left[M^2 + \gamma\alpha_1 + \frac{\alpha_1}{\alpha_2} \mp \sqrt{\left(M^2 + \gamma\alpha_1 + \frac{\alpha_1}{\alpha_2}\right)^2 - 4\frac{\alpha_1}{\alpha_2}M^2}\right].$$

The numerical results of the normalized velocities for the plane case are presented in Figs. 1 through 4. The effect of increasing the strength of the magnetic field (i.e., increasing the Hartmann number) is to flatten the velocity profile of the fluid phase as shown in Fig. 1, the same trend as found in one-phase MHD flow. Figures 3 and 4 show the effect of increasing the relaxation parameter α_1 on the velocity profiles of fluid and particle phases for a given Hartmann number. As shown in Fig. 3, the retarding effect on the fluid phase is carried over in the particle phase as well as indicated by the flattening of the velocity profile. A consequence of this effect will be a decrease in flow rate. The results in Figs. 3 and 4 also indicate that the particle phase is much more sensitive to changes in α_1 than the fluid phase is, even in the presence of a magnetic field. This behavior is the same as that noted by Yang [3] for ordinary two-phase flow. It is noteworthy that the effect of increasing α_1, while keeping M constant, is similar to that obtained by increasing M and keeping α_1 constant (compare Figs. 1 and 2 with Figs. 3 and 4). Variation of the slip coefficient, α_3, affects the velocity of the fluid phase only slightly, whereas it significantly changes the particle velocity. This fact can be found from the solution of Eqs. (11) and (12).

IV. TWO-DIMENSIONAL FLOW

In this section, flow is assumed to occur through a pipe of constant area. The same considerations apply as in the one-dimensional case except that the x-components (x-axis is assumed to be along the axis of the pipe) of the velocity fields u and u_p and the induced magnetic field are assumed to be functions of both y and z.

For two-dimensional pipe, Eqs. (4-6) apply directly and, because of their complexity, have been solved numerically using the finite difference successive over-relaxation scheme [7]. However, if the problem of interest has a strong magnetic field ($M \gg 1$), the implicit alternating direction method is recommended. The successive over-relaxation scheme is conditionally stable and not effective for high Hartmann-number problems.

Rectangular Pipe

The case considered is the flow through a rectangular pipe with insulated walls. It is assumed that the no-slip boundary condition applies to both fluid and particle phases. Figures 5 and 6 show the results for the dimensionless mean velocity of fluid phase $\overline{U}$ and particle phase $\overline{U}_p$, respectively, plotted against the Hartmann number M, for various values of the relaxation parameter α_1, with the height-to-width ratio k = 1 (square cross-section case). As α_1 approaches zero the results approach Shercliff's [6] exact solution for one-phase MHD flow. The corresponding variation of the dimensionless mean velocities $\overline{U}$ and $\overline{U}_p$, respectively, due to an increase in M, with the aspect ratio k = 2, is shown in Figs. 7 and 8. In this case, the results also approach

Shercliff's exact solution at $\alpha_1 = 0$. The qualitative nature of these results is the same as that for the one-dimensional case.

Pipe of Elliptical Cross Section

In this case, 2a is the length of the major axis (parallel to the z-axis) and 2b is the length of the minor axis (parallel to the y-axis) for the elliptical cross section. The technique used to treat the elliptical boundary grid points for the finite-difference approximation is similar to the one [7] used for the problem of torsion with a curved boundary.

Figures 9 and 10 show the results for the dimensionless mean velocity of fluid phase $\overline{U}$ and particle phase $\overline{U}_p$, respectively, plotted against the Hartmann number M, for various values of the relaxation parameter α_1, with the aspect ratio $k = (b/a) = 1$ (the circular case). As α_1 tends to zero, the results approach the exact results of Gold [8] for single-phase MHD flow. The corresponding results, with the aspect ratio $k = 0.5$, are shown in Figs. 11 and 12. Since the exact numerical solution of single-phase MHD flow for the elliptical case is not available, the results of the present case have been compared to Nei's [10] second approximation results obtained with the Ritz method; the difference is smaller than 3% in the range of $M \leq 10$. The behavior of the results for the elliptical cross-section case is similar to that for the rectangular cross-section case. Note that the exact solution for the case of elliptical cross section was obtained by Chang and Sahai [9] in terms of Mathieu functions. However, no numerical results were presented.

V. CONCLUSIONS

The magnetohydrodynamic two-phase Poiseuille flow problem with an electrically conducting fluid phase and a non-conducting solid phase has been successfully solved for the case of the flow of a suspension in which the particle cloud viscosity is not neglected.

The one-dimensional case -- flow between parallel insulated plates -- has been solved exactly. The effect that varying the Hartmann number, the relaxation parameter, and the particle density has on the mean velocities for the fluid and particle phases has been considered. Since the particles possibly may slip at the boundary, the effect of including a slip term (in an analogy with rarefied gasdynamics) in the boundary condition for the particle phase has also been tested.

The two-dimensional problem of MHD two-phase flow in insulated pipes of rectangular and elliptical cross-sections was solved numerically using an over-relaxation finite difference scheme. A particular contribution of this paper was the consideration of the case of an elliptical cross-section for which no numerical solutions (except for some approximate ones) were previously available even for the case of one-phase flow.

For simplicity, particle slip at the boundary was neglected in the two-dimensional case. The accuracy of the solution was demonstrated by comparing it to single phase MHD flow. The numerical solutions of rectangular and circular cross-sections were shown to be in good agreement with the known exact solutions for ordinary MHD flow. The exact numerical solution for the elliptical pipe of single phase MHD flow is not known, so the results were checked by comparing them with the known approximate results for one-phase MHD flow. Using all possible checks, it seems that the numerical technique used in the present study gives good results when applied to MHD two-phase pipe flow.

VI. ACKNOWLEDGMENTS

We thank Ms. Lois Craig for typing the manuscript and Ms. Aylene Kovensky for editing it.

VII. REFERENCES

1. Soo, S.L., Fluid Dynamics of Multiphase System, Blaisdell Publishing, Waltham, MA (1967).

2. Sense, K.A., and G.H. Gelb, AIAA J., 5, 862 (1967).

3. Yang, Y.C., M.S. Thesis, Tennessee Technological University (1971).

4. Yang, H.T., and J.V. Healy, Applied Scientific Research, 21, 387 (1973).

5. Marble, F.E., in Annual Reviews of Fluid Mechanics, Vol. 2, 397, Annual Reviews, Palo Alto, CA (1970).

6. Shercliff, J.A., Proc. Cambridge Phil. Soc. 49, 136 (1953).

7. Carnahan, B., H.A. Luther, and J.O. Wilkes, Applied Numerical Methods, John Wiley & Sons, New York (1969).

8. Gold, R.R., J. Fluid Mech. 11, 2192 (1962).

9. Chang, T.S., and V. Sahai, in Proc. Fifth U.S. Congress of Applied Mechanics, 741, ASME, New York (1966).

10. Nei, Y.W., M.S. Thesis, Tennessee Technological Univ. (1970).

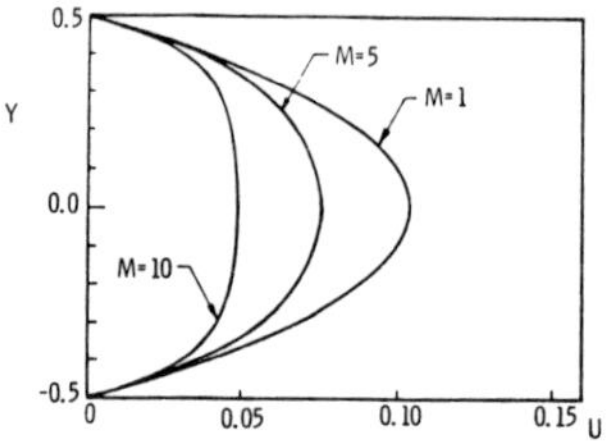

Figure 1. Dimensionless Velocity of Fluid vs. Hartmann Number $M(\alpha_2 = \alpha_3 = \gamma = G = 1,\ \alpha_1 = 5)$.

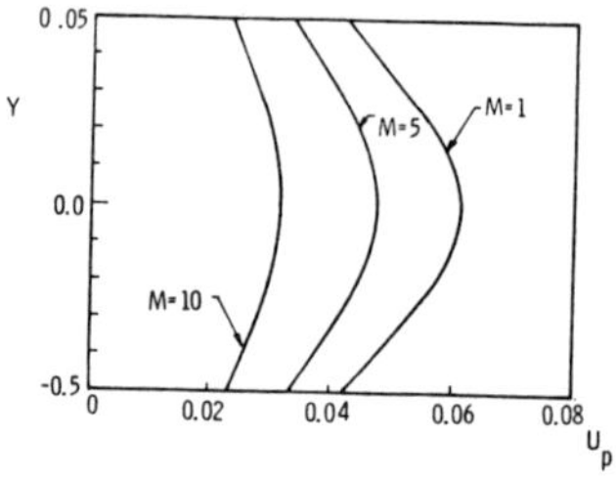

Figure 2. Dimensionless Velocity of Solid Phase vs. Hartmann Number $M(\alpha_2 = \alpha_3 = \gamma = G = 1,\ \alpha_1 = 5)$.

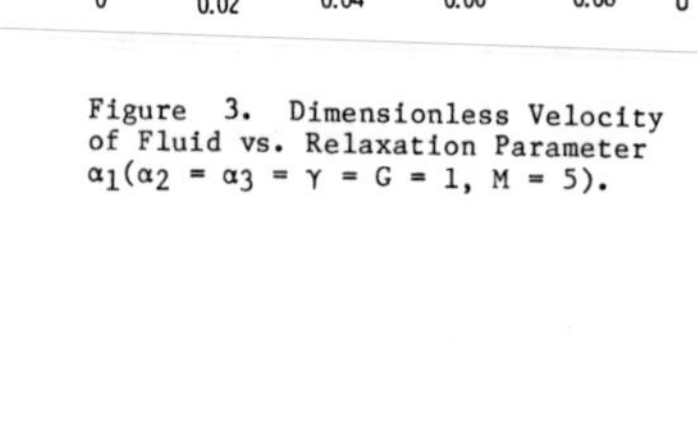

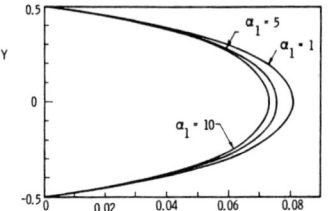

Figure 3. Dimensionless Velocity of Fluid vs. Relaxation Parameter $\alpha_1(\alpha_2 = \alpha_3 = \gamma = G = 1,\ M = 5)$.

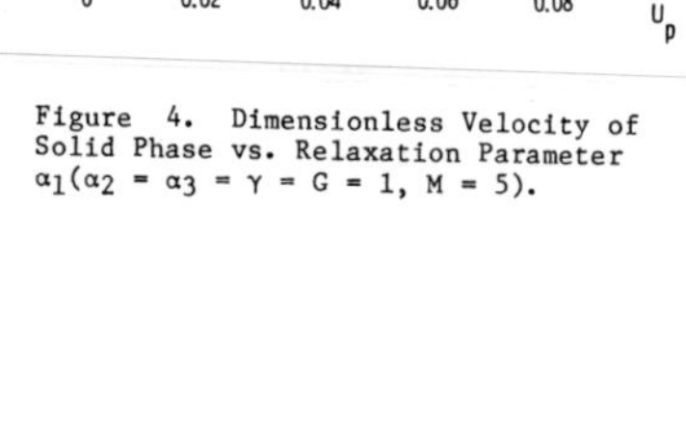

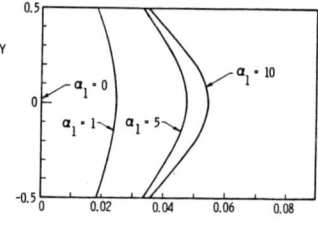

Figure 4. Dimensionless Velocity of Solid Phase vs. Relaxation Parameter $\alpha_1(\alpha_2 = \alpha_3 = \gamma = G = 1,\ M = 5)$.

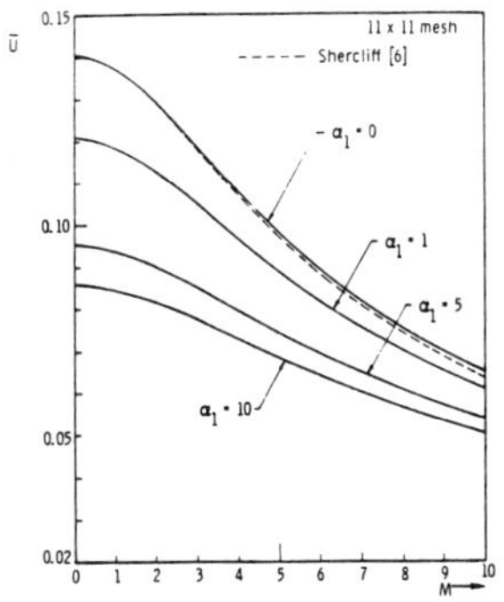

Figure 5. Dimensionless Mean Velocity of Fluid vs. Hartmann Number M for Square Pipe, ($\alpha_2 = \gamma = G = k = 1$, $\alpha_3 = 0$).

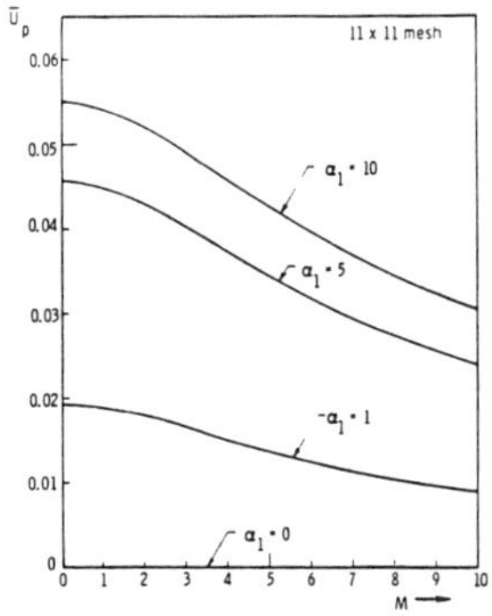

Figure 6. Dimensionless Mean Velocity of Particle Phase vs. Hartmann Number M for Square Pipe, ($\alpha_2 = \gamma = G = k = 1$, $\alpha_3 = 0$).

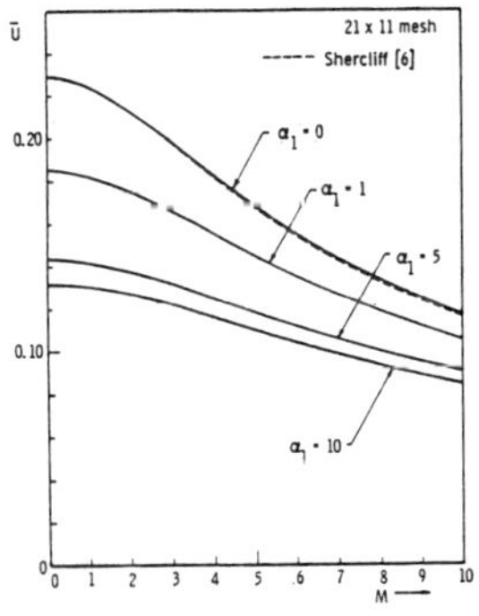

Figure 7. Dimensionless Mean Velocity of Fluid vs. Hartmann Number M for Rectangular Pipe, ($\alpha_2 = \gamma = G = 1$, $k = 2$, $\alpha_3 = 0$).

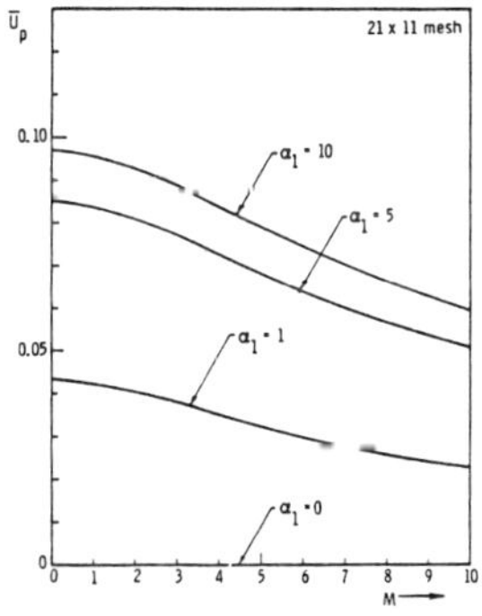

Figure 8. Dimensionless Mean Velocity of Particle Phase vs. Hartmann Number M for Rectangular Pipe, ($\alpha_2 = \gamma = G = 1$, $k = 2$, $\alpha_3 = 0$).

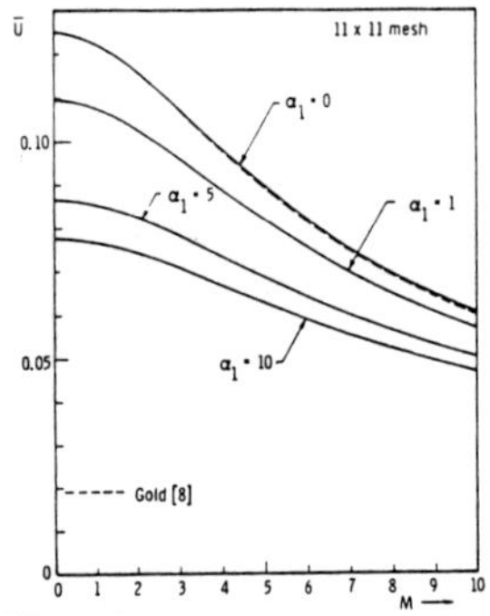

Figure 9. Dimensionless Mean Velocity of Fluid vs. Hartmann Number M for Circular Pipe, ($\alpha_2 = \gamma = k = G = 1$, $\alpha_3 = 0$).

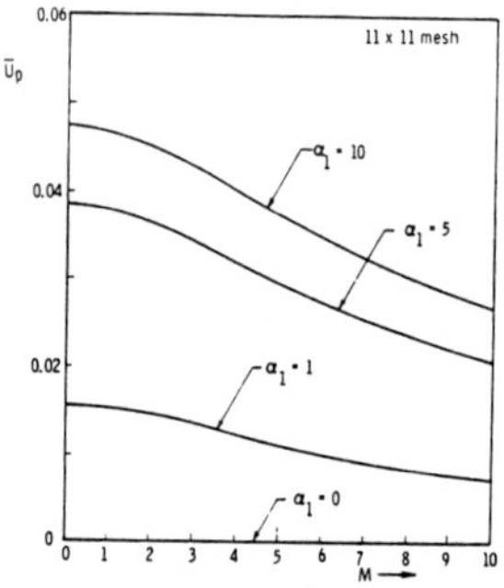

Figure 10. Dimensionless Mean Velocity of Particle Phase vs. Hartmann Number M for Circular Pipe, ($\alpha_2 = \gamma = k = G = 1$, $\alpha_3 = 0$).

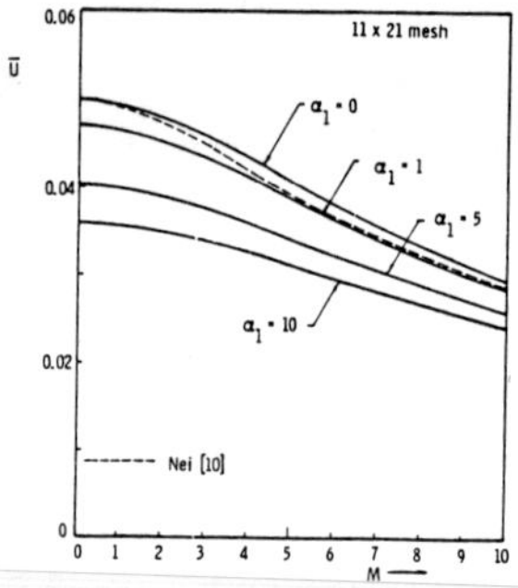

Figure 11. Dimensionless Mean Velocity of Fluid vs. Hartmann Number M for Elliptic Pipe, ($\alpha_2 = \gamma = G = 1$, $k = 0.5$, $\alpha_3=0$).

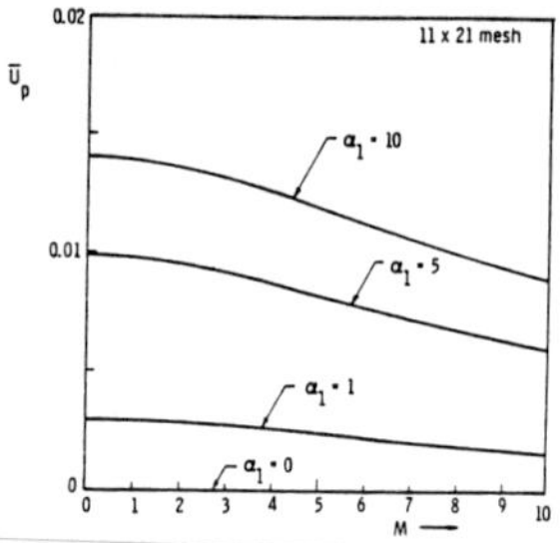

Figure 12. Dimensionless Mean Velocity of Particle Phase vs. Hartmann Number M for Elliptic Pipe, ($\alpha_2 = \gamma = G = 1$, $k = 0.5$, $\alpha_3 = 0$).

ON THE AUXILIARY EQUATION IN A GAS CENTRIFUGE

M. H. Berger
Oak Ridge Gaseous Diffusion Plant, Oak Ridge, Tennessee
operated by Union Carbide Corporation, Nuclear Division
for the U. S. Department of Energy

ABSTRACT

The auxiliary equation which uncouples temperature and angular velocity in linearized rotating gas flow is analyzed. This single, second order, anisotropic diffusion equation is carefully examined, and a boundary layer-like character for high speed conditions is identified through physical reasoning. Both exact and perturbation solutions are constructed for end cap, wall temperature, scoop and feed drives. The exact solutions are greatly simplified in the perturbation approximation of asymptotic expansions which reduce computational labor. As a result, the physics of this equation can be described by unusually simple expressions.

I. INTRODUCTION

One can show [1] that for linearized flow in a gas centrifuge, temperature and angular velocity can be combined to reduce the energy equation to an auxiliary equation of the form

$$- 4A^4 h_{xx} - h_{yy} = T(x,y) + 2(S-1) V(x,y). \tag{1.1}$$

Here the subscripts denote partial derivatives and V, T are the azimuthal momentum and energy sources, respectively. The geometry is illustrated schematically in Figure 1 below.

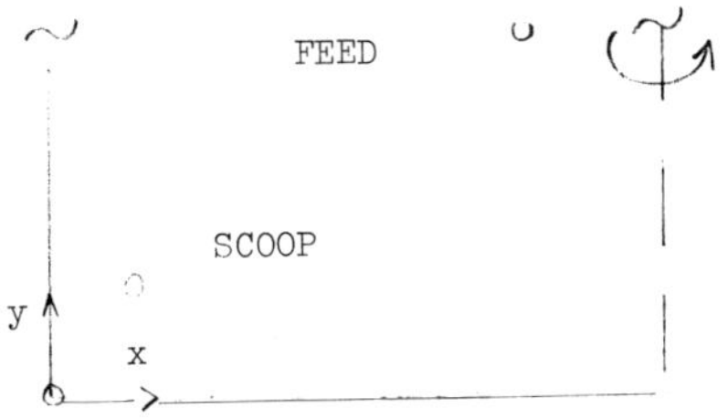

Fig. 1. Rotating Gas

Solutions to the auxiliary equation help to determine gas temperatures and/or heat fluxes. Non-continuum phenomena associated with large x are beyond the scope of this paper.

Anisotropic diffusion equations such as (1.1) occur elsewhere in centrifuge literature [2], as well as in heat transfer applications [3]. Here the tensor diffusion coefficient has an x-wise diffusion coefficient $4A^4$ times greater than in the y-direction. Although (1.1) is a second order, linear, elliptic partial differential equation on a rectangular domain and in theory can be solved exactly, it can be simplified by noting that for $A^4 >> 1$ it is a singular perturbation equation in the small, positive parameter ε, where $\varepsilon \equiv 1/(4A^4)$.

Rewriting (1.1),

$$h_{xx} + \varepsilon h_{yy} = - \varepsilon \left[T(x,y) + 2(S-1)\ V(x,y)\right] . \tag{1.2}$$

In the "high speed limit", $\varepsilon \to 0$,

$$h(x,y) = h(y), \tag{1.3}$$

and there is a complete loss of diffusion effects in both x and y directions. This special structure will be examined in Section 3.

Let us consider: 1) end cap (temperature or differential end cap rotation), 2) wall temperature, 3) scoop, 4) feed drives.

These drives can be treated individually and solutions superimposed in the general case. For the scoop we assume a two-dimensional delta function while the distributed feed is an assumed delta in x and arbitrarily distributed in y.

2. GENERALIZED FOURIER SERIES AND GREENS FUNCTIONS

Equation (1.2) can be transformed by the mapping $x \to X/\sqrt{\varepsilon}$ to a standard Poisson equation having known solutions.

$$h_{XX} + h_{yy} = - \left[T + 2(S-1)\ V\right] . \tag{2.1}$$

2.1 End Cap Drive

In the absence of internal sources and sinks we simply have the Laplacian

$$h_{XX} + h_{yy} = 0, \tag{2.2}$$

with the non homogeneous boundary conditions,

$$h(X,0) = \delta H_o,\ h(X,y_o) = h(0,y) = h_X(X_T,y) = 0. \tag{2.3}$$

This has the solution [3]

$$h(X,y) = \frac{4\ \delta H_o}{\pi} \sum_{n=0}^{\infty} \frac{1}{(2n+1)} \sin\left[\frac{(2n+1)\pi X}{2X_T}\right] \sinh\left[\frac{(y_o-y)(2n+1)\pi}{2X_T}\right] \operatorname{cosech}\left[\frac{(2n+1)\pi y_o}{2X_T}\right]. \tag{2.4}$$

Here a uniform end is assumed. A similar expression applies for the other end at y_o.

2.2 Wall Temperature Drive

Wall temperature drive is usually described by some prescribed rotor wall temperature profile matching the end caps. Thus,

$$h(X,0) = H(0),\ h(X,y_o) = H(y_o),\ h(0,y) = H(y),\ h_X(X_T,y) = 0. \tag{2.5}$$

The solution is obtained by superposition of three simpler solutions,

$$h(X,y) = \sum_{n=1}^{\infty} a_n \sin\left(\frac{n\pi y}{y_o}\right)\left\{\sinh\left[\frac{(2X_T-X)n\pi}{y_o}\right] + \sinh\left(\frac{n\pi X}{y_o}\right)\right\} \operatorname{cosech}\left(\frac{2n\pi X_T}{y_o}\right)$$

$$+ \frac{4\ H(0)}{\pi} \sum_{n=0}^{\infty} \frac{1}{(2n+1)} \sin\left[\frac{(2n+1)\pi X}{2X_T}\right] \sinh\left[\frac{(y_o-y)(2n+1)\pi}{2X_T}\right] \operatorname{cosech}\left[\frac{(2n+1)\pi y_o}{2X_T}\right]$$

$$+ \frac{4\ H(y_o)}{\pi} \sum_{n=0}^{\infty} \frac{1}{(2n+1)} \sin\left[\frac{(2n+1)\pi X}{2X_T}\right] \sinh\left[\frac{y\ (2n+1)\pi}{2X_T}\right] \operatorname{cosech}\left[\frac{(2n+1)\pi y_o}{2X_T}\right] \tag{2.6}$$

where the Fourier coefficient is

$$a_n = \frac{2}{y_o} \int_0^{y_o} H(y') \sin\left(\frac{n\pi y'}{y_o}\right) dy'. \tag{2.7}$$

2.3 Scoop Drive

For a point scoop we have a Greens function obeying

$$g_{XX} + g_{yy} = -\ \varepsilon^{\frac{1}{2}}\ \delta\,(X-X^*)\ \delta(y-y^*), \tag{2.8}$$

assuming a unit source.

This has the solution [3]

$$g(X,y) = \left(\frac{2}{2X_T}\right)\left(\frac{2}{y_o}\right) \sum_m \sum_n \frac{\varepsilon^{\frac{1}{2}}}{\pi^2\left[\left(\frac{m}{2X_T}\right)^2 + \left(\frac{n}{y_o}\right)^2\right]}$$

$$\sin\left(\frac{m\pi X^*}{2X_T}\right) \sin\left(\frac{n\pi y^*}{y_o}\right) \sin\left(\frac{m\pi X}{2X_T}\right) \sin\left(\frac{n\pi y}{y_o}\right) . \qquad (2.9)$$

Combining solutions by the method of images yields

$$g(X,y;\, X^*,y^*,\, 2X_T - X^*,y^*) = \frac{2\varepsilon^{\frac{1}{2}}}{X_T y_o} \sum_{\substack{m \\ odd}} \sum_{\substack{all \\ n}} \frac{1}{\pi^2\left[\left(\frac{m}{2X_T}\right)^2 + \left(\frac{n}{y_o}\right)^2\right]}$$

$$\sin\left(\frac{n\pi y^*}{y_o}\right) \sin\left(\frac{n\pi y}{y_o}\right)\left\{\sin\left(\frac{m\pi X^*}{2X_T}\right) \sin\left(\frac{m\pi X}{2X_T}\right) + \sin\left[m\pi\left(1 - \frac{X^*}{2X_T}\right)\right] \sin\left(\frac{m\pi X}{2X_T}\right)\right\} . \qquad (2.10)$$

which exhibits a logarithmic singularity at (X^*,y^*) [4]. Computationally, this series is slow to converge for $4A^4 >> 1$.

2.4 Feed Drive

Distributed feed solutions are derivable from the preceding scoop solution via a useful property of Greens function involving integration of the Greens function kernel over the given source distribution. Thus,

$$h(X,y) = \int_0^y \left[T(y^*) + 2(S-1)\ V(y^*)\right] g(X,y;X^*,y^*,2X_T - X^*,y^*)dy^* . \qquad (2.11)$$

If the integrand is sufficiently complicated such that the necessary term by term integrations cannot be done in closed form one could resort to numerical quadrature. Theoretically, equation (2.11) is an exact solution.

3. SINGULAR PERTURBATIONS

Let us consider an inner solution valid in region i near the end cap where axial diffusion is retained, i.e., boundary layer and an outer solution valid in region o away from the ends where axial diffusion is neglected, i.e., internal flow region. [By way of analogy with Prandtl's boundary layer theory [5], region i corresponds to the viscous boundary layer and region o corresponds to the purely inviscid core.] Then the following model results, for say, the end y = 0.

Region i:

PDE:

$$h_{xx} + \varepsilon h_{yy} = 0 \tag{3.1}$$

BC's:

$$h(x,0) = \delta H(0),\ h(x,\infty) = h_e(x,0),\ h(0,y) = H(y),\ h_x(x_T,y) = 0. \tag{3.2}$$

Here h_e is h external to the boundary layer.

Region o:

ODE:

$$h_{e_{xx}} = 0. \tag{3.3}$$

BC's:

$$h_e(0,y) = H(y),\ h_{ex}(x_T,y) = 0. \tag{3.4}$$

Here the partial differential equation (1.1) is reduced to an ordinary differential equation in the outer region. This physical, boundary layer approach can be formalized mathematically through the theory of perturbations. In particular the method of matched asymptotic expansions (M.A.E.) [6] is useful in obtaining approximate solutions. Gradients of h, which might be useful in computing heat fluxes, can be readily determined.

3.1 End Cap Drive

For $y_o/X_T >> 1$, the asymptote of equation (2.4) is

$$h(x,y) \sim \frac{4\ \delta H_o}{\pi} \sum_{n\ \mathrm{odd}} \frac{1}{n} e^{\left(-\frac{n\pi y}{2X_T}\right)} \sin\left(\frac{n\pi x}{2X_T}\right). \tag{3.5}$$

This same result could be obtained by M.A.E.

This is summable to [3]

$$h(x,y) \sim \frac{2}{\pi}\,\delta H_o \tan^{-1}\left[\frac{\sin\left(\frac{\pi x}{2X_T}\right)}{\sinh\left(\frac{\pi\sqrt{\varepsilon}\,y}{2x_T}\right)}\right]. \tag{3.6}$$

One can easily estimate the end boundary layer thickness Δ, where $h(x,\Delta) = .01\ \delta H_o$. From (3.6), after some algebra

$$\Delta(x) \simeq \frac{2x_T}{\pi\sqrt{\varepsilon}}\, \ell n \left[\frac{400}{\pi} \sin\left(\frac{\pi x}{2x_T}\right)\right] . \tag{3.7}$$

3.2 Wall Temperature Drive

For the outer expansion, $\varepsilon \to 0$, we have

ODE:

$$h^o_{xx} = 0 \tag{3.8}$$

BC's:

$$h^o(0,y) = h^o(2x_T,y) = H(y). \tag{3.9}$$

Here h^o is the first term in an asymptotic outer expansion for h.

Obviously then,

$$h^o \sim H(y). \tag{3.10}$$

For the inner expansion,

$$h^i_{xx} + h^i_{YY} = 0 \tag{3.11}$$

where $Y \equiv y/\varepsilon^{\frac{1}{2}}$. This is a simple coordinate stretching, and h^i is the first term in an asymptotic inner expansion for h.

BC's:

$$h^i(0,Y) = h^i(2x_T,Y) = H(0),\ h^i(x,0) = H(0). \tag{3.12}$$

By inspection,

$$h^i(x,Y) \sim H(0). \tag{3.13}$$

Similarly for $y = y_o$. Then the composite solution is given by the usual formula

$$h^c = h^o + h^i - (h^o)^i \tag{3.14}$$

which removes any redundance. This is trivial here.

$$h^c \sim H(y). \tag{3.15}$$

3.3 Feed Drive

In region i

$$h^i_{xx} + h^i_{YY} = 0 \tag{3.16}$$

with homogeneous boundary conditions

$$h^i(0,Y) = h^i_x(x_T,Y) = h^i(x,0) = 0. \tag{3.17}$$

This leads to the solution

$$h^i = a_1 Y \tag{3.18}$$

which one can readily discount on physcial grounds, hence

$$h^i = 0. \tag{3.19}$$

In region o let

$$h^o \sim \varepsilon\, \mathcal{h}^o + \text{h.o.t.} \tag{3.20}$$

valid for small ε. Thus,

$$\mathcal{h}^o_{xx} = -\left[T(y) + 2(S-1)\ V(y)\right] \delta(x-x^*) \tag{3.21}$$

with bc's

$$\mathcal{h}^o(0,y) = \mathcal{h}^o_x(x_T,y) = 0. \tag{3.22}$$

Here the x dependence of T and V has been specified.

This leads to the obvious solution

$$\mathcal{h}^o \sim \left[T(y) + 2(S-1)\ V(y)\right] \left[x - (x-x^*)\ U(x-x^*)\right] \tag{3.23}$$

where $U(x-x^*)$ is the Heaviside unit step function. One finds for small ε,

$$h^c \sim \varepsilon \left[T(y) + 2(S-1)\ V(y)\right] \left[x - (x-x^*)\ U(x-x^*)\right] \tag{3.24}$$

for the M.A.E. with $T(Y) = V(Y) = 0$. Physically this requires the feed to be outside the end boundary layers. This solution is the product of a ramp function in x with the feed distribution in y and exhibits no axial diffusion. For small ε, equation (3.24) is a good approximation to (2.11) for $T(y)$, $V(y)$ well behaved, smooth functions. However, considerable approximation errors may occur for strongly peaked distributions. Equation (3.24) breaks down for $T(y), V(y) = \delta(y-y^*)$. Thus when $T''(y)$ or $V''(y)$ are large this approximate solution breaks down locally due to non negligible axial diffusion. As long as the y extent of this breakdown region is small it should not lead to significant overall error.

3.4 Scoop Drive

Let us consider this as a limit of a feed drive such that the y-distribution goes to a point,

$$T(y),\ V(y) \to T_o\ \delta(y-y^*),\ V_o\ \delta(y-y^*). \tag{3.25}$$

In this limit the $O(\varepsilon)$ perturbation solution (3.24) breaks down. We will see that this is because the perturbation solution is really $O(\varepsilon^{\frac{1}{2}})$.

Replace the distributed source by a point source at (x^*,y^*), and rewrite equation (1.1) as

$$h_{xx} + \varepsilon\, h_{yy} = -\,\varepsilon\, \delta(x-x^*)\, \delta(y-y^*)\,. \tag{3.26}$$

This is subject to

$$h(0,y) = h_x(x_T,y) = h(x,0) = h(x,y_o) = 0 \tag{3.27}$$

on the closure of (x,y).

Equation (3.26), singular at (x^*,y^*), exhibits free shear layer character. That is to say, there are large local gradients in the absence of a boundary. Technically, mathematicians call this a free layer [7].

Assume a regular asymptotic expansion in ε,

$$h(x,y) \sim \mathcal{h}\, \varepsilon^{\frac{1}{2}} + \text{h.o.t.} \tag{3.28}$$

valid for small ε. Substitute (3.28) into the governing partial differential equation and obtain

$$\mathcal{h}_{xx} + \varepsilon\, \mathcal{h}_{yy} = -\,\varepsilon^{\frac{1}{2}}\, \delta(x-x^*)\, \delta(y-y^*) \tag{3.29}$$

after discarding the higher order terms.

For the outer expansion $\mathcal{h}^o$ to $O(\varepsilon^{\frac{1}{2}})$,

$$\mathcal{h}^o_{xx} = 0 \tag{3.30}$$

which has only the trivial solution

$$\mathcal{h}^o(x,y) = 0. \tag{3.31}$$

For the inner expansion we need the translation and stretching transformation

$$Y_1 = (y-y^*) \,/\, \sqrt{\varepsilon}\,. \tag{3.32}$$

Based on the integral property of Dirac's delta function and the coordinate stretching we have a classical Green's function problem in inner variables.

$$g_{xx} + g_{Y_1Y_1} = -\,\delta(x-x^*)\, \delta(Y_1) \tag{3.33}$$

with

$$g(0,Y_1) = g(2x_T,Y_1) = 0,\ \lim_{|Y_1| \to \infty} g(x,Y_1) \text{ finite} \tag{3.34}$$

Because Y_1 ranges from $+\infty$ to $-\infty$ the solution must exhibit exponential decay-like behavior and any end effects must be captured by the outer solution. This significantly different behavior from boundary layers is quite representative of interior layers.

From [8] the solution to equation (3.34) can be written as

$$g(x,Y_1) = \frac{1}{4\pi}\,\ell n\left[\frac{\cosh\left(\frac{\pi Y_1}{2x_T}\right) - \cos\left(\frac{\pi(x^*-x)}{2x_T}\right)}{\cosh\left(\frac{\pi Y_1}{2x_T}\right) - \cos\left(\frac{\pi(x^*+x)}{2x_T}\right)}\right] \tag{3.35}$$

which leads to

$$h^i(x,Y_1) \sim \frac{1}{4\pi}\left\{\ell n\left[\frac{\cosh\left(\frac{\pi Y_1}{2x_T}\right) - \cos\left(\frac{\pi(x-x^*)}{2x_T}\right)}{\cosh\left(\frac{\pi Y_1}{2x_T}\right) - \cos\left(\frac{\pi(x+x^*)}{2x_T}\right)}\right] + \ell n\left[\frac{\cosh\left(\frac{\pi Y_1}{2x_T}\right) + \cos\left(\frac{\pi(x+x^*)}{2x_T}\right)}{\cosh\left(\frac{\pi Y_1}{2x_T}\right) + \cos\left(\frac{\pi(x-x^*)}{2x_T}\right)}\right]\right\}. \tag{3.36}$$

Thus

$$h(x,y) = \frac{\varepsilon^{\frac{1}{2}}}{4\pi}\left\{\ell n\left[\frac{\cosh\left(\frac{\pi(y-y^*)}{2x_T\sqrt{\varepsilon}}\right) - \cos\left(\frac{\pi(x-x^*)}{2x_T}\right)}{\cosh\left(\frac{\pi(y-y^*)}{2x_T\sqrt{\varepsilon}}\right) - \cos\left(\frac{\pi(x+x^*)}{2x_T}\right)}\right] + \ell n\left[\frac{\cosh\left(\frac{\pi(y-y^*)}{2x_T\sqrt{\varepsilon}}\right) + \cos\left(\frac{\pi(x+x^*)}{2x_T}\right)}{\cosh\left(\frac{\pi(y-y^*)}{2x_T\sqrt{\varepsilon}}\right) + \cos\left(\frac{\pi(x-x^*)}{2x_T}\right)}\right]\right\}. \tag{3.37}$$

Clearly the greatest computational advantage of asymptotic expansions is for the feed case. The double series summations in equation (2.10) for the Greens function followed by numerical integration with the feed source distribution in (2.11) are avoided. Furthermore, the physics of the problem is greatly clarified.

4. NOMENCLATURE

A^2 = $\frac{MV_w^2}{2RT_w}$

g = "Greens" function

$H(y)$ = Lateral boundary data

h = $\theta + 2\,(S-1)\,\omega$

h_e = h external to the boundary layer

$\mathcal{h}$ = First term in some asymptotic expansions

M = Molecular weight

m,n = Summation indices

P_r = Prandtl number

R = Universal gas constant

R_e = Reynolds number

S = $1 + \frac{\gamma-1}{2\gamma}\,P_r\,A^2$

T_w = Reference temperature

U = "Heaviside unit step" function

$u,\omega,w,$ = Perturbation velocity components

V,T = Sources and sinks of angular momentum and energy

V_w = Wall velocity

X = $x/\sqrt{\varepsilon}$

x,y,z = Coordinates

x_T = Top of the atmosphere

x^*,y^* = Coordinates of delta function

Y,Y_1 = Inner y-variable

y_o = Rotor length

γ = Specific heat ratio

Δ = Boundary layer thickness

$\delta(x)$ = Dirac delta function

δH_o = End boundary data

ε = $1/(4A^4)$

θ = Perturbation temperature

Superscripts

c = Composite solution

i = Inner expansion

o = Outer expansion

Symbols

$\sim$ = Asymptotically equal

REFERENCES

1. Wood, H. G., and Morton, J. B., "Onsager's Pancake Approximation for the Fluid Dynamics of a Gas Centrifuge," Journal of Fluid Mechanics, Vol. 101, Part 1, pp. 1-31, November 1980.

2. Beardsley, R. C., Saunders, K. D., Warn-Varnas, A. C., and Harding, J. H., "An Experimental and Numerical Study of the Secular Spin-up of a Thermally Stratified Rotating Fluid," Journal of Fluid Mechanics, Vol. 93, Part 1, pp. 161-184 (1979).

3. Carslaw, H. S., and Jaeger, J. C., "Conduction of Heat in Solids," Oxford Press, Second Edition (1959).

4. Courant, R., and Hilbert, D., "Methods of Mathematical Physics," Interscience Publishers, Vol. 1 (1953).

5. Schlichting, H., "Boundary Layer Theory," McGraw-Hill, Sixth Edition (1968).

6. Van Dyke, M., "Perturbation Methods in Fluid Mechanics," Academic Press (1964).

7. Eckhaus, W., "Asymptotic Analysis of Singular Perturbations," North Holland (1980).

8. Carrier, G. F., Krook, N., and Pearson, G. E., "Functions of a Complex Variable Theory and Technique," McGraw-Hill (1966).

SOME ASPECTS OF THE AERODYNAMIC DRAG OF TRACTOR-TRAILER TRUCKS

by

Colin H. Marks
and
Frank T. Buckley, Jr.
Professors
University of Maryland
College Park, Maryland 20742

ABSTRACT

The aerodynamic drag of tractor-trailer trucks is complicated by the fact that such vehicles are made up of two blunt bodies in tandem, a tractor and a trailer, separated by a gap. Because of this complication, surprising results are often found. For example, a vehicle with a very unstreamlined tractor may have less drag than one with a streamlined tractor, or removal of exhaust stacks, which are high drag objects, may increase the drag.

This paper surveys the effects of tractor and trailer streamlining, trailer base drag and gap size on the aerodynamic drag, and shows the magnitude of the drag reductions which can be obtained by well-designed, add-on devices.

INTRODUCTION

The aerodynamic drag of a tractor-trailer truck is the sum of the drag on the tractor and the drag on the trailer. The drag on each of these can be divided into form drag due to flow separation at the front and rear of each body, parasite drag due to such protrusions as exhaust stacks, mirrors and wheels, and skin friction. Because of the flow interaction between the tractor and the trailer the drag around each of these is influenced by the geometry and proximity of the other.

Tractor-trailer trucks are unique among road vehicles in that they have a gap between the tractor and the trailer. As will be seen, the size of this gap is an important factor in the aerodynamic drag. In addition, whenever the wind has a component perpendicular to the axis of the truck, there is a horizontal flow through the gap which greatly affects the drag.

Because much of the fuel consumed by tractor-trailer trucks is used to overcome aerodynamic drag (approximately 50% at 50 mph), interest has been stimulated in finding ways of reducing the drag. Thus there have been a number of investigations whose object is to investigate schemes for reducing aerodynamic drag[1-9] and a number of commercially-available devices have appeared on the market which can be fitted to existing trucks.

Since the power required to overcome aerodynamic drag is exerted in the line of motion of the vehicle, the term drag in this paper shall refer to the net aerodynamic force along the vehicle's longitudial axis. This is known as the body-axis drag force, D_{BA}. Coefficients of aerodynamic drag will be based on the body axis drag force. Thus,

$$C_D = \frac{D_{BA}}{\frac{1}{2}\rho V_r^2 A_{ref}} \qquad (1)$$

for this equation, ρ is the free-stream air density, V_r is the free-stream air velocity relative to the vehicle, and A_{ref} is the reference area. (The reference area used varies with the investigator. There seems to be no convention followed by all investigators. We use the product of the height of the trailer above the ground and width of the trailer.)

YAW ANGLE

As with most blunt bodies, the aerodynamic drag of tractor-trailer trucks depends strongly on yaw angle. Figure 1 shows how the vector addition of the wind velocity, V_w, and the vehicle velocity, V, combine to yield the velocity of the air relative to the vehicle, V_r. The longitudinal axis of the truck is yawed with respect to V_r, and the yaw angle is ψ.

Figure 2 shows typical curves of C_D versus ψ for 1/8th-scale models of vehicles having a fairly streamlined conventional tractor. Note the large increase of C_D with yaw angle (a 43% at 20° over that at 0° for the vehicle having the round-edged trailer) and the substantial increase in the drag at all yaw angles due to the trailer having square front vertical edges rather than well-rounded ones. The amount of the increase of C_D with ψ depends very much on the shape of the tractor and the trailer and the interaction of the flow about the two. For example, Sherwood[10] found a 50% increase of C_D at 10° yaw over that at 0° for one of the vehicles he tested, but only a 28% increase for another.

It should be noted that there is no necessity for drag to increase with yaw angle. Indeed, Windsor[3] showed that a very streamlined tractor-trailer can have a body-axis drag which decreases as yaw angle increases, and that the body-axis drag can become negative! This unreasonable-appearing result is due to the fact that the body-axis drag is the vector sum of the wind-axis drag and the wind-axis side force, as shown in Figure 3. The component of the side force along the vehicle axis is a thrust. If the side force becomes large compared to the wind-axis drag force, the result is a net thrust along the axis of the vehicle. The ability to get thrust from the wind to move into the wind is utilized by sailboats, and so should not be unexpected.

WIND-AVERAGED DRAG COEFFICIENT

Because drag of a vehicle varies with yaw angle, and because the yaw angle of a vehicle constantly changes as its direction relative to wind changes, it is useful to have a single coefficient which will provide a measure of the average drag. Such a coefficient was suggested by Buckley and adopted by the DOT/SAE Truck and Bus Economy Measurement Conference in its report[11]. This coefficient is called the wind-averaged drag coefficient, $\overline{C}_D$. It is derived assuming that the wind is equally likely to occur from any direction, ϕ, (see Figure 1). The wind-averaged drag coefficient is calculated from the equation

$$\overline{C}_D = \frac{1}{2\pi} \int_0^{2\pi} C_D(\psi)\left[1+\left(\frac{V_w}{V}\right)^2 + 2\left(\frac{V_w}{V}\right)\cos\phi\right]d\phi \qquad (2)$$

The body-axis wind-averaged drag force, D_{BA}, is obtained from $\overline{C}_D$ by the equation

$$D_{BA} = \tfrac{1}{2}\rho V^2 \overline{C}_D A_r \qquad (3)$$

While it can be argued that the wind may not have an equal probability of occurrance for all angles, ϕ, for a given vehicle, it is very likely that this is so for the set of all vehicles of a given make and configuration. The wind-averaged drag coefficient is very useful when determining the average drag reduction brought about by modifications and various add-on devices whose purpose is to reduce aerodynamic drag. Generally, the effectiveness of these devices diminishes with yaw angle (some actually increase the drag of a vehicle at large yaw angles) and so the change of the wind-average drag coefficient, $\Delta\overline{C}_D$, due to the use of an add-on device, or due to the modification of a vehicle's shape, is a useful figure of merit.

For the vehicles whose C_D is shown in Figure 2, $\overline{C}_D$ for the one with the square-edged trailer is 0.98, whereas, for the one with the round edged trailer it is 0.85. Rounding the trailer's front vertical edges to a full-scale radius of 12 inches yielded a $\Delta\overline{C}_D$ of 0.13 (positive ΔC_D taken as a drag reduction). In this case, since the two C_D - ψ curves are very nearly parallel, $\Delta\overline{C}_D$ is about the same as ΔC_D at $\psi = 0$ ($\Delta C_D = 0.15$), but this is often not the case.

DISTRIBUTION OF DRAG

Most of the aerodynamic drag associated with tractor-trailers is due to forebody drag (pressure drag due to flow separation on the front of the trailer and on the tractor). Estimates of the skin friction[12] place it at about 10% of the total. The base drag (drag due to the pressure being less than free-stream pressure on the back of the trailer) varies with yaw angle. The base-drag coefficient C_{D_B} (C_{D_B} = base drag/$\frac{1}{2}\rho V_r^2 A_{ref}$) it is found to be very nearly 0.1 at zero yaw and between 0.23 and 0.28 at 20° yaw[13]. The former figure represents between 13% and 15% of the zero-yaw aerodynamic drag of a typical tractor-trailer truck and the latter represents 18% to 25%. Assuming that the skin friction remains constant at 10%, forebody drag is then about three-quarters of the drag at zero yaw and approximately two-thirds to three-quarters of the drag at 20°. Streamlining the front of the vehicle has no appreciable affect on the base drag, though it does increase its fraction of the total drag at small yaw angles due to the decrease of the total drag.

The fraction of a vehicle's aerodynamic drag due to the tractor and that due to the trailer depend strongly on the shape and size of each, and the size of the gap separating them. Flynn and Kyropolos[14] found that at zero yaw the drag of the tractor was about 36% of the total for a detailed model of a vehicle with a conventional tractor coupled to a chamferred-edge trailer. When used with a trailer with rounded front vertical edges, the tractor drag became 41% of the total. The tractor on a model having a trailer-height tractor was found to produce 112% of the total drag. This means that the pressure force on the front of the trailer and the trailer's skin friction was less than the pressure force on the rear, so that there was a thrust on the trailer. The low pressure on the trailer front was caused by air being entrained from the gap between the tractor and the trailer by air passing around the vehicle.

Marks and Buckley[15] found that a fraction of the drag due to the tractor generally decreased with increasing yaw angle up to about 15°, then started to increase again for tractor-trailers with conventional trailers or COE tractors (see Tables 1 and 2). In these tests the tractors were responsible for a higher percentage of the drag than found by Flynn and Kyropolos.

GAP BETWEEN TRACTOR AND TRAILER

It has long been known that the drag of two bodies in tandem at zero yaw depends on the distance separating them. It appears that Sherwood[10] and Windsor[3] were the first to show that the drag of tractor-trailer trucks generally increases with increasing separation between the tractor and the trailer, and that the effect of gap size increases as yaw angle increases. Windsor's data shows that for one vehicle tested the drag decreased by about 10% at zero yaw when gap size was decreased from 49.5 inches to 25.5 inches but decreased by 30% at 10° yaw for the same change in gap size. It was pointed out by Sherwood that the sensitivity of the drag to gap size depends to a large extent on the shape and size of the tractor. In one test that he made at zero yaw with a trailer-height cab, the drag only varied by 5% when the gap was increased from 25.5 inches to 49.5 inches, but with another, more normal tractor, the variation in drag was 20%. Other work [14,15,16] pretty well confirms that of Sherwood and Windsor, but some vehicles tested[4,15] have had a drag decrease at very large gaps (70 inches) over that of intermediate sized gaps (40 inches).

The increase of the magnitude of the effect of the gap with increased yaw angle is undoubtedly due to the strong cross flow which is observed to form in the gap. This appears to increase the flow separation on the leeward side of the trailer.

STREAMLINING

It was shown in Figure 2, and can be seen from Tables 1 and 2, that a vehicle having a trailer with well-rounded front edges has significantly less drag than one with square edges. Thus it appears that streamlining the trailer will diminish the drag of the tractor-trailer combination. Of interest is the amount of rounding of the forward edges of the trailer which is necessary to produce sugnificant drag reduction.

It was shown by Hoerner[17] that the aerodynamic drag of a long box whose axis is parallel with the wind decreases rapidly as the ratio of the radius of the forward edges to the vehicle width is increased. When this ratio reaches 0.1, no further significant decrease in the drag occurs due to further increases in edge radius.

Results similar to those given by Hoerner were found by Sherwood and by Windsor for trailers in wind-tunnel tests of 1/6th-scale tractor-trailer trucks. Windsor showed that there is very little difference in the drag reductions brought about by a 6-inch radius (radius/width = 0.063) and an 18-inch radius (radius/width - 0.19) on the side and top front edges of a trailer at yaw angles of up to 15°. At larger yaw angles (up to 30°), the 18 inch radius produced greater drag reductions.

The amount of the drag reduction due to rounding the exposed forward edges of the trailer depends to a large extent on the size and shape of the tractor. A large unstreamlined tractor would cause the trailer to be shielded from the oncoming air and thus the shape of the front of the trailer would be less important. This is pointed out by Sherwood, who found a 42% reduction in drag due to rounding the front edges of a trailer when used in combination with one tractor, but only a 2% reduction when used with a trailer-height tractor.

Streamlining the front of the tractor may result in increasing the drag of the tractor-trailer combination. This startling result is shown in Figure 4 in which C_D is plotted versus yaw angle for a COE tractor before and after its front vertical edges had been increased in radius from 4 inches to 10 inches below the windshield and 8 inches along the windshield. At zero yaw, C_D increased from 0.69 to 0.75. The vehicle with the more rounded tractor front continued to have a higher drag than the other until a yaw angle of about 4°. At above 4°, the vehicle with the more streamlined tractor had the lower drag. The key to understanding this effect is that of understanding the flow interaction between the tractor and the trailer. Observations of tufts on the front of the trailer showed that when the vertical edges on the front of the tractor were rounded, the stagnation point on the front of the trailer was lowered. Thus the trailer was less sheltered by the wake of the tractor at low yaw angles, and, because of its high drag shape, its drag increased more than that of the tractor decreased. At higher yaw angles, the trailer was less shielded by the tractor, and thus its drag was less affected by the tractor's flow.

The flow interaction between the tractor and the trailer is quite complicated, so there is no guarantee that streamlining the tractor will have the above effect, particularly at small yaw angles. Figure 5 shows a simplified 1/8th scale model of a tractor-trailer combination very similar to the COE vehicle whose characteristics were mentioned in the previous paragraph (see Figure II-I-2 of reference 4 for the tractor dimensions). Figure 6 shows how C_D of this vehicle changed as the radius of the front vertical edges of the tractor was increased. At zero yaw, the drag decreased, increased, decreased and then increased again as the full-scale radius was increased from zero to 18 inches. As yaw angle increased and the effect of the tractor on the flow about the trailer changed, the waviness of the $C_D - \psi$ curves diminished. At 10° yaw, the drag pretty much decreased with increasing radius up to 9 inches, and then (except for a small dip at 9 inches) remained constant.

Since the drag of tractor-trailers has been shown to be a function of gap size, a simple way of streamlining the vehicle is to make the distance separating the tractor from the trailer as small as possible. Another way, though not practical because it limits the vehicles maneuverability, is to fair the tractor into the trailer so that no gap exists between them and there is not a sudden change in height between the tractor and the trailer. Such a design eliminates cross flow in the gap and prevents separation on the top and sides of the trailer. This essentially changes the truck's configuration to that of a bus. The effectiveness of the fairing has been verified by a number of investigators [3,4,7,8]. Drag reductions are typically of the order of 25% at zero yaw.

A further attempt at streamlining tractor-trailer trucks has been to eliminate some of the protrusions which are sources of drag. Generally these contribute only a small amount to the total. Flynn and Kyropoulos[14] found that removing the mirrors from a conventional tractor reduced the drag by 1.5%, and that removing all of the details to form a simplified cab reduced the drag by 5% at zero yaw. Kirsch and Bettes[1] found that underbody details on the tractor and trailer are responsible for less than 1% of the total aerodynamic drag.

Exhaust and air inlet stacks are prominent protrusions on most tractor-trailers. Because of their high-drag shapes, it would be expected that their removal would decrease the vehicle's drag. However, it has been found[15] that removing these stacks increased the drag of a 1/8th-scale model of a vehicle with a COE tractor. The drag coefficient at zero yaw increased from 0.62 to 0.72 - a 16% increase. The difference between the drag with stacks on and off decreased as yaw angle increased. It was 5% at 10°. Measurements showed that removing the exhaust stacks decreased the drag of the tractor but increased the drag of the trailer. The cause of the large effects of the exhaust stacks is not known. It may have been the sheltering effect of their wake, or it may have been that the turbulence induced by them increased the turbulence in the boundary layer on the trailer so that it did not separate as easily.

Many ways of streamlining tractor-trailer trucks are not practical even though they would reduce the drag. Such modifications as skirts along the sides of the vehicle affect brake cooling and long boat tails at the rear of the trailer to prevent afterbody separation are not practical (and don't always reduce drag[3]). Despite its impracticability, Windsor[3] built and tested a "fully streamlined" wind tunnel model of a vehicle which had a well-rounded tractor, a complete fairing between the tractor and the trailer, skirts along the bottom of the tractor and the trailer and a beaver tail (pointed tail) at the back of the trailer. This vehicle had a drag coefficient at zero yaw which was only 0.17 (about 1/5th of that measured on a conventional tractor-trailer), and which, as mentioned earlier, became negative at high yaw angles.

DRAG REDUCING ADD-ON DEVICES

Recent years have seen the advent of a number of devices to be added onto tractor-trailer trucks to reduce their aerodynamic drag. Because most of the drag of these vehicles is due to forebody flow separation, all of these devices have been designed to reduce forebody drag. The two most popular classes of these devices are those which are mounted on the top of the tractor and which deflect or guide the air so that it does not impinge on the front of the trailer, and a pillow-shaped device which is fastened onto the front of the trailer to make the front of the trailer more streamlined. The latter device acts in the same way as well-rounded edges on the top and upper sides of the trailer and thus should produce drag reductions comparable to those produced by them. Other devices include air vanes (turning vanes) mounted on the front of the trailer, a rounded lip to be mounted on the front of the trailer, and a gap seal (a vertical plate mounted between the tractor and the trailer to stop the cross flow in the gap). Of these three, the first two act to prevent separation on square leading edges of the trailer and thus, at best, are comparable to rounding the forward edges of the trailer. Evaluations of several of these devices appear in the literature[1,4,18,19].

An example of a very effective drag-reducing device, one which we have developed, is a streamlined fairing which mounts on the top of a tractor[2,20]. This device, in combination with a gap seal, is shown in Figure 7. When combined with a gap seal it has produced reductions in the wind-averaged drag coefficient of as much as 30% ($\Delta C_D = 0.27$). As with all drag-reducing devices, the drag reductions produced vary significantly with vehicle configuration.

Because of its streamlined shape, the fairing is less sensitive to side winds and atmospheric turbulence than other roof-mounted devices which we have tested, and is unaffected by wind turbulence when used with a gap seal. The addition of a streamline fairing to the roof of the tractor increases the drag of the tractor (this includes the drag force on the fairing) but decreases the drag on the trailer to a much greater extent[15]. Because the top of the fairing is even with the top of the trailer, and because it guides the air smoothly around the trailer, the combination of tractor plus fairing is much like a trailer-height tractor. As in the case of the vehicle with the trailer-height tractor tested by Flynn and Kyropoulos, the pressure in the gap was greatly reduced and the drag on the trailer became zero at zero yaw.

MODEL TESTING AND TURBULENCE

Most of the results reported in this paper have been obtained from wind-tunnel tests of models of various sizes. Thus the question of scale effects must arise. In most cases reported herein the tests have been run at Reynolds numbers based on vehicle width of 10^6 or greater. A Reynolds number of 10^6 is about one-quarter of that of a full-scale truck at 50 mph. Comparison of wind tunnel data with full-scale data[1,2] indicates that wind tunnel results at a Reynolds number of 10^6 generally give results which agree with data from full-scale vehicles. There is still some question - yet unresolved - about the effect of local Reynolds number for flow about edges and small members. Testing at Reynolds numbers below 10^6 is not recommended [11].

The size of the model relative to the cross section of the test section of the wind tunnel is important if blockage effects are to be small. Blockage corrections for complex shapes such as tractor-trailer trucks are generally unknown and wind tunnel results are generally uncorrected. The recommendation of the DOT/SAE Truck and Bus Fuel Economy Measurement Conference[11] was that blockage should not exceed 4%.

The question of moving ground boards to eliminate boundary layer build-up under the models is often brought up. Most investigators feel that they are not necessary[3,20].

Atmospheric turbulence appears to affect the drag of tractor-trailer trucks. In particular, some drag reducing devices yield lower drag reductions when tested in a windy environment on full-scale vehicles than when tested in a relatively low turbulence wind tunnel[2]. The addition of a gap seal seems to help eliminate much of the detrimental effect of atmospheric turbulence by stabilizing the wake of the devices.

SUMMARY

The interaction of the flow between the tractor and trailer greatly complicates the aerodynamics of tractor-trailer trucks. Large drag reductions can be obtained by streamlining and by use of add-on drag reducing devices, but great care must be exercised or the drag may be increased instead of decreased. In particular, the yaw behavior of tractor-trailers and drag-reducing add-ons must be considered since road vehicles often are subject to side winds sufficient to produce substantial yaw angles.

REFERENCES

1. Kirsch, J.W. and Bettes, W.H., "Feasibility Study of the S^3 Air Vane and Other Truck Drag Reduction Devices", Proc. of the NSF/DOT Conference Workshop on Reduction of the Aerodynamic Drag of Trucks, Cal. Inst. of Technology, Pasadena, CA, Oct. 1974.

2. Buckley, F.T. Jr., Marks, C.H., and Walston, W.H., Jr., "Analysis of Coast Down Data to Assess the Aerodynamic Drag Reduction on Full-Scale Tractor-Trailer Trucks", SAE Trans., Vol. 85, 1976.

3. Windsor, R.I. Wind Tunnel Test of Trailmobile Trailers, 2nd Series, Univ. of Maryland Wind Tunnel Report No. 98, Univ. of Maryland, College Park, Md., 1953.

4. Buckley, F.T. Jr., Marks, C.H., and Walston, W.H. Jr., A Study of Aerodynamic Methods of Improving Truck Fuel Economy, Final Rpt. to NSF under Grant No. SlA-74-14843, Department of Mechanical Engineering, Univ. of Maryland, College Park, Md., Dec. 1978.

5. Lissaman, P.B.S., editor, Proc. of the NSF/DOT Conference/Workshop on Reduction of Aerodynamic Drag of Trucks, Cal. Inst. of Technology, Pasadena, CA, 1974.

6. Mason, W.T. Jr., "Wind Tunnel Development of the Dragfoiler - A System for Reducing Tractor-Trailer Aerodynamic Drag", SAE Paper No. 750705, West Coast Meeting, Seattle, WA, 1975.

7. Lissaman, P.B.S., and Lambie, J.H., "Reduction of Aerodynamic Drag of Large Highway Trucks", Proc. of the NSF/DOT Conference/Workshop on Reduction of the Aerodynamic Drag of Trucks, Cal Inst. of Technology, Pasadena, CA, 1974.

8. Stears, L.L., and Saltzmann, E.J., "Reduced Fuel Consumption Through Aerodynamic Design", AIAA Journal of Energy, Vol. 1, No. 5, 1977.

9. Marks, C.H., Buckley, F.T. Jr., and Walston, W.H, Jr., "An Evaluation of the Aerodynamic Drag Reduction Produced by Various Cab Roof Fairings and a Gap Seal on Tractor-Trailer Trucks", SEA Trans., Vo. 85, 1976.

10. Sherwood, W.A., Wind Tunnel Test of Trailmobile Trailers, Univ. of Maryland Wind Tunnel Report No. 85, Univ. of Maryland, College Park, MD., 1953.

11. DOT/SAE Truck and Bus Fuel Economy Measurement Conference Report, SAE Rpt. No. P-59.

12. Buckley, F.T. Jr., Marks, C.H., and Walston, W.L. Jr., "An Assessment of Drag Reduction Techniques Based on Observations of Flow Past Two-Dimensional Tractor-Trailer Models", Proc. of the NSF/DOT Conference/Workshop on Reduction of the Aerodynamic Drag of Trucks, Cal. Inst. of Technology, Pasadena, CA. 1974.

13. Marks, C.H. and Buckley, F.T. Jr., "A Study of the Base Drag of Tractor-Trailer Trucks", J. of Fluids Engr., Vol. 100, No. 4, 1978.

14. Flynn H. and Kyropoulos, P. "Truck Aerodynamics", SAE Trans. Vol. 73, 1963.

15. Marks, C.H., and Buckley, F.T. Jr., "The Effect of Tractor-Trailer Flow Interaction on the Drag and Distribution of Drag of Tractor-Trailer Trucks", SAE Trans., 1981.

16. Buckley, F.T. Jr., and Marks, C.H., "A Wind Tunnel Study of the Effect of Gap Flow and Gap Seals on the Aerodynamic Drag of Tractor-Trailer Trucks", J. of Fluids Engr. Vol. 100, No. 4, 1978.

17. Hoerner, S.F., Fluid Dynamic Drag, 2nd edition, published by author, Greenbrier, N.J., 1965.

18. Cooper, K.R., "A Wind Tunnel Investigation Into the Fuel Savings Available from the Aerodynamic Drag Reductions of Trucks", DME/NAE Quarterly Bulletin No. 1976(3) Ottowa, Canada, 1976.

19. Buckley, F.T. Jr., and Sekscienski, W.S., "Comparisons of Effectiveness of Commercially Available Devices for the Reduction of Aerodynamic Drag of Tractor Trailers", SEA Paper No. 750704, West Coast Meeting, Seattle, WA., 1975.

20. Sevran, G., Proc. of the Symposium on Aerodynamic Drag Mechanisms of Bluff Bodies and Road Vehicles, General Motors Research Laboratories, Warren, MI, 1976 (Plenum Press 1978).

TABLE 1

Breakdown of Drag, COE Tractor

C_D = Overall Drag Coefficient

C_{D_C} = Drag Coefficient of Tractor,

C_{D_T} = Drag Coefficient of Trailer,

Configuration	Yaw Angle (degrees)	C_D	C_{D_C}	C_{D_T}	C_D/C_D	C_{D_T}/C_D
1973 "IHC" COE tractor coupled to 12.5 ft-high trailer with rounded (12"-radius) front edges.	0	0.62	0.51	0.11	0.82	0.18
	5	0.76	0.54	0.22	0.71	0.29
	10	0.96	0.63	0.33	0.56	0.34
	15	1.17	0.67	0.50	0.57	0.43
	20	1.26	0.75	0.51	0.60	0.40
Same as above, but with square front-edged trailer	0	0.63	0.51	0.12	0.81	0.19
	5	0.79	0.54	0.25	0.68	0.32
	10	1.02	0.63	0.39	0.62	0.38
	15	1.23	0.67	0.56	0.55	0.45
	20	1.38	0.75	0.63	0.54	0.46
Same as 1st entry, but with 13.5 ft-high round-edged trailer	0	0.69	0.51	0.18	0.74	0.26
	5	0.82	0.54	0.28	0.66	0.34
	10	0.97	0.63	0.34	0.65	0.35
	15	1.17	0.67	0.50	0.57	0.43
	20	1.26	0.75	0.51	0.60	0.40
Same as 1st entry, but with 13.5 ft-high square-edged trailer	0	0.70	0.51	0.19	0.73	0.27
	5	0.81	0.54	0.27	0.67	0.33
	10	1.00	0.63	0.37	0.63	0.37
	15	1.22	0.67	0.55	0.55	0.45
	20	1.29	0.75	0.54	0.58	0.42

TABLE 2

Breakdown of Drag, Conventional Tractor

C_{D_C} = Drag Coefficient of Tractor

C_{D_T} = Drag Coefficient of Trailer

Configuration	Yaw Angle (degrees)	C_D	C_{D_C}	C_{D_T}	C_{D_C}/C_D	C_{D_T}/C_D
1974 IHC, Conventional tractor coupled to 12.5 ft-high round-edged trailer	0	0.75	0.43	0.22	0.57	0.43
	5	0.84	0.45	0.39	0.54	0.46
	10	0.98	0.47	0.51	0.48	0.52
	15	1.06	0.53	0.53	0.50	0.50
	20	1.12	0.53	0.69	0.47	0.53
Same as above, except with 13.5 ft-high round-edged trailer	0	0.74	0.39	0.35	0.53	0.47
	5	0.86	0.43	0.43	0.50	0.50
	10	0.99	0.49	0.50	0.49	0.51
	15	1.05	0.57	0.48	0.54	0.46
	20	1.12	0.59	0.53	0.53	0.47
Same as 1st entry, except with 13.5 ft-high square-edged trailer	0	0.88	0.40	0.48	0.45	0.55
	5	0.98	0.44	0.58	0.45	0.55
	10	1.10	0.45	0.65	0.41	0.59
	15	1.21	0.55	0.66	0.45	0.55
	20	1.27	0.60	0.67	0.47	0.63

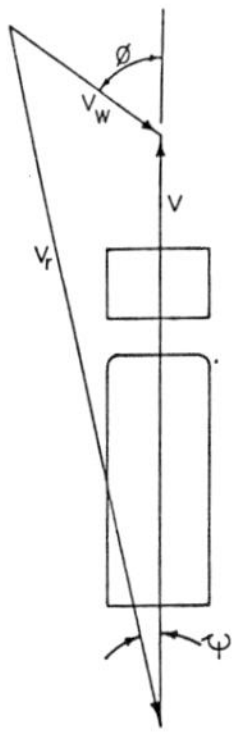

FIGURE 1 Relative Air Velocity Diagram

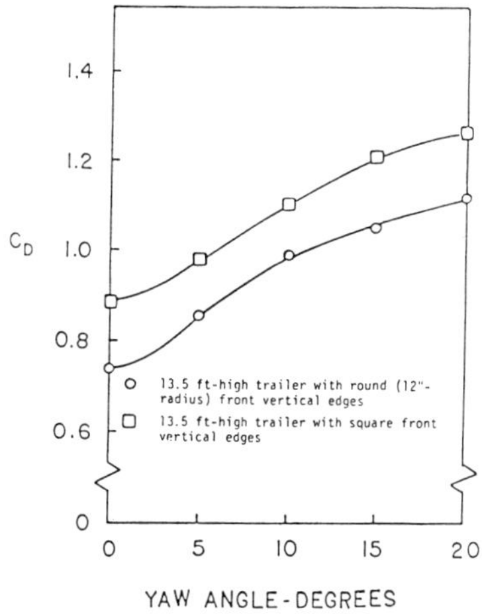

FIGURE 2 Body-Axis Drag Coefficient vs Yaw Angle for Tractor-Trailer Truck with a 1974 International Harvester Tractor. (1/8th scale, $Re=10^6$)

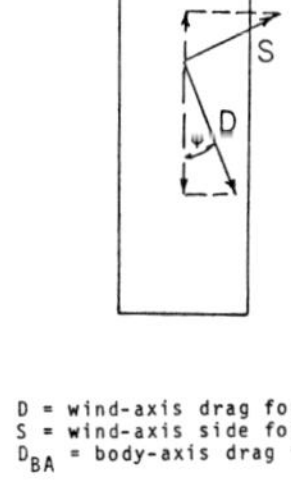

D = wind-axis drag force
S = wind-axis side force
D_{BA} = body-axis drag force

$D_{BA} = D \cos\psi - S \sin\psi$

FIGURE 3 Contribution of Wind-Axis Forces to Body-Axis Drag

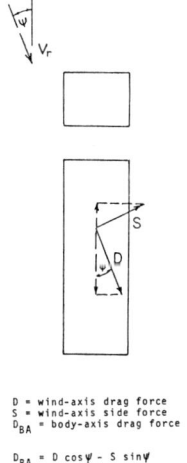

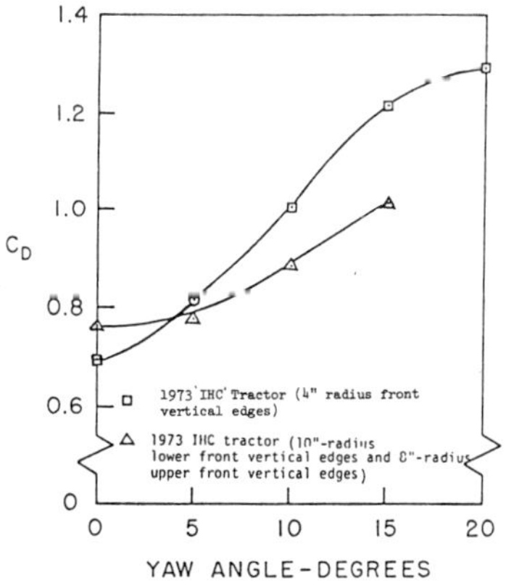

Figure 4 Body-Axis Drag Coefficient vs Yaw Angle for Tractor Trailer with a COE Tractor and a 13.5 ft. high Round-Edge Trailer (1/8th Scale, $Re = 10^6$)

Figure 5 1/8th Scale Tractor-Trailer Truck with Simplified Tractor

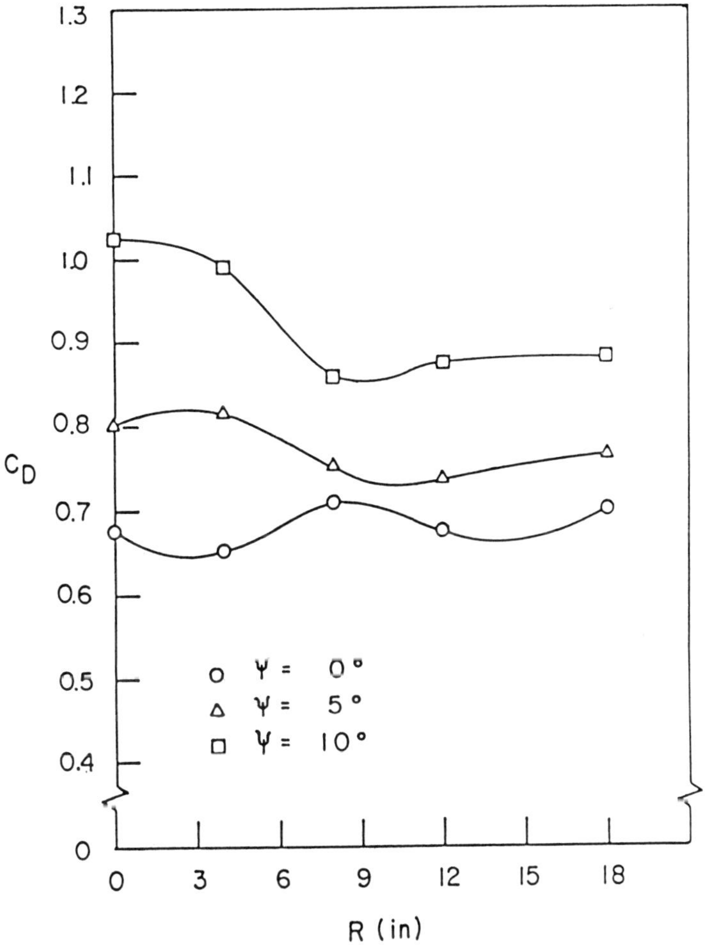

Figure 6 Effect of Tractor Front Vertical Edge Radius on Drag Coefficient

FLUID MECHANICS SIMULATION OF THE EFFECT OF COMBUSTION-RELATED POLLUTANTS ON THE FOG FORMATION

R. J. Hung*
The University of Alabama in Huntsville
Huntsville, Alabama

R. E. Smith**
NASA/Marshall Space Flight Center
Huntsville, Alabama

ABSTRACT

One of the most noticeable effects of air pollution on the properties of the atmosphere is the reduction in visibility. This paper reports the results of investigations of the fluid dynamical and microphysical processes involved in the formation of advection fog on aerosols from combustion-related pollutants, as condensation nuclei. The effects of a polydisperse aerosol distribution, on the condensation/nucleation processes which cause the reduction in visibility are studied. The results show that an aerosol population with a high particle concentration provides more favorable conditions for the formation of a denser fog that an aerosol population with a larger particle size, if the value of the mass concentration of the aerosols is kept constant . The effect of the aerosol particle size distribution on the wet removal nucleation processes is also discussed.

INTRODUCTION

A numerical simulation of advection fog formation on atmospheric aerosols from combustion-related pollutants, Hung and Liaw (1980) considered a

*Professor of Mechanical Engineering
**Deputy Chief, Division of Atmospheric Science

mono-disperse, multi-composition aerosol model for the initial aerosol population. Results show that the major contribution of combustion-related pollutants as condensation nuclei in the formation of fog comes from SO_x aerosols. However, NO_x also makes a considerable contribution.

In the present paper we generalize the analysis made by Hung and Liaw (1980) to include the effects due to a polydisperse distribution of atmospheric aerosols. The total water content in the atmosphere due to the condensation process is also investigated. This could provide an understanding of the acid rain problem, which is closely associated with combustion-related pollutants.

Acid rain is a major environmental problem on both sides of the Atlantic Ocean. Originally noticed and studied in the Scandinavian countries and in Canada, acid rain has been documented in this country; first in the Northeast and now throughout much of the United States. Increasing levels of acidity have already caused measurable damage to the environment. Many lakes are now totally devoid of fish population; monuments and other man-made structures are being degraded; and yields of agricultural crops and forests may be decreasing.

Acid rain, based on the definition of the terminology by the Environmental Protection Agency, includes the wet removal processes such as rain and snow; the dry deposition component such as particulate sedimentation and/or gaseous absorption and adsorption; and special events such as fog, frost and dew which also contribute to the overall acidic deposition in amounts that need to be quantified.

Strong evidence indicates that the major NO_x producer is the automobile. There is no doubt that stationaery fuel combustion sources also contribute a heavy fraction of the total NO_x emitted to the atmosphere. Fuel-bound nitrogen compounds are ammonia, pyridine, and many other amines (Glassman, 1977). Similar to NO_x production by combustion, SO_x is also produced through fuel combustion since S is an inherent impurity in both coal and petroleum. This SO_x and NO_x from energy-related fuel combustion is, in turn, transformed into aerosol particles through photochemical reactions (Council on Environmental Quality, 1973.)

Energy conversion processes account for 84% of the SO_x, 97% of the NO_x and 78% of the CO emissions based on the estimation made by the Council on Environmental Quality (1973). However, the figures are quite different for data chosen from anthropogenic source points of view. For example, volcanic eruption, such as ash coughed up by Mount St. Helens, Washington, on May 18, 1980, could account for more than 50% of SO_2, thunderstorms for more than 75% of NO_x, etc. Since the dispersal of air pollution is unaffected by national boundaries, the emission of pollutants often results in global problems and concerns.

It is most economical to use computer simulation techniques to study the characteristics of warm fog , which is the most stable and also accounts for about 95% of all fog occurrences worldwide (Sax et al, 1975), in particular, how the microstructure of advection fog is influenced by the saturation spectrum of the CCN in the polluted atmosphere. Recently, Baker (1977) and Eadie et al. (1975) proposed boundary layer models of advection fog. In these models, the effect of the cloud condensation nuclei (CCN) on water vapor nucleation was not taken into account.

Contrary to the boundary layer models treated by previous investigators, a one-dimensional Lagrangian model of the formation and evolution of droplet spectra in advection fog was proposed by Fitzgerald (1978). This model incorporates the physical processes governing the formation and evolution of fog droplets from a population of CCN contained in a parcel which is advected with the wind. The effects of vertical diffusion transfer, momentum transfer, and energy transfer are completely ignored in this model.

Recently, Hung et al. (1979) proposed a two-dimensional computer model of warm fog for studying fog formation and the evolution of droplet spectra. A description of the vertical and horizontal turbulent transfer of heat, moisture and momentum, together with the growth of fog droplets from a microphysical point of view, has been included in this model. Hung and Liaw (1980) used the mathematical model proposed by Hung et al (1979) to study the effect of a polluted atmosphere on advection fog formation and the evolution of droplet spectra. In this model, both advection and infrared effects are also taken into account. The driving force for the fog formation is a prescribed time rate of cooling of the ground.

The correlation of fog formation and CCN with different chemical compositions was investigated by Hung et al. (1979), and compared with the results obtained using NaCl nuclei. It was found that the value of the ratio of the Van't Hoff factor to the molecular weight plays a very important role in fog formation during the nucleation and condensation time period. Hung et al. (1979) concluded that hygroscopic chemicals with a high ratio of the Van't Hoff factor to the molecular weight provide more favorable conditions for denser advection fog than chemicals with lower values of the ratio.

In a computer simulation Hung and Liaw (1981) compared the formation of advection fog in a polluted atmosphere with the formation in a clean atmosphere. The results show that the condensation nuclei in a polluted atmosphere provide more favorable conditions for the production of a dense advection fog than the condensation nuclei in a clean atmosphere. It was also shown that attaining a certain degree of supersaturation is not necessarily required for the formation of advection fog when the condensation nuclei are monodisperse and the atmosphere is polluted.

To develop a mathematical model closer to the real atmosphere, a polydisperse condensation nuclei distribution may by necessary. In a study of fog optical properties and the size distribution of water droplets, Low et al. (1978) examined the effects of different spectral shapes and widths on gamma and lognormal distributions with the same liquid water content, and concluded that the drop size distribution greatly affects the results of visibility even though the liquid water content is the same. Observations show that particle size distributions for combustion-related aerosols are widely spread. In the present paper, we are interested in studying the effect of varying the particle size distributions of aerosols as condensation nuclei for fog formation while keeping the mass concentration fixed.

SO_x and NO_x are the major sources of acid rain through the transformation of gas molecules to aerosol particles via photochemical reactions and condensation processes in which the aerosol particles serve as condensation nuclei. The present paper, in particular, deals with the mathematical simulation of the formation of advection fog and the accumulation of liquid water content in the atmosphere through the condensation processes in which the effects of the initial polydisperse multi-composition aerosol distribution are taken into consideration.

MATHEMATICAL MODEL

A theoretical model is employed which describes the evolution of potential temperature, water vapor content, liquid water content, and horizontal and vertical winds as determined by the processes of vertical turbulent transfer and horizontal advection of momentum, energy and moisture, as well as radiation cooling, growth of water droplets based on microphysical processes, and drop sedimentation.

The mathematical model is two-dimensional in the X-Z plane , and the boundary layer model is assumed. All the quantities are uniform in the Y direction. The diffusivity coefficient is the same for liquid water droplets as for vapor. This is based on the assumption that eddy diffusion from the liquid water to vapor behaves similarly to eddy diffusion from vapor to liquid water.

The fundamental equations governing the macrophysical processes of the evolution of wind components, water vapor content, liquid water content and potential temperature under the influences of vertical turbulent diffusion transfer, turbulent momentum transfer and turbulent energy transfer can be expressed in the following three sets of conservation equations:

(a) Mass conservation equation of the air and diffusion equations for water vapor and liquid water contents

$$\frac{\partial U_x}{\partial x} + \frac{\partial U_z}{\partial z} = 0 \tag{1}$$

$$\frac{\partial C_{va}}{\partial t} = - U_x \frac{\partial C_{va}}{\partial x} - U_z \frac{\partial C_{va}}{\partial z} + \frac{\partial}{\partial z}\left(K_d \frac{\partial C_{va}}{\partial z} \right) - F_{cv} \tag{2}$$

$$\frac{\partial C_{wa}}{\partial t} = - U_x \frac{\partial C_{wa}}{\partial x} - U_z \frac{\partial C_{wa}}{\partial z} + \frac{\partial}{\partial z}\left(K_d \frac{\partial C_{wa}}{\partial z} \right)$$

$$+ \frac{\partial}{\partial z}(U_t \, C_{wa}) + F_{cw} \tag{3}$$

(b) Momentum conservation equations

$$\frac{\partial U_x}{\partial t} = - U_x \frac{\partial U_x}{\partial x} - U_z \frac{\partial U_x}{\partial z} - \frac{1}{\rho}\frac{\partial P}{\partial x} + \frac{\partial}{\partial z}\left(K_m \frac{\partial U_x}{\partial z} \right) + f(U_y - U_{gy}) \tag{4}$$

$$\frac{\partial U_y}{\partial t} = - U_x \frac{\partial U_y}{\partial x} - U_z \frac{\partial U_y}{\partial z} + \frac{\partial}{\partial z}\left(K_m \frac{\partial U_y}{\partial z} \right) + f(U_{gx} - U_x) \tag{5}$$

$$\frac{\partial U_z}{\partial t} = - U_x \frac{\partial U_z}{\partial x} - U_z \frac{\partial U_z}{\partial z} + \frac{\partial}{\partial z}\left(K_m \frac{\partial U_z}{\partial z} \right) + g\left(\frac{\theta - \theta_o}{\theta_o} \right) \tag{6}$$

(c) Energy conservation equation

$$\frac{\partial \theta}{\partial t} = - U_x \frac{\partial \theta}{\partial x} - U_z \frac{\partial \theta}{\partial z} + \frac{\partial}{\partial z}\left(K_h \frac{\partial \theta}{\partial z} \right)$$

$$+ \frac{1}{\rho C_p}\left(\frac{1000}{P}\right)^{2/7}\left(L \, F_{cv} \rho - \frac{\partial R_f}{\partial z} \right) \tag{7}$$

where

$$C_{va} = \frac{\rho_v}{\rho} \tag{8}$$

$$C_{wa} = \frac{4\pi}{\rho}\left[\rho_w \sum_i \int_0^\infty N_i(r) r^2 dr - \frac{1}{3} \sum_i \rho_{hi} N_i r_{oi}^3\right] \tag{9}$$

Here, U_x is the wind component in the horizontal x-direction; U_y, the wind component in the horizontal y-direction; U_z, the wind component in the vertical z-direction; C_{va}, water vapor-air mixing ratio; C_{wa}, liquid water-air mixing ratio; K_m, turbulent exchange coefficient for vertical turbulent transfer of momentum; K_d, turbulent exchange coefficient for turbulent transfer of water vapor diffusion and liquid water diffusion; K_h, turbulent exchange coefficient for turbulent transfer of heat flux; f, Coriolis parameter; U_{gx}, x-component of the velocity of geostrophic wind; U_{gy}, y-component of the velocity of geostrophic wind; F_{cv}, source function for condensation and evaporation for water vapor; F_{cw}, source function for condensation and evaporation for liquid water; U_t, mean termial velocity of fog drops; ρ, density of dry air; ρ_v, water vapor density; ρ_w, liquid water density; ρ_h, density of hygroscopic nucleus; N, number density of water droplet which is a function of droplet size; r, radius of water droplet; r_o, radius of nucleus; θ, the potential temperature; θ_o, potential temperature of dry adiabatic value; and R_f, net upward flux of infrared radiation.

The change of the water droplet radius, in Equation 9, due to condensation/evaporation is governed by microphysical processes and is given by (Carstens et al., 1974).

$$\frac{dr}{dt} = \frac{\rho_{sat}(\infty)\, D_{eff}}{\rho_w(r+\ell)} \left(S - 1 - \frac{a}{r} + \frac{c}{r^3}\right) \tag{10}$$

$$D_{eff}^{-1} = D^{-1} + \left(\frac{bL}{k} \right) \tag{11}$$

$$\ell = \left(\frac{\ell_\beta}{D} + \frac{\ell_\alpha bL}{k} \right) D_{eff}^{-1} \tag{12}$$

$$\ell_\alpha = \left(1 - \frac{\alpha}{2} \right) \frac{k}{\alpha p} \frac{\gamma-1}{\gamma+1} \left(\frac{8\pi T}{R} \right)^{\frac{1}{2}} \tag{13}$$

$$\ell_\beta = \left(1 - \frac{\beta}{2} \right) \frac{D}{\beta} \left(\frac{2\pi}{R_v T} \right)^{\frac{1}{2}} \tag{14}$$

$$a = \frac{2\sigma}{R_v T \rho_w} \tag{15}$$

$$c = \frac{3iM_w m_s}{4\pi M_s} \tag{16}$$

Here $\rho_{sat}(\infty)$ is the saturation vapor density at infinity; S, the saturation ratio of water vapor in air; r, the radius of the droplet; ℓ, characteristic length which is the weighted average of ℓ_α and ℓ_β; b, the slope of a linearized $\rho_{sat}(\infty)$ and T curve; D, the diffusion coefficient of water vapor in the air; k, the thermal conductivity of the mixture; α, the accommodation coefficient which is 0.67 in this case; β, the sticking coefficient which is 0.3536 in this case; R, the gas constant of dry air; σ, the surface tension of water; i, the Van't Hoff factor; M_w, the molecular weight of water; m_s, the mass of hygroscopic material dissolved; M_s, gram molecular weight of hygroscopic nucleus; and R_v, gas constant of water vapor

The surface tension of pure water decreases as the temperature increases. In addition, evaporation is greatly different over a surface with pure water than over a surface covered with a monolayer (Hoffer and Mallen, 1970). A Skylab experiment involving water droplets indicated that the surface tension of the water droplet decreased drastically from 72.7 dyne cm^{-1} to less than 50 dyne cm^{-1} at 20°C for droplets contaminated with grape drink and strawberry drink (Hung it al., 1976). Therefore, the decrease in the surface tension of water droplet exposed to a polluted atmosphere must be taken into account.

ADVECTION FOG IN A POLLUTED ATMOSPHERE BASED ON POLYDISPERSE, SINGLE-COMPOSITION MODEL

In the present study we consider a polydisperse particle size distribution aerosol population in contrast to the monodisperse aerosol distribution considered by Hung and Liaw (1980 and 1981).

Three types of particle size distribution are considered while keeping the total mass concentration at a fixed value. To simplify the calculation, and also to make the comparison of results easier, a combination of three step functions for each distribution is chosen. These three types of particle size distribution are expressed in Figure 1.

These distributions can be shown as follows:

(1) Type A Distribution:

$N = 8100$ particles/cm^3, for $r = 0.075$ μm;

$N = 4000$ particles/cm^3, for $r = 0.100$ μm;

$N = 1700$ particles/cm^3, for $r = 0.125$ μm;

(2) Type B Distribution:

$N = 3225$ particles/cm^3, for $r = 0.075$ μm;

$N = 4500$ particles/cm^3, for $r = 0.100$ μm;

$N = 2500$ particles/cm^3, for $r = 0.125$ μm;

(3) Type C Distribution:

$N = 2150$ particles/cm^3, for $r = 0.075$ μm;

$N = 3000$ particles/cm^3, for $r = 0.100$ μm;

$N = 3500$ particles/cm^3, for $r = 0.125$ μm;

where N stands for the number density, and r is the radius of the nucleus. The total initial value mass concentration of the aerosol particles as for Types A, B and C distributions is constant, 54.9 μg/m^3. With keeping the mass concentration fixed, the total number densities of Types A, B and C are 13,800, 10,225, and 8,650 particles/cm^3, respectively. The total aerosol concentration of number density varies widely, from as low as 10^3 particles/cm^3 in clean country air to over 10^5 particles/cm^3 in heavily polluted areas (Dennis, 1980). The examples are in the category of aerosols in polluted areas.

Direct identification of the nuclei actually contained in fog and cloud droplets indicated that the predominant particles were combustion nuclei and sea salt (Mason, 1971). In highly industrialized areas, large quantities of atmospheric aerosols of $SO_4^=$, NO_3^-, NH_4^+ and Cl^- were detected. In this study, the chemical species, $(NH_4)_2SO_4$, a typical aerosol particle observed in industrialized areas, was chosen. The mathematical simulation of the single-composition aerosol, NO_x, is not given here because the mass concentration of NO_x is only one-third that of SO_x (Hidy and Brock, 1971). In other words, NO_x, as a condensation nucleus for fog, may not be as effective as SO_x in the real atmosphere; however, the existence of NO_x in the real atmosphere may enhance the fog formation with SO_x as being the major contributor to the condensation nuclei produced by energy-related fuel combustion.

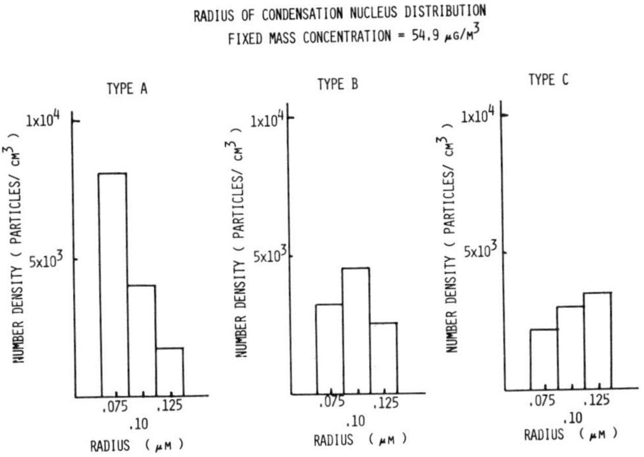

Figure 1. Three types of aerosol population distributions. Each distribution is a combination of three step functions in which the total mass concentration is conserved and equal to 54.9 μg/m^3.

The numerical techniques used in the application of the macroscopic equations, including the turbulent diffusion and other related physical parameters to the study of the dynamical behavior of warm fog, is given in Hung and Vaughan (1977). The method of implicit integration of the finite difference equation was used to solve the governing partial differential equations, a technique essentially similar to that of Richtmyer (1967). The utilization of the microscopic equations, as shown in Hung and Liaw (1980), in solving the initial stage of the nucleation and the growth of water droplets on aerosol particles is discussed in Hung et al. (1978).

In the present paper, the initial values of the wind profile, temperature profile, and the relative humidity are from the results of field measurements taken along the California coast by Mack et al. (1973, 1974 and 1975). The similar initial values from the field measurements have also been used in the fog formation studies, with detailed descriptions, given by Hung et al. (1979).

In this paper, the computations were carried out with a one-hour time coordinate after the fundamental equations were solved simultaneously in conjunction with the initial values adopted from the field measurements provided by Mack et al. (1973, 1974, and 1975). The one hour time coordinate as for the purpose of making all the physical parameters involved in the model self-consistent. In the horizontal geometrical x-coordinate, the initial temperature at the ground was 14°C everywhere. After a one hour time computation, the ground temperature decreased one degree linearly in one hour at the horizontal z-coordinates 3 km $\leq$ x $\leq$ 6 km. The results in this paper began at the time the ground temperature started to decrease linearly.

The densities of the fog, in terms of visibility, are computed for a variety of aerosol particle size distributions with the total mass concentration kept at a constant value. Ammonia sulfate is chosen in this paper as an example of a species of the air pollutant, SO_x, which dominates the combustion-related air pollution as indicates by Hidy and Brock (1971).

Figure 2 illustrates the relative humidity and visibility profiles at two horizontal locations, x = 3 and 6 km at z = 1 m during the formation of advection fog from a type A aerosol distribution, shown in Figure 1. Figure 3 shows similar relative humidity and visibility profiles at z = 4 m, under the same conditions as Figure 2. These two figures shown that advection fog with visibility below 1000 m was formed at time t = 52 min and relative humidity s = 98.6% at z = 1 m; and at time t = 86 min and relative humidity s = 97.8% at z = 4 m.

Similar computations were also made for aerosol particles $(NH_4)_2SO_4$ with size distribution types B and C, as shown in Figure 1. Figures 4 and 5 show similar relative humidity and visibility profiles at two different altitudes z = 1 and 4 m, respectively, with a type B aerosol size distirbution, with the rest of the conditions the same as that for Figure 2. These two figures show that advection fog with visibility below 1000 m was formed at time t = 53 min and relative humidity s = 99.0% at z = 1 m; and at time t = 92 min and relative humidity s $\simeq$ 98.1% at z = 4 m with a type B aerosol particle size distribution. Comparison between Figures 2 and 4, and Figures 3 and 5 indicate that a type A condensation nuclei distribution initiates the formation of advection fog at least more than 1 min earlier at z $\simeq$ 1 m, and 6 min eariler at z = 4 m than the type B condensation nuclei distribution. Furthermore, the relative humidity at which fog starts to form is lower for condensation nuclei aerosols with a type A distribution than a type B distribution. From Figures 1 to 5, it can be concluded that the effect of the aerosol particle size distribution is significant and also that the distribution with a higher number of particles and a smaller particle size is more favorable for dense fog formation than the distribution with a lower particle number and a larger particle size when the mass concentration is kept constant.

As to advection fog formation associated with a type C distribution of aerosols as condensation nuclei, Figures 6 and 7 show similar relative humidity and visibility profiles at two different vertical locations, z = 1 and 4 m, respectively, under the same conditions as Figures 2, and 3. These two figures again show that advection fog with visibility below 1000 m was formed at t = 55.5 min and relative humidity s = 98.5% at z = 4 m. A comparison of Figures 2, 4, and 6, shows that condensation nuclei with type A and type B distributions initiate the formation of advection fog at least more than 3.5 min and 2.5 min earlier, respectively, than a type C distribution at z = 1 m. For the altitude z = 4 m a comparison of Figures 3, 5 and 7 shows that condensation nuclei with type A and type B distributions initiate the formation of advection fog at least more than 8 min and 2 min earlier, respectively, than a type C distribution. Furthermore, the relative humidity at which fog starts to form is lower for aerosols with a type A distribution than a type B distribution, which is in turn lower than a type C distirbution.

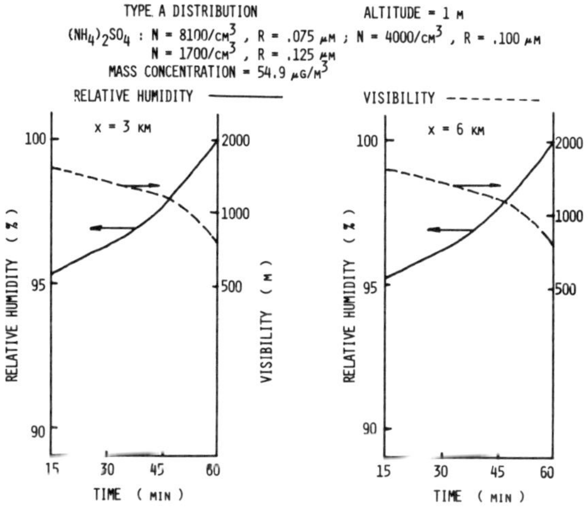

Figure 2. Visibility and relative humidity profiles of fog formation due to condensation nuclei, $(NH_4)_2SO_4$, with type A aerosol population distribution, as shown in Figure 1, at two horizontal locations, x = 3 and 6 km, and both at vertical location, z = 1 m.

LIQUID WATER CONTENT RESULTING FROM CONDENSATION

The mathematical simulation of the formation of advection fog based on a polydisperse aerosol distribution shows that the aerosol distribution with a higher number density, rather than greater size of aerosol nuclei, produces more favorable conditions for the formation of denser fog, if the value of mass concentration of the aerosol particles is kept constant. It is of considerable interest to study the liquid water content resulting from the condensation processes in which the aerosol particles with a variety of distribution functions are considered. The results of the liquid water content study are very important to the acid rain problem.

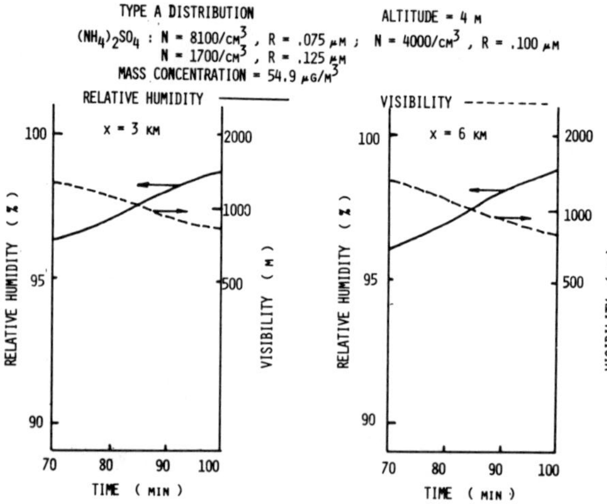

Figure 3. Visibility and relative humidity profiles of fog formation due to condensation nuclei, $(NH_4)_2SO_4$, with type A aerosol population distribution, as shown in Figure 1, at two horizontal locations, x = 3 and 6 km, and both at vertical location, z = 4 m.

A growing body of evidence suggests that acid rain may have substantial adverse effects on the environment. Such effects include acidification of lakes, rivers, and groundwaters, with resultant damage to fish and other components of the aquatic ecosystem; acidification and demineralization of soils; reduction of forest productivity; damage to crops; and deterioration of man-made materials. These effects may be cumulative or may result from peak acidity episodes. The study of liquid water content due to condensation, with SO_x and NO_x as the polydisperse condensation nuclei distribution may be beneficial to the study of potentially impacted ecosystems due to acid rain.

With the same initial and boundary conditions we considered in the previous section, the time dependent liquid water contents were calculated for types A, B and C particle size distributions, as shown in Figure 1, for $(NH_4)_2SO_4$ as a major component of the aerosol particles as condensation nuclei. Figure 8 illustrates the time change of the liquid water content at two horizontal locations, x = 3 and 6 km, both at z = 1 m, with a type A aerosol distribution. Figure 9 shows a similar time change of the liquid water content with a type A aerosol distirbution at z = 4 m, under the same conditions as Figure 8. These two figures show that the liquid water content becomes $102\mu g/m^3$ at time t = 40 min at z = 1 m; and 130 $\mu g/m^3$ at time t = 75 min at z = 4 m for a type A aerosol particle size distribution.

As for a type B aerosol size distribution, Figures 10 and 11 show a similar time change of liquid water content at two different altitudes, z = 1, and 4 m, respectively, with the remaining conditions the same as that of Figure 8.

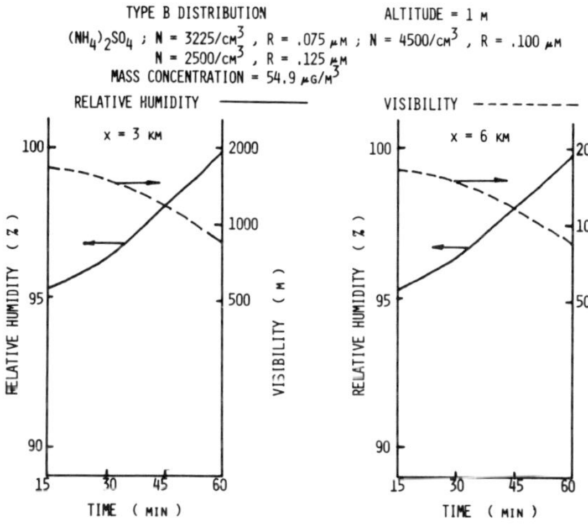

Figure 4. Visibility and relative humidity profiles of fog formation due to condensation nuclei, $(NH_4)_2SO_4$, with type B aerosol population distribution, as shown in Figure 1, at two horizontal locations, x = 3 and 6 km, and both at vertical location, z = 1 m.

A comparison between Figures 8 and 10, and Figures 9 and 11 indicates that the differences between the type A and B distributions make no difference in the liquid water content history. This is significant in that the process involved is similar to the wet process that removes acid from the atmosphere.

As to the time change of liquid water content with a type C distribution, Figures 12 and 13 show similar profiles at two different altitudes, z = 1 and 4 m, respectively. Comparison of Figures 8, 10, and 12, and Figures 9, 11 and 13 show that the aerosol population distributions have no effect on the liquid water content history so long as the mass concentration of the aerosol particles is the same value.

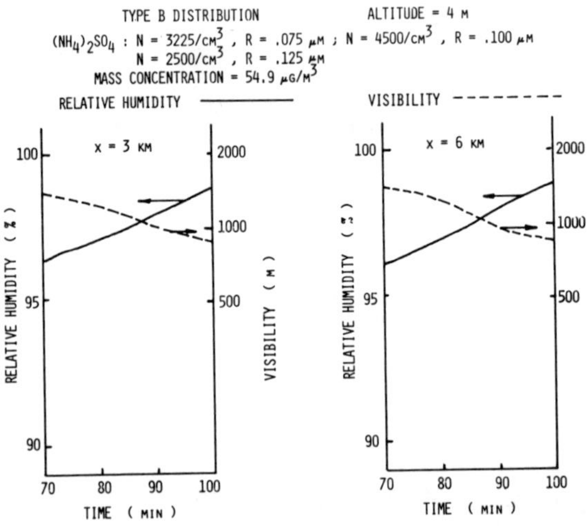

Figure 5. Visibility and relative humidity profiles of fog formation due to condensation nculei, $(NH_4)_2SO_4$, with type B aerosol population distribution, as shown in Figure 1, at two horizontal locations, x = 3 and 6 km, and both at vertical location, z = 4 m.

CONCLUSIONS AND DISCUSSIONS

Among the effects of air pollution on atmospheric properties, the most noticable is perhaps the reduction in visibility, with and without the occurrence of condensation. This study is particularly interested in how aerosols, produced by energy-related combustion, as condensation nuclei affect the formation of advection fog. Hung and Liaw (1980) showed that the greater the mass concentration, the particle concentration, and the size of the radius of condensation nuclei, the denser the fog that is formed. However, by using the monodisperse distribution model, Hung and Liaw (1980) indicate that the particle concentration, rather than the size of the radius of the nuclei makes a greater contribution to the formation of fog, if the value of mass concentration is kept constant.

Observations show that the particle size distributions for combustion related aerosols are broad rather than monodisperse. The results given in this study are based on a polydisperse distribution which is a combination of three step functions for each sdistribution, and a single composition aerosol model. The following conclusions resulted from the present study:

(1) The aerosol distribution with the higher particle concentration, rather than a larger aerosol nuclei size, makes a greater contribution to the formation of fog, if the value of mass concentration is kept constant.

(2) The relative humidity at which advection fog with a visibility below 1000 m is formed is lower for aerosols with a higher particle concentration with a smaller size than a lower particle concentration with a larger size, if the value of mass concentration is kept constant.

(3) The aerosol population with a high particle concentration-shifted distribution [see Figure 14(a)] provides more favorable conditions for the formation of dense fog than the aerosol population with low particle concentration-shifted distribution [see Figure 14(b) for description].

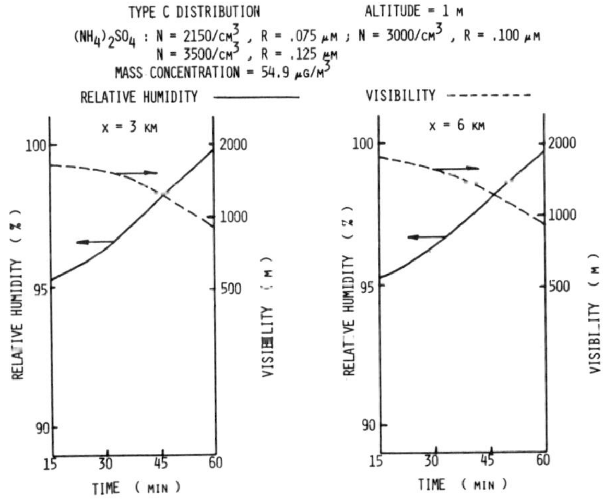

Figure 6. Visibility and relative humidity profiles of fog formation due to condensation nuclei, $(NH_4)_2SO_4$, with type C aerosol population distribution, as shown in Figure 1, at two horizontal locations, x = 3 and 6 km, and both at vertical location, z = 1 m.

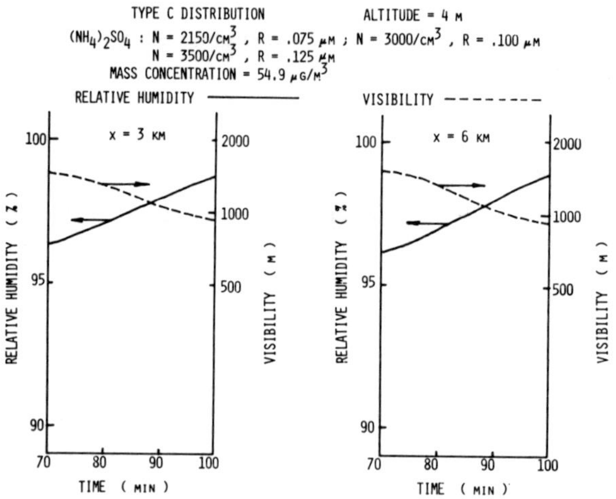

Figure 7. Visibility and relative humidity profiles of fog formation due to condensation nuclei, $(NH_4)SO_4$, with type C aerosol population distribution, as shown in Figure 1, at two horizontal locations, x = 3 and 6 km, and both at vertical location, z = 4 m.

Fresh water bodies in much of eastern North America and northern Europe which are adjacent to the areas of the highest acidic rains, are threatened by the continual deposition and further expansion of acid rain. The increasing acidity of lakes in North America and Europe has been documented, with the most tangible result being the decline in fish populations, and other aquatic organisms. The effect on condensation of combustion-related pollutants as condensation nuclei is considerable.

Results from the present study reveal that differences in aerosol particle size distributions make no difference in the amount of liquid water content resulting from the condensation and nucleation effects, so long as the mass concentration of the aerosols is kept constant. This also implies that the production of acid rain increases with an increase of the mass concentration of the combustion-related pollutants.

The results of the simulations have been compared with the field observations from the California coast made by Mack et al. (1973, 1974, and 1975) and others. Because most of the field observations were made as spotwise measurement, there was no single complete set of time-dependent vertical and horizontal simultaneous measurements of the wind profile, temperature profile, humidity profile, chemical constituent of condensation nuclei, and particle size distribution of the aerosols. The unavailability of a complete set of observations makes a detailed comparison between numerical simulation and field observation rather difficult. Nevertheless, the observed time-dependent evolution of the change of relative humidity with the change of visibility in the vertical and horizontal coordinates from the numerical simulation are in qualitatively good agreement.

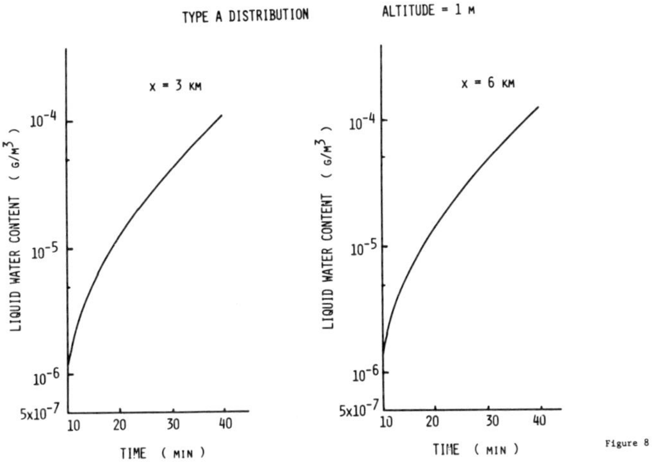

Figure 8. Time rate of change of the amount of liquid water content associated with condensation nuclei, $(NH_4)_2SO_4$ and $Ca(NO_3)_2$, with type A aerosol population distribution, as shown in Figure 1, at two horizontal locations, x = 3 and 6 km, and both at vertical location, z = 1 m.

ACKNOWLEDGEMENT

The authors appreciate the help of G. S. Liaw in numerical computations.

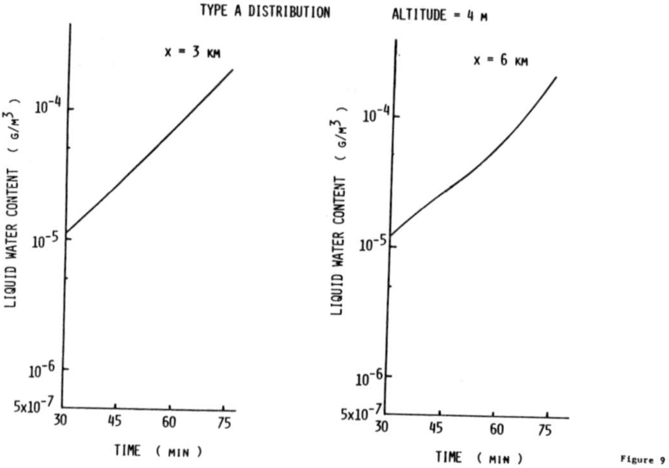

Figure 9. Time rate of change of the amount of liquid water content associated with condensation nuclei, $(NH_4)_2SO_4$ and $Ca(NO_3)_2$, with type A aerosol population distribution.

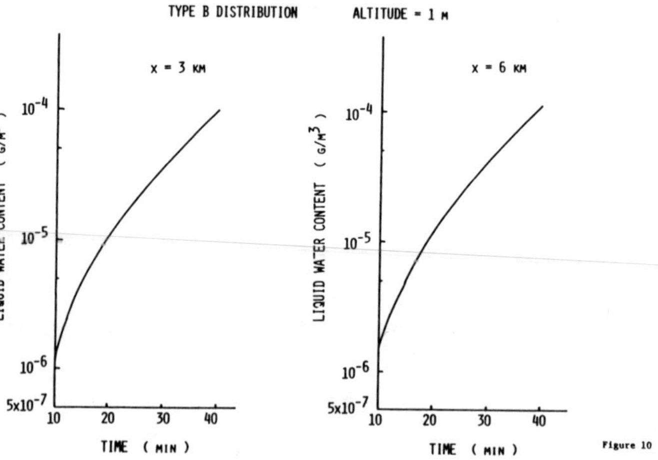

Figure 10. Time rate of change of the amount of liquid water content associated with condensation nuclei, $(NH_4)_2SO_4$ and $Ca(NO_3)_2$, with type B aerosol population distribution.

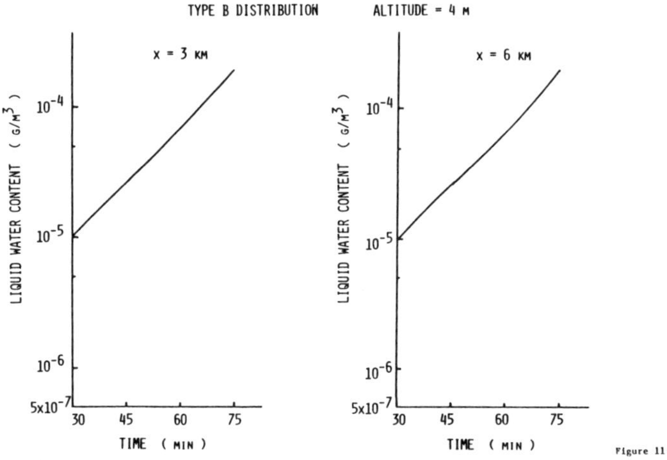

Figure 11. Time rate of change of the amount of liquid water content associated with condensation nuclei, $(NH_4)_2SO_4$ and $Ca(NO_3)_2$, with type B aerosol population distribution.

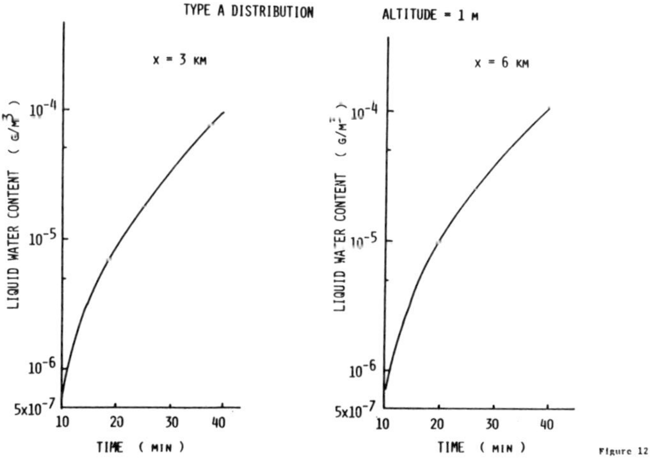

Figure 12. Time rate of change of the amount of liquid water content associated with condensation nuclei, $(NH_4)_2SO_4$ and $Ca(NO_3)_2$, with Type C aerosol population distribution.

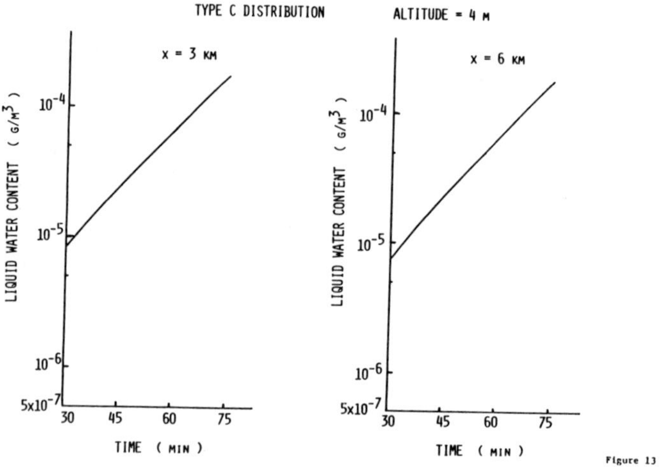

Figure 13. Time rate of change of the amount of liquid water content associated with condensation nuclei, $(NH_4)_2SO_4$ and $Ca(NO_3)_2$, with type C aerosol population distribution.

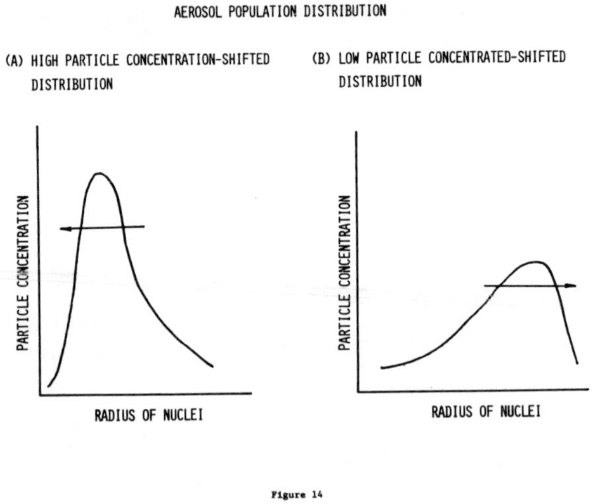

Figure 14. Aerosol population distribution with a fixed value of mass concentration: (a) High particle concentration - shifted distribution; (b) Low particle concentration - shifted distribution.

REFERENCES

Baker, E. H., Bound-Layer Meteor., 11, 267-294, 1977.

Carstens, J. C., Podzimek, J. & Sadd, A. N., J. Atmos. Sci., 31, 592-596, 1974.

Chagnon, S. A., Jr., Bull. Am. Meteorol. Soc., 49, 4-11, 1968.

Dennis, A. S., Weather Modification by Cloud Seeding, p. 267, Academic Press, New York, 1980.

Eadie, W. J., Rogers C. W., Katz, U. & Kocmond, W. C., Project fog drops. V. Part I: A numerical model of advection fog. Report NASA CR-2633, p. 54, Calspan Corp., New York., (1975).

Fitzgerald, J. W., J. Atmos. Sci., 35, 1522-1535, 1978.

Georgii, H. W., The effects of air pollution on urban climates, Bull. World Health Organ., 40, 624-632, 1969.

Hidy, G. M. & Brock, J. R., in Proc. Second International Clean Air Congress, ed. by H. M. Englund and W. T. Beery, Academic Press, New York, 1971.

Hoffer, T. E. & Mallen, S. C., J. Atmos Sci., 27, 914-918, 1978.

Hung, R. J. & Hunckle J. R., J. Rech. Atmos., 11, 165-178, 1977.

Hung, R. J. & Liaw, G. S., J. Air Pollut. Contr. Assoc., 31, 55-61, 1981.

Hung, R. J. & Vaughan, O. H., AIAA Paper No. 77-130, pp. 1-9, 1977.

Hung, R. J., Liaw, G. S. & Vaughan, O. H., in Proc. Atmospheric Environment of Aerospace Systems and Applied Meteorology, Amer. Meteor. Soc., Boston, Mass., pp. 169-173, 1978.

Hung, R. J., Liaw, G. S. & Vaughan, O. H., J. Rech. Atmos., 13, 37-61, 1979.

Hung, R. J, Vaughan, O. H. & Smith, R. E., Progr. in Astront. and Aeornaut., 48, 515-530, 1976.

Low, R. D. H., Duncan, L. D. & Gomez, R. B., The Microphysical Basis of Fog Optical Characterization, ASL-TR-0011, Atmos. Sci. Lab., White Sands Missile Range, New Mexico, p. 24, 1978.

Mack, E. J., Pilie, R. J. & Kockmond, W. C., An Investigation of the Microphysical and Micrometeorological Properties of Sea Fog, Calspan Corp., New York, p.33, 1973.

Mack. E. J., Katz, U., Rogers, C. W. & Pilie, R. J., The Microstructure of California Coastal Stratus and Fog at Sea, Calspan Corp., New York, p. 67, 1974.

Mack, E. J., Pilie, R. J. & Katz, U., Marine Fog Studied Off the California Coast, Calspan Corp., New York, p. 69, 1975.

Mason, B. J., The Physics of Clouds, 2nd Edition, Clarendon Press, Oxford, p. 671, 1971.

Richtmyer, R. D. Difference Methods for Initial-Value Problems, Interscience Publ., New York, p. 238, 1967.

Sax, R. I., Chagnon, S. A., Grant, L. O., Hitschfeld, W. F., Hobbs, P. V., Kahan, A. M. & Simpson, J., J. Appl. Meteor., 14, 652-672, 1975.

Seinfeld, J. H., Air Pollution- Physical and Chemical Fundamentals, McGraw-Hill, New York, p. 523, 1975.

SPIN-UP FLOWS IN A CYLINDER

Jae Min Hyun
Universities Space Research Association Scientist
ES82, NASA Marshall Space Flight Center, AL 35812

Abstract

Spin-up is the time-dependent process of adjusting a uniformly rotating fluid to a change in the rotation rate of the container. The model problem of spin-up flows, in a closed right circular cylinder, is considered in this paper. The previous theoretical and experimental work is reviewed in some detail, but a greater emphasis is placed on the recent findings stemming from the numerical solutions. The numerical results have been validated by checking them against the accurate laser-Doppler velocimeter measurements. Four cases of the spin-up flows are discussed: (1) linear spin-up of a homogeneous fluid, (2) spin-up from rest of a homogeneous fluid, (3) linear spin-up of a stratified fluid, and (4) spin-up from rest of a stratified fluid. As to (1), close agreement has been reported between theory and experiment, establishing the predominant mechanism due to the weak meridional circulation induced by the Ekman suction. Recent numerical solutions reveal the limitations of the classical analytical model on (2), pointing to the difficulty in formulating the nonlinear Ekman compatibility conditions in finite geometry. The correct location and the viscous structure of the moving front have been examined using the numerical results. The long-standing disagreement between theory and experiment on (3) has been resolved by numerical studies. It is shown that viscous diffusion in the interior, which was not included in the theory, is the cause of this discrepancy. No analytical models exist for (4). However, the results of the preliminary numerical studies on (4) are in qualitative agreement with the exploratory experimental observations on the bow-shaped moving front.

1. Introduction

Spin-up is a term to describe the transient adjustment process of a fully contained, uniformly rotating fluid to a change in the rotation rate of the container, i.e., from Ω_i to $\Omega = \Omega_i + \Delta\Omega$. The study of spin-up flows is of interest in certain applications, in particular, geophysical fluid systems. Further, the problem illuminates some of the salient features fundamental to rotating fluid dynamics.

This paper is confined to the model problem of the spin-up flows of a fluid in a closed, finite, right circular cylinder whose rotation axis coincides with its vertical symmetry axis. The choice of this cylindrical geometry allows analytical simplifications, and experimental verifications are relatively easy to carry out in the laboratory.

Modern work on spin-up dates from the linearized analysis of Greenspan and Howard [1] in which a homogeneous fluid, contained between two infinite discs and rotating about an axis perpendicular to the discs, is subjected to an infinitesimally small impulsive change in the rotation rate of the discs. Greenspan and Howard demonstrated that the relative azimuthal flow created by a rotation rate increase leads to the formation of Ekman layers on the boundary discs over a time scale $O(\Omega^{-1})$. These boundary layers exhibit an outward mass flux and draw fluid from the interior region (Ekman suction), which in turn leads to an inward meridional secondary circulation. This circulation fills the interior, where the direct effects of viscosity are negligible, and uniformly spins up the interior by angular momentum advection and the concomitant stretching of vortex lines. Further, the effect of the sidewall was shown to be dynamically unimportant for this problem. The fluid adjustment is substantially accomplished over the spin-up time scale $O(E^{-\frac{1}{2}}\,\Omega^{-1})$, which is much smaller than the viscous diffusion time scale $O(E^{-1}\,\Omega^{-1})$, where $E = \nu/\Omega h^2$ is the Ekman number, h is the characteristic dimension along the axis of rotation, and ν is the kinematic viscosity of the fluid. Subsequent studies dealt with various aspects of the spin-up process (see the review article by Benton and Clark [2]).

The primary purpose of this paper is to present a unified account of more recent studies of the spin-up flows in a cylinder based mainly on the numerical integrations of the full Navier-Stokes equations in axisymmetric form. In view of the inherent limitations of the previous analytical models, which will be discussed in later sections, there is a need to acquire accurate and complete flow data. The numerical studies were initiated to offer such data. The numerical results have been validated by checking them against the high-resolution, disturbance-free rotating laser-Doppler velocimeter (LDV) measurements. In Section 2, a brief description of the numerical model is given. The main body of the text is divided into four parts, reflecting the distinct characteristics of the spin-up process under a variety of conditions: (1) linear spin-up of a homogeneous fluid, (2) spin-up from rest of a homogeneous fluid, (3) linear spin-up of a stratified fluid, and (4) spin-up from rest of a stratified fluid.

2. The Numerical Model

2.a. Formulation

The numerical model adopted here was developed by Warn-Varnas, et al. [3]. Consider a right circular cylinder of radius a and height h ($\equiv$ 2H) filled with a fluid having kinematic viscosity ν, thermometric diffusivity K, and coefficient of volumetric expansion α. These physical properties of the fluid are assumed to be constant. In the case of a stratified fluid, the top and bottom boundaries are thermal conductors and are kept at constant temperatures to produce a stable stratification. The vertical temperature difference is $2\Delta T$ over $2H$. The sidewall boundary is a thermal insulator. The Froude number, $\Omega^2 H/g$, is much smaller than one, and the effect of the Sweet-Eddington circulation will be ignored [4].

The governing equations are the axisymmetric incompressible Navier-Stokes equations for a Boussinesq fluid written for cylindrical coordinates (r, θ, z) with respective velocity components (u, v, w). Viewed in a frame with the mid-depth plane at $z = 0$ and rotating with the final rotation rate Ω, these equations are

$$\frac{\partial u}{\partial t} = -\frac{1}{r}\frac{\partial}{\partial r}(ru^2) - \frac{\partial}{\partial z}(uw) - \frac{1}{\rho_o}\frac{\partial p}{\partial r} + \left(2\Omega + \frac{v}{r}\right) v + \nu\left(\nabla^2 u - \frac{u}{r^2}\right) , \qquad (1)$$

$$\frac{\partial v}{\partial t} = -\frac{1}{r}\frac{\partial}{\partial r}(ruv) - \frac{\partial}{\partial z}(vw) - 2\Omega\, u - \frac{vu}{r} + \nu\left(\nabla^2 v - \frac{v}{r^2}\right) , \qquad (2)$$

$$\frac{\partial w}{\partial t} = -\frac{1}{r}\frac{\partial}{\partial r}(ruw) - \frac{\partial}{\partial z}(w^2) - \frac{1}{\rho_o}\frac{\partial p}{\partial z} + \alpha\, g\, T + \nu\, \nabla^2 w , \qquad (3)$$

$$\frac{\partial T}{\partial t} = -\frac{1}{r}\frac{\partial}{\partial r}(ruT) - \frac{\partial}{\partial z}(wT) + K\, \nabla^2 T , \qquad (4)$$

$$\frac{1}{r}\frac{\partial}{\partial r}(ru) + \frac{\partial w}{\partial z} = 0 , \qquad (5)$$

where $\nabla^2 = \frac{1}{r}\frac{\partial}{\partial r} r \frac{\partial}{\partial r} + \frac{\partial^2}{\partial z^2}$, p is the reduced pressure, T the temperature such that the full temperature equals $(T_o + T)$, and the equation of state is

$$\rho = \rho_o(1 - \alpha T) , \qquad (6)$$

in which T_o and ρ_o are the reference values of temperature and density, respectively. Obviously, if the fluid is homogeneous, $\alpha g T$ is deleted from Eq. (3), and Eqs. (4) and (6) are removed from the system of equations.

The initial conditions for the fluid are

$$u = w = 0 , \quad v = -\Delta\Omega\, r, \qquad (7a)$$

$$\frac{\partial T}{\partial z} = \frac{\Delta T}{H} \qquad \text{at } t = 0 . \qquad (7b)$$

Clearly, the entire problem is symmetric about the mid-depth plane $z = 0$, and hence integration needs to be conducted for the bottom half of the cylinder, $-H \leq z \leq 0$, $0 \leq r \leq a$, only.

The boundary conditions at the bottom disc are

$$u = v = w = 0 \quad , \tag{8a}$$

$$T = -\Delta T \qquad \text{at} \quad z = -H \quad , \tag{8b}$$

and the symmetry conditions at the mid-depth plane are

$$\frac{\partial u}{\partial z} = \frac{\partial v}{\partial z} = w = 0 \quad , \tag{9a}$$

$$T = 0 \qquad \text{at} \quad z = 0 \quad . \tag{9b}$$

The boundary conditions on the cylinder sidewall are

$$u = v = w = 0 \quad , \tag{10a}$$

$$\frac{\partial T}{\partial r} = 0 \qquad \text{at} \quad r = a \quad . \tag{10b}$$

A thin solid cylinder of very small but finite radius ($r = r_i$) was inserted along the axis of symmetry to satisfy numerical stability requirements [3]. Symmetry conditions require that

$$u = v = \frac{\partial w}{\partial r} = 0 \quad , \tag{11a}$$

$$\frac{\partial T}{\partial r} = 0 \qquad \text{at} \quad r = r_i \quad . \tag{11b}$$

Again, it should be remarked that, if the fluid is homogeneous, all references to the temperature conditions, i.e., Eqs. (7b), (8b), (9b), (10b), and (11b), are not needed.

2.b. Numerical Simulation Technique

Equations (1) through (5) and the initial and boundary conditions were finite-differenced on a staggered mesh with nonuniform grid spacings. The resulting time-dependent difference equations were solved by a time marching procedure. The details of the numerical techniques are given in Warn-Varnas, et al. [3].

The stretching of the grid was accomplished to have better resolution of the boundary layers. A 42 x 42 grid for the full domain was used, and, owing to the grid stretching, for the values of the Ekman numbers used, there were approximately 8 points in each Ekman layer. In the case of spin-up from rest, a uniform grid in the r direction was used, since resolving the moving shear layer in the interior was of primary interest in that problem. The dependent variables were distributed over the staggered grid with azimuthal velocity, temperature, and pressure defined at the same grid points.

The pressure was found from a Poisson equation obtained by taking the divergence of Eqs. (1) and (3). Thus

$$\frac{\partial D}{\partial t} = - \nabla^2 P + C \quad , \tag{12}$$

where C denotes the combined divergence of the advection, Coriolis and buoyancy terms in the u and w equations. D is the divergence $(\nabla \cdot \vec{u})$ which is small but, due to machine round-off errors, not zero. This equation was solved by an ADI iterative approach.

3. Linear Spin-Up of a Homogeneous Fluid

If the Rossby number, $\varepsilon = \Delta\Omega/\Omega$, which measures the degree of nonlinearity, is very small, the problem can be linearized and certain powerful analytical techniques may be utilized. For a homogeneous fluid confined between two infinite parallel discs, Greenspan and Howard [1] showed that the interior azimuthal flow is essentially a time-dependent solid-body rotation. Although the geometry is of maximal simplicity, the interior fluid motion analyzed by Greenspan and Howard clearly demonstrates the predominant mechanism of the meridional circulation established by the horizontal Ekman boundary layers. A detailed description is given by Benton and Clark [2] of the formation of the Ekman layers and the subsequent development of the interior motion.

The overall decay of the interior azimuthal flow is given as

$$v(r,t) = - \Delta\Omega \; r \; [1 - \exp(- E^{\frac{1}{2}} \Omega \; t)] \quad . \tag{13}$$

Equation (13) indicates that the overall adjustment is substantially accomplished over the spin-up time scale $\tau = E^{-\frac{1}{2}} \Omega^{-1}$. Superposed on this overall decay are the inertial oscillations in the interior. These inertial oscillations are excited by the impulsive start, and they decay in amplitude as time progresses. Greenspan and Howard also showed that the effect of the sidewall boundary is unimportant; therefore, the principal qualitative picture that they obtained for two infinite discs is still valid for the finite cylinder as well.

Warn-Varnas, et al. [3] performed an extensive experimental (LDV measurements of the azimuthal flows) and numerical investigation of the impulsive spin-up of a homogeneous fluid in a closed cylinder. The numerical results were in excellent agreement with the LDV measurements, verifying the accuracy and reliability of the numerical model (see Figs. 8-16 of Warn-Varnas, et al. [3]). Both the experimental and numerical techniques had sufficient resolution to exhibit clearly the weak inertial oscillations. The agreement between the numerical and the experimental results is good not only for the overall decay of the azimuthal flow but also for the amplitudes and phases of the inertial modes. A comparison of the numerical results and the analytical predictions of Greenspan and Howard shows a good agreement for the overall decay of the azimuthal flow, but the inertial oscillations are not described well, either in phase or amplitude, by the Greenspan and Howard analysis. The dimensional angular frequencies of the linearized, inviscid, axisymmetric, and inertial-internal gravity modes about a state of rigid rotation in a cylinder are given by [5]

$$\beta = 2\Omega \left[\frac{1 + (2\alpha_n/\pi m A)^2}{1 + S^2\ (2\alpha_n/\pi m A)^2} \right]^{-\frac{1}{2}} , \tag{14}$$

where $A = a/H$, $S = (\alpha g \Delta T/H)^{\frac{1}{2}}/2\Omega$ is the stratification parameter ($S = 0$ for a homogeneous fluid), α_n are zeroes of the first-order Bessel function, and m is any integer. Now it is expected that most energy will be in the ($m = 2$, $n = 1$) mode since this mode describes the perturbation on the scale of the container and has symmetry about the mid-depth plane. The inertial mode frequency found in the numerical and the experimental results [3] indicates close agreement with the values given by Eq. (14) with $m = 2$, $n = 1$.

As was derived in Greenspan and Howard, the vertical mass flux into the Ekman layer (Ekman suction) requires a radial inflow in the bulk of the interior inviscid fluid to satisfy continuity. The magnitude of the radial inflow is $u \sim E^{\frac{1}{2}} \varepsilon r \Omega$; therefore, it is important to note that, over the spin-up time scale, the radial distance travelled by a fluid particle is only $O(\varepsilon r)$.

In summary, for the case of the well-established problem of linear spin-up of a homogeneous fluid, agreement is good between the analytical predictions and the numerical solutions in describing the overall adjustment process of the interior fluid motion.

4. Spin-Up from Rest of a Homogeneous Fluid

When the Rossby number $\varepsilon = \Delta\Omega/\Omega$ is not small, due consideration has to be given to nonlinear effects. The weak meridional circulation induced by Ekman boundary layers is still the dominant mechanism controlling spin-up even in the nonlinear regime. Therefore, the spin-up time scale is still $O(E^{-\frac{1}{2}}\ \Omega^{-1})$. However, an important distinction has to be noted between nonlinear spin-up from an already established rotation rate and spin-up from rest. The former case is a direct extension of the linear problem discussed in Section 3, and only quantitative changes are observed. In the latter case, however, the bulk of the interior fluid is not rotating initially; therefore, the angular momentum advection, which controls linear spin-up, is not effective for this problem. As will be seen later, the interior fluid is first flushed into the Ekman boundary layers, and while the fluid is transported radially outward in the Ekman boundary layers it acquires angular momentum and spins up from rest.

A classical analytical model of spin-up from rest of a homogeneous fluid in a cylinder was proposed by Wedemeyer [6]. Here the velocity components (u, v', w) on an inertial framework are used (note $v' = v + r\Omega$). As is evident from a systematic scaling argument [5], the meridional flow in the interior is $O(E^{\frac{1}{2}} v')$, and the azimuthal velocity is independent of z to the leading order. Then we obtain an equation for v' in the interior

$$\frac{\partial v'}{\partial t} + u \left(\frac{\partial v'}{\partial r} + \frac{v'}{r} \right) = \nu \left[\frac{\partial^2 v'}{\partial r^2} + \frac{\partial}{\partial r} \left(\frac{v'}{r} \right) \right] , \tag{15}$$

which is referred to as Wedemeyer's equation. The formulation is complete if an explicit relationship between u and v' is derived. An approximate expression for this relation was given by Wedemeyer [6], which is based on a boundary layer analysis. To determine the local mass flux into the Ekman boundary layers for the unsteady problem of spin-up from rest, Wedemeyer used a linear fit to the steady Ekman flux data of Rogers and Lance [7] which were computed for fluid in solid-body rotation above an infinite rotating disc. Then, Wedemeyer obtained

$$u(r,t) = - k E^{\frac{1}{2}} (r\Omega - v') , \tag{16}$$

where k is a constant which is taken to be unity [5].

Substitution of Eq. (16) into Eq. (15) yields a first-order partial differential equation for v'. Furthermore, if $E^{\frac{1}{2}} A^{-2} \ll 1$ so that the viscous diffusion term can be neglected, Eq. (15) has an analytic solution

$$\begin{aligned} v' &= 0 \quad \text{for} \quad r < r_o , \\ v' &= \left[\Omega r - \Omega \frac{a^2}{r} \exp(-2E^{\frac{1}{2}}\Omega t) \right] / [1 - \exp(-2E^{\frac{1}{2}}\Omega t)] , \quad \text{for} \quad r > r_o , \end{aligned} \tag{17}$$

where

$$r_o = a \exp(-2E^{\frac{1}{2}}\Omega t) ,$$

which is termed as Wedemeyer's inviscid solution. Equation (17) indicates that a cylindrical shear discontinuity front at $r = r_o$ divides the interior into two regions. Ahead of the front, the fluid is not rotating. Behind the front, the fluid motion is a combination of a time-dependent solid-body rotation and a vortex.

The Wedemeyer model has since been tested and extended by a number of authors (e.g., Venezian [8,9]; Watkins and Hussey [10,11]; Weidman [12]), revealing various features of spin-up from rest. All these investigations, however, necessarily depend on several crucial assumptions which were incorporated in the Wedemeyer model.

Kitchens [13,14] solved the unsteady Navier-Stokes equations for the transient spin-up flow in a cylinder. Several calculations of the decay of the azimuthal flow and the plots of the meridional streamfunctions were presented. The papers of Kitchens, however, do not contain comprehensive flow data. No discussions on the moving front were given by Kitchens.

Hyun, et al. [15] recently carried out extensive numerical simulations for spin-up flows from rest using the numerical model of Warn-Varnas, et al. [3]. The numerical results were verified by checking them against the rotating LDV measurements, and close agreement was found.

The z-independence of the azimuthal velocities is quite apparent in the numerical results. In order to examine the detailed structure of the moving front, the radial profiles at mid-depth are displayed by the five terms on the right side of Eq. (2), representing respectively the radial advection, vertical advection, Coriolis acceleration, curvature effect, and viscous diffusion. Figure 1 is typical of these diagnostic studies. The top plot depicts the first four terms on the right side of Eq. (2). The middle plot compares the sum of the first four terms, representing the inviscid effects, with the viscous diffusion term. The bottom plot demonstrates the discrepancy in azimuthal velocity between the numerical results and Wedemeyer's inviscid solution, Eq. (17). Several points emerge from the diagnostic studies. The nonlinear effects are dominant at early times, but as spin-up progresses, the Coriolis acceleration, which is a linear term, increases in magnitude. As was pointed out by Venezian [9], the shear discontinuity front of Wedemeyer's inviscid solution implies that it is a region of strong flow gradients in which viscous diffusion is significant. The middle plot clearly depicts the interior region of pronounced viscous diffusion effects. This is identified as the moving shear layer front, which scales with $O(E^{\frac{1}{4}}h)$. Far ahead of the front, both the inviscid acceleration terms and the viscous diffusion term are zero. The moving front may also be identified as the region in which the nonlinear advection terms undergo rapid variations. The middle plot also shows that there is a stationary boundary layer very close to the sidewall to adjust the flow to the sidewall conditions. As seen in the bottom plot, Wedemeyer's inviscid solution becomes a poorer representation of the azimuthal flow at large times.

The approximate relation between u and v', Eq. (16), was derived under the assumption that the boundary layer flow for the configuration of two infinite discs can still be used for a finite cylinder. This assumption implies that the vertical velocity of the edge of the Ekman layer, w_∞, is independent of r [5]. Figure 2 shows the numerical results for typical profiles of w in the lower half of the cylinder. Far ahead of the front where the fluid is not rotating, there is a uniform suction into the Ekman layers.

Furthermore, the linear dependence on z of w in the interior is apparent. However, as the radial location is moving closer to the front, w exhibits a strong r-dependence. In the core of the front, w becomes very small. Behind the front, the fluid is ejected into the interior (w_∞ is positive). The magnitude of w_∞ behind the front decreases with increasing time since mass continuity in the vertical direction has to be satisfied.

Another important assumption in Wedemeyer's model is that the fluid is locally in solid-body rotation. This is equivalent to saying that the Ekman mass flux depends only on the local difference in vorticity across the Ekman layer [12]. In view of strong radial flow gradients in the vicinity of the front, as shown in Fig. 1, there arises a question as to the validity of this assumption, since the Ekman flux may also be a function of radial gradient of flow variables.

The results of the numerical simulations reported here put the classical Wedemeyer model of spin-up from rest in proper perspective. The Wedemeyer model describes the crude qualitative features of the fluid motion in the interior; however, the basic assumptions embodied in the model have to be carefully evaluated before any quantitative significance can be given to the model results. The numerical results also point to the difficulty of formulating the nonlinear Ekman compatibility conditions in finite geometry.

5. Linear Spin-Up of a Stratified Fluid

The linear spin-up of a stratified fluid in a cylinder with insulated sidewall [see Eq. (10b)] is now considered. The parameter S is assumed to be order one. One of the most important consequences of stable stratification is that the vertical motion is inhibited, although the structure of the Ekman boundary layers is substantially unchanged. When the sidewall is insulated, the sidewall boundary layer cannot transport the Ekman boundary layer mass flux and, consequently, the Ekman mass flux enters into the interior directly from the corner regions [16]. This flow is known as the corner jet. In the interior it forms a meridional circulation which is restricted to a region closer to the Ekman layers. The weak oscillations excited by the impulsive start are now modified by stratification, resulting in inertial-internal gravity modes [see Eq. (14)].

Walin [16] and Sakurai [17] demonstrated that in stratified spin-up, the effect of the meridional circulation, in general, is to bring about only a partial spin-up of the interior. Walin (and Sakurai) assumed that the direct effect of viscous diffusion was confined to the boundary layers. In line with the approach of Greenspan and Howard [1] for linear homogeneous spin-up, Walin (and Sakurai) obtained the analytical expression for linear stratified spin-up for the decay of the azimuthal velocity in the interior [16,17]

$$\left|\frac{v}{\Delta\Omega r}\right| = 1 + \frac{H}{r} \Sigma C_n \frac{\cosh S K_n z/H}{\cosh s K_n} J_1(K_n r/H) \times [1 - \exp(-t/\tau_n)] \quad , \qquad (18)$$

where $C_n = 2A/\alpha_n J_o(\alpha_n)$, $K_n = \alpha_n/A$, $A = a/H$, $\tau_n = E^{-\frac{1}{2}} \Omega^{-1} (\tanh S K_n)/SK_n$, and α_n are zeroes of the first-order Bessel functions.

On the experimental side, three carefully controlled laboratory experiments on linear stratified spin-up have been reported: Buzyna and Veronis [4], Saunders and Beardsley [18], and Lee [19]. The first two experiments employed conventional measurement techniques, and Lee [19] used the rotating LDV system. In general, the comparisons of the experimental results against the linearized theoretical predictions of Eq. (18) revealed that even within the homogeneous spin-up time scale $E^{-\frac{1}{2}} \Omega^{-1}$, the measured azimuthal velocities decay faster than the theoretical predictions. Further, the discrepancies in the decay rates are a function of the stratification parameter and the location in the fluid.

Barcilon, et al. [20] solved the Navier-Stokes equations in axisymmetric form by means of finite-difference techniques on a uniform grid. Barcilon, et al. confirmed that the model results decayed faster than the theoretical predictions. By comparing results for different values of the Rossby number, Barcilon, et al. established that the effects of nonlinearity were not the cause of the discrepancy. From an examination of the meridional streamfunction plots, Barcilon, et al. argued that, due to the oscillating corner jet, the fluid ejected into the interior from the vertical boundary layer yields its newly acquired angular momentum to the bulk of the interior, which then appears to have a faster spin-up time than predicted theoretically.

In order to resolve this long-standing discrepancy between theory and experiment, Hyun, et al. [21] performed comprehensive numerical simulations for linear stratified spin-up using the numerical model of Warn-Varnas, et al. [3]. The numerical results of Hyun, et al. [21] were compared against the LDV measurement data of Lee [19], and excellent agreement was observed.

The numerical results of Hyun, et al. [21] clearly indicate that in stratified spin-up the Taylor-Proudman column is no longer applicable and the scaled azimuthal velocity, $|v/\Delta\Omega r|$, is not uniform in the interior. Figure 3 illustrates the decay of the azimuthal velocity. A systematic examination of the data leads to the finding that agreement between the theory and the numerical results is good when dealing with locations and times for which the meridional circulation flow is strong. Correspondingly, if this meridional flow is weak, agreement is poor. Hyun, et al. [21] concluded that the observed discrepancy between the theory and the numerical results for the time span $E^{-\frac{1}{2}}\Omega^{-1}$ is due to viscous diffusion in the interior. As was mentioned earlier, viscous diffusion in the interior was not included in the theory. The effects of viscous diffusion in the interior are enhanced for stratified spin-up over homogeneous spin-up since the effect of the meridional circulation in the stratified case is to produce radial and vertical gradients of the azimuthal velocity in the interior, whereas in the homogeneous case the azimuthal velocity in the interior decays essentially as solid-body rotation. Hyun, et al. [21] also pointed out the difficulty with the suggestion of Barcilon, et al. [20] concerning the transport of angular momentum by the oscillating corner jet from the sidewall boundary layer into the bulk of the interior. Note that the inward radial distance moved by a fluid particle during the meridional circulation phase of spin-up is only $O(\varepsilon r)$. For the small Rossby number flows considered by Barcilon, et al., this distance is small and, hence, the fluid ejected into the interior from the vertical boundary layer will not reach the bulk of the interior.

6. Spin-Up from Rest of a Stratified Fluid

Recent exploratory experiments by Greenspan [22] suggest that fluid instability is a principal mechanism by which a stratified fluid spins up from rest. After the impulsive start, Ekman layers form and transport the fluid to the developing sidewalls. The buoyancy effects produce instabilities at the sidewall with vigorous action in the corner regions. Unlike the vertical front that appears in spin-up from rest of a homogeneous fluid, the front in the stratified fluid is bow-shaped. Since stratification restricts the meridional flow closer to the Ekman boundary layers, the magnitude of the radial velocities is larger near the horizontal boundaries, resulting in a bow-shaped propagation front separating rotating from non-rotating fluid. After the edges of this advancing shear layer meet at the central axis, vortex generation-penetration combines with Ekman layer suction to produce substantial spin-up of the fluid.

Preliminary runs of the numerical model have been made for spin-up from rest of a stratified fluid over a time span of $E^{-\frac{1}{2}}\,\Omega^{-1}$. The numerical results confirm the bow-shaped propagating front. There is, however, a question as to the effectiveness of the present axisymmetric numerical model to study fluid motions after the edges of the advancing shear layer meet at the axis. The vortex generation-penetration process was described by Greenspan [22] to involve non-axisymmetric elliptical cross-sections. In order to acquire a full understanding of stratified spin-up from rest, a three-dimensional numerical model will have to be used.

Acknowledgments: It has been a pleasure to collaborate on this work with Drs. W. Fowlis and F. Leslie of Marshall Space Flight Center and Dr. A. Warn-Varnas of Naval Ocean Research and Development Activity. I am also grateful to NASA, Office of Space and Terrestrial Applications for supporting this research.

References

1. Greenspan, H. P. and Howard, L. N., 1963, J. Fluid Mech. 17, 385.
2. Benton, E. R. and Clark, A., 1974, Ann. Rev. Fluid Mech. 6, 257.
3. Warn-Varnas, A., Fowlis, W. W., Piacsek, S. and Lee, S. M., 1978, J. Fluid Mech. 85, 609.

4. Buzyna, G. and Veronis, G., 1971, J. Fluid Mech. 50, 579.
5. Greenspan, H. P., 1969, The theory of rotating fluids. Cambridge U. Press.
6. Wedemeyer, E. H., 1964, J. Fluid Mech. 20, 383.
7. Rogers, M. H. and Lance, G. N., 1960, J. Fluid Mech. 7, 617.
8. Venezian, G., 1969, Topics in Ocean Engr. 1, 212.
9. Venezian, G., 1970, Topics in Ocean Engr. 2, 87.
10. Watkins, W. B. and Hussey, R. G., 1973, Phys. Fluids, 16, 1530.
11. Watkins, W. B. and Hussey, R. G., 1977, Phys. Fluids, 20, 1596.
12. Weidman, P. D., 1976, J. Fluid Mech. 77, 685.
13. Kitchens, C. W., 1979, Tech. Rept. ARBRL-TR-02193, Aberdeen Proving Ground, Maryland.
14. Kitchens, C. W., 1980, AIAA J. 18, 929.
15. Hyun, J. M., Leslie, F. W., Fowlis, W. W. and Warn-Varnas, A., 1981, To be submitted to J. Fluid Mech.
16. Walin, G., 1969, J. Fluid Mech. 36, 289.
17. Sakurai, T., 1969, J. Fluid Mech. 37, 689.
18. Saunders, K. D. and Beardsley, R. C., 1975, Geophys. Fluid Dyn. 7, 1.
19. Lee, S. M., 1975, MS Thesis, Florida State Univ.
20. Barcilon, A., Lau, J., Piascsek, S. and Warn-Varnas, A., 1975, Geophys. Fluid Dyn. 7, 29.
21. Hyun, J. M., Fowlis, W. W., and Warn-Varnas, A., 1981, To appear in J. Fluid Mech.
22. Greenspan, H. P., 1980, Geophys. Astrophys. Fluid Dyn. 15, 1.

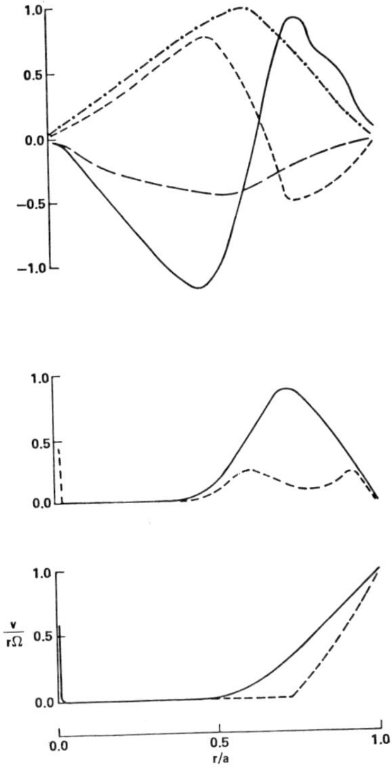

Fig. 1. Radial profiles at mid-depth of the magnitudes of each term on the right side of Eq. (2). In the top plot, —— radial advection; ---- vertical advection; -·- Coriolis acceleration; - - curvature effect. In the middle plot, —— the sum of the above four terms; ---- viscous diffusion term. The ordinates of the top and the middle plots are in the units of $E^{\frac{1}{2}}\ \Omega^2$ a. In the bottom plot, —— the numerical results, ---- Wedemeyer's inviscid solution. The ordinate of the bottom plot shows the scaled azimuthal velocity on the inertial frame. Conditions are a = 10.14 cm, H = 5.02 cm, Ω = 0.211 rad s^{-1}, ν = 0.01 $cm^2\ s^{-1}$. The time is $E^{\frac{1}{2}}\ \Omega$ t = 0.32.

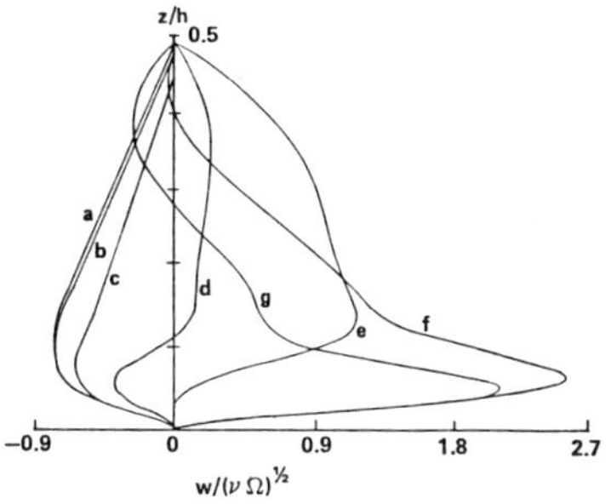

Fig. 2. Plots of the vertical velocity w. The ordinate shows the vertical distance from the bottom (z/h = 0.0), where h is the height of the cylinder. Flow parameters are the same as in Fig. 2. The time is $E^{\frac{1}{2}}\ \Omega$ t = 0.16. The radial locations, r/a, of the curves are (a) 0.65, (b) 0.70, (c) 0.75, (d) 0.80, (e) 0.85, (f) 0.90, (g) 0.95.

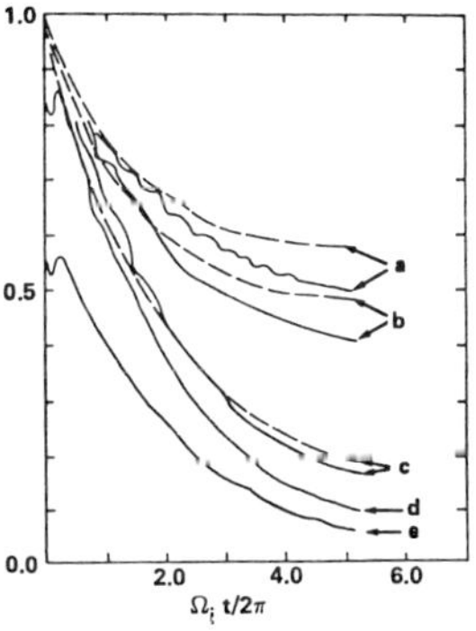

Fig. 3. Stratified spin-up results for S A^{-1} = 0.49, E = 7.24 x 10^{-4}, ε = 0.02. Solid lines are the numerical results, and broken lines the theoretical predictions of Walin. The radial location is at mid-radius (r/a = 0.5). The vertical locations, z/H, of the curves are (a) 0.0, (b) -0.42, (c) -0.82, (d) -0.93, (e) -0.97.

PROPAGATION OF SMALL-AMPLITUDE DISTURBANCES IN A PARTICULATE SUSPENSION

Tongyou Kanaruksunti[1] and John Peddieson, Jr.[2]
Department of Engineering Science and Mechanics
Tennessee Technological University
Cookeville, Tennessee 38501

ABSTRACT

A continuum theory of particulate suspensions is applied to the problem of one-dimensional acoustic non-harmonic wave propagation in a semi-infinite duct. Both gas-liquid drop and liquid-gas bubble suspensions are considered. The problem is solved by Laplace transforms and the inversions are found in closed form by an approximate technique. Some typical results are presented graphically and used to illustrate some interesting properties of the solutions. Attention is focused on illustrating the differences in the behavior of the two types of suspensions and on illustrating the significance of a finite particulate volume fraction.

INTRODUCTION

This paper is concerned with small-amplitude non-harmonic wave propagation in fluid-particle suspensions. This subject is of interest in connection with a variety of engineering problems including sound attenuation in fogs, acoustic cavitation, loss-of-coolant accidents in nuclear reactors, and attenuation of combustion instabilities in rocket motors.

The analysis to be discussed subsequently is based on a continuum theory in which the suspension is treated as a mixture of two interacting continua. This approach is appropriate for suspensions in which representative volume elements, with dimensions that are small compared to the characterisitic dimensions of the flow field, contain amounts of each phase sufficient to allow the formulation of meaningful averages of the properties of each phase within the volume elements. The volume elements are treated as differential volumes (points) and the average properties are treated as continuous variables.

---

[1]Research Assistant

[2]Professor

The primary purpose of the present work is to compare certain aspects of the behaviors of two types of gas-liquid suspensions. These are (1) a suspension of drops of incompressible liquid in a compressible gas (subsequently referred to as a gas-liquid drop suspension) and (2) a suspension of bubbles of compressible gas in an incompressible liquid (subsequently referred to as a liquid-gas bubble suspension).

An outline of this paper is as follows. First, linearized equations for one-dimensional flow of each type of suspension are developed. Next, these equations are employed to determine certain limiting forms of the speeds of propagation of small disturbances. Then, both sets of equations are analyzed using Laplace transforms for a unit step increase in fluid velocity at the end of a semi-infinite duct containing an initially quiescent suspension. The corresponding inversions are carried out approximately by the method discussed by Rasmussen [1], [2]. This procedure, which is valid for moderate and large times, results in convolution integrals which, for the situations considered herein, must be evaluated numerically. Finally, a representative set of numerical results is presented graphically and used to illustrate various interesting parametric trends. It is shown that important differences exist between the responses of the gas-liquid drop suspension and the liquid-gas bubble suspension to the same excitation.

GOVERNING EQUATIONS

For simplicity, attention will be confined to suspensions for which interphase mass transfer, changes in particle dimensions, viscous stresses, external body forces, and external body heating are negligible. Then the continuum formulation provides the one-dimensional balance laws

$$\partial_t\rho + \partial_x(\rho u) = 0, \qquad \partial_t(\rho u) + \partial_x(\rho u^2) = -\partial_x p - L$$

$$\partial_t(\rho(e + u^2/2)) + \partial_x(\rho u(e + u^2/2)) = -\partial_x(up) - E - u_p L \tag{1}$$

for the fluid phase and

$$\partial_t\rho_p + \partial_x(\rho_p u_p) = 0, \qquad \partial_t(\rho_p u_p) + \partial_x(\rho_p u_p^2) = -\partial_x p_p + L$$

$$\partial_t(\rho_p(e_p + u_p^2/2)) + \partial_x(\rho_p u_p(e_p + u_p^2/2)) = -\partial_x(u_p p_p) + E + u_p L \tag{2}$$

for the particle phase. The symbols ρ, p, u, and e denote in-suspension density (mass of phase per unit volume of suspension), in-suspension pressure (force on phase per unit area of suspension), velocity, and specific internal energy respectively. They are unsubscripted when they refer to the fluid phase and carry the subscript p when they refer to the particulate phase. The symbols x, t, L, and E denote position along the duct, time, interphase linear momentum transfer per unit volume, and interphase energy transfer per unit volume respectively. In both (1) and (2) the equations result from balances of mass, linear momentum, and energy in that order.

If the particulate volume fraction is denoted by ϕ, it follows that

$$\rho = (1 - \phi)\bar{\rho}, \qquad \rho_p = \phi\bar{\rho}_p, \qquad p = (1 - \phi)\bar{p}, \qquad p_p = \phi\bar{p}_p \tag{3}$$

In (3) the symbols $\bar{\rho}$ and $\bar{p}$ denote true density (mass of phase per unit volume of phase) and true pressure (force on phase per unit area of phase) respectively.

Constitutive equations are required to supplement the balance laws and render the problem determinate. These can be formulated on a variety of levels of complexity and a multitude of proposals exist in the literature (see, for instance, Kanaruksunti and Peddieson [3] for a partial summary). In the present paper the equations will later be linearized and it can be shown that linearization wipes out most of the differences between the various proposed formulations. Thus the following representative constitutive equations (which are objective and which lead to non-elliptic field equations) will be employed without further comment.

$$e = cT, \qquad e_p = c_p T_p$$

$$L = p_s \partial_x \phi + \bar{\rho}_p \phi (L_1(u-u_p) + L_2(\partial_t u - \partial_t u_p + u_p \partial_x u - u \partial_x u_p))$$

$$E = c_p \bar{\rho}_p \phi (E_1(T-T_p) + E_2(\partial_t T - \partial_t T_p + u_p \partial_x T - u \partial_x T_p))$$

$$\bar{p}_p = p_s, \qquad p_s = \bar{p} - P\bar{\rho}\phi(u - u_p)^2 \tag{4}$$

where the symbols c and T denote specific heat at constant volume and temperature respectively and the subscript **s** indicates that the variable to which it is attached represents the average over a typical volume element of the values of that variable evaluated at the particle surfaces. The quantities L_1, L_2, E_1, E_2, and P are constitutive coefficients (the specific values of which depend on the dimensions, shapes, and distributions of the particles). The first term on the right-hand side of (4c) accounts for buoyancy. The terms multiplied by L_1 and E_1 in (4c) and (4d) account for steady-state drag and heat transfer respectively while those multiplied by L_2 and E_2 account roughly for unsteady effects (such as added mass). The term multiplied by P in (4f) accounts for the Bernoulli effect. It should be noted that the forms of (1), (2), (3), and (4) are independent of which phase is the gas and which the liquid.

For the case of liquid drops suspended in a perfect gas the final constitutive equation can be written as

$$\bar{p} = (\gamma - 1)c\bar{\rho}T \tag{5}$$

where γ is the ratio of specific heats for the gas. Combining (1) - (5), making the usual acoustic linearizations about a uniform state characterized by

$$\bar{\rho} = \bar{\rho}_o, \qquad \phi = \phi_o, \qquad u = u_p = 0, \qquad T = T_p = T_o \tag{6}$$

and rearranging the results produces a set of three equations governing the dimensionless dependent variables

$$\overset{*}{u} = u/u_o, \qquad \overset{*}{u}_p = u_p/u_o, \qquad \overset{*}{T} = T/T_o - 1; \qquad u_o = (\gamma(\gamma-1)cT_o)^{1/2} \tag{7}$$

where u_o is the speed of sound in a clear gas. These equations involve the dimensionless independent variables

$$\overset{*}{x} = L_{10}x/u_o, \qquad \overset{*}{t} = L_{10}t \tag{8}$$

and the dimensionless parameters

$$\overset{*}{\kappa} = (\bar{\rho}_p\phi_o)/(\bar{\rho}_o(1 - \phi_o)), \qquad \overset{*}{\varepsilon} = \bar{\rho}_o/\bar{\rho}_p, \qquad \overset{*}{\mu} = c/c_p$$
$$\overset{*}{L}_{20} = L_{20}, \qquad \overset{*}{E}_{10} = E_{10}/L_{10}, \qquad \overset{*}{E}_{20} = E_{20} \tag{9}$$

where the subscript o indicates the initial state. The governing equations can be written as

$$\partial_{tt}u + (\partial_{xt}T - (\partial_{xx}u + \kappa\varepsilon\partial_{xx}u_p))/\gamma + \kappa(\partial_t u - \partial_t u_p + \varepsilon M_o(\partial_{tt}u - \partial_{tt}u_p)) = 0$$
$$\partial_{tt}u_p + \varepsilon(\partial_{xt}T - (\partial_{xx}u + \kappa\varepsilon\partial_{xx}u_p))/\gamma + \partial_t u_p - \partial_t u + \varepsilon M_o(\partial_{tt}u_p - \partial_{tt}u) = 0$$
$$\partial_{tt}T + \delta(1 + \kappa/\mu)\partial_t T + (\gamma - 1)(\partial_{xt}u + \delta(\partial_x u + \kappa\varepsilon\partial_x u_p)) = 0 \tag{10}$$

where the stars have been dropped for simplicity and the substitutions

$$L_{20} = \varepsilon M_o, \qquad E_{20} = 0 \tag{11}$$

(M_o being an added-mass coefficient) have been made.

For the case of perfect gas bubbles suspended in a liquid the final constitutive equation can be written as

$$\bar{p}_p = (\gamma_p - 1)c_p\bar{\rho}_p T_p \tag{12}$$

where γ_p is the ratio of specific heats for the gas.

Combining (1) - (4) and (12), linearizing about (6), and rearranging produces a set of three equations governing the dimensionless dependent variables

$$\overset{*}{u} = u/u_{po}, \qquad \overset{*}{u}_p = u_p/u_{po}, \qquad \overset{*}{T}_p = T_p/T_o - 1; \qquad u_{po} = (\gamma_p(\gamma_p-1)c_pT_{po})^{1/2} \tag{13}$$

where u_{po} is the sound speed associated with the gas. The above-mentioned equations contain the dimensionless independent variables given by (8) with u_o replaced by u_{po} and the dimensionless parameters given by (9) with $\bar{\rho}_p$ replaced by ρ_{po} and ρ_o replaced by $\bar{\rho}$. These equations can be written as

$$\partial_{tt}u + (\partial_{xt}T_p - \partial_{xx}u/(\kappa\varepsilon) - \partial_{xx}u_p)/(\gamma_p\varepsilon) + \kappa(\partial_t u - \partial_t u_p$$
$$+ \varepsilon M_o(\partial_{tt}u - \partial_{tt}u_p)) = 0$$

$$\partial_{tt}u_p + (\partial_{xt}T_p - \partial_{xx}u/(\kappa\varepsilon) - \partial_{xx}u_p)/\gamma_p + \partial_t u_p - \partial_t u$$
$$+ \varepsilon M_o(\partial_{tt}u_p - \partial_{tt}u) = 0$$
$$\partial_{tt}T_p + (1+\kappa/\mu)\delta\partial_t T_p + (\gamma_p - 1)(\partial_{xt}u_p + (\delta/\mu)(\kappa\partial_x u_p + \partial_x u/\varepsilon)) = 0 \tag{14}$$

where the stars have again been dropped and use has been made of (11).

It can be seen that (10) and (14) appear to have similar mathematical forms. It will be demonstrated subsequently that, despite this fact, gas-liquid drop suspensions and liquid-gas bubble suspensions can respond quite differently to the same excitation.

LIMITING CASES

In this section certain limiting forms of (10) and (14) will be discussed. From these will be deduced some corresponding limiting forms of the speed of propagation of small disturbances. These calculations will be carried out both to establish connections with previous work and to aid in the interpretation of the numerical results to be presented later.

The behavior of the gas-liquid drop suspension is governed by (10). The next three paragraphs present some interesting limiting forms of these equations and discuss the physical significance of each one.

First, consider the case of a finite particle loading κ and a small volume fraction ϕ_o. From (9a,b) it is apparent that this corresponds to the limit of (10) as $\varepsilon \to 0$ with κ remaining finite. This yields

$$\partial_{tt}u + (\partial_{xt}T - \partial_{xx}u)/\gamma + \kappa(\partial_t u - \partial_t u_p) = 0$$
$$\partial_t u_p + u_p - u = 0$$
$$\partial_{tt}T + \delta(1 + \kappa/\mu)\partial_t T + (\gamma - 1)(\partial_{xt}u + \delta\partial_x u) = 0 \tag{15}$$

where (15b) is obtained by integrating the simplified form of (10b) once with respect to t. Equations (15) have been extensively studied (see, for instance, Marble [4]). For the special case of $\gamma = 1$ (15c) is satisfied by T = 0 and (15a,b) become

$$\partial_{tt}u - \partial_{xx}u + \kappa(\partial_t u - \partial_t u_p) = 0, \qquad \partial_t u_p + u_p - u = 0 \tag{16}$$

These are useful model equations for the qualitative study of particulate attenuation effects and have been used as such by Rochelle and Peddieson [5], [6] and Rasmussen [2]. The usefulness of (16) is increased by the fact that these equations are mathematically analogous to certain equations arising in viscoelasticity and in the study of chemically reacting flows. This makes it possible to adopt solutions already obtained in these fields to the study of wave propagation in suspensions.

Next, consider a suspension exhibiting no steady-state drag or heat transfer. The appropriate limiting forms of (10) are

$$\partial_{tt}u + (\partial_{xt}T - (\partial_{xx}u + \kappa\varepsilon\partial_{xx}u_p))/\gamma + \kappa\varepsilon M_o(\partial_{tt}u - \partial_{tt}u_p) = 0$$

$$\partial_{tt}u_p + \varepsilon(\partial_{xt}T - (\partial_{xx}u + \kappa\varepsilon\partial_{xx}u_p))/\gamma + \varepsilon M_o(\partial_{tt}u_p - \partial_{tt}u) = 0$$

$$\partial_t T + (\gamma-1)\partial_x u = 0 \tag{17}$$

where (17c) is obtained by integrating the simplified form of (10c) once with respect to t. It can be seen that for the small volume fraction-finite particle loading situation ($\varepsilon \to 0$) there will be no coupling between the motions of the phases. For this reason the term frozen flow is attached to this limiting case. It should be noted that this terminology is not strictly accurate for finite volume fractions. Equations (17) can be combined into the wave equation

$$\partial_{tt}u - u_{fg}^2\partial_{xx}u = 0; \quad u_{fg}^2 = (1+\kappa\varepsilon^2/\gamma + M_o\varepsilon(1+\kappa\varepsilon)(1+\kappa\varepsilon/\gamma))/(1+M_o\varepsilon(1+\kappa)) \tag{18}$$

where u_{fg} is the frozen wave speed for the gas-liquid drop suspension. It can be seen that this quantity is independent of δ and μ. If added mass is neglected ($M_o = 0$) (18b) reduces to

$$u_{fg}^2 = 1 + \kappa\varepsilon^2/\gamma \tag{19}$$

This corresponds to a situation in which the only coupling between the phases is the geometric one due to the finite volume fraction. For small volume fractions ($\varepsilon \to 0$) (19) reduces to

$$u_{fg} = 1 \tag{20}$$

Which is the wave speed in a clear gas.

Finally, consider a suspension exhibiting almost complete interphase momentum and energy transfer. The appropriate governing equations are found by multiplying (10b) by κ, adding the result to (10a), taking the limit of (10c) as $\delta \to \infty$, and setting $u_p = u$ in both of these results to get

$$(1+\kappa)\partial_{tt}u + (1+\kappa\varepsilon)(\partial_{xt}T - (1+\kappa\varepsilon)\partial_{xx}u)/\gamma = 0$$

$$(1+\kappa/\mu)\partial_t T + (\gamma-1)(1+\kappa\varepsilon)\partial_x u = 0 \tag{21}$$

Equations (21) have a form appropriate for a single gas with modified density and compressibility. Since the suspension behaves as a single-phase material, the name equilibrium flow is associated to this limiting case. Equations (21) can be combined to yield

$$\partial_{tt}u - u_{eg}^2\partial_{xx}u = 0; \quad u_{eg}^2 = (1+\kappa\varepsilon)^2(\gamma+\kappa/\mu)/(\gamma(1+\kappa)(1+\kappa/\mu)) \tag{22}$$

where u_{eq} is the equilibrium wave speed for the gas-liquid drop suspension. This quantity is independent of M_o and δ. For small volume fractions (22b) reduces to

$$u_{eg}^2 = (\gamma + \kappa/\mu)/(\gamma(1 + \kappa)(1 + \kappa/\mu)) \tag{23}$$

The behavior of the liquid-gas bubble suspension is governed by (14). The next three paragraphs will discuss the limits corresponding to those introduced above.

For a liquid-gas bubble suspension the case of small volume fractions ($\varepsilon \to 0$ with κ remaining finite) has no physical meaning. In a suspension of this type the only real compressibility derives from the gas contained in the bubbles. This can be transformed into effective compressibility of the liquid phase only through the geometric effect associated with a finite volume fraction (see (3a)). Without this disturbances cannot be transmitted through the liquid phase and wave motion is impossible. For this reason, no equations appropriate for this limit will be presented.

For frozen flow operations analogous to those discussed for the gas-liquid drop suspension result in

$$\partial_{tt}u - u_{f\ell}^2\partial_{xx}u = 0; \qquad u_{f\ell}^2 = (1+1/(\kappa\gamma\varepsilon^2)+M_o(1+\kappa\varepsilon)(1+1/(\kappa\gamma\varepsilon)))/(1+M_o\varepsilon(1+\kappa)) \tag{24}$$

where $u_{f\ell}$ is the frozen sound speed for the liquid-gas bubble suspension. It is independent of δ and μ. With added mass neglected (24b) reduces to

$$u_{f\ell}^2 = 1 + 1/(\kappa\gamma\varepsilon^2) \tag{25}$$

For equilibrium flow operations analogous to those discussed for the gas-liquid drop suspension lead to

$$\partial_{tt}u - u_{e\ell}^2\partial_{xx}u = 0; \qquad u_{e\ell}^2 = (1+\kappa\varepsilon)^2(\gamma+\mu/\kappa)/(\kappa\gamma\varepsilon^2(1+\mu/\kappa)(1+\kappa)) \tag{26}$$

where $u_{e\ell}$ is the equilibrium wave speed of the liquid-gas bubble suspension. It is independent of δ and M_o. For $\gamma = 1$ (26) reduces to

$$u_{e\ell}^2 = ((1 + \kappa\varepsilon)/(\kappa\varepsilon))^2\kappa/(1 + \kappa) = \kappa/((1 + \kappa)\phi_o{}^2) \tag{27}$$

where (9a) and (9b) have been used to obtain

$$\phi_o = \kappa\varepsilon/(1 + \kappa\varepsilon) \tag{28}$$

which was, in turn, used to obtain (27b) from (27a). For $\kappa << 1$ (9b) reduces to

$$u_{e\ell}^2 = \kappa/\phi_o{}^2 \tag{29}$$

which is identical to the result given by Van Wijngaarden [7] who considered equilibrium flow of a bubbly liquid under isothermal conditions for small particle loadings. Equation (29) provides a very simple illustration of the fact, mentioned earlier, that no wave propagation is possible in the limit of $\phi_o \to 0$ (since $u_{e\ell} \to \infty$).

The previous discussion reveals that both the frozen and equilibrium limits are associated with nondispersive wave propagation. This is true for both types of suspensions under discussion. In general the frozen and equilibrium wave speeds are distinct. There are some cases, however, in which they are the same. As an example, the parametric values $\varepsilon = 1$, $\mu = 1$ produce

$$u_{fg}^2 = u_{eg}^2 = 1 + \kappa/\gamma, \qquad u_{f\ell}^2 = u_{e\ell}^2 = 1 + 1/(\kappa\gamma) \tag{30}$$

regardless of the values of the other parameters. This suggests the possibility that nondispersive wave propagation may occur even if the conditions for equilibrium or frozen flow are not satisfied. That this is, in fact, the case is demonstrated by Kanaruksunti and Peddieson [8].

SOLUTION PROCEDURE

Consider the problem of wave motion in an initially quiescent suspension in a duct caused by a step increase in fluid velocity at the end. This problem will be solved by Laplace transform analysis with s denoting the Laplace transform parameter and a superposed bar indicating a Laplace transform.

Because of the relatively large number of dimensionless parameters appearing in (10) and (14), it was decided to fix those which are not central to describing the two-phase nature of the system; namely γ and μ. These were assigned values of 7/5 and 1 respectively. This is reflected in the following equations.

If the Laplace transform of (10) is taken with respect to time and the results are rearranged one can obtain algebraic equations relating $\bar{u}_p$ to $\bar{u}$ and $\bar{T}$ to $\bar{u}'$ together with the differential equation

$$\bar{u}'' - p^2\bar{u} = 0; \qquad p = as((1 + \alpha_1 s + \alpha_2 s^2)/(1 + \beta_1 s + \beta_2 s^2))^{1/2} \tag{31}$$

where

$$a = (1 + \kappa)(1 + \kappa\varepsilon)/(1 + 5\kappa/7)^{1/2}$$

$$\alpha_1 = 1/(\delta(1+\kappa)) + (1+\varepsilon M_o(1+\kappa))/(1+\kappa), \quad \alpha_2 = (1+\varepsilon M_o(1+\kappa))/(\delta(1+\kappa)^2)$$

$$\beta_1 = (1+\varepsilon(M_o(1+2\kappa\varepsilon)+\kappa\varepsilon(1+\kappa\varepsilon M_o)))/(1+\kappa\varepsilon)^2+(1+5\kappa\varepsilon/7)/(\delta(1+\kappa\varepsilon)(1+5\kappa/7))$$

$$\beta_2 = (1+\varepsilon(M_o(1+\kappa\varepsilon)+5\kappa\varepsilon(1+M_o(1+\kappa\varepsilon)))/7)/(\delta(7/5+\kappa)(1+\kappa\varepsilon)^2) \tag{32}$$

The solution of (31a) subject to

$$\bar{u}(0) = 1/s, \quad \bar{u}(x) \to 0 \text{ as } x \to \infty \tag{33}$$

is

$$\bar{u} = \exp(-ps)/s \tag{34}$$

No closed-form inversion of (34) was apparent to the present authors. For this reason use was made of the approximate inversion technique discussed by Rasmussen [1], [2]. (It was felt that this method would produce results appropriate for comparison of the behaviors of the two types of suspensions without requiring extensive numerical work.) This yields (see [1] and [2] for details)

$$u = (abx/\pi^{\frac{1}{2}}) \int_0^t H(t-t_1)\exp(-(b(t_1-ax))^2/t_1)dt_1/t_1^{3/2} \tag{35}$$

where H denotes the unit step function, t_1 is a dummy variable and

$$b = 1/(2(\beta_1-\alpha))^{\frac{1}{2}} \tag{36}$$

The convolution integral (35) was evaluated numerically by Simpson's rule to obtain the numerical results to be presented in the next section. Similar convolution integrals were determined for u_p and T (these are omitted for the sake of brevity) and were evaluated in a similar way.

The analysis of the liquid-gas bubble suspension is similar to that discussed above. Rearrangement of the Laplace transforms of (14) produces algebraic equations relating $\bar{u}_p$ to $\bar{u}$ and $\bar{T}_p$ to $\bar{u}'$ together with (31) with (32) replaced by

$$a = \varepsilon(1 + \kappa)/((1 + \kappa\varepsilon)(1 + 5/(7\kappa)^{\frac{1}{2}})$$

$$\alpha_1 = (1+\delta(1+\varepsilon M_o(1+\kappa)))/(\delta(1+\kappa)), \quad \alpha_2 = (1+\varepsilon M_o(1+\kappa))/(\delta(1+\kappa)^2)$$

$$\beta_1 = (1+7\kappa\varepsilon/5)/(\delta(1+7\kappa/5)(1+\kappa\varepsilon)) + (1+\varepsilon(M_o+\kappa\varepsilon(1+M_o(2+\kappa\varepsilon))))/(1+\kappa\varepsilon)^2$$

$$\beta_2 = (1+\varepsilon(M_o+7\kappa\varepsilon(1+M_o(12/7+\kappa\varepsilon))/5))/(\delta(1+7\kappa/5)(1+\kappa\varepsilon)^2) \tag{37}$$

The inversion process is the same as that discussed previously and leads to (35).

For reasons discussed in [1] and [2], the inversion process employed herein is not valid for small times. This should be kept in mind when viewing the numerical results to be presented in the next section.

RESULTS AND DISCUSSION

In the course of the work reported on herein a large number of cases were considered. Only a small representative sample of these will be discussed in this section. Figures 1-5 pertain to the gas-liquid drop suspension while Figures 6-9 pertain to the liquid-gas bubble suspension. Since the objective of this work is to compare the behavior of two idealized models of fluid-particle suspensions, no attempt was made to choose sets of parameters corresponding to any actual suspension or to make detailed comparisons with experimental results. Instead, the parametric values were assigned in such a way as to facilitate the determination of parametric trends.

Figure 1 presents some typical results for u plotted as a function of x for various values of t. It can be seen that the wave speed decreases and the initially sharp wave front spreads out as t increases. This behavior is typical of wave propagation in a dissipative medium. In this case the dissipation is provided by steady-state drag and heat transfer.

Rochelle and Peddieson [6] and Rasmussen [2] showed that according to the model equations (16) the wave front travels with the frozen wave speed for small times and with the equilibrium wave speed (in an average sense) for large times. Figure 1 is consistent with this behavior but it should be recalled that the present method of analysis cannot be employed for small values of t.

In Figure 2 the same data presented in Figure 1 is replotted versus x/t. It can be seen that on this scale a sharp wave front reappears in the limit as $t \to \infty$. Presenting the data in this way makes it clear that this wave front travels with speed u_{eg} (= 0.72 for the parametric values associated with Figures 1 and 2). All subsequent results will be presented in the form of Figure 2. It is felt that this is the most efficient way to display the data.

Figures 3-5 show the influence of κ (and, thus, of ϕ_o) on the behavior of u, u_p, and T. These results clearly show the influence of particulate matter on the flow structure. For $\kappa = 0$ the motion of the gas phase is independent of that of the drop phase and nondispersive wave propagation occurs in the gas (the wave speed being that of the clear gas). The solutions corresponding to this case can be found in closed form to be

$$u = H(t - x), \qquad T = 2H(t - x)/5$$

$$u_p = (1 - (1 - \varepsilon)\exp(-(t-x)/(1+\varepsilon M_o))/(1+\varepsilon M_o))H(t-x) \tag{38}$$

As κ increases from 0.01 to 10 (with a corresponding increase, as computed from (23), of ϕ_o from 0.009 to 0.5) the sharp wave front begins to spread out and eventually disappears altogether. At the same time the average speed of the wave front decreases. These results are qualitatively similar to those obtained by solving (15) (appropriate for small volume fractions and finite particle loadings). This data is not presented graphically for the sake of brevity. Quantitatively, on the other hand, the two sets of predictions are quite different. This was found to be true for a wide variety of situations which were investigated.

Figure 6 shows some typical results for u plotted versus x/t for various values of t for the liquid-gas bubble suspension. The small value of κ employed is typical of liquid-gas bubble suspensions. It can be seen that the time required for the liquid-gas bubble suspension to reach equilibrium conditions will normally be much longer than for the gas-liquid drop suspension (compare Figure 2). In Figure 6 the sharp wave front traveling at the frozen wave speed

(u_{ef} = 0.98) can still be discerned in all the curves presented.

Figures 7-9 illustrate the influence of κ on u, u_p, and T_p for the liquid-gas bubble suspension. Considerable overshoot can be observed in the values of u_p and T_p behind the wavefront. This behavior is typical of a large number of cases investigated in the course of the work reported on herein. This phenomenon was not observed in any of the cases investigated for the gas-liquid bubble suspension. These observations represent a significant qualitative difference in the response of the two suspensions to the same excitation. With ε fixed, variations in κ are equivalent to variations of ϕ_o (see (23). It can be seen from Figures 7-9 that the value of the volume fraction has a significant quantitative effect on the results.

CONCLUSION

In the present paper the governing equations for one-dimensional acoustic wave propagation in an idealized gas-liquid drop suspension and in an idealized liquid-gas bubble suspension were developed. Certain limiting forms of these equations were investigated and employed to deduce some limiting forms of the acoustic wave speeds. The complete equations were then applied to the problem of wave propagation in a semi-infinite duct due to a step increase in fluid velocity at the end. The analysis was performed by the Laplace transform method and an approximate method, valid for moderate and large times, was used to determine the inverse transforms of the dependent variables in the form of convolution integrals. These were evaluated numerically and some typical results of these calculations were presented graphically.

It was found that both gas-liquid drop suspensions and liquid-gas bubble suspensions are capable of supporting non-harmonic small-amplitude wave propagation with a finite wave speed. In both cases a step fluid velocity excitation produces a wave having an initially sharp front which spreads out as time increases. In both cases the speed of the initial sharp wave front is that associated with the frozen flow limit while the average wave speed of the final diffused wave front is that associated with the equilibrium flow limit. The major difference in the behaviors of the two suspensions appeared to be the presence of overshoot in some of the variables for the liquid-gas bubble suspension and the absence of this phenomenon in the gas-liquid drop suspension.

REFERENCES

[1] RASMUSSEN, M. L., J. Engr. Math 9, 261 (1975).

[2] RASMUSSEN, M. L., J. Appl. Mech. 44, 354 (1977).

[3] KANARUKSUNTI, T. and PEDDIESON, J., "Comparison of Constitutive Equations for One-Dimensional Two Phase Flow," presented at the 16th Annual Meeting of the Society of Engineering Science (1979).

[4] MARBLE, F. E., Ann. Rev. Fluid Mech. 2, 397 (1970).

[5] ROCHELLE, S. G. and PEDDIESON, J., "Small Oscillations of a Particulate Suspension in a Tube," presented at the 11th Annual Meeting of the Society of Engineering Science (1974).

[6] ROCHELLE, S. G. and PEDDIESON, J., "One Dimensional Wave Propagation in Particulate Suspensions," presented at the 13th Annual Meeting of the Society of Engineering Science (1976).

[7] VAN WIJNGAARDEN, L., Ann. Rev. Fluid Mech. 4, 369 (1972).

[8] KANARUKSUNTI, T. and PEDDIESON, J., "Harmonic Wave Propagation in a Finite Volume Fraction Particulate Suspension," presented at the Fall Meeting of SIAM (1979).

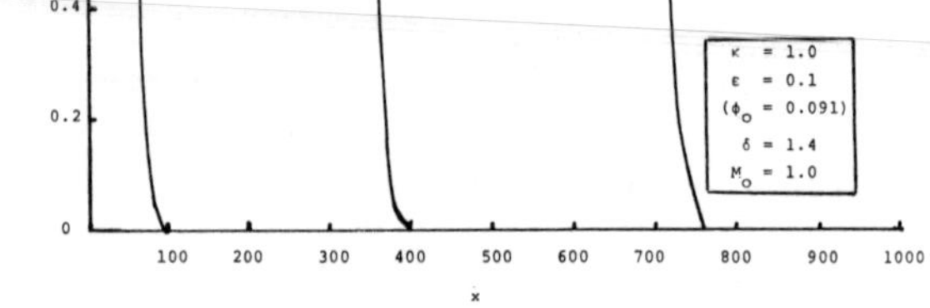

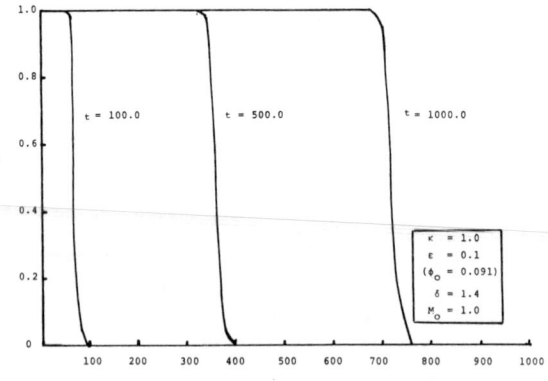

Figure 1. Gas Velocity vs. Distance

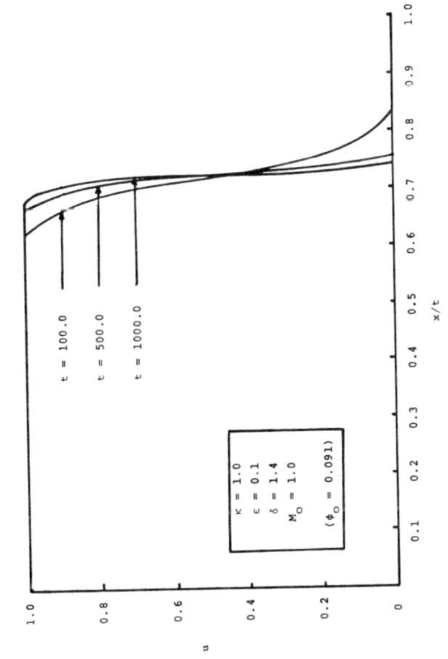

Figure 2. Gas Velocity vs. Distance

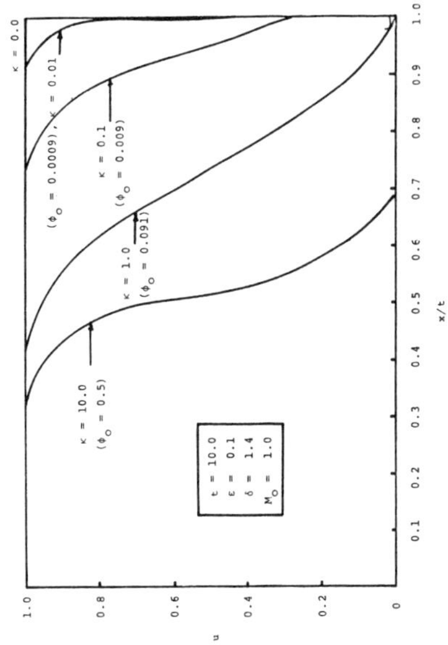

Figure 3. Gas Velocity vs. Distance

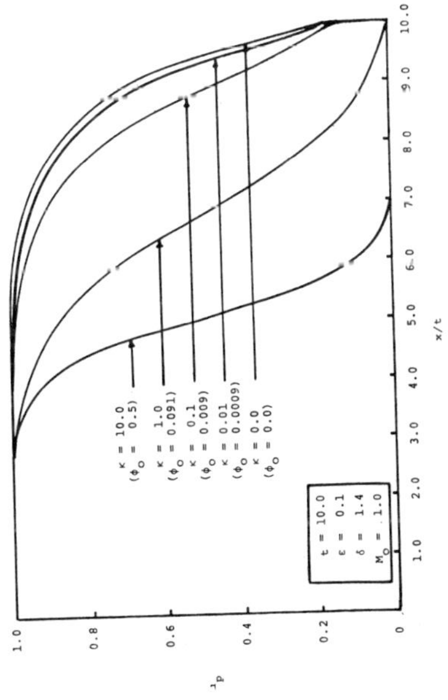

Figure 4. Particle Velocity vs. Distance

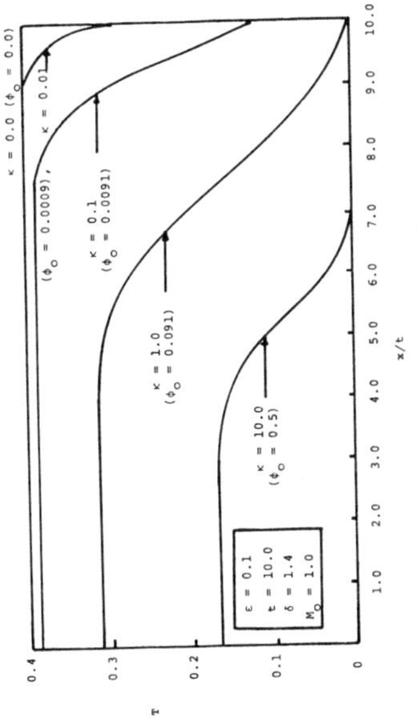

Figure 5. Gas Temperature vs. Distance

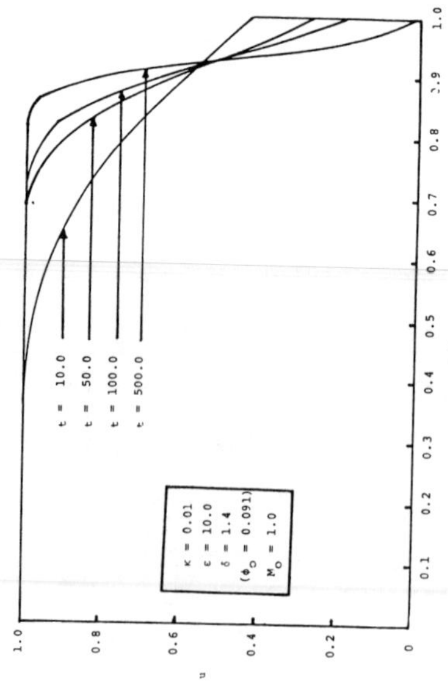

Figure 6. Liquid Velocity vs. Distance

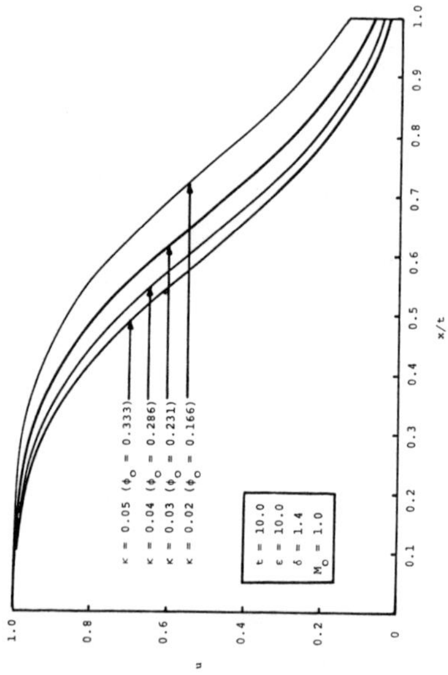

Figure 7. Liquid Velocity vs. Distance

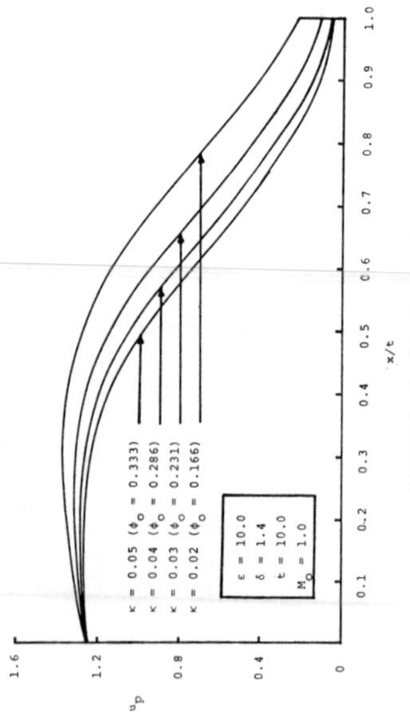

Figure 8. Gas Velocity vs. Distance

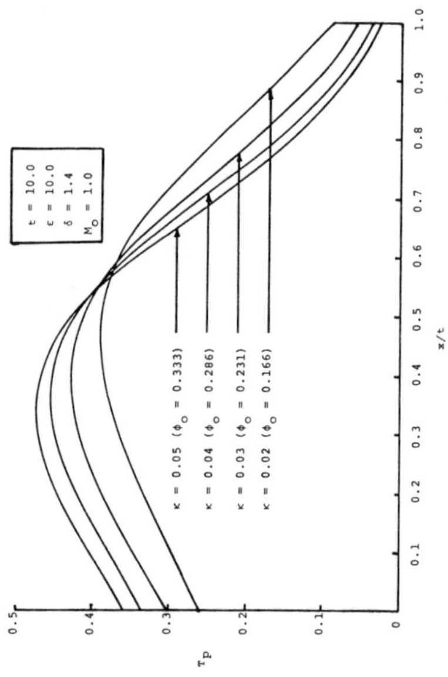

Figure 9. Gas Temperature vs. Distance

SLUG FLOW HEAT TRANSFER IN A CYLINDER WITH A RING OF TUBES*

A. K. Naghdi
Professor of Aeronautical-Astronautical Engineering and Mathematical Sciences, Purdue University School of Engineering and Technology at Indianapolis, Indiana

ABSTRACT

The problem of slug flow heat transfer in a cylinder with a ring of equally spaced circular tubes is investigated. It is assumed that the outer surface of the cylinder is insulated, and that the common temperature of the inner tubes is uniform. The problem is solved with a new technique of combining a closed form particular solution of the governing equation in polar coordinates with complementary solutions in multi-bipolar coordinate systems. The particular solution with singularities at the centers of the inner tubes, and the complementary solutions are designed such that they automatically satisfy the zero radial heat flux condition on the outer boundary of the cylinder. The unknown constants involved in the combination of these functions are evaluated through the point-by-point satisfaction of the inner boundary condition and the method of least square error. Numerical values of temperature at various points, bulk temperature, and Nusselt number are presented for different cases.

Introduction

In recent years a few authors have investigated problems similar to the one studied in this article. Among these authors are Sparrow and Loeffler, El-Saden, Snyder, and Rowley and Payne. Sparrow and Loeffler [1,2] employed polar coordinates in order to obtain solutions for a fully developed laminar flow and the related fully developed heat transfer in an array of tubes arranged in triangular and square forms. Since they assumed infinitely long and wide cross-sectional area for the arrays, their solutions were reduced to those for the cases of flow and heat transfer around one tube. The continuity condition along each plane separating the regions was then satisfied through the point-by-point technique. El-Saden [3] using bipolar coordinate system, derived a solution for the problem of steady-state heat conduction in an eccentrically hollow cylinder with internal heat generation. Snyder [4] employed the same technique to obtain the solution for slug flow heat transfer in an eccentric annulus. Some of the results found in this investigation for the particular case of one inner tube shall be compared with those given by Snyder. Rowley and Payne [5] used Howland functions [6] to derive a solution for the problem of a heat generating cylinder cooled by a ring of circumferential holes.

The technique employed in this investigation is entirely different from those used by the previous authors. It combines a closed form particular integral of the governing equation in polar coordinates with complementary solutions in a system of multi-bipolar coordinates. The particular solution has singularities at the center of each inner tube, and satisfies automatically the zero radial heat flux condition at the outer boundary of the cylinder. It should be noted that such a solution in closed-form cannot be written in a multi-bipolar coordinate system. It shall be shown by numerical results that the present technique produces remarkably accurate solutions as compared to the computer time it requires.

Method of Solution

Consider a fluid flow with uniform velocity U in the axial direction in a cylinder having a row of N equally spaced internal tubes (see Fig. 1). It is assumed that the properties of the fluid do not change with temperature, and that the rate of change of temperature in the axial direction $\frac{\partial T}{\partial Z}$ remains constant. It is also assumed that the outer boundary is insulated and the boundary temperatures of the inner tubes are all the same and uniform. The governing differential equation for the temperature field T in the cross-section is given by:

$$\nabla^2 T = \frac{\rho C_p U \frac{\partial T}{\partial Z}}{K} \tag{1}$$

in which ρ, C_p, k are respectively the mass density, specific heat at constant pressure, and thermal conductivity of the fluid.

The Particular Solution The differential equation (1) in dimensionless polar coordinate form is written as

$$\left.\begin{aligned} &\frac{\partial^2 \overline{T}}{\partial \overline{r}^2} + \frac{1}{\overline{r}}\frac{\partial \overline{T}}{\partial \overline{r}} + \frac{1}{\overline{r}^2}\frac{\partial^2 \overline{T}}{\partial \theta^2} = \overline{P}, \\ &\overline{T} = \frac{T}{T_1}, \quad \overline{r} = \frac{r}{R}, \quad \overline{P} = \frac{\rho C_p R^2 U \frac{\partial \overline{T}}{\partial Z}}{K}, \end{aligned}\right\} \tag{2}$$

in which T_1 is the common uniform temperature of all inner tubes. The zero radial heat flux condition at the outer boundary of the cylinder dictates that all of the heat absorbed or rejected by the fluid must cross the inner boundaries. For the steady-state case considered, this means that there have to be heat sources or sinks located inside the inner tubes.

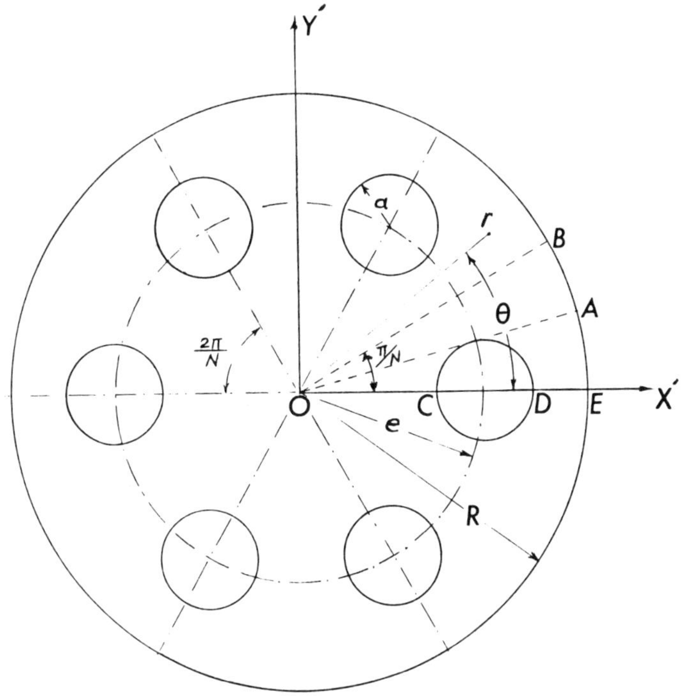

Fig. 1. Circular Cylinder with N Circular Inner Tubes

Due to this fact, a set of N segmental line sources confined in central angle $\overline{\theta}$ and having the intensity $\overline{S}$ per unit length are placed inside the inner tubes (see Fig. 2). Thus, the differential equation (2) is written in a new form:

$$\left.\begin{aligned} &\frac{\partial^2 \overline{T}}{\partial \overline{r}^2} + \frac{1}{\overline{r}} \frac{\partial \overline{T}}{\partial \overline{r}} + \frac{1}{\overline{r}^2} \frac{\partial^2 \overline{T}}{\partial \theta^2} = \overline{P} + \\ &\delta(\overline{r} - \overline{r}_0) \left[S_0^* + S_1^* \operatorname{Cos}\theta + ---- \right] , \\ &\overline{r}_0 = \frac{r_0}{R} , \end{aligned}\right\} \qquad (2')$$

where δ is the unit impulse function, and S_0^* , S_1^* ,...S_n^* are the Fourier coefficients for N equally-spaced segmental line sources at the circumference $\overline{r} = \overline{r}_0$. They are given by

$$S_0^* = \frac{N\,\overline{S}\,\overline{\theta}}{\pi R} , \qquad (3)$$

$$S_n^* = \frac{2N\bar{S}\bar{\theta}}{\pi R} \cdot \text{Cos}\, nN\theta_o \quad . \tag{3}$$

It should be noted that the solutions of equation (2′) also satisfy equation (2) in the region between the outer and inner tubes as the sources are located inside the inner circles. The solution of equation (2′) subject to the boundary condition

$$\frac{\partial \bar{T}_p}{\partial \bar{r}} = 0 \qquad \text{at} \qquad \bar{r} = 1 \tag{4}$$

is sought in the form

$$\bar{T}_p = g(\bar{r}) + \sum_{n=o}^{\infty} f_n(\bar{r}) \, \text{Cos}Nn\theta \quad . \tag{5}$$

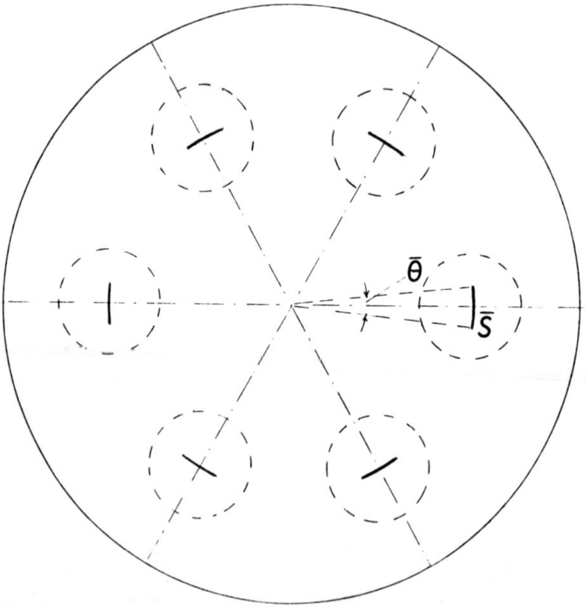

Fig. 2. Circular Region with N Segmental Line Sources

Substituting expression (5) in equation (2′), and employing the technique of variation of parameters [7], it is found

$$
\left.
\begin{aligned}
& g(\bar{r}) = \frac{1}{4} P \bar{r}^{2} , \\
& f_o(\bar{r}) = f_{op}(\bar{r}) + f_{oc}(\bar{r}) \qquad \text{for } \bar{r} \geq \bar{r}_o , \\
& f_o(\bar{r}) = f_{oc}(\bar{r}) \qquad \text{for } \bar{r} < \bar{r}_o , \\
& f_n(\bar{r}) = f_{nc}(\bar{r}) + f_{np}(\bar{r}) \qquad \text{for } \bar{r} \geq \bar{r}_o , \\
& f_n(\bar{r}) = f_{nc}(\bar{r}) \qquad \text{for } \bar{r} < \bar{r}_o , \\
& f_{oc}(\bar{r}) = A_o , \qquad f_{op}(\bar{r}) = S_o^* \bar{r}_o \ln\left(\frac{\bar{r}}{\bar{r}_o}\right) , \\
& f_{nc}(\bar{r}) = A_n \cdot (\bar{r})^{Nn} , \qquad n = 1,2,3,\ldots , \\
& f_{np}(\bar{r}) = \frac{S_n^*}{2Nn} \left[(\bar{r}_o)^{1-Nn} \cdot (\bar{r})^{nN} - (\bar{r}_o)^{1+Nn} \cdot (\bar{r})^{-nN} \right]
\end{aligned}
\right\} \qquad (6)
$$

It should be noted that the functions $f_{op}(\bar{r})$ are continuous at $\bar{r} = \bar{r}_o$. Applying relation (4) to the solution (5), and making use of equations (6), the following result is obtained:

$$A_n = - \frac{S_n^*}{2n} \left[(\bar{r}_o)^{1-Nn} + (\bar{r}_o)^{1+Nn} \right] . \qquad (7)$$

As the value of $\bar{\theta}$ tends to zero while the intensity $\bar{S}$ of the segmental line sources tends to infinity such that $\bar{S}\theta\bar{r}_o R$ tends to S, the values of S_o^*, S_1^*, ...S_n^* for the concentrated sources at the center the inner circular tubes are found:

$$
\left.
\begin{aligned}
& S_o^* = \frac{NS}{\Pi R^2 \bar{r}_o} , \\
& S_n = \frac{2NS}{\Pi R^2 \bar{r}_o} \qquad \mathrm{Cos} Nn\theta_o .
\end{aligned}
\right\} \quad \text{for concentrated source} \qquad (8)
$$

Here in equation (8) θ_o is the polar angular position of the first of N concentrated sources. Since the center of one of the circles is assumed to be on the negative side of the X′ axis (see Fig. 1.) the value of θ_o shall be set equal to Π.

Now, considering that the heat gained or lost by the fluid must be equal to that produced or absorbed by the sources, it is not too difficult to show that

$$S = -\frac{1}{2}\,\frac{\Pi R^2\,\bar{P}}{N} \tag{9}$$

Substituting (9) in (8) and utilizing the results in (6), finally the particular solution satisfying the zero radial heat condition at $\bar{r} = 1$ is found:

$$\left.\begin{aligned}
\bar{T}_p &= \frac{1}{4}\bar{P}\,\bar{r}^2 + A_o - \frac{1}{2}\bar{P}\ln\left(\frac{\bar{r}}{\bar{r}_o}\right) + \frac{\bar{P}}{4N}\sum_{n=1}^{\infty}\frac{1}{n}\left[(\bar{r}.\bar{r}_o)^{nN} + \left(\frac{\bar{r}_o}{\bar{r}}\right)^{nN}\right]\cdot\\
&\quad\cdot\left[\text{Cos}\,nN\,(\theta+\theta_o) + \text{Cos}\,nN\,(\theta-\theta_o)\right] \qquad \text{for } \bar{r} \geq \bar{r}_o, \text{ and}\\
\bar{T}_p &= \frac{1}{4}\bar{P}\,\bar{r}^2 + A_o + \frac{\bar{P}}{4N}\sum_{n=1}^{\infty}\frac{1}{n}\left[\left(\frac{\bar{r}}{\bar{r}_o}\right)^{nN} + (\bar{r}.\bar{r}_o)^{nN}\right]\cdot\left[\text{Cos}\,nN\,(\theta+\theta_o) + \text{Cos}\,nN\,(\theta-\theta_o)\right]\\
&\qquad \text{for } \bar{r} < \bar{r}_o \quad .
\end{aligned}\right\} \tag{10}$$

It is observed that each of the various series in relation (10) can be written written in the form $\sum_{j=1}^{\infty}\frac{e^{-j\xi}}{j}\,\text{Cos}\,j\theta$, in which $\xi > o$. According to the the previous works [8,9,10] these series have closed-form sums:

$$\sum_{j=1}^{\infty}\frac{e^{-j\xi}}{j}\,\text{Cos}\,j\phi = \frac{1}{2}\ln\left[\frac{\text{Cosh}\,\xi-1}{\text{Cosh}\,\xi-\text{Cos}\,\phi}\right] - \ln\,(1-e^{-\xi}) \quad . \tag{11}$$

Therefore, the particular solution can be written in the following closed-form:

$$\bar{T}_p = \frac{1}{4}\bar{P}\,\bar{r}^2 + A_o - \frac{1}{2}\bar{P}\ln\left(\frac{\bar{r}}{\bar{r}_o}\right) + G(\theta) \qquad \text{for } \bar{r} \geq \bar{r}_o, \tag{12}$$

$$\bar{T}_p = \frac{1}{4}\bar{P}\,\bar{r}^2 + A_o + G(\theta) \quad \text{for} \quad \bar{r} < \bar{r}_o,$$

$$\begin{aligned}
G(\theta) = \frac{\bar{P}}{4N}\Bigg[& \frac{1}{2}\ln\left(\frac{\cosh(N\ln[\bar{r}.\bar{r}_o]) - 1}{\cosh(N\ln[\bar{r}.\bar{r}_o]) - \cos N(\theta+\theta_o)}\right) \\
& - 2\ln\left(1 - e^{N\ln[\bar{r}.\bar{r}_o]}\right) + \frac{1}{2}\ln\left(\frac{\cosh(N\ln[\bar{r}.\bar{r}_o]) - 1}{\cosh(N\ln[\bar{r}.\bar{r}_o]) - \cos N(\theta_o-\theta)}\right) \\
& - \frac{1}{2}\ln\left\{\frac{\cosh(N\ln\frac{\bar{r}}{\bar{r}_o}) - 1}{\cosh(N\ln\frac{\bar{r}}{\bar{r}_o}) - \cos N(\theta_o+\theta)}\right\} \\
& + 2\ln\left(1 - e^{-N\ln\frac{\bar{r}}{\bar{r}_o}}\right) \\
& - \frac{1}{2}\ln\left\{\frac{\cosh(N\ln\frac{\bar{r}}{\bar{r}_o}) - 1}{\cosh(N\ln\frac{\bar{r}}{\bar{r}_o}) - \cos N(\theta_o-\theta)}\right\}\Bigg].
\end{aligned} \tag{12}$$

Complementary Solutions It is considered, for the time being, that there is only one eccentric circular tube in the cylinder [see Fig. 3.]. Employing bipolar coordinates [11], it is found that the governing equation for the complementary solution $\bar{T}_c$ takes the form:

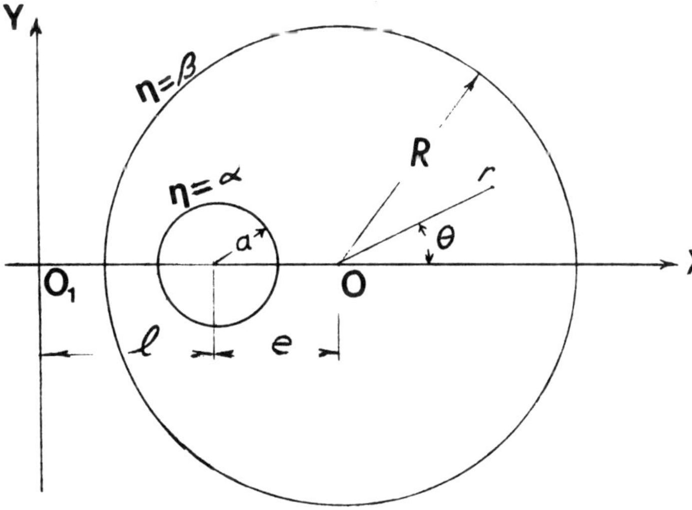

Fig. 3. Bipolar Coordinate System.

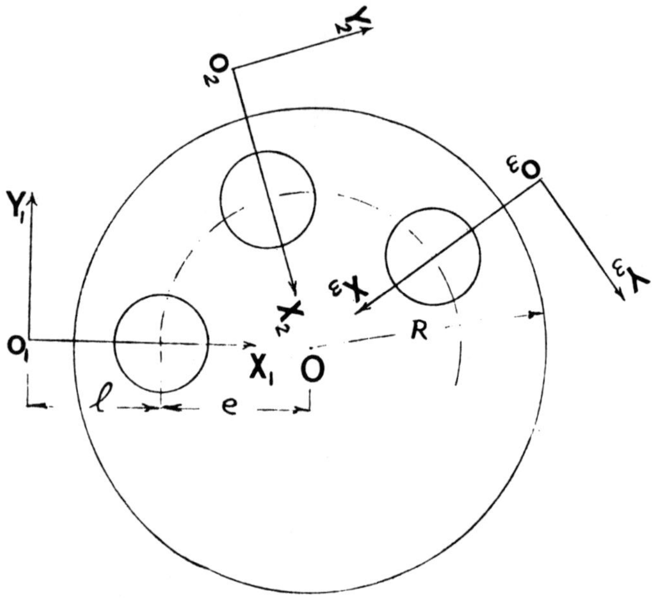

Fig. 4. Multi-Bipolar Coordinate System

$$
\left.
\begin{aligned}
&\nabla^2 \bar{T}_c = \frac{\partial^2 \bar{T}_c}{\partial \xi^2} + \frac{\partial^2 \bar{T}_c}{\partial \eta^2} = 0 \,, \\
&\xi = \text{Arctan}\, \frac{2\bar{c}\,\bar{y}}{\bar{x}^2 + \bar{y}^2 - \bar{c}^2} \,, \quad \bar{x} = \frac{x}{R} \,, \quad \bar{y} = \frac{y}{R} \,, \\
&\bar{c} = \frac{c}{R} = \bar{a}\sinh\alpha \,, \quad \eta = \frac{1}{2}\ \ln \left[\frac{(\bar{x} + \bar{c}\,)^2 + \bar{y}^2}{(\bar{x} - \bar{c}\,)^2 + \bar{y}^2}\right] \,, \\
&\bar{a} = \frac{a}{R} \,, \qquad \alpha = \cosh^{-1}\left[\frac{1 - \bar{a}^{\,2} - \bar{r}_o^{\,2}}{2\,\bar{a}\,.\,\bar{r}_o}\right] \,, \qquad \bar{r}_o = \bar{e} = \frac{e}{R} \,.
\end{aligned}
\right\} \qquad (13)
$$

The coordinate lines η = constant correspond to eccentric circles. In particular $\eta = \alpha$, and $\eta = \beta = \cosh^{-1}[\bar{a} \cosh \alpha + \bar{r}_o]$ respectively represent the inner and outer circular boundaries of the region. The solutions of equation (13) are obtained in the usual way. Thus, it is found:

$$\left.\begin{aligned} (\bar{T}_c^*)_o &= A_o^* \eta + B_o^* , \\ (\bar{T}_c^*)_n &= \left[A_n^* e^{n\eta} + B_n^* e^{-n\eta} \right] \cdot \cos n\xi , \end{aligned}\right\} \qquad (14)$$

in which A_o^*, B_o^*, A_n^*, B_n^* are constants of integration. Applying the condition of no radial heat flux at the outer boundary $\frac{\partial \bar{T}_c^*}{\partial \eta} = 0$ at $\eta = \beta$, the following results are obtained:

$$A_o^* = 0, \qquad B_n^* = A_n e^{2n\beta} , \quad n = 1,2,3,\ldots. \qquad (15)$$

It should be noted that B_o^* may be set equal to zero since an unknown constant A_o already is taken into account as a part of the particular solution. Utilizing (15) into (14) it is found for a region with one tube

$$\bar{T}_c^* = \sum_{n=1}^{\infty} A_n^* \left[e^{n\eta} + e^{n(2\beta-\eta)} \right] \cdot \cos n\xi \quad . \qquad (16)$$

The solution (16) suggests that for the case of a multiple tube cylinder the following eigenfunctions may be utilized:

$$\left.\begin{aligned} \bar{T}_c = \sum_{n=1}^{\infty} \bar{A}_n \Big\{ & \left[e^{n\eta_1} + e^{n(2\beta-\eta_1)} \right] \cdot \cos n\xi_1 + \\ & \left[e^{n\eta_2} + e^{n(2\beta-\eta_2)} \right] \cdot \cos n\xi_2 + \\ & \ldots + \left[e^{n\eta_N} + e^{n(2\beta-\eta_N)} \right] \cdot \cos n\xi_N \Big\} , \end{aligned}\right\} \qquad (17)$$

in which η_i and ξ_i for i = 1, 2, ...N are bipolar coordinates corresponding to $\bar{x}_i$ and $\bar{y}_i$ coordinate system (see Fig. 4.). Note that the solution (17) automatically satisfies the condition $\frac{\partial \bar{T}_c}{\partial \eta} = 0$ at $\eta = \beta$, and that it is periodical in the circumferential direction with the period of $\frac{2\Pi}{N}$.

Numerical Results

It was seen in the previous section that the particulr solution (12), and the eigenfunctions (17) automatically satisfy the zero radial heat flux condition at the outer boundary. It remains to fulfill the condition

$$\overline{T} = \overline{T}_p + \overline{T}_c = 1. \tag{18}$$

at the inner boundaries. Due to the geometrical symmetry and the periodicity of the solution, it is sufficient to satisfy this condition on one hald of one of the circular inner tubes. In order to achieve this goal, M_1 terms in the series solution (17) are retained and the boundary condition (18) is satisfied at $M_2 > M_1$ points of the mentioned inner boundary. This procedure leads to a system of M_2 linear algebraic equations with M_1 unknowns. The system is normalized and solved approximately by the method of least square error [12].

The solutions so obtained are remarkably accurate. For example, in a case of 4 inner tubes, and choosing $M_1 = 24$, $M_2 = 35$, the maximum value of relative error in satisfying the inner boundary condition is of the order of 10^{-11}. This outstanding accuracy shows how rapidly the series solution (17) converges. Thus, for practical engineering applications much smaller sets of linear equations can be utilized.

To ensure the validity of the solution presented, another technique for the particular case one inner tube has been employed. This technique uses a solution of Laplace's equation in polar coordinates which automatically satisfy the inner boundary condition $\overline{T} = 1$. The outer boundary condition is then satisfied by the employment of the method of least square error. The numerical results obtained from this method are in full agreement with those found by the technique presented in this article.

For the presentation of some of the numerical results,the following dimensionless parameter is introduced:

$$\sigma = \frac{K\,(T - T_1)}{\rho c_p\, c^2\, U \frac{\partial T}{\partial Z}} = \frac{\overline{T} - 1}{\overline{c}^{\,2} \cdot \overline{P}} \tag{19}$$

The values of

$$\sigma^* = \sigma_o - (\sigma_o)_{Ave.} = \frac{1}{\overline{c}^2 \cdot \overline{P}} \left[\overline{T}_o - (\overline{T}_o)_{Ave.} \right] \tag{20}$$

versus angle θ have been presented in Table 1 for different cases. Here in relation (20) σ_o and $\overline{T}_o$ are the quantities at the outer boundary, and the subscript "Ave." denotes the average values. The numerical values in the first two rows of Table 1 are for cylinders geometrically identical with those considered by Snyder [4]. For the case $\overline{a} = 0.28571$, $\overline{e} = 0.57143$ the graphs given in Ref. [4]show the values of σ^* starting approximately from -3.7 and ending with 1.3. These values obviously are not in agreement with the results presented in Table 1. However, since our results have been checked with the aforementioned independent technique of polar coordinates, it is believed that they are correct.

In Tables 2 and 3 the values of dimensionless fluid temperature $\overline{T}$ are presented for various points of the cross-section.

The heat flux per unit length of the cylinder is written as

$$Q = \rho\ U\ \Pi\ (R^2 - N\ a^2)\ C_p\ \frac{\partial T}{\partial Z} = 2\Pi aNh\ (T_B - T_1) \tag{21}$$

in which h is the convection heat transfer coefficient, and T_B is the bulk temperature of the fluid. Defining a Nusselt number on the basis of the radius of the inner tubes:

$$(NU)_a = \frac{ha}{K}\ , \tag{22}$$

it is easily concluded from (21) that

$$\left.\begin{aligned} (NU)_a &= \frac{(1 - N\overline{a}^{\,2})\ \overline{P}}{2N\ (1 - \overline{T}_B)}\ , \\ \overline{T}_B &= \frac{T_B}{T_1} \end{aligned}\right\} \tag{23}$$

The values of $\overline{T}_B$ are determined by a highly accurate eight order polynomial approximation for numerical intergration [12].

The values of $(NU)_a$, $\overline{T}_B$ and $(\overline{T}_o)_{ave}$ for different cases are presented in Table 4.

Conclusion

With the solution of a small set of linear algebraic equations, the presented technique produces results with acceptable engineering approximations. The technique can also be utilized for the solutions of other heat transfer problems concerning the same geometry. These problems include the cases of shape factors, cylinders with internal heat generation, and the case of a cylinder with a convective inner boundary condition.

Table 1. The Values of σ^* Versus θ for Various Cases. Center of One of The Tubes at $\theta_o = \pi/N$

Angles in Radians		0	$\pi/5N$	$2\pi/5N$	$3\pi/5N$	$4\pi/5N$	π/N
$\bar{a}$ = 0.2671, e = 0.57143, N = 1	σ^*	-2.0912	-1.8548	-1.1146	0.21909	2.1438	3.2834
$\bar{a}$ = 0.28571, e = 0.42857, N = 1	σ^*	-0.47230	-0.41247	-0.22955	0.07951	0.46420	0.66862
$\bar{a}$ = 0.125, e = 0.5, N = 3	σ^*	-0.07722	-0.06466	-0.02920	0.02080	0.06781	0.08772
$\bar{a}$ = 0.1, e = 0.6, N = 4	σ^*	-0.11440	-0.09581	-0.04334	0.03075	0.10052	0.13014

Table 2. The Values of $\bar{T}$ along Lines OA, OB, OC, and DE (See Fig.1) for The Case of $\bar{a}$ = 0.125, e = 0.5, N = 3, P = 0.75.

$\bar{r}$	0	1/6	1/3	1/2	2/3	5/6	1
$\bar{T}$ along OA	0.90893	0.91740	0.95534	0.10002	0.92940	0.87919	0.86470
$\bar{T}$ along OB	0.90893	0.90971	0.89799	0.87042	0.83928	0.81823	0.81098
$\bar{r}$	0	0.0625	0.1250	0.1875	0.2500	0.3125	0.3750
$\bar{T}$ along OC	0.90893	0.90990	0.91376	0.92207	0.93675	0.96077	1.0000
$\bar{r}$	0.625	0.6875	0.7500	0.8125	0.8750	0.9375	0.10000
$\bar{T}$ along DE	1.0000	0.95105	0.91954	0.89866	0.88538	0.87806	0.87577

TABLE 3. The Values of $\overline{T}$ along Lines CA, OB, OC, and DE (See Fig. 1) for The Case of $\overline{a} = 0.1$, $\overline{e} = 0.6$, $N = 4$, $\overline{P} = 0.75$.

$\overline{r}$	0	1/6	1/3	1/2	2/3	5/6	1
$\overline{T}$ along OA	0.89102	0.89667	0.91904	0.97323	0.97970	0.92272	0.90622
$\overline{T}$ along OB	0.89102	0.89567	C.90319	0.90029	0.88524	0.87009	0.86415
$\overline{r}$	0	.5/6	1/6	1.5/6	1/3	2.5/6	0.5
$\overline{T}$ along OC	0.89102	0.89236	0.89680	0.90565	0.92136	0.94867	1.0000
$\overline{r}$	0.7	0.75	C.80	0.85	0.90	0.95	1
T along DE	1.0000	0.96576	0.94368	0.92903	0.91967	0.91450	0.91287

TABLE 4. The Values of Bulk Temperature, Average Surface Temperature, Nusselt Number for Various Cases, and $\overline{P} = 0.75$.

	$\overline{T}_B$	$(\overline{T}_o)_{ave}$	$(NU)_a$
$\overline{a} = 0.125$, $\overline{e} = 0.5$, $N = 3$	0.87555	0.84131	0.98101
$\overline{a} = 0.1$, $\overline{e} = 0.6$, $N = 4$	0.90773	0.88694	0.97540
$\overline{a} = 0.28571$, $\overline{e} = 0.57143$, $N = 1$	0.52449	0.52441	0.72441
$\overline{a} = 0.28571$, $\overline{e} = 0.42857$, $N = 1$	0.63128	0.60556	0.93402

REFERENCES

[1] Sparrow, E. M., and Loeffler, A. L., Jr., "Longitudinal Laminar Flow Between Cylinders Arranged in Regular Array," A.I.Ch.E. Journal, Vol. 5, No. 3, pp. 325-330, 1959.

[2] Sparrow, E. M., Loeffler, A. L., Jr., and Hubbard, H. A., "Heat Transfer to Longitudinal Laminar Flow Between Cylinders," Journal of Heat Transfer, 83, No. 4, pp. 415-422, 1961.

[3] El-Saden, M. R., "Heat Conduction in an Eccentrically Hollow, Infinitely Long Cylinder with Internal Heat Generation," Diags. Trans. Ser. C 83, pp. 510-512, 1961.

[4] Snyder, W. T., "An Analysis of Slug Flow Heat Transfer in an Eccentric Annulus," A.I.Ch.E. Journal, Vol. 9, No. 4, pp. 503-506, 1963.

[5] Rowley, J. C. and Payne, J. B., "Steady State Temperature Solution for a Heat Generating Circular Cylinder Cooled by a Ring of Holes," bibliog. diags. Trans. Ser. C 86, J. Heat Transfer, pp. 531-536, 1964.

[6] Howland, R. C. J., "Potential Functions Periodicity in One Coordinate," Proc. of the Cambridge Phil. Soc., London, England, Vol. 30, pp. 315-326, 1935.

[7] Miller, F. H., Partial Differential Equations, 11th Edition, John Wiley, 1965.

[8] Kantorovich, L. V., and Krylov, V. I., Approximate Methods of Higher Analysis, Interscience Publishers, Inc. New York - The Netherlands, 1964.

[9] Naghdi, A. K., and Gersting, J. M., Jr., "The Effect of Transverse Shear Acting on the Edge of a Circular Cutout in a Simply Supported Circular Cylindrical Shell," Ing. - Arch., Vol. 42, No. 2, pp. 141-150, 1973.

[10] Naghdi, A. K., "On the Convergence of Series Solution for a Short Beam," Trans. ASME Vol. 96. J. Appl. Mech. Vol. 41, Series E, No. 2, 1974.

[11] Bateman, H. Partial Differential Equations of Mathematical Physics, Dover Publications, New York, N. Y., 1944.

[12] Hildebrand, F. B., Introduction to Numerical Analysis, McGraw-Hill Book Company, 1956.

ACKNOWLEDGEMENT

* The author wishes to thank the Department of Computing Services of IUPUI for providing CDC 6600 computer time for this investigation.

ON A TWO-DIMENSIONAL MEAN FLOW OF GRANULAR MATERIALS

M. Shahinpoor, Mechanical and Industrial Engineering Department
J.S.S. Siah, Civil and Environmental Engineering Department
Clarkson College of Technology
Potsdam, N.Y. 13676

Abstract

Presented is a reformulation of Bagnold's theory for the rapid flow of granular materials in the inertia dominated regions. This reformulation is free of ambiguities regarding the angle of pairwise collisions α_i. It will be shown that if α_i is defined as a random angle of collision between pairs of particles with respect to the direction of flow, then Bagnold's equations for dispersive and shear forces must be multiplied by $\sin\alpha_i$. The dispersive forces arise due to random pair collisions of grains. This correction is shown to be important for mean flows when statistical ensemble phase averages are computed for realistic 2-D flows of granular materials in which pair collisions are random and particle velocity fluctuations are present.

Introduction

The fact that cohesionless granular materials and "bulk solids" behave much like the non-Newtonian microfluids in rapid flow regimes has recently been established by Shahinpoor and Siah [1], and Shahinpoor [2]. They have derived a new set of constitutive equations for the dispersive stresses developed due to particle collisions in such rapid flows, which depend strongly and rather nonlinearly on the spatial gradients of both the particle velocity and the particle microrotation. The interest on rapid flow of granular materials and rapid transport of bulk solids, however, dates back to Hagen [3], and Bagnold [4],[5]. Recent interests on such rapid flow theories can be traced in the works of Brown and Richards [6], Savage [7], Jenkins and Cowin [8], Ackermann and Shen [9], McTigue [10], Blinowski [11], Kanatani [12], Ogawa, Umemura and Oshima [13], Shahinpoor and Lin [14], Jenkins and Savage [15], and Jenkins and Shahinpoor [16].

In the present paper we present a reformulation of Bagnold's granular materials fast flow theory for inertia dominated regions. This reformulation is free of ambiguities regarding the angle of pairwise collisions α_i for the ith pair of particles. This reformulation is important in computing the statistical ensemble phase averages on the dispersive stresses that arise due to continuous random collisions of particles in a simple two dimensional shear flow, in the presence of particle velocity fluctuations.

Referring to Fig. 1 below, we consider a simple unidirectional shear motion of a body of granular materials comprising of rigid particles of

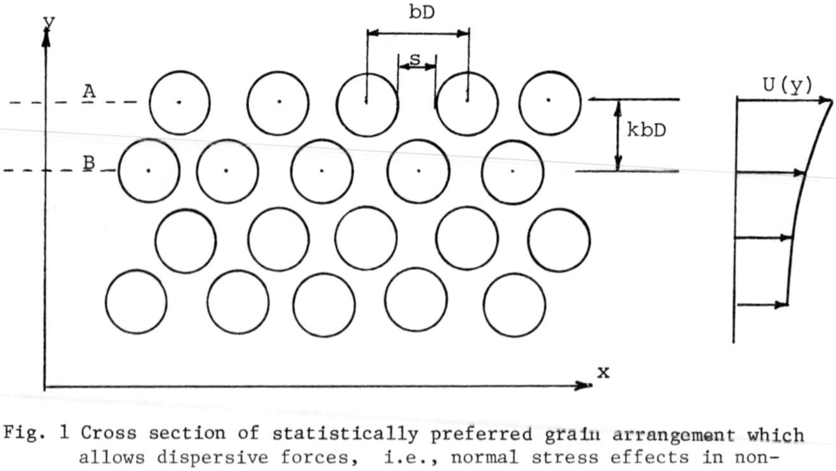

Fig. 1 Cross section of statistically preferred grain arrangement which allows dispersive forces, i.e., normal stress effects in non-Newtonian fluids, to be developed.

diameter D, and density γ_0, which are randomly separated in each layer by parametric distances bD, s, and kbD, respectively.

Bagnold [4] has shown that for such flows the dispersive grain forces arising from collision of pairs of grains do give rise to a normal stress P_{yy} and a shear stress T_{xy} such that

$$P_{yy} = a_i \gamma_o \lambda f(\lambda) D^2 \left(\frac{dU}{dy}\right)^2 \cos \alpha_i \ , \tag{1}$$

$$T_{xy} = P_{yy} \tan \alpha_i \ , \tag{2}$$

where $f(\lambda)$ is an unknown function related to the collision frequency, λ is defined as the linear concentration and is given by $\lambda = (D/s)$, and α_i according to Bagnold [4] is "some unknown angle determined by the collision conditions, including grain rotation, in this inertia state of grain motion." This interpretation of α_i, specifically in reference to grain rotations which have not been considered by Bagnold, makes the definition of α_i rather ambiguous. Here we define α_i as the instantaneous angle of random collision pertaining to the ith pair of particles as shown in Fig. 2. Obviously, α_i now becomes a statistical quantity possessing a frequency distribution which is most likely Gaussian under such random collision conditions. Furthermore, all other parameters such as b, s, and k also become statistical with defined frequency distributions. Thus, we proceed to present a reformulation of dispersive and shear forces arising in such random two-dimensional flows which are dominated by random pair-wise collisions between pairs of grains with negligible effects due to particle velocity fluctuations superimposed on particle relative mean velocity δU.

In the next section we shall present the details of calculations leading to correct form of P_{yy} and T_{xy} for such simple cases.

Governing Equations:

Consider an arbitrary pair of grains colliding elastically as shown in Fig. 2. Note that δU is the relative velocity at a pair collision and is given by

$$\delta U = \left(\frac{dU}{dy}\right) kbD \ . \tag{3}$$

We follow the line of thought of Bagnold [4] in assuming the A grains to form a rigid granular surface of infinite inertia being collided randomly by the B grains at a relative velocity of δU. Let each B grain make $f(\lambda)\frac{\delta U}{s}$ collisions with an A grain in unit time. The change of momentum, as shown in Fig. 2, is

$$(\delta mv) = 2m\delta U \sin \alpha_i \ , \tag{4}$$

and is in the direction of implusive force, i.e., in the direction perpendicular to the tangent between the colliding spheres. This is where we fundamentally differ from Bagnold's results because his expression for this change of momentum is given as ($2m\delta U$) only.

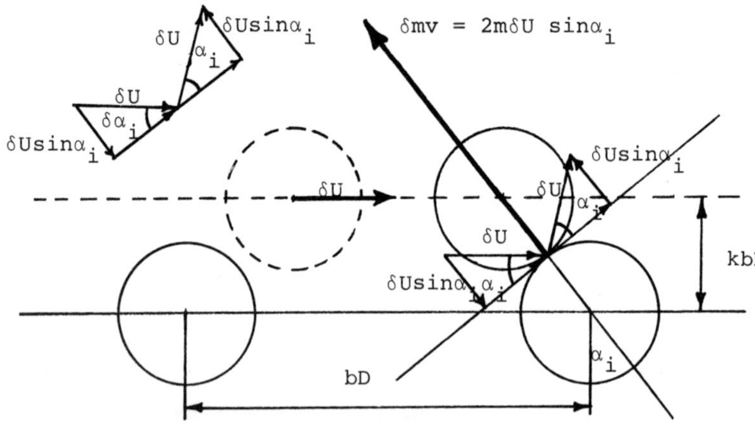

Fig. 2 Geometry of a random collision between a pair of spheres as evisaged by Bagnold [11].

Thus, the total repulsive force per unit area of plane xz may be readily shown to be

$$P = N_{xz}\, f_c(\delta mv) \ , \tag{5}$$

where N_{xz} is the two dimensional grain concentration in the xz plane, i.e., number of grains per unit area of xz plane, f_c is the number of collisions per unit time or the collision frequency.

Let us consider a steady fast flow. We note that within certain period of time T, the particles of layer A have travelled an additional distance, s, compared to the particles of layer B. Thus, every A grain will have a chance for pair collisions with (s/bD) B grains in a time interval of $(\frac{s}{\delta U})$ seconds. In other words the collision frequency f_c must be proportional to $(\frac{s}{bD})/(\frac{s}{\delta U})$ or $\frac{\delta U}{(1+\lambda)s}$. In reality the collisions are random and at best we may write

$$f_c = f(\lambda)\,\frac{\delta U}{s} \ , \tag{6}$$

where $f(\lambda)$ is an unknown function of linear concentration λ. Furthermore, the two-dimensional concentration N_{xz} is clearly proportional to the local number of grains per unit area and this is given by

$$N_{xz} = \beta_i \frac{n_1 n_2}{n_1 bD \, n_2 bD} = \beta_i \left(\frac{1}{b^2 D^2}\right) , \tag{7}$$

where n_1 and n_2 are the number of grains in the x and z directions for any arbitrary rectangular area $\ell_1 x \ell_2$ of layer A, and β_i is a local concentration proportionality constant. Thus, from equations (3), (4), (5), (6), and (7) we find that

$$P = \frac{\beta_i}{b^2 D^2} \frac{f(\lambda)\delta U}{s} (2m\delta U \sin \alpha_i)$$

$$= \frac{\beta_i}{b^2 D^2} \frac{f(\lambda)}{(D/\lambda)} \, 2 \, \frac{\gamma_0 \pi D^3}{6} (k_i bD)^2 \left(\frac{dU}{dy}\right)^2 \sin \alpha_i$$

$$= \beta_i \frac{\pi k_i^2}{3} \gamma_0 D^2 \lambda f(\lambda) \left(\frac{dU}{dy}\right)^2 \sin \alpha_i , \tag{8}$$

or finally

$$P = a_i \gamma_0 D^2 \lambda f(\lambda) \left(\frac{dU}{dy}\right)^2 \sin \alpha_i , \tag{9}$$

where $a_i = \frac{\pi}{3} \beta_i k_i^2$.

The components of this dispersive stress in the y and x directions are, respectively, given by

$$P_{yy} = P \operatorname{Cos} \alpha_i , \; T_{xy} = P \sin \alpha_i = P_{yy} \tan \alpha_i , \tag{10)a,b}$$

so that the final equations for P_{yy} the normal and T_{xy} the shear stresses read

$$P_{yy} = a_i \gamma_0 D^2 \lambda f(\lambda) \left(\frac{dU}{dy}\right)^2 \sin \alpha_i \operatorname{Cos} \alpha_i , \tag{11}$$

$$T_{xy} = P_{yy} \tan \alpha_i = a_i \gamma_0 D^2 \lambda f(\lambda) \left(\frac{dU}{dy}\right)^2 \sin^2 \alpha_i . \tag{12}$$

These equations differ from the ones given by Bagnold [4], [5] in that they have been multiplied by a sin α_i. This modification is quite significant in a number of ways. First, since both α_i and a_i are random statistical quantities they possess probability distributions which, when taken into account, will definitely affect the values of both P_{yy} and T_{xy}. Thus, the absence of a sin α_i will become significant in calculating the statistical averages. Second, as experimentally verified by Bagnold [11] the average value of the ratio of the shear stress T_{xy} to the normal stress P_{yy}, namely tan α_i was about 0.32, based on this he calculated the value of a constant α_i in his equation (3 of [4]) equal to 0.042. So that for $\lambda<12$, for which $f(\lambda)$ is proportional to λ, he found that (equation 6 of [4]):

$$P_{yy} = 0.042 \, (\lambda D)^2 \left(\frac{dU}{dy}\right)^2 \operatorname{Cos} \alpha \tag{13}$$

Taking the sin α_i correction into account we find that the above equation should read

$$P_{yy} = 0.1378\ (\lambda D)^2 \left(\frac{dU}{dy}\right)^2 \mathrm{Sin}\ \alpha_i\ \mathrm{Cos}\ \alpha_i\ , \qquad (14)$$

in which the new value for a_i is equal to 0.1378.

Statistical Ensemble Averages:

Considering equations (11) and (12) we can compute statistical ensemble phase averages on the dispersive stresses T_{xy} and P_{yy} if we define a probability frequency distribution density $p(\alpha_i)$ on α_i such that

$$\int_0^{\frac{\pi}{2}} p(\alpha_i)d\alpha_i = 1 \qquad (15)$$

$$\langle T_{xy}\rangle = \int_0^{\frac{\pi}{2}} p(\alpha_i)a_i\gamma_o D^2\lambda f(\lambda)\left(\frac{dU}{dy}\right)^2 \mathrm{Sin}\ \alpha_i\ \mathrm{Cos}\ \alpha_i\ d\alpha_i\ , \qquad (16)$$

$$\langle P_{yy}\rangle \quad \int_0^{\pi} p(\alpha_i)a_i\gamma_o D^2\lambda f(\lambda)\left(\frac{dU}{dy}\right)^2 \mathrm{Sin}^2\ \alpha_i\ d\alpha_i\ . \qquad (17)$$

Here we have chosen α_i to vary from 0 to $\frac{\pi}{2}$. However, this may prove to be a rather arbitrary assumption although it makes a lot of sense to consider such a variation range for α_i. Furthermore, since we are considering a random distribtion for α_i , then $p(\alpha_i) = (2/\pi)$. Thus,

$$\langle T_{xy}\rangle = \frac{a_i}{\pi} \gamma_o D^2\lambda f(\lambda)\left(\frac{dU}{dy}\right)^2 , \qquad (18)$$

$$\langle P_{yy}\rangle = \frac{a_i}{2} \gamma D^2\lambda f(\lambda)\left(\frac{dU}{dy}\right)^2 . \qquad (19)$$

Finally, if we can define a frequency distribution density for a_i in terms of known random variables such as s_i and k_i we may further simplify equations (18) and (19) for further computations.

Acknowledgement

This research was partially supported by CCT - Division of Research Grant No. 357-226-79-80, and the NSF Grant No. CME-8021032. The authors thank Professor N.L. Ackermann for valuable discussions.

References

1. Shahinpoor, M., and Siah, J.S.S., "New Constitutive Equations for the Rapid Flow of Granular Materials," J. Non-Newtonian Fluid Mechanics, vol. 9, pp. 147-156, (1981).

2. Shahinpoor, M., "On Rapid Flow of Bulk Solids," Int. J. Bulk Solids Handling, vol. 1, no. 3, pp. 487-500, (1981).

3. Hagen, G., "Flow of Sand in Tubes," Berlin Monatsber., Akad. Wiss., pp. 35-42, (1852).

4. Bagnold, R.A., "Experiments on a Gravity-Free Dispersion of Large Spheres in Newtonian Fluid Under Shear," Proc. Roy. Soc., Lond., A., vol. 225, pp. 49-63, (1954).

5. Bagnold, R.A., "The Shearing and Dilatation of Dry Sand and the Singing Mechanism," Proc. Roy. Soc. Lon., A295, pp. 219-232, (1966).

6. Brown, R.L., and Richards, J.C., "Principles of Powder Mechanics," Pergamon Press, Oxford, 1st edition, (1970).

7. Savage, S.B., "Gravity Flow of Cohesionless Granular Materials in Chutes and Channels," J. Fluid Mech., vol. 92, part 1, pp. 53-96, (1979).

8. Jenkins, J.T., and Cowin, S.C., "Theories for Flowing Granular Materials," The joint ASME-CSME Appl. Mech. Fluid Engng. and Bioengng. Conf., AMD. vol. 31, pp. 79-89, (1979).

9. Ackermann, N.L., and Shen, H.T., "Flow of Granular Materials as a Two Component System," Proc. US-Japan Seminar "Continuum Mechanical and Statistical Approaches in the Mechanics of Granular Materials," edited by S.C. Cowin and M. Satake, Sendai, Japan, pp. 258-265, (1978).

10. McTigue, D.F., "A Model for Stresses in Shear Flow of Granular Materials," Proc. US-Japan Seminar (see ref. 9), pp. 266-271, (1978).

11. Blinowski, A., "On the Dynamic Flow of Granular Media," Archivum Mech. Stoso., vol. 30, no. 1, pp. 27-34, (1978).

12. Kanatani, K.I., "A Micropolar Continuum Theory for the Flow of Granular Materials," Int. J. Engng. Sci., vol. 17, pp. 419-432, (1979).

13. Ogawa, S., Umemura, A., and Oshima, N., "On the Equations of Fully Fluidized Granular Materials," ZAMP, vol. 31, pp. 483-493, (1980).

14. Shahinpoor, M., and Lin, S.P., "Rapid Couette Flow of Cohesionless Granular Materials," ACTA Mechanica, to appear (1981).

15. Jenkins, J.T., and Savage, S.B., "The Mean Stress Resulting From Interparticle Collisions in a Rapid Granular Shear Flow," Proc. 4th Int. Conf. on Continuum Models of Discrete Systems, Stockholm, June, (1981).

16. Jenkins, J.T., and Shahinpoor, M., "On Plane and Cylindrical Poiseuille Flow of Granular Materials," ASME J. Appl. Mech., to appear (1982).

TURBULENCE

ON THE STRUCTURE OF BOUNDED TURBULENT SHEAR FLOW: A PERSONAL VIEW

By

James M. Wallace*
Department of Mechanical Engineering
The University of Maryland
College Park, MD 20742

ABSTRACT

Evidence from a wide variety of research carried out during the last thirty years is presented in support of a specific view of the structure of bounded turbulent flows. This view holds that vortex lines which diffuse from the wall are lifted and stretched to form "hairpin"-like vortices which are fundamental to the generation and maintenance of turbulence in both transitional and fully developed flows. The special structural characteristics of pipe and channel flows when compared to boundary layers are discussed.

I. INTRODUCTION

This paper is intended as a personal view of the structure of boundary layer, parallel channel and pipe flows. It is not a review or survey of all the literature on this subject. The space available would not allow that, and there already exist very good reviews by Kovasznay [1], Mollo-Christensen [2], Laufer [3], Willmarth [4], Mathieu and Charnay [5] and Cantwell [6]. Rather, I want to draw on some of the results cited in these reviews as well as newer ones to gather support for a specific interpretation of the structure of these flows. My interpretation will not, in every case, agree completely with that of the authors since it is in the nature of the results of these investigations that they are often subject to more than one interpretation. It remains for the reader to judge which interpretation seems more plausible.

Laufer [4] and Cantwell [6] have pointed out that the ways of seeing these flows have largely been determined by the experimental devices and analysis techniques which were developed at the time and Comte-Bellot [7] has provided a thorough review of these. The renewed use, in the late fifties and early sixties, of flow visualization in both fixed and moving frames of reference marked a watershed in the understanding of these flows. This work indicated

* *Written while on sabbatical leave at the Laboratoire de Mécanique des Fluides, Ecole Centrale de LYON, FRANCE.*

that, far from being completely chaotic processes, the flows have a clearly observable structure which evolves in space and time in a quasi-repeatable manner. As with all developments of understanding, this insight was not entirely new. The finite values of the principal Reynolds stress component, $\overline{uv}$, implies a certain structure. Indeed the mixing length theories of the thirties can be interpreted as a kind of crude structural assumption. By the beginning of the seventies however, it was understood that, to properly characterize these structures statistically, phase information had to be retained. As a result of the visual observations, quantitative techniques using conditional sampling were quickly developed. Free shear flows, including the outer intermittency region of the boundary layer, was more easily accessible to these techniques; the requirement of a means to discriminate the events near the wall of bounded flows proved to be much more difficult. Very imaginative data analysis techniques were conceived with the growing availability of small digital computers. However, the limited number of Eulerian spatial points at which data could reasonably be obtained simultaneously has proven to be a limiting factor in the experiments. The emergence of image process analysis in recent years when combined with flow visualization may provide a significant new advance in our ability to quantitatively characterize these flows.

At this point, despite the limitations mentioned above, it seems beyond doubt that the processes responsible for the transport of momentum and kinetic energy as well as heat and mass are, to a significant degree, due to the motions of coherent, convecting structures which are generated, evolve, dissipate and are regenerated by means which I hope this paper will help clarify. A word of caution is necessary, however. Even the kinematic description of these processes and their associated structures is still somewhat sketchy and their dynamics is largely a matter of speculation. These dynamical questions will certainly not be answered directly by experiments. The most that experiments can provide is an adequate kinematic description (something I hope to show is slowly emerging) and also, perhaps, a critical confirmation of a dynamical theory. Such a theory is most likely to emerge however, when a convincing physical model of the dynamical processes has been given.

II. A MODEL OF BOUNDED TURBULENT FLOW

I wish to argue for a dynamical model which has its origins with Willmarth and Tu [8] and Willmarth [5] following Lighthill [9] and has a number of common features with, among others, models put forth by Theodorsen [10] and later by Black [11] with little experimental data to draw on. If we take the curl of the Navier-Stokes equation, an equation of motion in terms of vorticity, Ω_i, is obtained

$$\frac{\partial \Omega_i}{\partial t} + U_j \frac{\partial \Omega_i}{\partial X_j} = \Omega_j \frac{\partial U_i}{\partial X_j} + \nu \frac{\partial^2 \Omega_i}{\partial X_j \partial X_j} \qquad (1.)$$

When evaluated at the wall the component equations

$$\frac{\partial \Omega_x}{\partial t} = \nu \nabla^2 \Omega_x \quad (2.1), \quad \frac{\partial \Omega_y}{\partial t} = 0 \quad (2.2) \text{ and } \frac{\partial \Omega_z}{\partial t} = \nu \nabla^2 \Omega_z \quad (2.3)$$

illustrate that the rate of change in vorticity at a point on the wall is simply due to diffusion within the plane of the wall and normal to it. Expressions for the two terms giving the flux of vorticity components normal to the wall can be obtained by evaluating the Navier-Stokes equation itself there

$$\frac{1}{\rho}\frac{\partial P}{\partial X} = -\nu\frac{\partial \Omega_z}{\partial y} \quad (3.1) \quad \text{and} \quad \frac{1}{\rho}\frac{\partial P}{\partial Z} = \nu\frac{\partial \Omega_x}{\partial y} \quad (3.2)$$

where the instantaneous transport of Ω_z and Ω_x out of the wall is related to the instantaneous pressure gradients over it. Emmerling et al. [12] have shown that there are intense, small scale pressure gradients occurring instantaneously at the wall under bounded turbulent flows. Thus the lines of vorticity, which diffuse out of the wall into the shearing flow, although having a mean orientation in the spanwise direction, can be expected to be severely kinked at many locations. The kinked loops pointing downstream will rise because of their mutually induced velocities, whereas those loops pointing upstream will move toward the wall and effectively become attached (see Willmarth [5]). As the rising loops pass through regions of higher velocity, their "legs" will stretch, as described by the first term on the right of equation (1), and come closer together intensifying the vorticity contained in the tubes and also their induced upward velocity. Their transverse "heads" will be compressed in this process reducing their vorticity and increasing their scale. Thus "hairpin"-like vortices form which lift upward through the mean shear layer while convecting downstream. They are responsible for large local momentum transport and will create high shear layers on their upstream sides in the vertical planes between and on either side of their legs as well as high shear layers in the horizontal planes. They are also responsible for much of the subsequent large pressure variations at the wall which in turn give rise to new diffusing, contorted vortex lines entering the flow. Thus a continuous generation of "hairpin" vortex structures occurs which, because of the basic shear, have a preferred direction at an angle to the wall inclined downstream. Near the wall, in the region of intense mean shear, the loops are primarily oriented with their axes in the streamwise direction because they are stretched before they rise. This stretching, however, induces the strong upward motion, increasing the angle of the axes of the "legs" to the wall. This process, for any vortex loop, will continue until viscous diffusion so reduces its vorticity so that it is no longer identifiable. During this diffusion process one would expect to see the scale of the vortices increase. A sketch, attempting to illustrate these elements of the process (albeit with difficulty because of its spatio-temporal nature) is given in Figure 1.

III. EVIDENCE FOR THE MODEL

A. DEVELOPED FLOWS

Townsend [13], based on Eulerian correlation measurements had, in the early fifties, suggested that long cylindrical eddies, with their axes in the direction of mean flow and with the plane of circulation normal to the principal axis of the positive rate of strain, were the main organized motion in shear flows. These eddies were thought to draw energy from the mean flow and dissipate it through turbulent friction and viscous effects at the wall. Superimposed on this system he conceived of a whole range of more

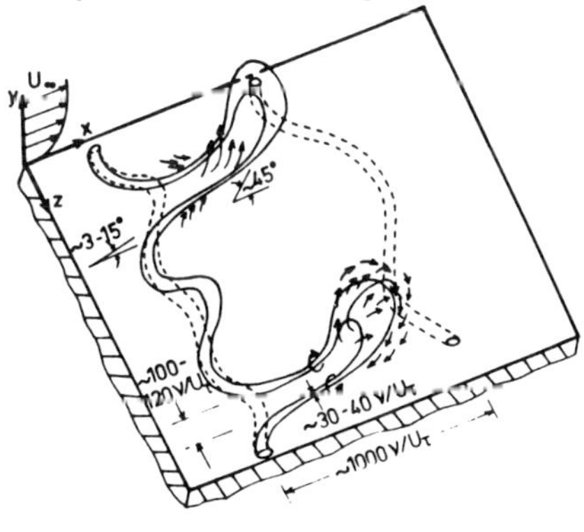

Fig. 1 - Conceptual model of diffusion, stretching and lifting of vortex tubes.

disorganized eddies which transferred energy down the scale of sizes.

An eigen-function decomposition of the streamwise fluctuating velocity components obtained from correlation functions in a pipe flow of gylcerin (Re_d = 8 700) with a thick wall layer enabled Bakewell and Lumley [14] to obtain mean streamlines of the organized structure near the wall as seen in Figure 2. The structure consists of counter-rotating pairs of eddies with a gradual influx of fluid in the spanwise direction and a strong outflow in the direction normal to the wall. The centers of rotation are at $y^+ \sim 35$. They hypothesize that these eddies occur randomly over the wall and originate with their planes of circulation tilted upstream although their origin is not known. The subsequent intensification and ultimate destruction of the eddies was believed to be governed by vortex stretching.

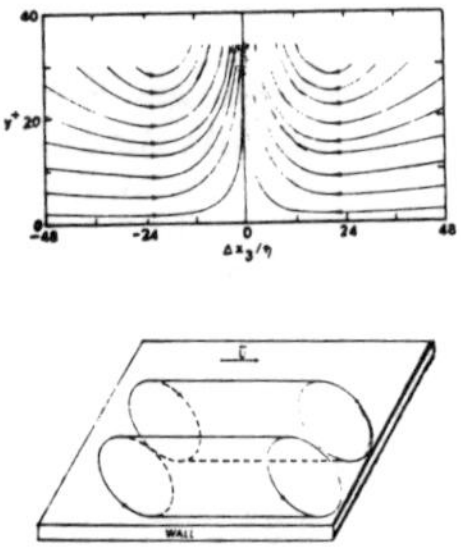

Fig. 2 - From Bakewell and Lumley [14] courtesy of the American Inst. of Phys.

Based on extensive space-time correlation measurements between velocity components in the flow and the pressure at the wall, Willmarth and Tu [8] proposed a model represented in Figure 3. The vortex lines are inclined downstream and away from the wall due to the induced motion upward, with the upstream apex anchored at the wall. The sweptback shape of spatial correlation measurements, R_{pv}, in planes close to and parallel to the wall as well as the correlation contours of R_{pw} (Fig. 3) which show a change in sign with distance from the wall and downstream of the pressure transducer in planes perpendicular to the wall and the mean flow, provided strong evidence for the type of structure postulated. Measurements which were later analyzed by Willmarth and Lu [15], were made by the same authors using a detecting probe near the wall at $y^+ \sim 15$ which sensed "ejection" of low speed fluid from the wall and a probe sensitive to streamwise vorticity placed downstream, above and to either side of the detecting probe. They indicated that, when these "ejections" were detected, large values of streamwise vorticity were found on either side with opposite signs.

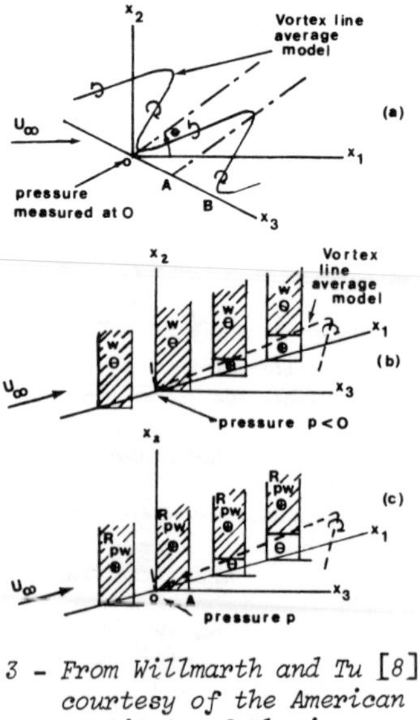

Fig. 3 - From Willmarth and Tu [8] courtesy of the American Institute of Physics.

A series of papers giving results of flow visualization in low Re_θ water flow using dye and/or hydrogen bubbles have greatly illuminated our understanding of the flow structures. The most important results of Kline et al. [16] was to establish that elongated regions of alternating high and low speed fluid with a relatively well defined spanwise spacing of $z^+ \sim 100$ were quite evident in and just above the sublayer. The regions of low speed appeared to gradually lift up from the wall, oscillate and finally break up into chaotic motion. This sequence was called a "burst" (A ubiquitously used name which can be traced at least back to Corrsin (17). Later Kim et al 18 determined that a large percent of the turbulent energy production is associated with the process. Kline et al [16] interpreted these observations with a partial dynamical model which has much in common with that of Willmarth and Tu [8]. A sketch indicating the elements of this model is shown in Figure 4 where a spanwise vortex element is shown being lifted, rotated and stretched downstream setting up a dynamically unstable local shear layer. Corino and Brodkey [19] observed many of these same features in a pipe flow seeded with neutrally bouyant particles and photographed with a moving camera system. In addition, they identified the large contribution to the stress by inflows of larger scale, high speed motions which they called "sweeps" as was also found by Grass [20] in a visual study using hydrogen bubbles over smooth and rough walls and subsequently quantitatively studied by Wallace et al. [21] and Willmarth and Lu [15] using an analysis based on the signs of the velocity components.

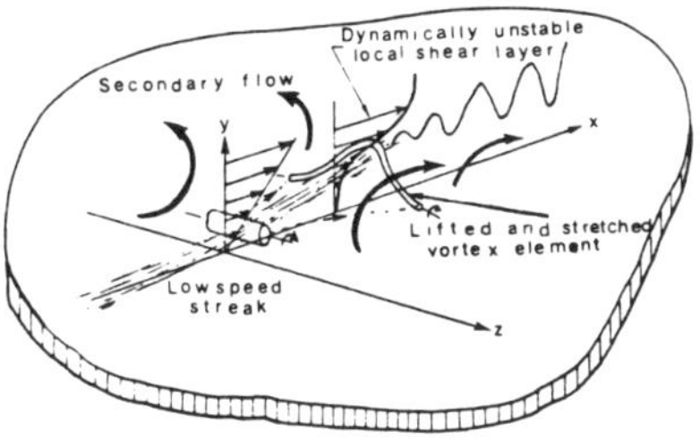

Fig. 4 - From Kline et al. [16] courtesy of Cambridge University Press.

In addition to quantifying the contribution of "bursts" to turbulence production, Kim et al. [18] were also able to establish the mean rate of passage of "bursts" per unit span past a fixed observer. Using a hydrogen bubble wire perpendicular to the wall, they made observations of the evolution of the lifting fluid as seen in Figure 5. Here quite clearly a structure shown in sequential frames with streamwise vorticity is evident; it lifts away from the wall and grows in scale with its progress downstream. Also evident in the picture are distortions in the time lines indicating spanwise vorticity of relatively larger scale above and downstream of the streamwise vortex. Motion of fluid down toward the wall and in front of the streamwise vortex is apparent. One may conjecture that the vertical plane view is capturing a section of the lifting, stretching process of one "leg" of a newer vortex and the spanwise "head" of an older vortex which originated upstream. The growth in scale of the vortex structures is consistent with observations made from two point correlation measurements of Grant [22] and Tritton [23] and with the space-time isocontigency curves of Dumas et al [24]. However, the reported increase in strength of the vortices is difficult to account for.

Using combinations of up to four hydrogen bubble wires oriented parallel

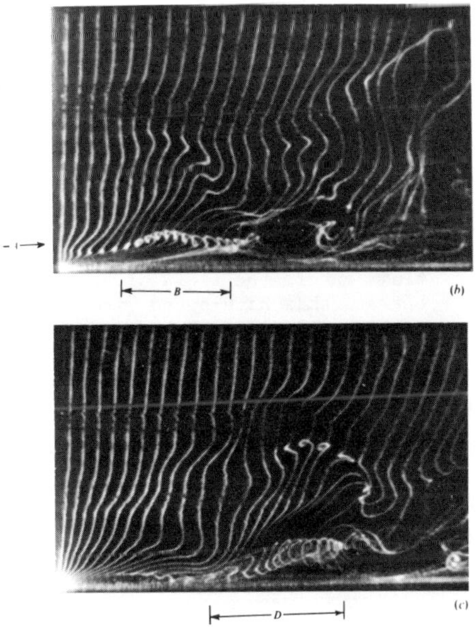

Fig. 5 - From Kim et al. [18] courtesy of Cambridge University Press.

and normal to the wall in a low Reynolds number water channel, Clark and Markland [25] have identified counter-rotating streamwise vortices in the region $7 \lesssim y^+ \lesssim 70$ inclined at small angles ($\sim 3°-7°$) downstream and away from the wall. These vortices convect at a velocity greater than the local mean near the wall ($0.65\ U_\infty$) and at nearly equal to it near the edge of the buffer layer. The vortex radii grow to about 30-40 viscous lengths and the rate of spin decreases over their average convected lengths of $\Delta x^+ \sim 450$. Transverse vortices, which proceeded the counter-rotating pairs were also observed in the same region and were of about the same scale. They also increased in size with downstream evolution, and there was some, although not conclusive evidence, that these were the connecting segments to the counter-rotating pairs. Results suggesting the counter-rotating vortex pair picture also have been obtained by Lee et al. [26] and Kastrinakis et al. [27].

Blackwelder and Eckelmann [28] have found evidence for counter-rotating pairs of vortices using combinations of hot-film sensors in the flow and flush with the wall of an oil channel at low Reynolds number with a thick viscous sublayer. Using a conditional analysis based on the sign of the four measured signals as well as one using the VITA detection technique (See Blackwelder and Kaplan [29]), they conceived a partial model of the flow near the wall as seen in Figure 6. A low speed streak is created by the counter-rotating vortex pairs which are separated by the half streak wavelength and which "pump" the low-speed fluid away from the wall. A high shear layer, reflected in the inflextional velocity profile seen, is produced by this momentum transfer. They find the middle of the vortex structures lie approximately at $y^+ \sim 20-30$, and they appear to have a streamwise extent $\Delta x^+ > 1{,}000$, in agreement with visual results of Oldaker and Tiedermann [30]. The strength of the vortex is estimated to be an order of magnitude less than the mean spanwise vorticity at the wall.

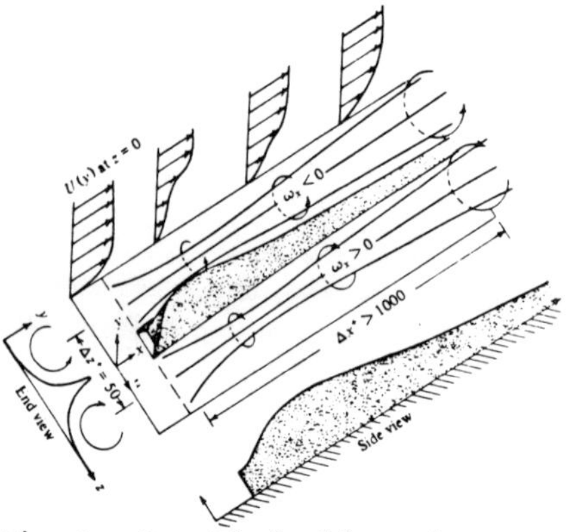

Fig. 6 - From Blackwelder and Eckelmann [28] courtesy of Cambridge University Press.

A highly innovative visualization of a boundary layer was carried out by Praturi and Brodkey [31] using stereoscopic motion pictures of suspended particles in a low-speed water flow with a moving camera system. Many events were observed, and these are summarized in Figure 7. A large region of high-speed fluid enters the field of view (its origin unknown), and, on the interface between this high-speed "front" and the downstream flow, a transverse vortex occurs. This is interpreted as arising from an instability of a Helmholtz type earlier suggested by Nychas et al. [32]. The vortex grows as it is convected downstream and slightly away from the wall and appears to be associated with wall layer activity of ejecting fluid and smaller scale streamwise vortices. The transverse vortex gives rise to the "bulge" in the rotational/irrotational interface at the edge of the boundary layer, and a large scale inflow of irrotational fluid occurs in the developing interstice. Although the authors were never able to observe connected "hairpin" vortices, all their elements were observed. The observations were also not able to clarify how the vortex lines associated with these elements are connected, but this is a problem common to

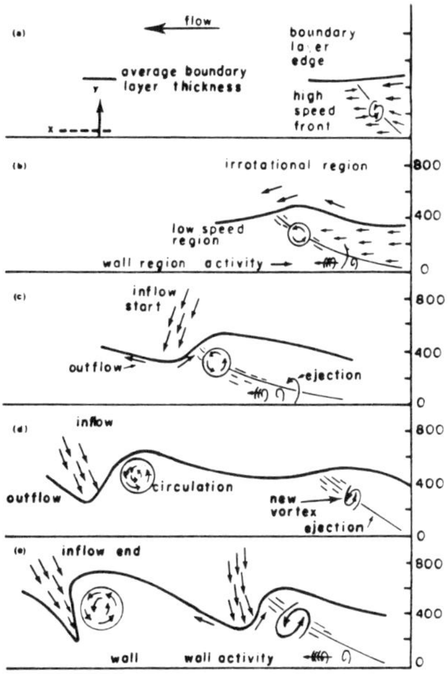

Fig. 7 - From Praturi and Brodkey [31] courtesy of Cambridge University Press.

most of the models described herein. The streamwise vortices observed near the wall have diameters of the order of 50 ν/u_τ, in substantial agreement with Blackwelder and Eckelmann [28].

Using a rake of hot-wire probes across a boundary layer at Re_θ = 10,160 and a flush mounted shear sensor at the wall, Brown and Thomas [33] obtained a picture of the large scale structure of the flow, given in Figure 8, in a frame of reference convected at $0.8U_\infty$. Here a large inclined "hairpin"-like vortex structure with streamwise extent near the wall of about 2δ is seen in two planes. There is a stagnation point on the "back" of the outer interface of the structure with irrotational fluid flowing down the intercises between successively occurring structures. The high frequency component of the wall shear, which appeared to be correlated with the large scale motion, was interpreted to be associated with the wall layer "bursting". They speculate that the origin of the streamwise vorticity near the wall is due to a Taylor-Goertler instability resulting from the convex curvature of the streamlines near the wall and perform a calculation to show that the center of instability is at a height above the wall of $y^+ \sim 50$. This, then, is an alternative suggestion to the proposal made in Section II for the origin of the wall layer streamwise vortices. The structure observed appears, from their correlation measurements, to be inclined downstream at an oblique angle of about 18° across the boundary layer. From results to be described below, it will be apparent that this angle is in some dispute.

In the outer part of the layer the figure of Brown and Thomas [33] looks remarkably similar to the results obtained by Blackwelder and Kovasznay [34] and by Falco [35]. Using space-time correlation methods and conditional sampling Blackwelder and Kovasznay [34] obtained an average qualitative picture of the "bulges" in the outer intermittency region of the boundary layer which showed the same stagnation point and large scale rotation within the bulges. Falco [35] confirms this and, in addition, observes a finer scale "mushroom"-like structure on the "backs" of these "bulges" with scales of the order of 100-200 ν/u_τ which he calls "typical eddies" as seen in Figure 9.

Chen and Blackwelder [36] used a slightly heated wall under a turbulent boundary layer (Re_δ = 29,000) as a passive marker and discovered well defined "fronts" of sharp temperature change extending from deep in the wall

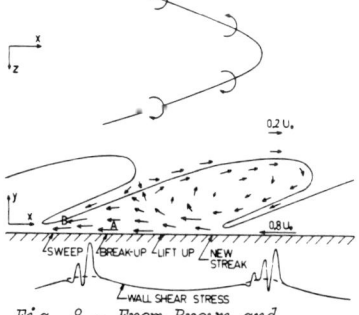

Fig. 8 - From Brown and Thomas [33] courtesy of the American Inst. of Physics.

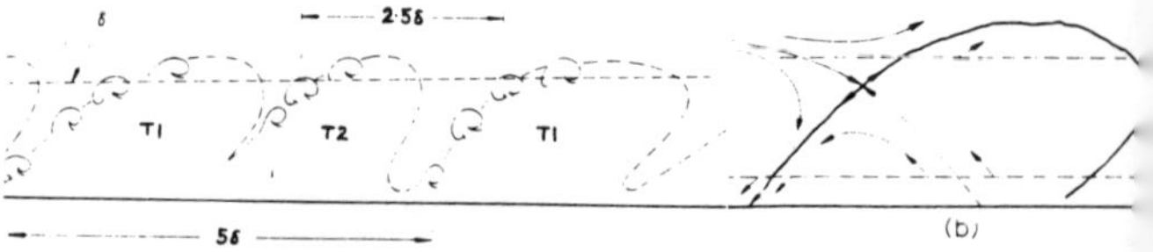

Fig. 9 - From Falco [35] courtesy of the American Institute of Physics

layer into the intermittent zone. This provides direct evidence that a relation between the large scale outer structure and the events occurring near the wall exists, as has long been conjectured. The average angle of this "front" can be estimated to be about 48°. The mean frequency of passage of this "front" was seen to be remarkably constant across the entire layer except very near the wall and to be similar to the mean frequency of passage of the large scale outer bulges. A similar constancy has been obtained from measurements by Lu and Willmarth [37] and of Sabot and Comte-Bellot [38], and a correlation of all the available data provided by Fleischmann and Wallace [39]. Using a three-sensor probe made up an X-array and a temperature sensor, the two-dimensional flow field associated with these "fronts" can be observed and is shown in Figure 10. The front is associated with intense Reynolds stress contributions and spanwise vorticity. Shear layers similar to these have also been found by Wallace et al. [40] and Eckelmann and Wallace [41] at locations across a channel flow. If the "front" is assumed to be convected at the local mean velocities, the spatial extent of the disturbance of the velocity field associated with it is about 60-120 ν/u_τ in the mean flow direction. This scale is similar to the "typical eddy" scales of Falco [35] and much smaller than the "bulge" scales.

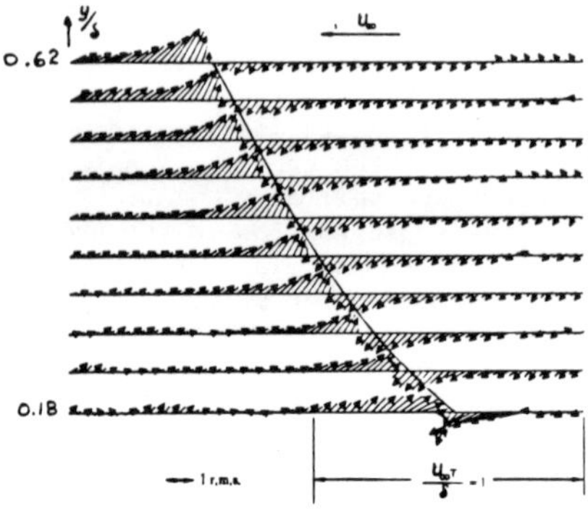

Fig. 10 - From Chen and Blackwelder [36] courtesy of Cambridge University Press.

Kreplin and Eckelmann [42] have also detected a disturbance "front" in the sublayer and buffer layer of the oil channel flow described above. Space-time correlation measurements between the streamwise fluctuating velocity component in the flow and the fluctuating spanwise vorticity component at the wall show positive values all the way out to y/b = 0.5 where b is the half channel width. The "fronts" are inclined downstream at an angle varying from approximately 5° at the wall to 14° at $y^+ \sim 50$. The celerity of the "front" also varies from approximately 0.6 U_c at the wall to 0.88 U_c at $y^+ \sim 50$. Two-point spatial correlation measurements at the wall show positive values for $\Delta X^+ \sim 1{,}300$ and a quasi-periodicity over $\Delta Z^+ \sim 120$. The space-time correlations of the fluctuating spanwise velocity component are antisymmetric for $y^+ \gtrsim 40$ which the authors explain in terms of the counter-rotating vortex pair model.

The most convincing evidence for the model of Figure 1 has been recently provided in a paper by Head and Bandyopadhyay [43]. They use a smoke filled turbulent boundary layer over a relatively wide range of Reynolds numbers and photographed various laser illuminated planes across the flow. The most revealing picture of the lifting "hairpin" vortices were obtained with light planes inclined either upstream or downstream at 45° to wall as seen in Figure 11. These photographs were taken at low Reynolds number Re_θ = 600; similar results were obtained at higher Reynolds numbers up to Re_θ = 9,400. The absolute scale of the vortex loops in the downstream light plane and the "mushroom" shaped cross-section in the upstream light plane decreases with Re_θ, but, when normalized

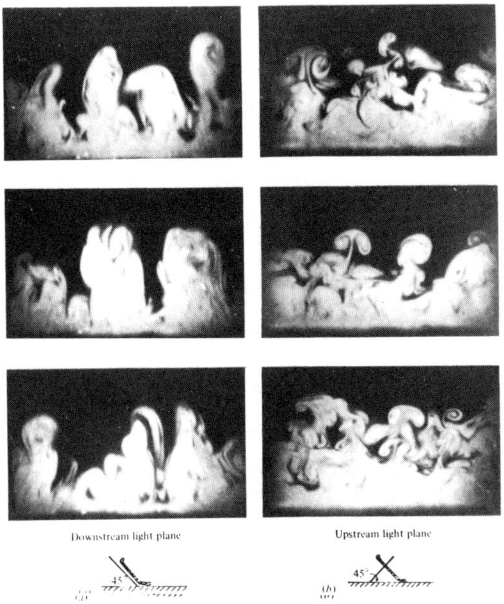

Fig. 11 - From Head and Bandyopadhyay [43] courtesy of Cambridge University Press.

they are consistently about 100 ν/u_τ independent of Re_θ. Additional visualization in the near wall flow revealed counter-rotating vortex pairs quite clearly. The loops lifted outward, in many cases are seen all the way across the boundary layer, at a characteristic angle of about 45°. This does not disagree with the smaller angles found by Clark and Markland [25] and Kreplin and Eckelmann [42] because their observations were much closer to the wall. At the lowest Reynolds numbers these "hairpins" appear to be the only scale of motion in the flow. At larger Reynolds numbers, Re_θ > 2,000, larger scale structures seem to be agglomerations of these hairpins with a slow spanwise rotating motion, an observation earlier made by Smith [44] with a moving hydrogen bubble wire in a low Reynolds number water flow. Occasionally the "backs" of these agglomerations exhibit an angle of inclination to the wall of about 20°, similar to the results of Brown and Thomas [33] but the basic angle of the "hairpin" structures themselves is of the order of 45°. It should be mentioned, however, that these pictures were obtained using a rather larger serrated trip which raises questions about the effects on the subsequent structure. It is regrettable that so little information is available about these effects in the literature.

Figure 12 shows the distribution of the flux of mean spanwise vorticity, Q(y), for a turbulent channel flow numerically obtained by Bernard and Berger [45] using a closure to the mean vorticity equation which utilizes a Lagrangian time expansion technique. Spanwise vorticity diffuses into the flow in the region $y^+ \lesssim 2$ and is unaltered in this lower part of the viscous sublayer. In the region $2 \lesssim y^+ \lesssim 14.6$, beneath the streamwise vortex pairs, the flux of spanwise vorticity increases rapidly (becomes more negative) due to the intense production of mean vorticity by spanwise stretching. In the region $14.6 \lesssim y^+ \lesssim 39.7$, the flux decreases due to the rotation of the spanwise vorticity into the y direction by the lifting process.

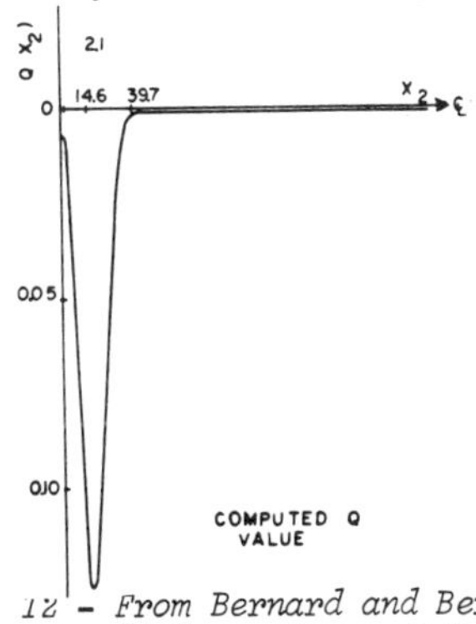

Fig. 12 - From Bernard and Berger [45] courtesy of the SIAM Journal of Appl. Math.

B. TRANSITION

Kleabanoff et al. [46] and Hama et al.

[47] in studies of the latter stages of transition found that the breakdown of two dimensional Tollmien-Schlichting waves into fully turbulent spots was accompanied by a stage where "hairpin" type vortices occurred. Fleischmann and Wallace [39] have shown that the mean frequency of passage of these transition vortices is very close to the passage frequencies of the organized structures in the fully turbulent layer as observed in a large number of different experiments. Purtell et al. [48] speculate that the "hairpin"-like vortices occurring in the transition process may be the same type structure which occurs in the fully developed layer.

Perry et al. [49] have beautifully photographed developing "hairpin"-vortices behind a vibrating trip wire as seen in Figure 13 and have convincingly shown that the characteristics of a transitional turbulent spot (as given by the ensemble average results of studies like those of Wygnanski et al. [50] and Cantwell et al. [51]) are due to an internal hierarchy of "hairpin" vortices with the vortices nearer the wall being generated by the presence of the previously created ones which have lifted further away. The vortex filaments, in this view, propagate outward and set up conditions laterally and below for **new** vortex loops to occur. Additionally, they performed experiments showing the effects of such vortex loops on a sheet of smoke beneath the "heads" of the loops. The results are shown in Figure 13 where the form of the smoke pattern is quite similar to the patterns in a turbulent boundary layer obtained by Falco [52], with a similar sheet of smoke. One can conjecture that these patterns which Falco called "pockets" and which are compared to Perry et al's. in the figure, are, in fact, the "footprints" of the lifted segments of the vortex loops in the flow above them.

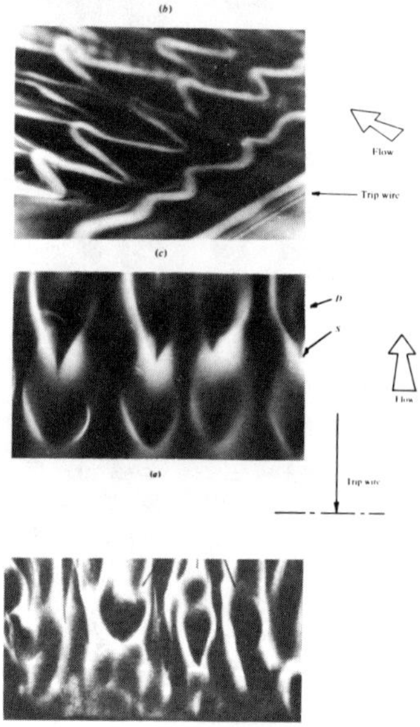

Fig. 13 - From Perry et al. [49] and Falco [52] courtesy of Cambridge Univ. Press and the American Inst. of Aeronautics and Astronautics.

In a remarkable numerical experiment Leonard [53] has tracked the Lagrangian evolution of vortex filaments, initially oriented in the spanwise direction to make up a Blassius boundary layer, after they are perturbed in a local region. The tracking is accomplished by numerically solving the equations of motion in time steps, and Figure 14 shows the evolution of the computed vortex lines. Here the lifted vortex structure is evident and occurs simply from the vortex interactions in a sheared flow environment. Taking ensemble averages of cross-sections in the x - y plane, Leonard [53] obtains a picture of the average turbulent spot which is strikingly similar to the conditional averaged results obtained by Wygnanski et al. [50] and Cantwell et al. [51]. It now seems clear that there are finer scale interior vortices in these turbulent spots that have not been revealed by the conditional sampling measurements, perhaps because of transverse "jitter" in the techniques used. Antonia et al. [54] concur with this observation. Finally,

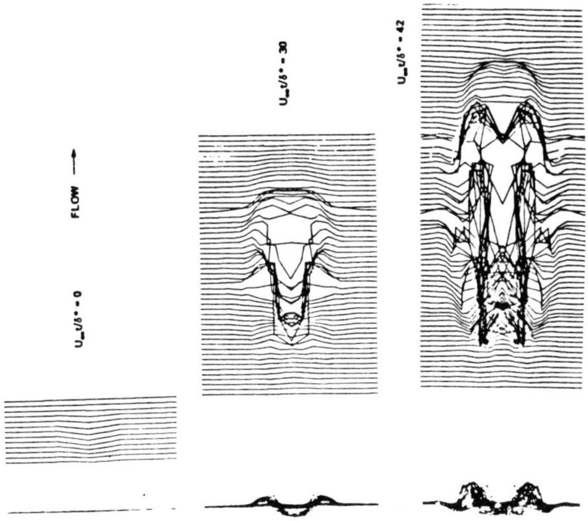

Fig. 14 - From Leonard [53] courtesy of Springer Verlag.

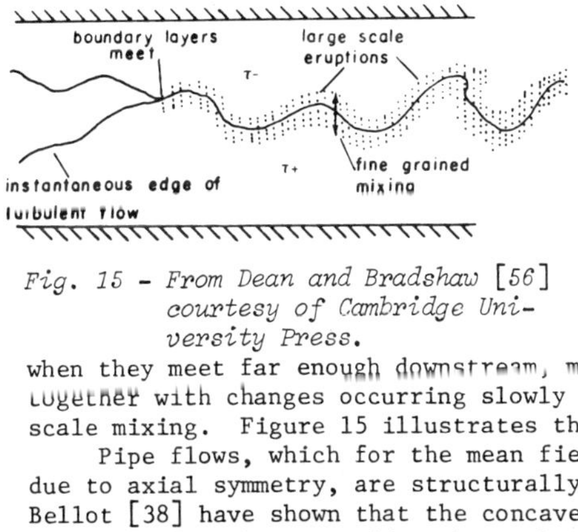

Fig. 15 - From Dean and Bradshaw [56] courtesy of Cambridge University Press.

Zilbermann [55] has tracked a turbulent spot induced in a laminar layer as it merges with surrounding spots which form the developed boundary layer downstream. They were able to follow this spot because it retains identifiable features up to 70δ downstream demonstrating their resilience.

IV. ADDITIONAL CHARACTERISTICS OF CHANNEL AND PIPE FLOWS

For fully developed channel and pipe flows there is no free boundary beyond which the flow is irrotational. The entire flow is turbulent, and the structure is similar to that of boundary layers. At the centerline, the turbulence is far from random. Dean and Bradshaw [56] have found that the large scale structures from the two boundary layers growing on two of the parallel walls of a high aspect ratio channel fit together when they meet far enough downstream, much like a zipper. They remain locked together with changes occurring slowly at the interface through viscous fine scale mixing. Figure 15 illustrates their conception of the process.

Pipe flows, which for the mean field are the most simply describable due to axial symmetry, are structurally the most complex. Sabot and Comte-Bellot [38] have shown that the concave pipe walls "focus" the large scale structures at the centerline where they fit together in a manner similar to, but more complicated than, the channel flow. They find, in fact, three zones: rotating structures coming from either side and "remnants" of older structures. They believe this result applies to channels as well. The zero value of the Reynolds stress at the centerline, $\overline{u_x u_r}$, as for the channel, is due to symmetry and not to isotropy.

V. CONCLUSION

A considerable body of evidence has been gathered to support the view that the interior structure of turbulent spots in transition and of the organized motion in the fully developed bounded flows are identical and indeed, result from one continuous process. This structure is a "hairpin"-like vortex which takes its shape from the stretching and lifting of vortex lines by the shearing flow as they diffuse from the wall. The process begins in the last stage of transition with the emergence of the spots and regenerates itself through the integral effects of the vortices on the pressure field at the wall. Some mean features of spots and developed flow structures are different and are as yet unaccounted for by this model.

An outstanding question which remains unanswered by this model is how and shy the "hairpins" agglomerate to form the observed larger scale "bulges" in the outer part of the layers at higher Reynolds numbers. The effects of trips on the observed structures also needs to be thoroughly investigated.

ACKNOWLEDGEMENTS

This work was made possible by the hospitality of Professors J. MATHIEU, G. COMTE-BELLOT AND J. P. SCHON and by the partial support of the D.G.R.S.T., France. Partial support from NSF Grant CME 79-19835 and NASA Grant NAG1-24 is also gratefully acknowledged.

REFERENCES

1. Kovasznay, L.S.G. 1970 Ann. Rev. Fl. Mech. 2, 95.
2. Mollo-Christensen, E. 1971 AIAA Jour. 9, 1217.
3. Laufer, J. 1975 Ann. Rev. Fl. Mech. 7, 307.
4. Willmarth, W.W. 1975 Adv. Appl. Mech. 15, 159.
5. Mathieu, J. and Charnay, G. 1981 Lect. Notes in Phys. 136, 146.
6. Cantwell, B.J. 1981 Ann. Rev. Fl. Mech. 13, 457.
7. Comte-Bellot 1979 AGARD Conf. Proc. No. 271, 1.
8. Willmarth, W.W. and Tu, B.J. 1967 Phys. Fl. Supp. 10, S134.
9. Lighthill, M.J. 1963 Laminar Boundary Layers ed. by L. Rosenhead, Oxford Univ. Press, Chap. 2.
10. Theodorsen, T. 1952 Proc. 2nd Midwestern Conf. on Fl. Mech., Ohio State University.
11. Black, T.J. 1966 Proc. Heat Transfer and Fl. Mech. Inst., Stanford University Press.
12. Emmerling, R., Meier, G.E.A. and Dinkelacker, A. 1973 AGARD Conf. Proc. n° 131, pap. n°.24.
13. Townsend, A.A. 1950 Proc. Cambridge Phil. Coc. 47, 375.
14. Bakewell, H.P. and Lumley, J.L. 1967 Phys. Fl. 10, 1880.
15. Willmarth, W.W. and Lu, S.S. 1972 J. Fl. Mech. 55, 65.
16. Kline, S.J., Reynolds, W.C., Schraub, F.A. and Bunstadler, P.W. 1967 J. Fl. Mech. 30, 741.
17. Corrsin, S. 1957 Symp. on Naval Hydrodyn., Publ. 515 NAS-NRC, 373.
18. Kim, H.T., Kline, S.J. and Reynolds, W.C. 1971 J. Fl. Mech, 50, 133.
19. Corino, E.R. and Brodkey, R.S. 1969 J. Fl. Mech. 37, 1.
20. Grass, A.J. 1971 J. Fl. Mech., 50, 233.
21. Wallace, J. M. Eckelmann, H. and Brodkey, R.S. 1972 J. Fl. Mech. 55, 65.
22. Grant, H.L. 1958 J. Fl. Mech. 4, 149.
23. Tritton, D.J. 1967 J. Fl. Mech. 28, 439.
24. Dumas, R., Arzoumanian, E. and Fulachier, L. 1977 C.R. Acad. Sc. Paris, t. 284, série B, 487.
25. Clark, J.A. and Markland, E. 1971 J. Hydr. Div. ASCE HY10, 1653.
26. Lee, M.K., Eckelman, L.D. and Hanratty, T.J. 1974 J. Fl. Mech. 66, 17.

27. Kastrinakis, E.G., Wallace, J.M., Willmarth, W.W., Ghorashi, B. and Brodkey, R.S. 1978 Lect. Notes in Phys. 75, 175.
28. Blackwelder, R.F. and Eckelmann, H. 1979 J. Fl. Mech 94, 577.
29. Blackwelder, R.F. and Kaplan, R.E. 1976 J. Fl. Mech. 76, 89.
30. Oldaker, D.K. and Tiederman, W.G. 1977 Phys. Fl. Supp. 20, S133.
31. Praturi, A.K. and Brodkey, R.S. 1978 J. Fl. Mech. 89, 251.
32. Nychas, S.G., Hershey, H.C. and Brodkey, R.S. 1973 J. Fl. Mech. 61, 513.
33. Brown, G.L. and Thomas, A.S.W. 1977 Phys. Fl. Supp. 20, S243.
34. Blackwelder, R.F. and Kovasznay, L.S.G. 1972 Phys. Fl. 15, 1545.
35. Falco, R.E. 1977 Phys. Fl. Supp. 20, S124.
36. Chen, C.H.P. and Blackwelder, R.F. 1978 J. Fl. Mech. 89, 1.
37. Lu, S.S. and Willmarth, W.W. 1973 J. Fl. Mech. 60, 481.
38. Sabot, J. and Comte-Bellot, G. 1976 J. Fl. Mech. 74, 767.
39. Fleischmann, S. and Wallace, J.M. 1981 Report of Mech. Engr. Dept. Univ. of Md.
40. Wallace, J.M., Brodkey, R.S. and Eckelmann, H. 1977 J. Fl. Mech. 83, 673.
41. Eckelmann, H. and Wallace, J.M. 1981 Lect. Notes in Phys. 136, 292.
42. Kreplin, H.P. and Eckelmann, H. 1979 J. Fl. Mech. 95, 305.
43. Head, M.R. and Bandyopadhyay 1981 J. Fl. Mech. 107, 297.
44. Smith, C.R. 1978 Proc. of Workshop on Coherent structure of turbulent boundary layers, Lehigh University, 48.
45. Bernard, P. S. and Berger, B.S. 1981 to be published in SIAM J. Appl. Math.
46. Klebanoff, P.S., Tidstrom, K.D. and Sargent, L.M. 1960 J. Fl. Mech. 12, 1.
47. Hama, F.R. and Nutant, J. 1963 Proc. Heat Transfer and Fl. Mech. Inst., Stanford Univ. Press, 77.
48. Purtell, L.P., Klebanoff, P.S. and Buckley, F.T. 1981 Phys. Fl. 24, 802.
49. Perry, A.E., Lim, T.T. and Teh, E.W. 1981 J. Fl. Mech. 104, 387.
50. Wygnanski, I., Sokolov, M. and Friedman, D. 1976 J. Fl. Mech. 78, 785.
51. Cantwell, B., Coles, D. and Dimotakis, P. 1978 J. Fl. Mech. 87, 641.
52. Falco, R. 1980 AIAA - 80 - 1356.
53. Leonard, A. 1980 Lect. Notes in Phys. 136, 119.
54. Antonia, R.A., Sokolov, M., Van Atta, C.W. 1981 J. Fl. Mech., 108, 317.
55. Zilmerman, M., Wygnanski, I. and Kaplan, R.E. 1977 Phys. Fl. Supp. 20, 5258.
56. Dean, R.B. and Bradshaw, P. 1976 J. Fl. Mech. 78, 641.

THE THREE DIMENSIONAL MEAN VORTICITY AND COVARIANCE

TURBULENCE CLOSURE

BY

PETER S. BERNARD
ASSISTANT PROFESSOR
DEPT. OF MECHANICAL ENGINEERING
UNIVERSITY OF MARYLAND
COLLEGE PARK, MD
20742

BRUCE S. BERGER
PROFESSOR
DEPT. OF MECHANICAL ENGINEERING
UNIVERSITY OF MARYLAND
COLLEGE PARK, MD
20742

ABSTRACT

The MVC, mean vorticity and covariance, turbulence closure has evolved from the vorticity transport theory of G.I. Taylor and the coarse grained approximations of A.J. Chorin. It utilizes a Lagrangian time expansion technique to effect closure to the exact equations for the vorticity mean and covariances. This aspect of the closure reflects the central role which has been allotted to vorticity dynamics in the closure's formulation. In this respect it differs from previous turbulence closures which are most often given to the averaged momentum or Reynolds stress equations. The MVC closure is also distinguished in that the approximations used in closing the exact equations are, in principle, experimentally verifiable. Such tests, however, await the refinement of current experimental techniques.

Solutions for two-dimensional channel flow have been obtained which display a drag crisis at low Reynolds numbers and a critical Reynolds number separating laminar and turbulent solutions. The closed mean vorticity equation for the case of three-dimensional channel flow has been solved in which pairs of streamwise counter-rotating vortices were used to model anisotropy in the wall region. The computed values of the mean velocity field agreed with experimental values across the whole flow field. Solutions to the complete set of coupled PDE predicted by the MVC closure for channel flow have also been obtained. These are consistent with the numerical results found previously.

The closed equations are given for a class of two-dimensional flows in generalized cylindrical coordinates. The form of the equations corresponding to circular cylindrical coordinates is shown to be a special case.

Introduction

The MVC, mean vorticity and covariance, turbulence closure is derived through the expression of unclosed correlations which occur in the mean vorticity and covariance equations, as functions of the mean vorticity, double correlations of velocity fluctuations and double correlations of gradients of velocity fluctuations. Through the integration of the exact vorticity equation along the path of a fluid particle, a relationship is obtained between the vorticity fluctuation at a point and various quantities which contributed to it in the immediate past. The resulting expression is substituted into the unclosed terms of the mean vorticity and covariance equations which yield terms composed of the sums of various correlations. Those correlations which are of a higher order, vanish or are uncorrelated because of the randomizing nature of turbulence are omitted. The resulting set of nine non-linear partial differential equations are supplemented by integrals which relate the global properties of the vorticity and velocity moments. The closed forms of the mean vorticity transport and covariance equations are given, in curvilinear coordinates, by equations (1) and (2).

$$\langle W_i\rangle_{,t} = \underset{1}{-\langle U^j\rangle \langle W_i\rangle|_j} + \underset{2}{(1/R) \langle W_i\rangle|^j{}_j} + \underset{3}{\langle W^j\rangle \langle U_i\rangle|_j} + \underset{4}{(\langle u_m u_n\rangle T^{mn}_{jk} \langle W_i\rangle|^k)|^j} + \underset{5}{g^{jk} \langle u_q|_r u_s|_p\rangle S^{qrsp}_{ijkm} \langle W^m\rangle} \qquad (1)$$

$$\begin{aligned} \zeta_{ij,t} = {} & -\langle U^k\rangle \zeta_{ij}|_k + (1/R) \zeta_{ij}|^k{}_k + (\langle U_i\rangle|^k \zeta_{kj} \\ & + \langle U_j\rangle|^k \zeta_{ik}) + \langle u_m u_n\rangle T^{mn}_{k\ell} (\langle W_j\rangle|^\ell \langle W_i\rangle|^k \\ & + \langle W_i\rangle|^\ell \langle W_j\rangle|^k) + g^{sr} g^{kn} (T^{mq}_{kr} \langle u_m u_q\rangle \zeta_{ij}|_s)|_n \\ & - 2 \zeta_{ij}/(\gamma^2 R) + \langle u_q|_r u_s|_t\rangle[(\langle W^n\rangle\langle W^k\rangle + \zeta^{nk}) \\ & (S^{qrst}_{ikjn} + S^{qrst}_{jkin}) + \zeta^n{}_j g^{m\ell} S^{qrst}_{im\ell n} + \\ & \zeta^n{}_i g^{m\ell} S^{qrst}_{jm\ell n}] \end{aligned} \qquad (2)$$

where $U_i \equiv$ covariant component of the velocity vector, $\langle(\ )\rangle \equiv$ ensemble average of ( ), $u_i \equiv$ velocity fluctuation, $W_i \equiv$ vorticity vector, $w_i \equiv$ vorticity fluctuation vector, $R \equiv$ Reynold's number, $g_{ij} \equiv$ metric tensor, $(\ )|_i \equiv$ covariant derivative of ( ) with respect to X_i, $\zeta_{ij} \equiv \langle w_i w_j\rangle$, $\gamma^2 \equiv$ vorticity microscale, $T^{k\ell}_{ij}$ and $S^{qrst}_{ijk\ell}$ are integral time scales. A complete derivation of (1) and (2) is given in [4]. The present form of the MVC closure evolved from the coarse grained approximation of A.J. Chorin, [1] and its extension by P.S. Bernard, [2,3].

In the following, (1) and (2) are expressed in rectangular Cartesian coordinates and applied to the numerical study of channel flows. An extensive comparison of numerical solutions with experimental results is given for previous computations. Current studies of three dimensional turbulent channel flows are discussed.

The closed equations, (1) and (2), are specialized for a restricted class of two dimensional flows in generalized cylindrical coordinates. The resulting equations provide a computationally tractable system of P.D.E. which model physically significant cases. The case of the circular cylindrical coordinate system is examined in particular. Current studies of turbulent external flows around circular cylinders are discussed.

Channel Flows

A stringent test of the validity of equations (1) and (2) can be made by comparing their computed solutions for channel flow with experimental data. Application of the MVC closure to channel flow has followed a progression of increasing generality. Initially, a preliminary form of the MVC closure valid only for planar flows, was applied to the idealized case of two-dimensional channel flow, see [2]. Such flows have no vortex stretching. This omission was strikingly reflected in the numerical computations where it was not possible to obtain physically meaningful solutions in the wall region of the flow. In this region vortex stretching plays a fundamental role in the dynamics of coherent vortical structures, e.g., see [5-7]. In spite of this limitation it was still possible to match measured values of the mean velocity distribution in the central region of the channel. In addition, the computed solutions displayed a drag crisis and a bifurcation at a critical Reynolds number separating laminar and turbulent solutions.

For the case of fully developed three-dimensional channel flow equation (1) gives

$$\frac{1}{R} \langle W_3 \rangle_{,22} + (\langle u_2^2 \rangle T \langle W_3 \rangle_{,2})_{,2} + S\, P_{33} \langle W_3 \rangle = 0 \tag{3}$$

for the mean spanwise vorticity $\langle W_3 \rangle$ where $P_{ij} \equiv \langle u_{i,1}\, u_{1,j} + u_{i,2}\, u_{2,j}$ $u_{i,3}\, u_{3,j} \rangle$, X_1 is the streamwise, X_2 the normal, and X_3 the spanwise directions. Similarily (2) gives

$$\frac{1}{R} \zeta_{11,22} + 2\langle U_1 \rangle_{,2}\, \zeta_{12} + (\langle u_2^2 \rangle T\, \zeta_{11,2})_{,2} - 4\, \zeta_{11}/(R\gamma^2)$$

$$+ 2S(\langle u_{1,1}^2 \rangle \zeta_{11} + \langle u_{1,2}^2 \rangle \zeta_{22} + \langle u_{1,3}^2 \rangle (\zeta_{33} + \langle W_3 \rangle^2))$$

$$+ 4S\langle u_{1,1}\, u_{1,2} \rangle \zeta_{12} + 2S\, P_{11}\, \zeta_{11} + 2\, S\, P_{12}\, \zeta_{12} = 0 \tag{4}$$

$$\frac{1}{R}\zeta_{22,22} + (\langle u_2^2\rangle T\,\zeta_{22,2})_{,2} - 4\,\zeta_{22}/(R\gamma^2)$$

$$+ 2S(\langle u_{2,1}^2\rangle\zeta_{11} + \langle u_{2,2}^2\rangle\zeta_{22} + \langle u_{2,3}^2\rangle(\zeta_{33} + \langle W_3\rangle^2))$$

$$+ 4S\langle u_{2,1}u_{2,2}\rangle\zeta_{12} + 2S\,P_{22}\,\zeta_{22} + 2S\,P_{21}\,\zeta_{12} = 0 \qquad (5)$$

$$\frac{1}{R}\zeta_{33,22} + 2\,T\langle u_2^2\rangle\langle W_3\rangle_{,2}^2 + (\langle u_2^2\rangle\,T\zeta_{33,2})_{,2} - 4\zeta_{33}/(R\gamma^2)$$

$$+ 2S\,(\langle u_{3,1}^2\rangle\zeta_{11} + \langle u_{3,2}^2\rangle\zeta_{22} + \langle u_{3,3}^2\rangle(\zeta_{33} + \langle W_3\rangle^2))$$

$$+ 4S\,\langle u_{3,1}u_{3,2}\rangle\zeta_{12} + 2S\,P_{33}\,\zeta_{33} = 0 \qquad (6)$$

$$\frac{1}{R}\zeta_{12,22} + \langle U_1\rangle_{,2}\,\zeta_{22} + (T\langle u_2^2\rangle\zeta_{12,2})_{,2} - 4\zeta_{12}/(R\gamma^2)$$

$$+ 2S(\langle u_{1,1}u_{2,1}\rangle\zeta_{11} + \langle u_{1,2}u_{2,2}\rangle\zeta_{22} + \langle u_{1,3}u_{2,3}\rangle(\zeta_{33} + \langle W_3\rangle^2))$$

$$+ 2S(\langle u_{1,1}u_{2,2}\rangle + \langle u_{1,2}u_{2,1}\rangle)\zeta_{12} + S(P_{11} + P_{22})\,\zeta_{12} + S\,P_{21}\,\zeta_{11}$$

$$+ S\,P_{12}\,\zeta_{22} = 0 \qquad (7)$$

for the vorticity covariances ζ_{11}, ζ_{22}, ζ_{33} and ζ_{12} respectively. Equations for ζ_{13} and ζ_{23} are not needed since these quantities are identically zero for channel flow. The equations used in the two-dimensional study consisted of (3) and (6) with the stretching terms eliminated, i.e., $S = 0$.

Prior to solving the complete coupled system of equations (3) - (7), separate solutions to just the mean vorticity equation (3) were obtained. This simplified the task of later solving the coupled system by giving an indication of how a balance of physical process could be achieved in the complete system. To solve (3) separately values for $\langle u_2^2\rangle$ and P_{33}, ordinarily coming from the solution to (4) - (7), had to be supplied externally. $\langle u_2^2\rangle$ was obtained from measurements of Eckelmann [8], at R = 8200. P_{33} was computed from a model of the coherent structures in a turbulent boundary layer. In this model pairs of streamwise counter rotating vortices were assumed to exist in the wall region. P_{33} was computed as an average of the values of $u_{3,1}u_{1,3} + u_{3,2}u_{2,3} + u_{3,3}^2$ across the spanwise period of the vortices. The solutions to (3) computed in this way gave turbulent-like velocity

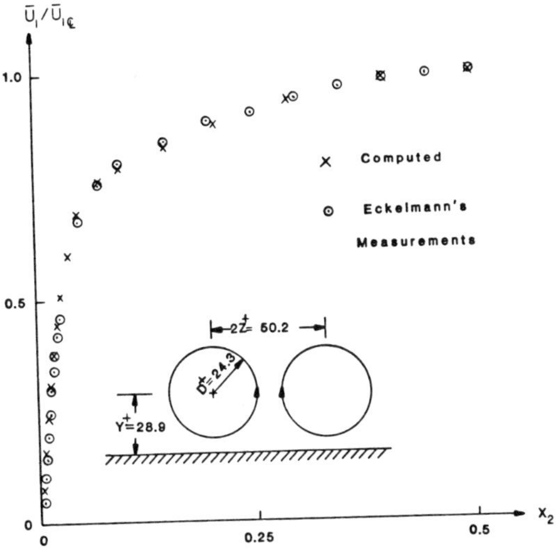

Fig. 1

Comparison of predicted and computed values of $\bar{U}_1$

profiles over a significant range of the dimensions of the vortex pairs. An optimal solution was found for which the computed mean vorticity profile agreed almost exactly with experimental values across the whole channel, see fig. 1. For this solution streamwise vortices with diameters of 24.3 (in wall units) centers at $X_2^+ = 28.9$ and streak spacing = 100.4 were used. These values are remarkably close to those observed in various experimental studies [5 - 7].

The computed values of the terms in (3) provided a picture of the vorticity dynamics in the turbulent boundary layer. Vorticity generated at the wall diffuses by molecular and turbulent means towards the center of the channel. This flux of vorticity is first enhanced by a vortex stretching process beneath the vortex pairs adjacent to the wall. Subsequently, near the centers of the vortex pairs, it is reduced by rotation of the spanwise vorticity into the component of vorticity perpendicular to the wall. The latter process is in agreement with the concept of uplifting "hairpin" or "horseshoe" vortices.

In the most recent computations solutions have been sought to the complete system of equations (3) - (7). An iterative method, based on using the non-steady form of (3) - (7), has been used. Convergent solutions have been obtained with properties very similar to those found in the earlier solutions to (3) above. It still remains to conduct a comprehensive study of the effects of all of the various parameters in (3) - (7) on these results. This will disclose how closely the computed solutions can match experimental values for the mean velocity field and the Reynolds stresses.

Cylindrical Geometries

The expansion of (1) and (2) for a general curvilinear geometry leads to systems of non-linear partial differential equations of considerable complexity. However a significant simplification occurs for crossflows over a circular cylinder which may be approximately modeled as 2-D in the mean. Consider the restricted two dimensional flow for which the velocity vector, U_i, mean velocity, $<U_i>$ and velocity fluctuations, u_i, are not functions of X_3 nor have an X_3 component. Let Z_i be a rectangular Cartesian coordinate system and X_i an orthogonal curvilinear coordinate system such that $X_1 = X_1(Z_1,Z_2)$, $X_2 = X_2(Z_1,Z_2)$ and $Z_3 = X_3$. Assume that the integral time scales T^{mn}_{jk} and $S^{qrst}_{jm\ell n}$ may be expressed as $T^{mn}_{jk} = \delta^m_j\delta^n_k T$, $S^{qrst}_{jm\ell n} = \delta^q_j\delta^r_m\delta^s_\ell\delta^t_n S$ where T and S are constants. The expansions of the integral time scales as products of Kronecker deltas and the scalars T and S is discussed in [4]. Under the foregoing assumptions, the forms taken by (1) and (2) will be derived. Since U_i, $<U_i>$ and u_i are not functions of X_3, have no X_3 component and X_i is cylindrical and orthogonal it follows from $W^k = \varepsilon^{k\ell m} U_{m|\ell}$, $\varepsilon^{k\ell m} = e^{k\ell m}/\sqrt{g}$ that $<W_1> = 0$, $<W_2> = 0$, $w_1 = 0$, $w_2 = 0$ and $\zeta_{ij} = 0$ except for $i = j = 3$.

Substituting the Kronecker delta expansion of the integral time scale $S^{qrst}_{jm\ell n}$ into term 5 of (1) and expanding gives $Sg^{jk} <u_i|_j u_k|_m><W^m> =$ $S[g^{11} <u_i|_1 u_1|_3> + g^{22} <u_i|_2 u_2|_3> + g^{33} <u_i|_3 u_3|_3>]<W^3> = 0$ since $u_i|_3 = 0$, $i = 1,2,3$. Note that $u_i|_j = u_{i,j} - \left\{ {\ell \atop ij} \right\} u_\ell$, $u_{i,3} = 0$ and $\left\{ {\ell \atop i3} \right\} = 0$, [9], where $u_{i,j} \equiv \partial u_i/\partial X_j$.

Expanding term 1 of (1) for $i = 3$, noting that the term vanishes for $i = 1,2$ gives $<U^j><W_3>|_j = <U^1><W_3>|_1 + <U^2><W_3>|_2 = <U^1><W_3>_{,1} +$ $<U^2><W_3>_{,2}$ since $<W_3>|_i = <W_3>_{,i}$. Define the stream function ψ such that $\psi_{,2} = \sqrt{g}\ <U^1>$ and $\psi_{,1} = -\sqrt{g}\ <U^2>$ where $g \equiv \det.\ g_{ij}$. Then term 1 becomes

$$<U^j><W_3>|_j = (1/\sqrt{g})(\psi_{,2}<W_3>_{,1} - \psi_{,1}<W_3>_{,2}) \tag{8}$$

For $i = 3$ and omitting the constant multiplier, term 2 of (1) is $<W_3>|^j_j = g^{jn}(W_3|_n)|_j = g^{11}W_{31}|_1 + g^{22}W_{32}|_2 + g^{33}W_{33}|_3$ where for simplicity of notation $<W_3> \equiv W_3$ and $<W_3>|_k \equiv W_{3k}$. From the expansion of $W_3|_k$ it follows that $W_3|_k = W_{3,k}$. The expression of $W_{3\ell}|_m$ is given by, [9],

$$W_{3\ell}|_m = W_{3\ell,m} - \left\{ {3 \atop 3\ m} \right\} W_{3\ell} - \left\{ {n \atop \ell\ m} \right\} W_{3n} \tag{9}$$

From (9) it follows that $W_{31}|_1 = W_{3,11} - \left\{ {1 \atop 1\ 1} \right\} W_{3,1}$ $- \left\{ {2 \atop 1\ 1} \right\} W_{3,2}$, $W_{32}|_2 = W_{3,22} - \left\{ {1 \atop 2\ 2} \right\} W_{31} - \left\{ {2 \atop 2\ 2} \right\} W_{32}$ and $W_{33}|_3 = W_{3,33}$. It follows that

$$\langle W_3\rangle|^j{}_j = (1/g_{11})(\langle W_3\rangle_{,11} - C111\langle W_3\rangle_{,1} - C211\,\langle W_3\rangle_{,2}) + (1/g_{22})(\langle W_3\rangle_{,22} - C122\langle W_3\rangle_{,1} - C222\langle W_3\rangle_{,2}) + \langle W_3\rangle_{,33} \tag{10}$$

where $Cijk \equiv \left\{ {i \atop j\ k} \right\}$.

Summing with respect to j in term 3 of (1) and recalling that $\langle W^j\rangle = 0$ for $j = 1,2$ it follows that $\langle W^j\rangle\langle U_i\rangle|_j = \langle W^3\rangle\langle U_i\rangle|_3 = 0$ since $\langle U_i\rangle|_3 = 0$ for all i.

Substituting the Kronecker delta expansion of T^{mn}_{jk} into term 4 of (1), setting $i = 3$ and omitting the scalar T gives $(\langle u_j u_k\rangle\langle W_3\rangle|^k)|^j$. Letting $A_{jk} \equiv \langle u_j u_k\rangle$ and lowering the contravariant indices yields $g^{mk}(A_{jk}W_3|_m)|_n$ g^{nj} for this term. Carrying out the indicated differentiation, expanding, substituting into term 4 and replacing terms 1, 2, 3, 4 and 5 of (1) by their expanded forms gives

$$\begin{aligned} W_{,t} &= -1\sqrt{g}\ (\psi_{,2}\, W_{,1} - \psi_{,1}\, W_{,2}) + (1/(Rg_{11})) \\ &(W_{,11} - C111W_{,1} - C211W_{,2}) + (1/(Rg_{22})(W_{,22} - C122W_{,1} - C222W_{,2}) \\ &+T((g^{11})^2((A_{11,1} - 2\,C111A_{11} - 2\,C211A_{21})W_{,1} + A_{11}(W_{,11} - C111W_{,1} - C211W_{,2})) \\ &+ (g^{11}g^{22})\ ((A_{21,2} - C122A_{11} - C222A_{21} - C112A_{21} - C212A_{22})W_{,1} + A_{12}(W_{,12} \\ &- C112W_{,1} - C212W_{,2}) + (A_{12,1} - C111A_{12} - C211A_{22} - C121A_{11} - C221A_{12})W_{,2} \\ &+ A_{12}(W_{,12} - C112W_{,1} - C212W_{,2})) + (g^{22})^2((A_{22,2} - 2C122A_{12} - 2C222A_{22}) \\ &W_{,2} + A_{22}(W_{,22} - C122\,W_{,1} - C222W_{,2}))) \end{aligned} \tag{11}$$

For i = 1 and 2(1) is identically zero. Thus it is seen that the system of three partial differential equations given by (1) reduces, under the stated assumptions, to the single P.D.E. given by (11).

Similarly the covariance equations, (2) reduce, under the stated assumptions to a single P.D.E. corresponding to i = j = 3. The expression of (2) in the given orthogonal curvilinear cylindrical coordinate system involves computations which are similar to the foregoing and which will therefore be omitted. The expanded form of (2) is

$$\zeta,_t = (1/\sqrt{g})\,(\psi,_1\zeta,_2 - \psi,_2\zeta,_1) + (1/R)\,(g^{11}\zeta,_{11} + g^{22}\zeta,_{22} - (g^{11}\,C111 + g^{22}\,C122)\zeta,_1 - (g^{11}\,C211 + g^{22}\,C222)\zeta,_2) + 2T\,((g^{11})^2A_{11}(W,_1)^2 + 2g^{11}g^{22}A_{12}W,_1W,_2 + (g^{22})^2A_{22}(W,_2)^2) + T((g^{11})^2((A_{11,1} - 2(C111A_{11} + C211A_{21}))\zeta,_1 + A_{11}(\zeta,_{11} - C111\zeta,_1 - C211\zeta,_2)) + g^{11}g^{22}((A_{21,2} - C122A_{11} - C222A_{21} - C112A_{21} - C212A_{22})\zeta,_1 + A_{21}(\zeta,_{12} - C112\zeta,_1 - C212\zeta,_2)) + g^{11}g^{22}((A_{12,1} - C111A_{12} - C211A_{22} - C121A_{11} - C221A_{12})\zeta,_2 + A_{12}(\zeta,_{21} - C112\zeta,_1 - C212\zeta,_2)) + (g^{22})^2((A_{22,2} - 2(C112A_{12} + C222A_{22}))\zeta,_2 + A_{22}(\zeta,_{22} - C122\zeta,_1 - C222\zeta,_2))) - 2\zeta/(R\gamma^2) \qquad (12)$$

The expanded forms of the closed mean vorticity equation, (11), and covariance equation, (12), may be readily expressed in specific cylindrical coordinate systems appropriate to various internal and external two-dimensional turbulent flows. For the case of circular cylindrical coordinates $Z_1 = X_1 \cos X_2$, $Z_2 = X_1 \sin X_2$, $Z_3 = X_3$, $g_{11} = g^{11} = g_{33} = g^{33} = 1$, $g_{22} = 1/g^{22} = X_1^{\,2}$, $C212 = C221 = 1/X_1$, $C122 = -X_1$ and all other $Cijk = 0$.

Substitution into (11) and (12) gives the closed mean vorticity and covariance equations for circular cylindrical coordinates.

A finite difference solution to these equations for the case of the circular cylinder in a uniform flow has been constructed. Numerical studies are currently underway which, when compared to experimental results, will determine the validity of the closure.

Acknowledgement

The authors wish to express their appreciation to Dr. Oscar Manley and Dr. James Wallace for their encouragement. Those aspects of the study related to channel flows were supported by the National Science Foundation, under project No. CME79-19835, while those related to flows about cylinders were supported by the Department of Energy, under project No. DE-AS05-81ER10809.A000. All computations were performed at the Computer Science Center, University of Maryland.

References

1. Chorin, A.J., "Lectures on Turbulence Theory", Publish/Perish, Boston (1975).

2. Bernard, P.S., "A Method for Computing Two-Dimensional Turbulent Flows", SIAM J. Applied Math., 38 (1980), pp. 81-92.

3. Bernard, P.S., "A Mean Vorticity and Covariance Turbulence Closure with Application to Channel Flow", University of Maryland, Dept. of Mech. Eng. Report, Jan. 1980.

4. Bernard, P.S. and Berger, B.S., "A Method for Computing Three-Dimensional Turbulent Flows", SIAM J. Applied Math., (To appear).

5. Blackwelder, R. and Eckelmann, H., "Streamwise Vortices Associated with the Bursting Phenomenon", J. Fluid Mech., 94 (1979), pp. 577-594.

6. Smith, C.S. and Abbot, D.E., Eds. Workshop on Coherent Structure of Turbulent Boundary Layers, Lehigh U., Bethlehem, Pa., 1978.

7. Willmarth, W.W., "Structure of Turbulence in Boundary Layers", Advances in Applied Mechanics, Vol. 15, Academic Press, N.Y., 1975.

8. Eckelmann, H., "The Structure of the Viscous Sublayer and The Adjacent Wall Region in a Turbulent Channel Flow", J. Fluid Mech., 65 (1974) pp. 439-459.

9. Lovelock, D. and Rund, H., "Tensors, Differential Forms, and Variational Principles", John Wiley and Sons, New York, 1975.

A VISUAL STUDY OF THE CHARACTERISTICS, FORMATION, AND REGENERATION OF TURBULENT BOUNDARY LAYER STREAKS

C.R. Smith and S.P. Metzler

Department of Mechanical Engineering and Mechanics
Lehigh University
Bethlehem, PA 18015

ABSTRACT

Using hydrogen bubble-wires and negatively buoyant glass beads for flow visualization in water, a series of visual studies have been done of the characteristics of the low-speed streaks which occur in the near-wall region of turbulent boundary layers. Using a two camera, high-speed video system both single view and multiple view studies have been completed. The results indicate that low-speed streaks are regions of extreme streamwise extent, demonstrating an apparent persistance for times much longer than accepted bursting times associated with wall region turbulence production. The use of a fiber optic lens for simultaneous top-end view studies has revealed the low-speed streaks to be regions of significant outward flow resulting from the apparent presence of counter-rotating streamwise vortices. Further dual views (top-side) indicate a symbiotic relationship between the occurrence of low-speed streaks and the formation of vortex loops in the near-wall region. Using these studies as a basis, a cyclic process for low-speed streak formation and reinforcement is hypothesized.

INTRODUCTION

Probably the most universally distinguishing flow structures of a turbulent boundary layer are the sublayer "streaks" [1] of alternately high and low-speed flow, which are always observed near the surface beneath a turbulent boundary layer. However, despite substantial documentation of the statistical characteristics of the streak structure, the origin of streaks is still not understood. This is due to the three-dimensional, transient behavior of the streaks, which precludes detailed probe measurement, and has necessitated primarily flow visualization studies using dye injection, hydrogen bubble-wires, or smoke. However, most of these studies have been limited to motion picture recording of a single view of the streak phenomena (generally the top-view), which allows only limited interpretation of three-dimensional effects. The present paper describes flow visualization results obtained with a unique two-camera, video flow visualization system which circumvents many of the limitations of previous photographic techniques with regard to three-dimensional effects and view perspective. The selected results of a series of recent studies [2,3,4,5] are presented which reveal some key characteristics of low-speed streaks and provide substantial insight into the process of low-speed streak formation. On the basis of these studies, a model of the flow process giving rise to momentum transfer and low-speed streak formation in the wall region of a turbulent boundary layer is hypothesized.

EXPERIMENTAL APPARATUS

The present studies were all performed in a free-surface facility with a 6m working section, 0.9m wide by 0.3m deep as described in [2,3].

Visualization of the streak structure was done using a uniquely designed hydrogen bubble-wire probe (described in [2]) which allows a 25 μm dia. platinum wire, 20 cm in length, to be located parallel to the floor of the channel and transverse to the flow direction. The wire, used as the cathode in an electrolysis process, generates hydrogen bubbles by employing a specially designed generator unit. Using this probe, lines and sheets of hydrogen bubbles can be introduced parallel to the floor of the channel at any desired height and frequency.

The video viewing and recording system is a two-camera, high-speed video system (manufactured by the Video Logic Co.) which incorporates synchronized strobe lights to provide 120 frame/s with effective frame exposure times of 10^{-4}s. A split-screen capability allows two different fields of view to be simultaneously displayed and recorded. All recorded data can be played in flicker-free slow motion (both forward and reverse), as well as single-framed for detailed data analysis and hard copy output. For some studies, a specially designed fiber optic lens, 1 cm in diameter and 1 m long, was incorporated with one of the video cameras to yield end-on views of the hydrogen bubble-wire.

Once a video sequence is recorded, conventional still photographs of individual stop-action frames, such as appear in this paper, can be taken directly from the video screen.

RESULTS

A. Top View Studies

Using top-view sequences of flow behavior in the near-wall region ($1<y^+<40$), the characteristics of the low-speed streaks which occur in the near wall region of turbulent boundary layers have been examined for a Reynolds number range of $740 \leq Re_\theta \leq 5830$. As shown in Figure 1, low speed streaks are found to be essentially identical in appearance and character for all Reynolds numbers examined. Using visual counting techniques for long video sequences, the statistics of non-dimensional spanwise streak spacing were determined and shown to be essentially invariant with Reynolds number, exhibiting consistent values of $\overline{\lambda^+} \cong 100$ and remarkably similar probability distributions conforming to lognormal behavior [3]. Further studies show that streak spacing increases with distance from the wall due to a merging and intermittency process which occurs for $y^+ \gtrsim 5$. An additional observation was that although low speed streaks are not fixed in time and space, they demonstrate a tremendous persistence, often maintaining their integrity and reinforcing themselves for time periods up to an order of magnitude longer than the observed bursting times associated with wall region turbulence production.

An investigation of streak persistence and longevity was carried out through an experiment that capitalized upon the observation that stray particulate matter on the channel bottom appeared to resolve itself into long streamwise aggregations. This appraoch involved the injection via syringe of a thin slurry of 100 μm glass beads (s.g. $\cong$ 2 and $d^+ \cong 1$) along the surface beneath a fully turbulent boundary layer. Within a few seconds after injection the beads collected into streamwise concentrations of width $\Delta z^+ \cong 35$ which extended essentially uninterrupted the length of the channel with a nearly uniform spanwise spacing of $\Delta z^+ \cong 80$. This pattern appeared stable and quasi-stationary, with the only movement being a quasi-periodic buffeting and a gradual downstream migration of the beads, apparently in response to localized "sweeps" of high speed fluid. Further investigation with a horizontal hydrogen bubble wire disclosed that the bead concentrations coincided with the low-speed streaks, as can be seen in Figure 2.

Since the beads used were not light enough to move freely with small velocity fluctuations, the results of this experiment should be interpreted with some caution. While it is possible that the bead concentrations tended to mark the "mean" locations of very long, naturally occurring low-speed streaks, it is more probable that the wakes induced by the streamwise aggregations of the beads acted to cause their further concentration downstream. However, since the hydrogen bubble flow patterns occurring directly above the lengthy, coherent bead concentrations were those of low-speed streaks, this implies that the flow structures forming above the beads (and causing their continued concentration) must be the same structures which occur naturally, only more focussed by the presence of the beads.

B. Split-Screen Studies: Top-end Views

The bead concentration studies suggested that some streamwise mechanism which occurs in a repetitive fashion must be responsible both for the extreme streamwise length of the beads concentrations and for the formation of a low-speed streaks. In an attempt to reveal both the presence and the characteristics of the streamwise behavior, a dual-view study [4,5] was conducted using a fiber optic lens to obtain simultaneous top-end views of the near-wall region within which low-speed streaks are observed.

Figures 3 and 4 show two typical dual (top-end) view sequences of the streak behavior taken for $Re_\theta = 1700$. The two sequences, both taken with the bubble-wire at $y^+ = 14$, are to the same scale and correspond to conventional orthographic projections (i.e., the lower picture is the end-view projection of the corresponding top-view above). The flow is top to bottom in the top-views and out of the picture in the end-views; the bubble-wire appears at the upper extreme in the top-view and about mid-picture in the end-view. The two dark rectangles appearing near the top of each picture are scene and frame indicators inserted electronically by the video system.

It was found that bubble-lines generated in close proximity to each other tended to obscure one another when viewed in end-view. Thus, the sequences presented here employ single-realization time-lines (1 Hz pulse frequency, 3% on and 97% off) in order to clarify the local flow dynamics. For the end-view pictures, the fiber optic lens was located 12 cm downstream of the wire at a viewing angle of about 2° to the surface. Thus, the bubble lines in the end-views will appear to move slightly downward as they progress towards the lens. In addition, the lighting and lens aperature limited the end-view depth of field to about 3 cm. So the bubbles nearer the lens frequently appear defocussed.

The top-view in Figure 3a illustrates the very strong span-wise velocity variations that exist in the wall region, with the corresponding end-view sequence showing the strong vertical movement and the extreme span-wise discreteness of the low-speed regions (which correspond to the low-speed streaks illustrated by Figure 1). Close examination of the end-view of Figure 3b reveals the beginning of a "mushrooming" effect for each of the low-speed regions. This indicates a spreading of the low-speed fluid in a spanwise direction as it moves upward. Notice that the low-speed upwelling on the far left has developed a definite "hook" in the bubble-line which was observed to be the result of a strong streamwise vortex adjacent to the line.

In the top-view of Figure 4a, discrete low-speed regions (two appear in this sequence) again appear as the dominant characteristic. The corresponding end-view again shows these to be well-defined upwellings of fluid. In the following picture, the bubble-line has passed out of the top-view, but in the end-view the low-speed upwelling on the right has developed from a double-hook appearance in Figure 4a into a double-loop in Figure 4b (this pattern was also observed for the left upwelling, but it is not as clearly shown by the still photographs). This behavior was determined from the video sequence to be the result of a pair of counter-rotating vortices with dimensionless diameter and vorticity (during the sequence) of $D^+ = Du_\tau/\nu \cong 30$ and $\omega^+ = \omega_x \nu/u_\tau^2 \cong 0.4$. Detailed examination of a series of dual view sequences over a wide range of Reynolds numbers and y^+ values have shown the low-speed streaks to always be regions of vertical motion, with the bubble line patterns, frequently revealing associated streamwise rotation (40% of observations) or counter-rotating vortex pairs (15-20% of observations). Generally, the degree and characteristics of observed counter-rotating structures varied with wire location from the wall [5]. However, the most important observation was that whenever counter-rotating structures were detected in end-view, they were always observed in the corresponding top-view to evolve either from or in conjunction with a low-speed streak. Thus, it was apparent that the presence of counter-rotating, streamwise vortices is either necessary for or the result of low-speed streak formation.

C. Split-Screen Studies: Top-Side Views

In order to further examine the three-dimensional aspects of the near-wall flow behavior, and the origin of the counter-rotating vortices which top-end view studies showed to be intimately involved in low-speed streak formation, a series of studies was done employing simultaneous side and top views of a horizontal hydrogen bubble-wire. A slightly oblique positioning of the side view camera at a small downwards angle (7°-9°) yielded a perspective of the horizontal bubble wire improved over that which would be obtained if the line of sight was parallel to the wire. The cameras were carefully adjusted to obtain equal magnification scales and coincident streamwise fields of view.

A problem that inevitably accompanies use of split-screen viewing is the difficulty in identifying the same spanwise location in the side and top views. In the present study, this problem was overcome by insulating all but a one centimeter span of the bubble wire, thus producing a narrow bubble sheet that is entirely within the viewing depth of field. For the flows examined, the width of this bubble sheet was of the order of one streak spacing ($\Delta z^+ \simeq 110$). This approach permits (1) a direct association to be made between the top and side-views, and (2) all bubble lines produced to be kept in focus.

Basically, several different types of flow structure were observed to occur in the near-wall region. The most dominant was the narrow, well defined low speed streak, which was always observed to be a region of upwelling motion. Depending on the position of the wire, these low-speed streaks often were observed to undergo a strong transverse moment (in top-view) causing the streak to become "kinked" or wavy as shown in Figure 5a. Alternatively, the streak would be observed to degenerate into one or more spanwise bulges (almost a bisymmetric "kink") as shown in Figure 5b. These bulges appear to be the "pockets" described in previous studies [6,7] . In both cases, a series of vertical undulations in the streak could frequently be observed in side view to form transverse and sometimes streamwise rotational motions. The formation of these undulations and the breakdown into rotational elements was determined to be the consequence of the apparent bursting behavior originally observed and described by Kim [8].

Less frequently, streaks were observed to break down in a more symmetrical fashion into a series of 2-5 loop-like vortex structures. This was normally due to a fortuitous coincidence of wire location and streak breakdown, but was observed for bubble-wire locations from $5<y^+<35$. Figure 6 is a sequence showing the evolution of such a vortex (marked L.V. on the Figure). The remnants of the legs of previously generated loop vortices can be observed in this figure (although they are difficult to identify in these still pictures). Figure 7 is a schematic representation of this type of flow structure with associated characteristics. Note that the vorticity associated with the upstream-pointing legs is oriented similarly to that of the counter-rotating vortices discussed earlier in conjunction with streak formation. The head of the loop appears as a transverse vortex aligned with the mean strain.

Velocity characteristics of the loops were measured from video footage of a stationary horizontal bubble wire. It was generally determined that the trajectories of individual loop "heads" typically did not contain a substantial upwards component until reaching $y^+ \doteq 15$. Locating the bubble wire at $y^+ = 15$ resulted in loop head trajectories that attained an average height of $y^+ = 78$ after traveling a distance $\Delta x^+ = 377$ downstream. At $y^+ \simeq 40$, the stream wise convection velocity was $U_c = 13.1\ u_\tau$, and the upwards normal component was $v = 2.2\ u_\tau$. The former value (U_c) is close to that measured by Kline et al. [1] for the convection velocity associated with the ejection phase of his bursting cycle i.e. $13.8\ u_\tau$ at $y^+ \simeq 40$. Where discrete arrays of typically 2-5 loop-like ejections could be observed, the mean spacing between loops was typically $\Delta x \simeq 200$. Separate arrays of loops, kinks, or bulges were observed to appear at time intervals between the arrays commensurate with the empirical expressions for the bursting frequency based on inner variables ($f^+ = f\nu/u_\tau^2 = .005$), or outer variables ($U_\infty T_B/\delta = 5$), which for moderate Reynolds numbers are nearly coincident.

In addition to the formation of an array of loop-like ejections during the "bursting" of a streak, it was observed that a given streak will frequently persist and give rise to multiple bursts. In numerous cases a low-speed streak was observed to persist for times as long as $4\text{-}5T_B$, resulting in the generation of several loop arrays. In essence this observation is consistent with the observations of the previous bead slurry study, with the implication being that the formation of the vortex loop arrays may be the mechanism causing both low speed streak formation - perpetuation and the great streamwise length of the bead concentrations.

DISCUSSION AND HYPOTHESIS

In the results presented in this paper, it is clear that the structures associated with both the streak formation and breakdown display a wide variation in appearance. Although this is in part due to the stochastic nature of the flow, it was also observed that small changes in the location of the bubble-wire and the phase of development at which it intersected a given structure has a tremendous effect on the features emphasized. For example, to be well defined and discernible when viewed in top-side views, it is necessary for a loop vortex to begin its formation essentially at or slightly downstream of the bubble-wire. If a loop formed upstream or well downstream of the wire, the vorticity contained in the loop would not be marked by the bubble lines; only the effect of the loops presence would appear impressed on the visualization medium. Thus, the observation of entire loop structures (in top-side view) or the counter-rotating legs of the loop (in top-end view) occurs less frequently than the observation of impressed motions such as kinked streaks and upwellings.

Despite the limitations of the visualization process, a number of new aspects of streak behavior have been observed which suggest a cylical process for streak formation, persistence, and bursting. The following is an attempt to describe and define that process.

Basically, the formation of a low-speed streak requires that a spanwise perturbation in the mean vorticity be present very near the surface. This could be due to the presence of an existing streak or a surface irregularity (as caused by the bead concentrations), or as a consequence of the transition process. At some point in time, through either viscous diffusion or interaction with a high-speed sweep from the outer region of the boundary layer, the vorticity sheet in the near-wall region becomes sufficiently perturbed such that it becomes locally unstable, precipitating the formation of three-dimensional flow structures (vortex loops) via a Kelvin-Helmholz type instability. Note that the wavelength of the instability would be the distance between individual vortex loops of an array, which the present study indicates to be $\Delta x^+ \simeq 200$ (see Figure 7). Note that this value is consistent with wavelengths discussed by Blackwelder [9] in reference to a potential bursting mechanism.

The number of concentrated zones of vorticity or vortex loops formed or shed as a result of this breakdown process is dependent on local conditions, but 2 to 5 are commonly observed. As shown in Figure 7, the vorticity concentrations are quickly stretched and distorted by the local velocity gradient into a series of "nested" vortex loops, with the front of the loop moving away from the wall and the legs trailing back and terminating in a spanwise vortex sheet in the high-speed regions which flank the low-speed streak (note that the legs do not actually terminate, since the vortex lines comprising the loops must be continuous; this term implies only that the actual rotational vortex tubes cease to be observed).

The stretching of the upstream-pointing legs of the vortex loop results in a strong local pressure gradient which causes retarded, low-speed fluid at or near the surface to be swept in and pumped up between the legs of the vortex as it convects downstream. This accretion of fluid results in either the formation of a new low-speed streak or the reinforcement of a previous streak which spawned the vortex loop. The passage of more than one loop, as in the arrays of loops observed, merely enhances the process. The streak thus formed or reinforced becomes the spanwise perturbation in vorticity which yields the site for generation of subsequent vortex loop arrays during the burst process.

An important effect which bears on the above description of streak formation is the observation of an apparent streamwise coalescence process whereby the stretched legs of the multiple, nested vortex loops continually reinforce the counter-rotating behavior which yields the persistence of the streak. As shown in Figure 7, the streamwise legs of multiple loops appear to coalesce in a three-dimensional fashion, with the legs of the older loops (which have been weakened by viscous effects) tending to orbit the more recent ones (see detail B-B in Figure 7). The result is a continued agglomeration of the streamwise vorticity from the legs of individual vortex loops (which indivudually would be stretched and dissipated rapidly) into larger concentrations which will maintain their integrity and coherence in the presence of extreme stretching, external buffeting, and visious dissipation.

The hypothesized streak formation process described above is thus a self-sustaining, cylical process requiring only the presence of a velocity gradient and an external stimulus of suffucient amplitude to perpetuate the process of streak formation and turbulence production near a surface. It is important to point out that the process described and the model shown in Figure 7 are idealized and represent only a part of the overall turbulence production-dissipation process. However, the proposed process contains all the elements observed in the 6 hours of video tape data which comprise the series of near-wall studies described here, and it is felt to provide at least a tentative explanation of the origin and persistence of low-speed streaks.

ACKNOWLEDGEMENTS

The authors would like to gratefully acknowledge the continuing support of this research by the Fluid Mechanics Division of the Air Force Office of Scientific Research.

REFERENCES

1. Kline, S.J., Reynolds, W.C., Schraub, F.A.,and Runstadler, P.W. "The Structure of Turbulent Boundary Layers," J. Fluid Mech., 30, part 4, 1967, p. 741.
2. Metzler, S.P. "Processes in the Wall Region of a Turbulent Boundary Layer", M.S. Thesis, Dept. of Mech. Engr. & Mech., Lehigh University, 1980.
3. Smith, C.R. and Metzler, S.P. "The Characteristics of Low-Speed Streaks in the Near-Wall Region of a Turbulent Boundary Layer," under review, J. Fluid Mechanics.
4. Smith C.R., Schwartz, S.P., Metzler, S.P., and Cerra, A.W. "Video Flow Visualization of Turbulent Boundary Layer Streak Structure," Flow Visualization II, W. Merzkirch, ed., Hemisphere Pub. Co., Washington, 1981, p. 605.
5. Schwartz, S.P. "Investigation of Vortical Motions in the Inner Region of a Turbulent Boundary Layer," M.S. Thesis, Dept. of Mech. Engr. & Mech. Lehigh University, 1981.
6. Smith C.R. "Visualization of Turbulent Boundary Layer Structure Using a Moving Hydrogen Bubble-Wire Probe," Coherent Structure of Turbulent Boundary Layers, Smith and Abbott, ed., AFOSR/Lehigh University Workshop, Dept. of Mech. Engr. & Mech., Bethlehem, PA., 1978, p. 50.
7. Falco, R. "The Role of Outer Flow Coherent Motions in the Production of Turbulence Near a Wall," ibid ref. 6, p. 448.
8. Kim, H.T., Kline, S.J., and Reynolds, W.C. "The Production of Turbulence Near a Smooth Wall in a Turbulent Boundary Layer," J. Fluid Mech., 50, part 1, 1971, p. 133.
9. Blackwelder, R.F., "The Bursting Process in Turbulent Boundary Layers," ibid ref. 6, p. 211.

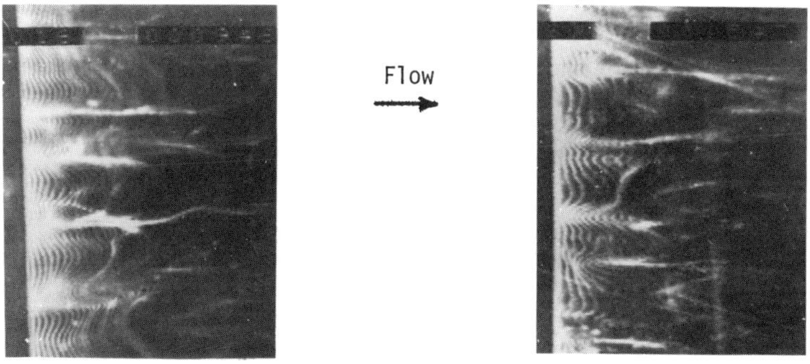

a) Re_θ=1490 b) Re_θ=5930

Figure 1 Low Speed Streaks, y^+=5

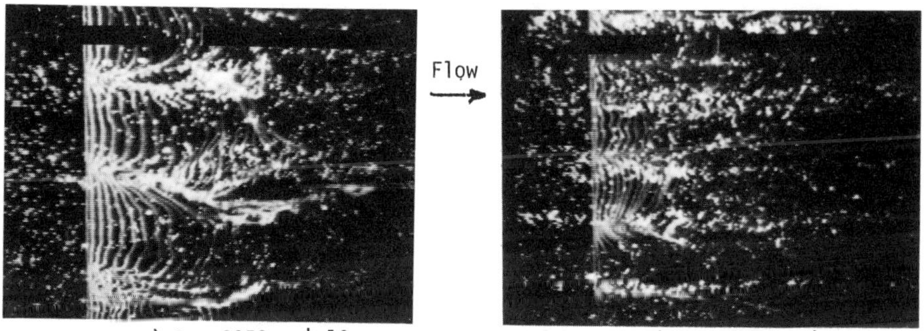

a) Re_θ=2250, y^+=13 b) Re_θ=2070, y^+= 7

Figure 2 Bead Aggregations Due to Streaks

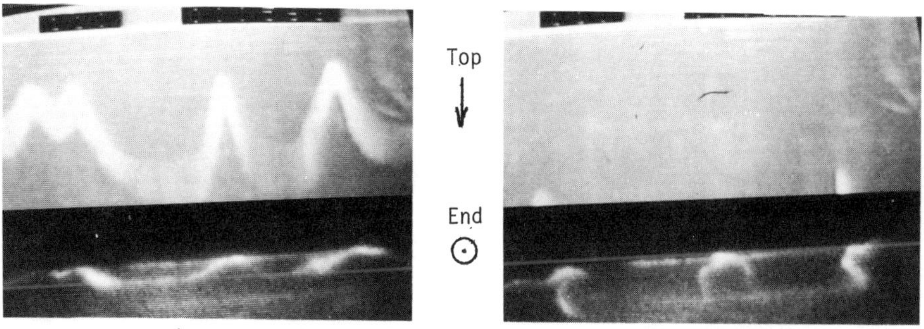

a) t^+ = 0 b) t^+ = 14

Figure 3 Top-End View of Streak Upwelling: Re_θ=1700, y^+=14

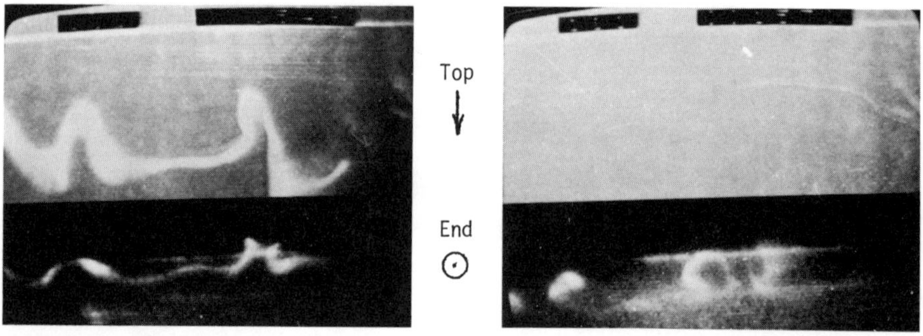

a) $t^+ = 0$ b) $t^+ = 16$

Figure 4 Top-End Views of Counter-Rotating Vortices: Re_θ=1700, y^+=14.

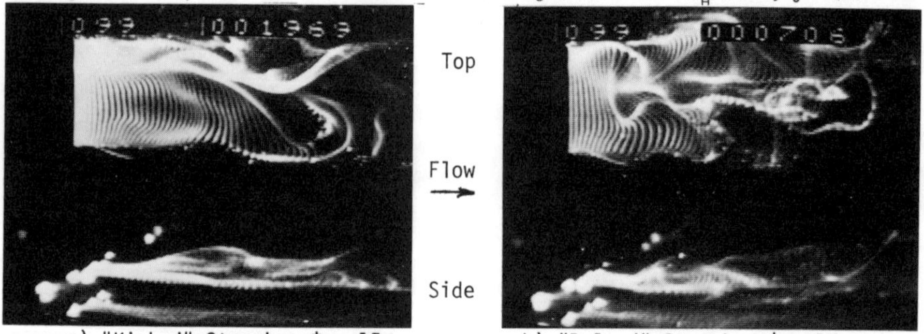

a) "Kinked" Streak, $y^+ = 15$ b) "Bulged" Streak, $y^+ = 15$

Figure 5 Top-Side Views Illustrating Streak Behavior During Breakdown. Re_θ=1910.

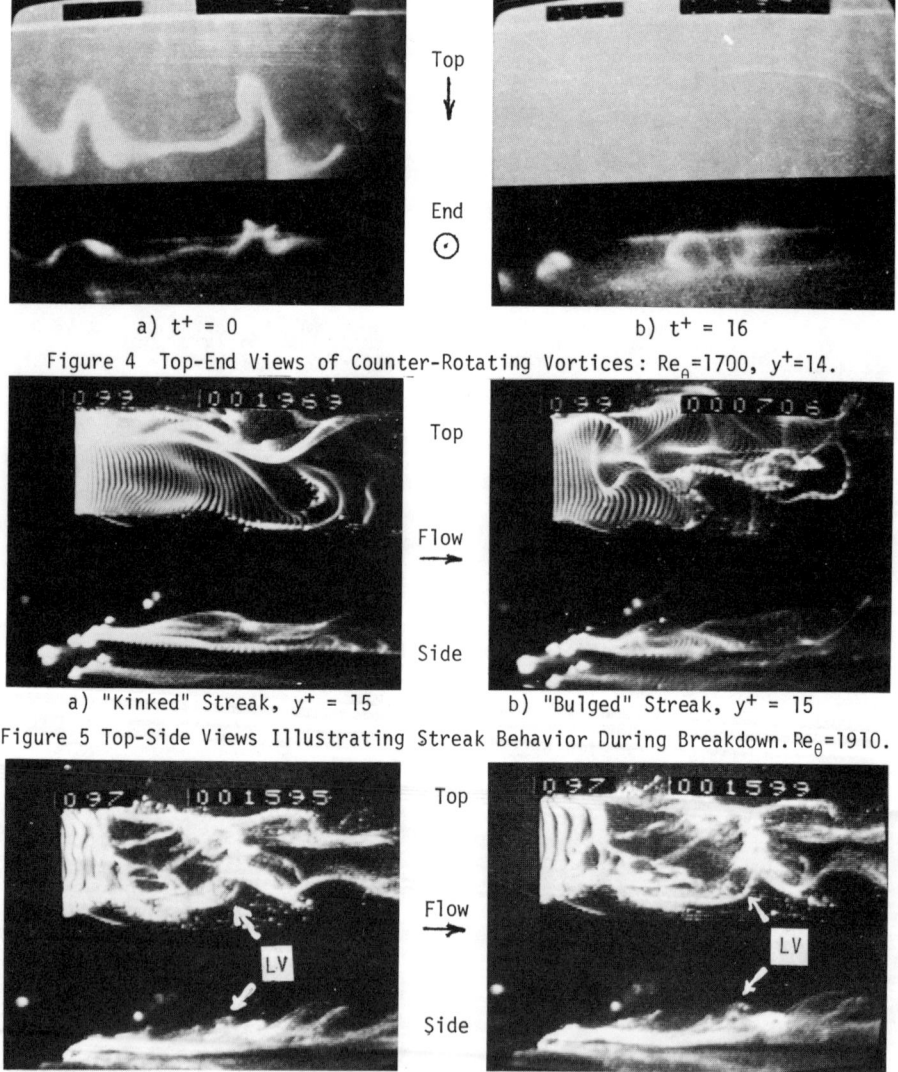

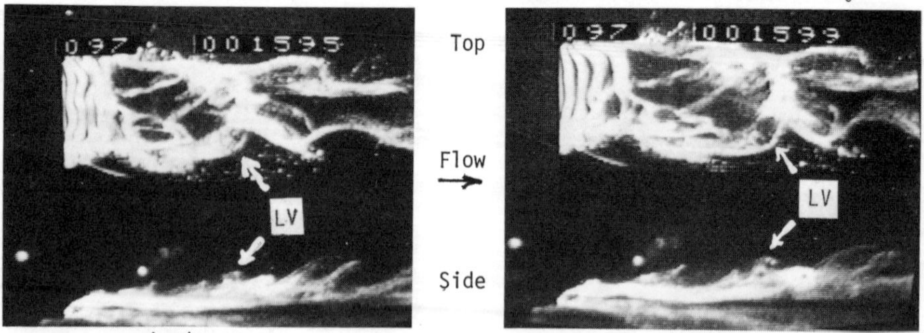

a) $t^+ = 0$ b) $t^+ = 4$

Figure 6 Top-Side View Sequence of "Loop" Formation-Appearance: Re_θ=1910, y^+=5.

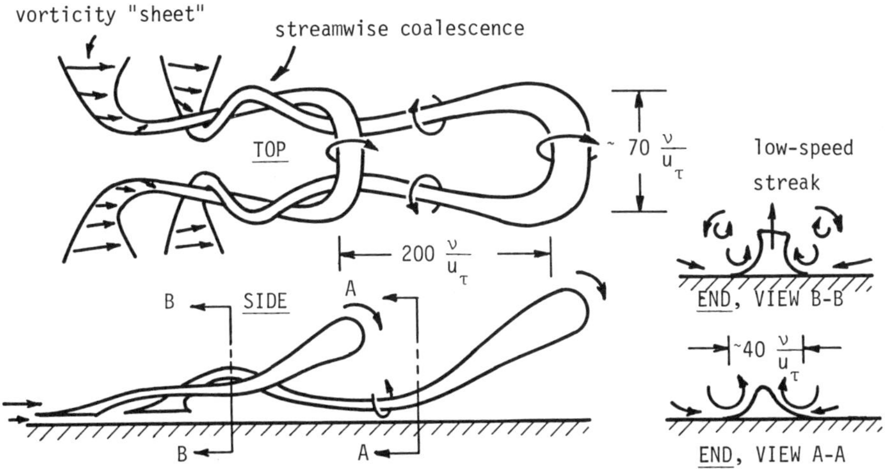

Figure 7. Orthographic Projection of Hypothesized Model of Vortex-Loop Flow Structure causing Low-Speed Streak Formation. A Two-loop array (not to scale) is shown.

THE APPLICATION OF PHOTON CORRELATION LASER VELOCIMETRY
TO TURBULENT FLOW FIELD INVESTIGATIONS

by

G. D. Catalano
Louisiana State University
Baton Rouge, Louisiana

H. E. Wright
Air Force Institute of Technology
Wright Patterson AFB, Ohio

Abstract

A laser velocimeter with photon correlation processing scheme used to investigate various turbulent flow fields. Naturally occurring contaminant serves as scattering centers for the experiments. The photon correlation technique is tested for its sensitivity to laser incident intensity, and response to flow unsteadiness. In addition, a test configuration which features high shear regions and significant reverse flow is also investigated. An improved method for the caluclation of the turbulent intensities from the correlation curves is presented and examined. A direction for future interpretation of photon correlation data is proposed. Actual real world engineering applications where turbulent velocity data is required are discussed

Introduction

Within the past few years, the commerical availability of fast digital correlators has led to new technologies of measurement based upon the quantum resolved properties of low light level. The digital or photon correlation processing technique used in laser velocimetry applications has remarkable power in that it covers a wide dynamic range, is sensitive to extremely low scattered light intensities and does not require a continuous signal. The requirement for low light levels permits the researcher to employ relatively small lasers (e.g. 5-15 mW) and the use of naturally occurring contaminant as the scattering centers. However, there are drawbacks to the correlation technique such as serious problems with flare and ambient light, the photon pile-up phenomenon, little control over the size distribution of the scattering centers, only ensemble average data retrievable, and the mean velocity and the turbulent intensity information alone calculable from the correlation function. The purpose of this report is to document the experiences gained by the authors with the photon correlation technique as it applies to the measurement of turbulent flow fields. In concert with the experiences, several new data analysis algorithms are presented.

Experimental Approach

In order to establish the credibility of the photon correlation processing technique using the Malvern 50 ns unit, the first segment of the experimental investigation focussed on measuring the turbulent flow field of a rectangular nozzle free jet [1]. The rectangular jet was chosen because considerable hot wire anemomentry obtained data already existed and thus a comparison was realizable. The exit plane velocity varied from 130 m/sec to 260 m/sec. Measurements of the mean velocities and the turbulence intensities were made at various downstream locations. Examples of the data are shown in Figures 1 and 2. For this experiment, the naturally occurring contaminant found in the laboratory compressed air supply was used as marking particles. The data presented in Figure 2 points out one of the drawbacks of the 50 ns unit. The experimenter has the choice of being able to measure high subsonic speeds in the central region of the jet or measuring turbulent intensities in high shear regions. The existence of this delemma is due to required beam spacing for the inclusion of the phase modulator in the laser velocimeter optical set-up. The spacing then limits to a large extent the angle of intersection between the two laser beams.

The second step involved measuring the flow field entering the combustion chamber of an actual jet engine [2]. Once again, comparison was made to hot wire anemometry data when available. In addition to the practical engineering significance, the long inlet duct provided regions of relatively large flow accelerations and decelerations (Figure 3). This would enhance the understanding of the sensitivity of the photon correlation scheme to particle size biases. In fact, this portion of the investigation led to effort being concentrated in the area of particle dynamics [3]. For the inlet duct flow, the air supply was the frequently not-too-clean Dayton atmosphere present outside the jet engine test facility. No additional scattering centers were introduced. Typical results are shown in Figures 4 and 5.

Difficulties in properly seeding flows behind blunt or bluff bodies led to the next investigation [4]. The flow field chosen was a subsonic jet impinging upon a flat disc. Laboratory compressed air without additional scattering particles was used. Velocity measurements were then made behind the disc in the wake region at various downstream locations. Figures 6 and 7 present examples of the mean velocity and turbulent intensities measured.

A quantification of the sensitivity of the photon correlation scheme to incident laser intensity was the next goal [5]. The approach taken was to vary the actual size of the laser beam by means of a beam expander/aperture arrangement. The $1/e^2$ beam diameter was varied from 0.5 mm to 10 mm. The resultant varying sized control volumes are located in the flow field of the previously discussed rectangular nozzle turbulent free jet. Mean velocities and turbulent intensities were then calculated from the resultant correlation curves (Table 1).

Mean velocities and turbulent intensities in the near field region of the wake flow behind a circular disc immersed in a turbulent jet flow are obtained. Figure 6 presents mean velocity profiles for various downstream locations.

With the application of the photon correlation processing technique to a flow field located in the test section of the wind tunnel, problems concerning ambient and/or flare light and photon pileup arose. The manifestation of these problems often is that the expected damped cosinusoidal correlation function is skewed and/or distorted. This makes it very difficult to obtain credible mean velocity or turbulent intensity data. An approach [6] was devised based on polynomial curve fitting procedure and applied to a two dimensional turbulent wake flow field. Figures 8 through 11 indicate the correlation functions prior to and after the correction procedure has been employed. Once the credibility of the correction procedure had been established, the near field of the two dimensional wake was documented in detail [7].

Typical results are presented in Figures 12 and 13. For this investigation, the free stream velocity and the Reynolds number based on cylinder diameter are 5.9 m/sec and approximately 50,000 respectively. The cylinder was tested with a stationary (ω = 0 RPM) and a rotating (ω = 500 RPM) initial condition.

A flapping (oscillating) jet allowed the sensitivity of the photon correlation scheme to an unsteady flow field to be documented. The nozzle design employed consists of a modified fluidic element with a feedback mechanism [8]. For this experiment, a centrifugal blower was employed with the naturally occurring contaminant serving as the light scatterers. Mean and turbulent velocity profiles are presented in Figures 14 and 15.

The most recent application of the photon correlation laser velocimeter has been in the turbulent flow field associated with a thrust augmenting ejector wing design. The two dimensional model consists of two subsonic airfoils with an ejector nozzle/constant area duct [9] (Figure 16). The intent of the investigation was to determine the effects of an ejector on the flow characteristics around the airfoil(s) specifically with respect to supercirculation and, thus, increased lift. A thorough documentation was performed [10]. Typical results are shown in Figures 17 through 19. The ratio of the ejector velocity to free stream velocity was kept constant and equal to 2.0. Here, once again, naturally occurring contaminant present in the tunnel/ejector flow served as scattering centers.

Experimental Results and Discussion

The investigation of the rectangular nozzle turbulent jet both established the credibility of using the photon correlation processing technique with its inherent advantages of low laser intensity and no artificial seeding but also it pointed out one significant drawback (Figures 1 and 2). Whereas with a counter or tracker processor, a Bragg cell can be used to shift the frequency of the laser light up to 40 MHz, the photon correlator requires the use of a phase modulator which has an upper limit frequency shift equal to approximately 2 MHz. Thus, a serious problem arises when the flow is both high in mean velocity and in turbulent intensity. The mean velocity data can be obtained but at the expense of any turbulence information.

Figure 4 presents mean velocity profiles for the flow location downstream from the venturi. Note the uniformity of the flow in the center of the duct. The flow field measured by the laser velocimeter seems to be somewhat wider than that measured by the hot wire. This trend is apparent in several additional profiles not presented and is principally due to the optical arrangement with the collecting lens being in the same plane as the beam intersection.

An additional comment should be made concerning the comparison of the mean velocity profiles downstream from the venturi. Though the shapes for the LV and hot wire data are similar, the absolute velocities at the centerline are 3.6 and 6.4 m/sec lower using the hot wire anemometer for 50 percent and 70 percent throttle respectively. These variances represent approximately 10 percent of the measured mean velocity, and are due to the relatively large size particles in the air (i.e. approximately 5-10 microns in diameter).

The lateral distributions of the turbulence intensities, $\overline{u^2}^{\frac{1}{2}}/U$, are plotted versus lateral displacement from the centerline, y/r_o, in Figure 5. Note that near the centerline, the LV seems to indicate zero turbulence level. In fact, the photon correlation scheme is not well suited for determining turbulence intensities for nearly laminar flow. Again note that the LV measured flow fields seem wider for the different locations.

For the disc immersed in a turbulent jet (Figures 6 and 7), it is possible to use the natural scatteners and obtain meaningful data. The local mean velocity, U, is plotted versus non-dimensionalized lateral displacement from the disc center, Y/R where R is the disc radius. At the centerline of the flow field (i.e. Y/R=0), the flow initially accelerates in a negative X direction, then decelerates and finally accelerates in a positive downstream location. Note that the maximum negative mean velocity is not reached until the downstream location X/R=1.50. The maximum width of the recirculation region is approximately 3.5 diameters. The local turbulence intensity varies between 0.20 and 0.40 with the largest magnitude being reached at X/R=1.50 near the flow centerline (Figure 7).

The non-dimensionalized wake momentum thickness, $\Theta_o/2R_o$, calculated by integrating the momentum equation over the flow cross section is plotted versus non-dimensionalized downstream location, $X/2R_o$ in Figure 12. The values Θ_o plotted for the rotating case, which are consistently larger for $X/2R_o > 0.75$, represent the sum of the momentum thickness above the cylinder centerline (i.e. rotation in the direction of the flow) and below the cylinder centerline (i.e. opposite ot the rotation direction). The momentum thickness is asymetric for $\omega = 500$ rpm. Consider a kinetic perspective. The rotation creates an asymmetric pressure distribution around the cylinder, which coupled with the uniform flow produces a Magnus-type or lift force. The associated increment in the drag is manifest in the increase in the value of Θ_o.

The mixing width, $Y_{1/2}$, non-dimensionalized by Θ_o is plotted versus non-dimensional distance downstream, X/Θ_o, in Figure 13. As is the case for the momentum thickness, $Y_{1/2}$, for $\omega = 500$ rpm, is consistently larger after the initial flow development. The mixing width is a possible length scale with which to describe the flow field and is related to the slope of the mean velocity profile. In the recirculation region, the cross stream momentum advection turn $(V \partial V/\partial y)$ is much larger than for the fully developed case. The associated transverse flow strain rate would tend to compress eddies in the cross-stream direction and thus result in a decrease in the magnitude of a characteristic length scale. This is a possible explanation for the initial decrease in $Y_{1/2}$.

The photon correlation processing technique was shown to be quite insensitive to incident laser intensity (Table 1). The unfocussed laser beam diameter is 1.1 mm. For diameters up to 7 mm, the measured mean velocity varies by less than 5%. The error begins to grow more rapidly for the larger beam diameters. An interesting occurrence is that the measured mean velocity has the greatest magnitude for the largest size beam diameter. This phenomenon which is repeatable is directly opposite to the trend observed when tracking or counting signal processing is used. The data for the half-widths shows a similar behavioral pattern.

In Figure 14(a) the value of the mean velocity, U, in the x direction is plotted versus oscillation frequency, f, for several downstream locations. The mean velocities shown are taken at the jet/nozzle centerline (i.e. y = z = 0). Notice that as the oscillation frequency is increased from f=4 hz to f=18 hz, the measured mean velocity also increases. In fact, as the frequency increases from 8 hz to 18 Hz, U varies almost linearly with f.

In Figure 14(b) the mean velocity, U, at the vertical location, y = 1.90 cm or $y/2r_o=1.5$ is plotted for the various frequencies and downstream locations. Note that while there still is a nearly direct relationship between increasing f and thus resulting in the net increase in U, the functional dependence is not as strong as is the case for $y/2r_o=0$.

Turbulent intensities in the longitudinal direction, $(\overline{u^2})^{\frac{1}{2}}/U$, plotted versus oscillation frequency are shown in Figures 15(a) and 15(b) for several different downstream locations. For both $y/2r_o=0$ (Figure (15(a)) and for $y/2r_o=1.50$ (Figure 15(b)), the value of the turbulence intensity is clearly not a simple function of the oscillation frequency. It is evident that the potential core region which typifies the near field of a classical turbulent jet does not exist for the oscillating jet. In fact at $x/2r_o=2$ and $y/2r_o=0$, the turbulence intensity ranges from 0.10 to 1.58 as the oscillation frequency is increased.

In Figure 16, the location of the mean velocity and turbulent intensity data obtained are shown for the ejecter wing experiment. Note that in all cases, x is measured, longitudinally, from the leading edge and z is measured vertically from the airfoil surfaces.

Figure 17 shows mean velocity profiles upstream of the ejector wing. The effect of the ejector in the mean velocity profiles is to accelerate the mean flow above the upper surface and to decelerate the mean flow beneath the ejector wing. This effect is quite pronounced immediately upstream of the leading edge.

Mean velocity and turbulent intensity profiles are shown for the downstream location x/c = 0.2, in Figure 18. the mean flow is consistently faster in the ejector powered case. The value of the turbulent intensities reduce to the free stream value closer to the wing surface with the ejector working. This would indicate a shift of the potential flow down toward the upper surface.

The mean and turbulent velocity field downstream of the ejector nozzle is examined in Figure 19 for x/c = 0.58. Note that the flow for both cases actually accelerates after it enters the constant area mixing duct. Also consider the relatively high turbulent intensities in the confining duct for the ejector powered case. Value of u_{rms}/U equal to 0.30 are measured which is indicative of jet mixing rather than characteristic of duct type flow.

Bendat and Piersol [12] have developed the notion of partial coherence functions which can help establish the cause of a linear dependence indicated by an ordinary coherence function. Consider the application of partial coherence functions to photon correlation spectroscopy.

Defining R_1 as the finite fourier transform (FT) of the autocorrelation function r_{11} where

$$r_{11}(\tau) = \lim_{T\to\infty} \frac{1}{T}\int_{o}^{T} u_1\ (x,t)\ u_1\ (x,t+\tau)\ dt$$

and

$$R_2 = FT\left[r_{22}\right]$$

$$R_c = FT\left[r_{12}\right]$$

then

$$R_{2\cdot 1} = R_2 - L_{12} R_1 = R_2 - \left(\frac{S_{12}}{S_{11}}\right) R_1$$

$$R_{c\cdot 1} = R_c - L_{1c} R_1 = R_c - \left(\frac{S_{1c}}{S_{11}}\right) R_1$$

where $R_{2\cdot 1}$ denotes the finite fourier transform over long record length of input r_{22} with linear effects of r_{11} removed from r_{22}. Here S(f) denotes the autospectral or cross-spectral density function. The partial coherence function can then be defined as

$$\gamma^2_{c\cdot 1} = \frac{S_{c\cdot 1}{}^2}{S_{22\cdot 1}\, S_{cc\cdot 1}}$$

where

$$S_{22\cdot 1} = \frac{\text{Expected Value of } [R^*_{2\cdot 1}\, R_{2\cdot 1}]}{T}$$

and similarly for $S_{cc\cdot 1}$.

The implications of this type of analyses are quite straightforward. The photon correlation spectroscopy technique permits the experimenter to calculate the classical turbulence scalar correlation functions. The use of partial coherence will then help establish the relationship between the scalar functions. Any degree of non-linearity (i.e. noise) in the transformation will be clearly identified.

Conclusions

The photon correlation processing scheme proved to have several major advantages and disadvantages. The use of naturally occurring contaminant as the light scattering centers permitted experiments to be set up and data obtained fairly rapidly since no artificial seeding was required. Additionally, the technique was shown to be fairly insensitive to laser incident intensity as the control volume size was varied by more than an order of magnitude with no discernable effect on the data. Problems can and did arise with the response of the naturally occurring contaminant specifically in regions of high flow acceleration and deceleration, such as in the jet engine inlet duct. However, the photon correlation scheme permits the application of laser velocimetry to flow fields such as the near field wake region, and internal aerodynamic configurations which would have been exceedingly difficult to investigate using a processing scheme requiring artificial scatterers.

References

[1] Catalano, G. D., Wright, H. E., and Cerrulo, "Photon Correlation Laser Velocimeter Measurements in Highly Turbulent Flow Fields," AIAA 18th Aerospace Sciences Meeting, Pasadena, CA, 14-16 January 1980.

[2] Catalano, G. D., Wright, H. E., Rogers, H., Rivir, R., Viets, H., and Pratt, M., "Steady and Unsteady Turbulence Measurements Using a Photon Correlation Laser Velocimeter," 4th International Conference on Photon Correlation Techniques in Fluid Mechanics, Standord, CA, 24-27 August 1980.

[3] Catalano, G. D., "Marker Particle Velocity Perturbations in Compressible Flows Over a Wavy Wall," AIAA Journal, Vol. 18, No. 10, pp. 1270-1272, Oct. 1980.

[4] Catalano, G. D., Wright, H. E., "Turbulence Measurements Behind a Bluff Body," AIAA/AFIT Minisymposium, Wright-Patterson AFB, Ohio, 25 March 1981.

[5] Catalano, G. D., Hamid, I., Wright, H. E., "Effect of Test Rhombus Size on Photon Correlation Laser Velocimeter Data," AIAA/AFIT Minisymposium, Wright-Patterson AFB, Ohio, 25 March 1981.

[6] Catalano, G. D., Walterick, R. E., adn Wright, H. E., "Improved Measurement of Turbulent Intensities by Use of Photon Correlation," to be published in AIAA Journal, March or April 1981.

[7] Catalano, G. D., Walterick, R. E., and Wright, H. E., "The Near Field Two Dimensional Wake with Rotation and a Turbulent Ambient Stream" AIAA 14th Fluid and Plasma Dynamics Conference, Palo Alto, CA, 23-25 June 1981.

[8] Catalano, G. D., Elrod, W. C., Viets, H., and Wright, H. E., "Photon Correlation Laser Velocimeter Measurements in an Unsteady Jet," presented at 53rd Meeting of the Supersonic Tunnel Association, NASA-Ames, CA, 26-28 Mar. 1980.

[9] Viets, H., "Flip-Flop Jet Nozzle," AIAA Journal Vol. 13, No. 10, Oct. 1975, pp. 1375-1379.

[10] Catalano, G. D., Nagaraja, K., Walterick, R. E., and Wright, H. E., "Turbulence Measurements in an Ejector Wing Design," AIAA Aircraft Systems and Technology Meeting, Dayton, Ohio, 11-13 August 1981.

[11] Pike, E. R., "The Application of Photon Correlation Spectroscopy to Laser Doppler Measurements," Journal of Physics D: Applied Physics, Vol. 5, 1972, L23-25.

[12] Bendat, J. S., and Piersol, A. G., Engineering Applications of Correlation and Spectral Analyses, Wiley & Sons, New York, 1980, pp. 220-233.

Table 1: Effect of Laser Beam Diameter on Mean Velocity and Half-Width Measurements

Laser Beam Diameter (mm)	U/U_{exit}	$Y_{\frac{1}{2}}/Y_{\frac{1}{2}}$ @ 1.1 mm
1.1	.607	1
0.5	.606	1.09
1.0	.620	.99
3.0	.630	1.07
5.0	.580	.99
7.0	.563	1.27
10.0	.687	1.30

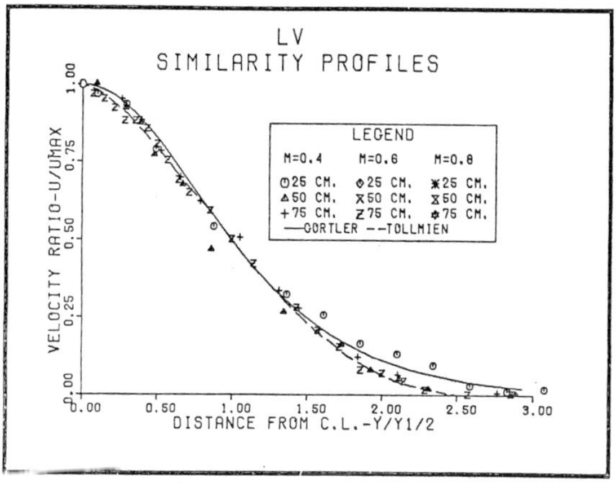

FIG. 1 LV MEAN VELOCITY SIMILARITY PROFILES

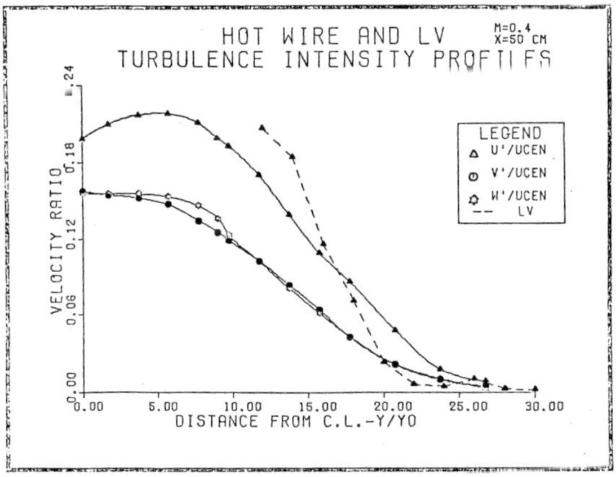

FIG. 2 HOT WIRE AND LV TURBULENCE INTENSITY AT M=0.4 AND 50 CM

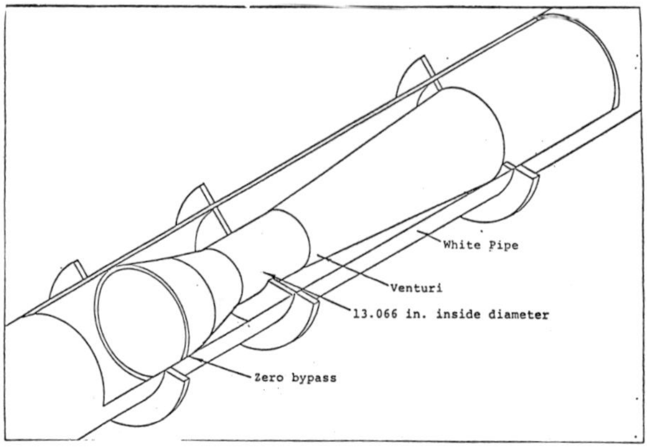

Figure 3. Inlet Duct Flow Configuration

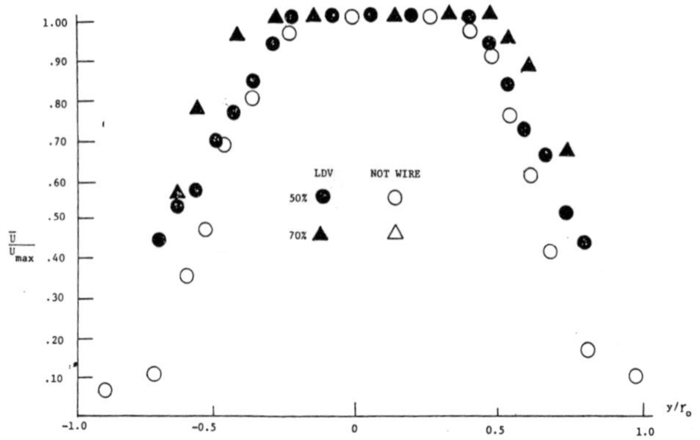

Figure 4. Mean Velocity Profile Comparison (Position II), Downstream of Venturi

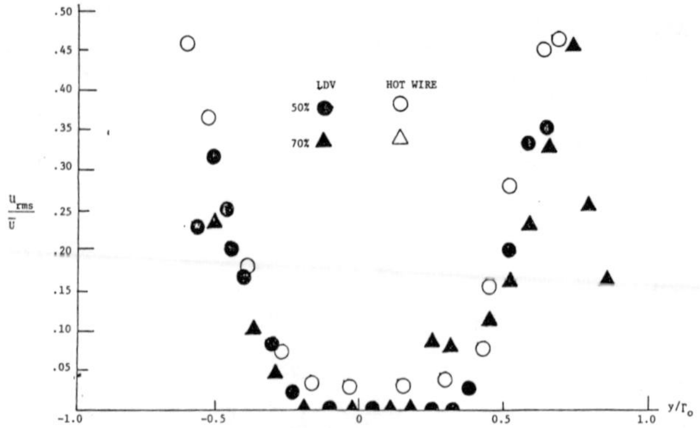

Figure 5. Turbulent Intensity Profiles Comparison (Position II), Downstream of Venturi

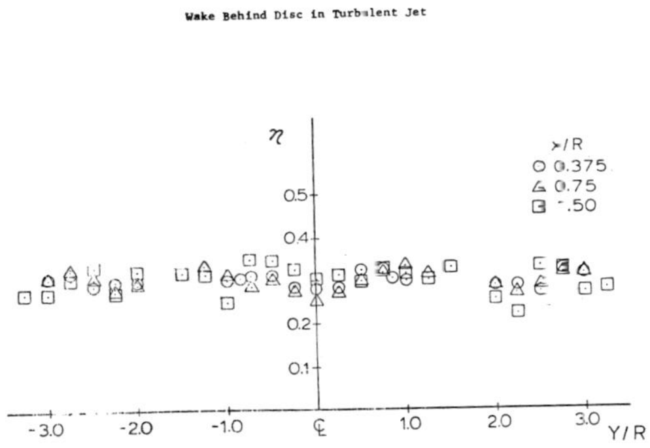

Figure 7. Turbulent Intensity Profile Summary, 2 inch Diameter Disc

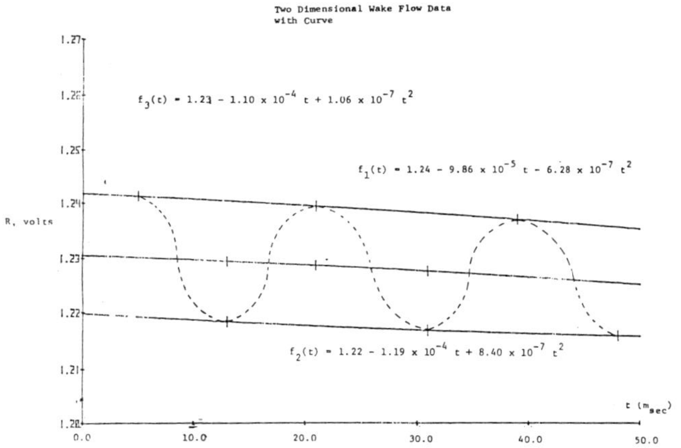

Figure 8. Autocorrelation Function (x= -2.15D, y= -.085D)

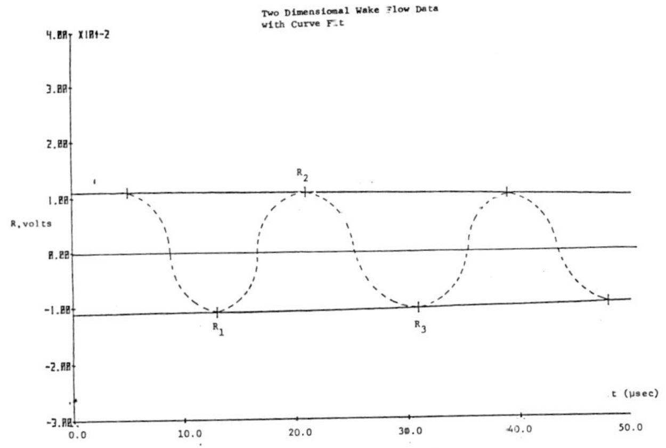

Figure 9. Modified Autocorrelation Function (x= 2.15D, y= -.0851D)

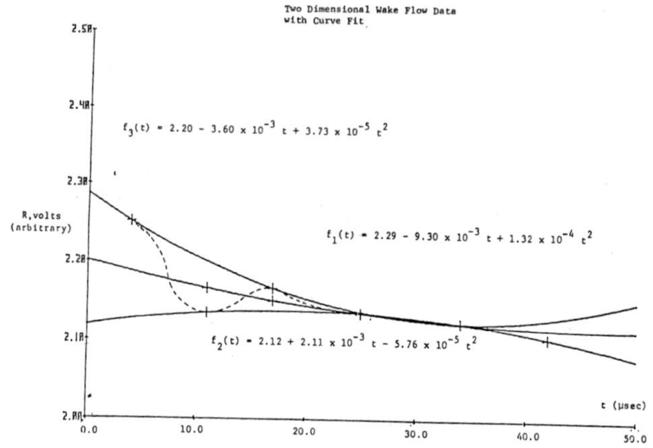

Figure 10. Autocorrelation Function (x= .5D, y= 0D)

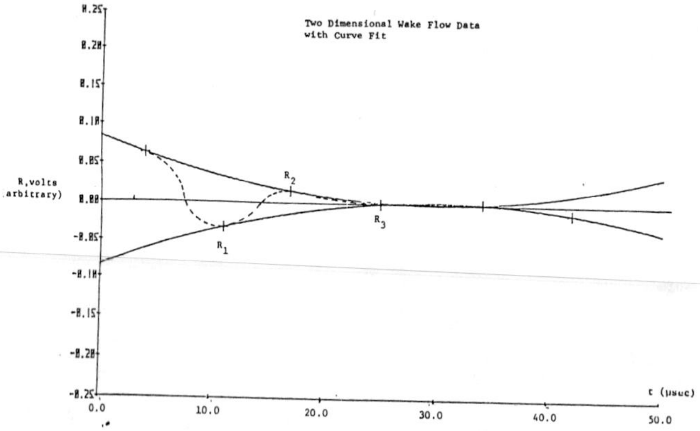

Figure 11. Modified Autocorrelation Function (x= .5D, y= 0D)

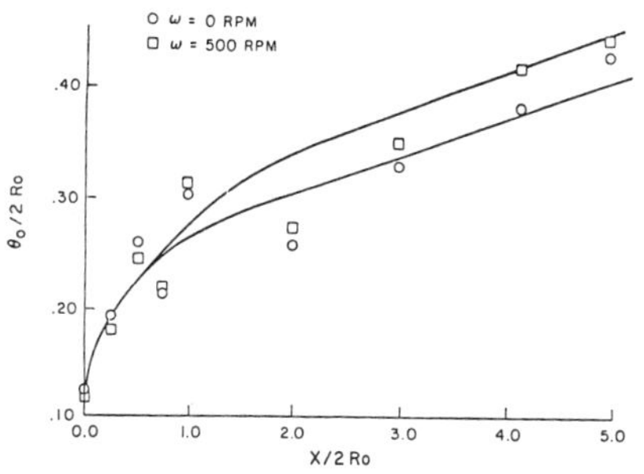

Figure 12. Momentum Thickness Growth with Downstream Wake Development for Fixed and Rotating Cylinder at Wake Centerline

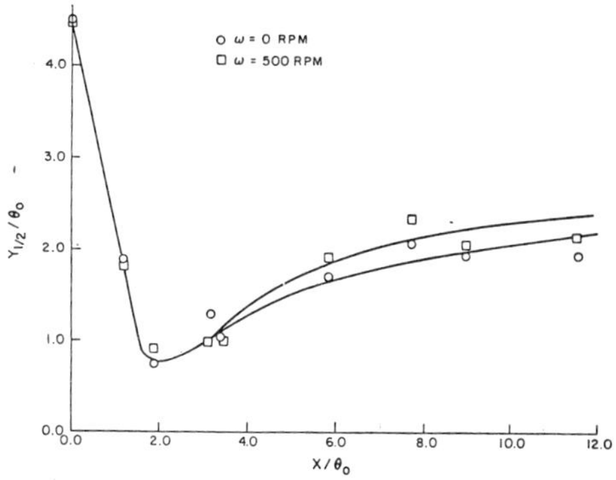

Figure 13. Growth of Mixing Half Width with Downstream Wake Development for Fixed and Rotating Cylinder at Wake Centerline

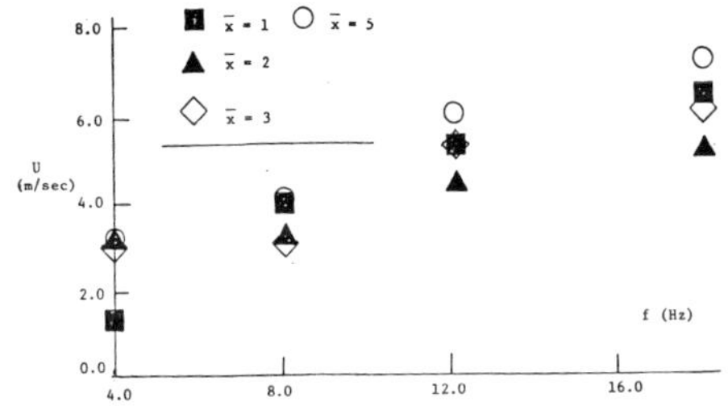

Figure 14a. Mean Velocity Development (y/D=0) where $\bar{x}$ = X/D

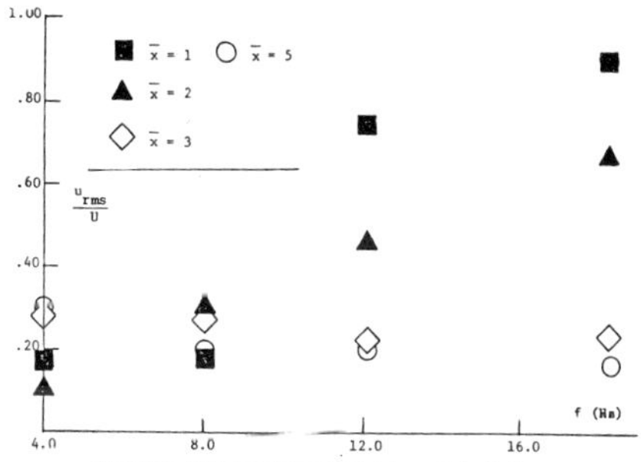

Figure 15a. Turbulent Intensity Development (y/D=0) where $\bar{x}$ = X/D

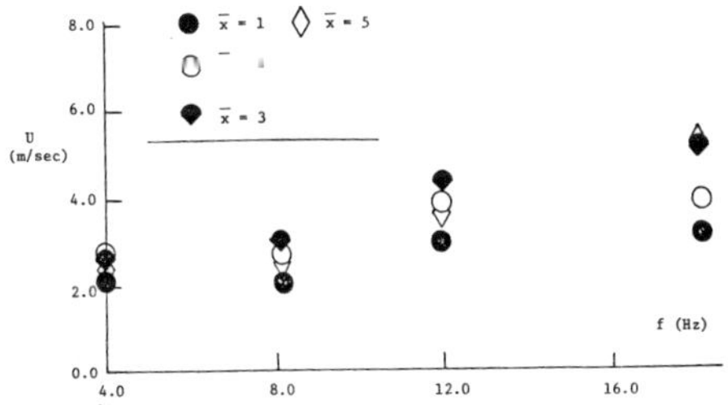

Figure 14b. Mean Velocity Development (y/D=1.50) where $\bar{x}$ = X/D

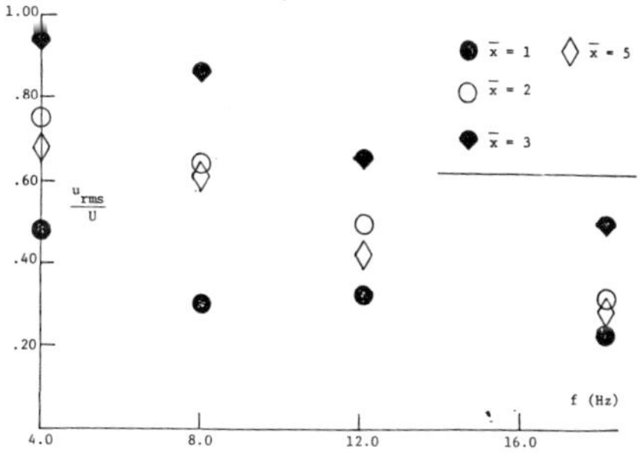

Figure 15b. Turbulent Intensity Development (y/D=1.5) where $\bar{x}$ = X/D

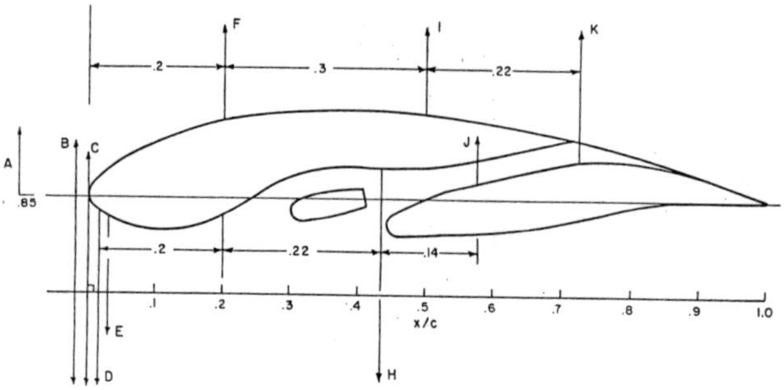

Figure 16. Two-Dimensional Ejector Wing Model with Measurement Locations

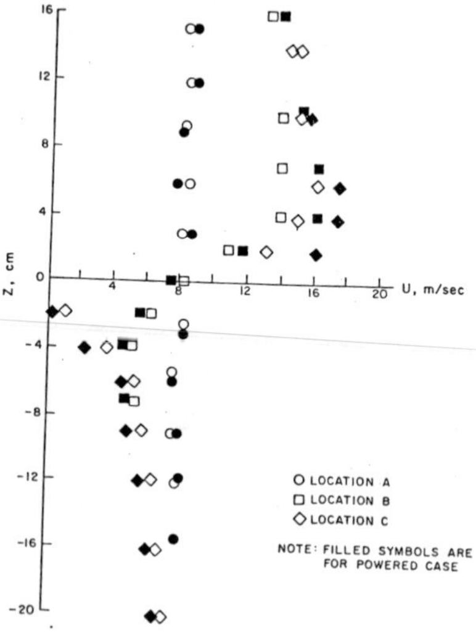

Figure 17. Mean Velocity Profiles Near the Leading Edge

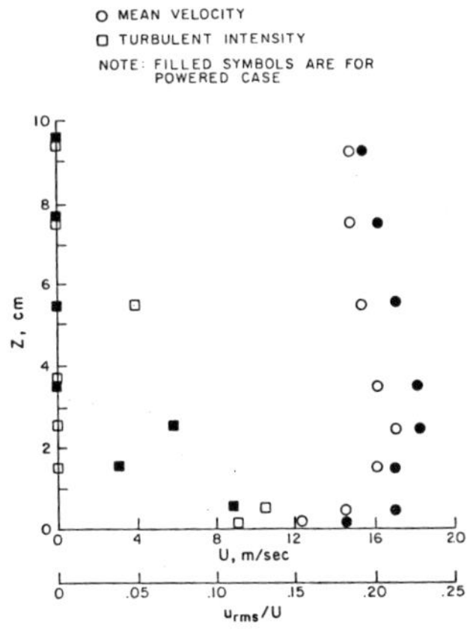

Figure 18. Mean Velocity and Turbulent Intensity Profiles above Upper Airfoil Section

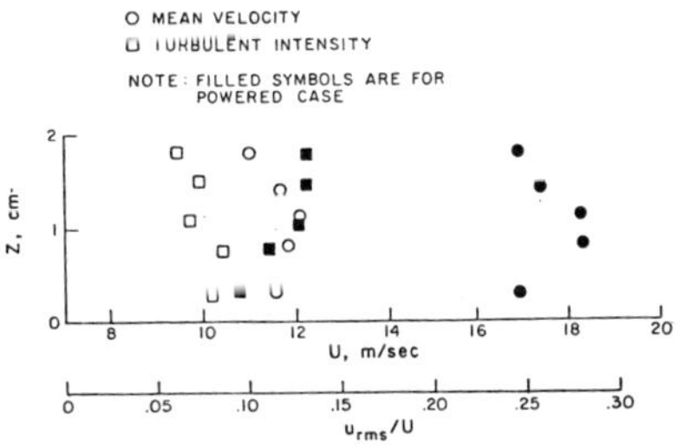

Figure 19. Mean Velocity and Turbulent Intensity Profiles in Constant Area Duct

MULTIPHASE FLOW

STATE OF MULTIPHASE INSTRUMENTATION*

S. L. Soo
Department of Mechanical and Industrial Engineering
1206 West Green Street
University of Illinois at Urbana-Champaign
Urbana, IL 61801

ABSTRACT

Local measurements of the dynamic behavior of multiphase flow consist of measuring two of three inter-related quantities: density, mass flow, and velocity of phases. The phases may include various combinations of gas, liquid, and solid. Where the configuration of a phase is undefined, its measurement is further needed. Instrumentation for multiphase flow can be categorized into mechanical, optical, nuclear, acoustical, and electrical methods. Some of them have reached their limits of usefulness; others, like holography, call for systematic ways of evaluating and quantifying the results. Many challenges remain.

*Study supported in part by the National Science Foundation.

INTRODUCTION

Multiphase flow systems include suspensions of solid particles in a gas or liquid, and mixtures of liquid and vapor or a non-condensable gas in various flow regimes. The latter include bubbles in liquid, slug flow, froth flow, annular flow, and liquid droplets in vapor. Mixtures of immisible liquids may also exist in various flow regimes but will not be dealt with in the present discussion.

The organizer of this session should be complimented for choosing experimental studies in multiphase flow rather than theories for presentation. Almost everyone involved in the disagreements of formulation of multiphase flow can agree that resolution of controversies in theories rests on experimental evidences. Such an agreement, however, is not yet supported by a sufficient amount of basic instrumentation as in the case of mechanics of a single phase fluid. One soon finds that primary standards are few and secondary standards of measurements tend to be complicated and inaccurate. If one expects only accurate, direct reading instruments for multiphase flow, this discussion will be very short indeed.

Several reviews on measurement techniques on multiphase flow are based on many thousands of reports (Hewitt, 1982), among them, Hewitt (1982) and Hsu (1980) on gas-liquid flow and Lieberman (1982) and Soo (1967,1979) on gas-solid flow, with Lieberman emphasizing aerosols or fine particles. These reviews include both local and overall measurements. Typical methods will be outlined in the following; an exhaustive survey is not intended.

To take stock of the present state of availability of instrumentation in multiphase flow, we shall restrict ourselves to measurements of momentum transfer parameters. Once the phase configuration is identified (such as particle sizes of solids ϕ or flow regimes of a liquid-vapor liquid system) basic measurements can be grouped among the three parameters constituting a triangular relation: density (or volume fraction) of phases, mass flow of phases, and the velocity of phases as shown in Fig. 1. In general, measurement of any two will give the third. Where velocities of phases are identical, such as micron-sized aerosols, density of a particle phase can be deduced from its mass flow.

One readily notes that some of the methods for gas-solid flow are applicable to droplets in a vapor or bubbles in a liquid, while special devices might be needed for each flow regime of liquid-gas systems. An attempt will be made to cut across these multiphase systems to identify the commonality, specialty, and areas of challenge among their instrumentation.

We shall attempt to group the current techniques for measuring multiphase flow under mechanical, optical, nuclear, acoustical, and electrical methods. In so doing, an outline of options of instrumentation will be made available and cross-reference can be made. Their limitations and promises can be noted and the needs for further development can be identified.

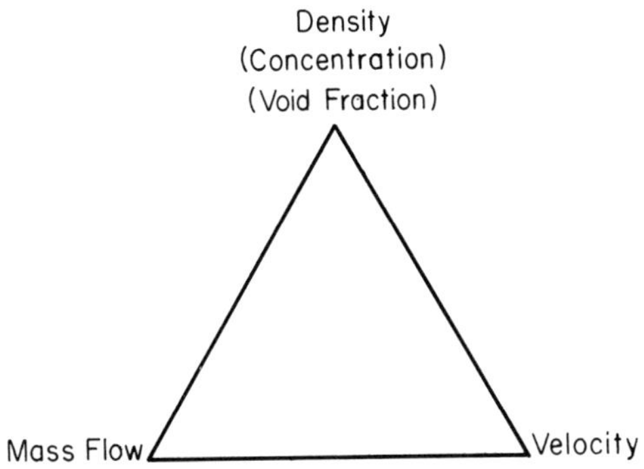

Fig. 1 Triangular Relation of Basic Measurements of Quantities of a Phase Relating to Momentum Transfer in Multiphase Flow

For a Phase:

$$\text{Velocity} = \frac{\text{Mass Flow}}{\text{Density}}$$

VELOCITY OF PHASES

For predicting events in flow of suspension of solid partaicles, the equation of motion of a suspended solid particle in a flowing fluid needs to be validated. This prompted many to measure the velocity of the solid particles or droplets or bubbles as outlined in Table 1. Under the subtrifuge of a dilute suspension, the fluid velocity can be measured with a probe with a bleed for the mean air velocity or the random velocity can be determined with a hot film probe to withstand the particle impact. Measurement of the velocity of the particle phase first relied on various ways of taking and evaluating photography with successive exposures of the particle path. In recent times, laser-Doppler anemometry has been used to determine particle velocity or to use tracer particles below 4 μm diameter to determine the velocity of the fluid phase, the latter use being more frequent. Applications to droplet flow have been successful (Hewitt, 1978).

For dense suspensions such as a fluidized bed, local particle velocity was measured using a radioactive tracer particle with a basic group of three scintillation counters with output linked to a computer. The output gives the instantaneous location and velocity of the tracer particle (Lin, 1981). A tracer with pulsed neutron activation has been used in a reactor system (Rochau, 1979).

Measurements for liquid-gas mixtures based on the mechanical approach are directly applicable to pipe flow; velocity has to be deduced from signals in mass flow.

Several electrical measurements in liquid vapor mixture measure the velocity of passage of the liquid-gas interface. The mean local velocity of each phase remains to be deduced.

In general, local mean velocity of a particulate phase, when measured optically, has to be averaged properly to give the local mean velocity of that phase. Determination of local mean velocity from locally measured mass flow and density of a phase is often more convenient than optical methods.

MASS FLOW

The primary standard for measuring and calibrating local mass flow is via isokinetic sampling, called by different names for different multiphase mixtures and different flow ranges. Table 2 gives an outline of means for measuring mass flow.

Methods based on the rate of change of momentum of a bubble of fluid are mainly applicable to pipe flow or overall measurements.

Radioactive tracers, when grouped under mass flow measurment, are calibrated in reference to velocity.

Table 1 Measurement of Phase Velocity

Methods of Approach	Gas-Solid Flow	Liquid-Solid Flow	Gas-Liquid Flow Droplets-----------Bubbles
Mechanical	← Shapiro Probe for Fluid Phase (Cheng, et al, 1970) →		← Spool Piece → (Turnage and Davis, 1979)
			Film Probe (Cousins, et al., 1965)
			Stagnation Probe (Banerjee, et al., 1978b)
Optical	← Successive Exposure Photography (Soo, et al., 1960) →		↔ (Vanoni, 1950)
	← Laser-Doppler Anemometer →		(Hewitt, 1978b)
Radioactive	↔ Tracer (Lin, 1981)		↔ Tracer with Pulsed Neutron Activation (Rochan, 1979)
Electrical	↔ Hot film probe for fluid phase (Chao, et al., 1979)		↔ Electrodes with Signal Analyzer

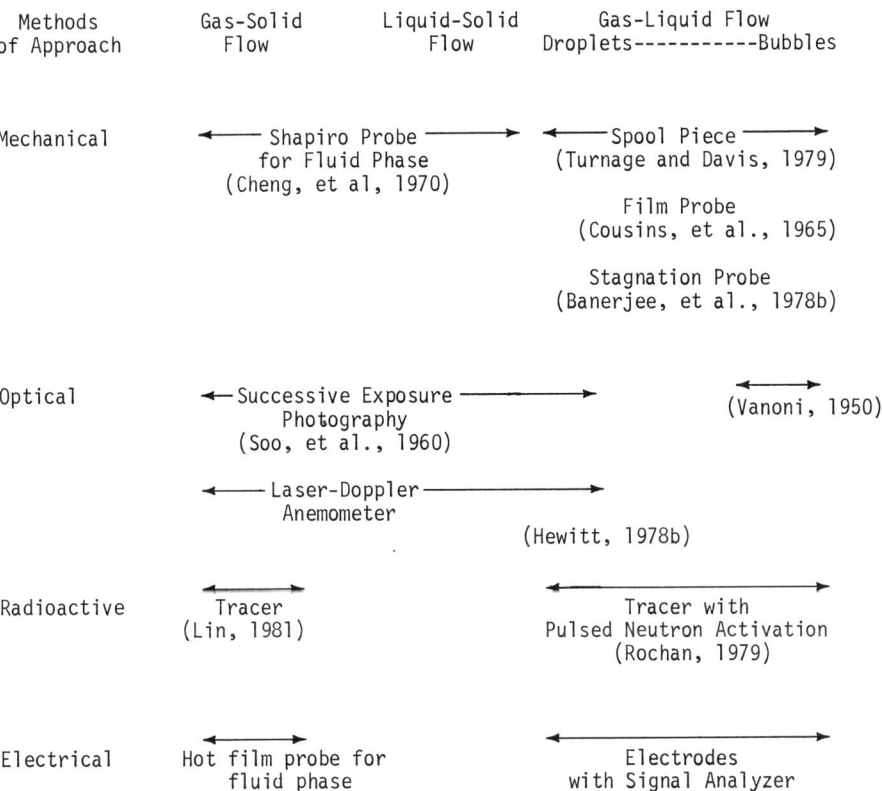

Table 2 Measurement of Mass Flow

Method of Approach	Gas-Solid Flow	Liquid-Solid Flow	Gas-Liquid Flow Droplets--------Bubbles
Mechanical	Isokinetic Sampling	Flow Divider (Newitt, et al., 1962) ←	← Isokinetic Flow Separation → (Alia, et al., 1968)
		(For pipe flow only)	Differential Pressure (Chisholm, 1972); Momentum Method (Ryley and Kirman, 1967); Pool Piece (Turnage and Davis, 1979); Drag Body (Anderson, 1979); Turbine Method (Turnage, et al., 1979) (Claus, et al., 1979)
Optical	(All optical methods for density measurements when sampling by extraction is used.)		
Radioactive	Tracer (Lin, 1981)		Tracer (Cousins and Hewitt, 1968)
Electrical	Ball Probe		← Needle Contact → (Hewitt, et al., 1962)
	Counter Probe (Saunders, 1981)		Double Needle Contact (Wicks and Dukler, 1966)
Thermal			Phase Indicating Thermocouple (Delhayer, 1969)

(All measurements of mass flow gives density when phase velocities are equal.)

The electrostatic probes for solid particles in gas suspensions have the advantage of instantaneous measurements and remain to be a secondary standard of measurements except for monodispersed suspensions using a counter probe.

For liquid gas mixtures, electrical probes based on needle contact actually record the passage of interfaces, so is the phase indicating thermocouple. Observations of phase configurations need to be made or assumed. Even in the case of spherical bubbles, interpretation of results call for an assumption of average bubble diameter.

DENSITY OR VOID

Intuitively, an absolute measurement of density of a phase can be made by closing trap doors at two ends of a tube and measuring its fraction in the mixture. This is the basic rationale of all the devices using quick closing valves. Table 3 is an outline of devices for measuring density or void of a phase.

True density measurements must not involve sampling over a time duration. Localized density measurements have to be made in situ. They include the use of fiber optics and photodiodes. Both suffer from degradation due to deposition of particles when measuring fine dusts. Fiber optics, when used with liquid-gas mixtures, call for different interpretations for different flow regimes. Again spherical shapes of particulate phases are preferred whether bubbles or droplets.

Laser holography promises an accurate means of measuring interfaces and volumes of various flow regimes of liquid-gas mixture. A data system remains to be formulated to link to the basic relations of multiphase flow such as the interface transfer integrals (Soo, 1981).

The use of x-ray (Rowe, et al. 1962) gave the picture of events inside of a fluidized bed, but quantitative evaluation remains to be done.

Measurements based on nuclear power sources such as gamma rays or neutron beams have been intensively studied in relation to reactor safety. For other than pipe flow, these measurements are difficult to interpret. An alternative is using an ultrasonic source. Basic problems of interpretation remain.

Electrical measurements have been applied. The effect of presence of different phases need to be carefully calibrated and the calibration is strongly influenced by deposition. In all cases of calibration of void probes for liquid gas mixture, a knowledge of phase configuration is needed. Here the void is really given in terms of fraction of residence time of a phase.

DISCUSSION

As can be seen from the above review, phase configuration remains an important information in the data reduction of mutliphase measurements. Table 4 summarizes various means of determining phase configurations. Solid

Table 3 Measurement of Density or Void

Methods of Approach	Gas-Solid Flow	Liquid-Solid Flow	Gas-Liquid Flow Droplets--------Bubbles
Mechanical	←	Quick Closing Valve	→ (Johnson and Abu Sabe, 1952)
			Film Thickness (Anular Flow) (Coney and Fisher, 1976)
Optical	Fiber Optics (Soo, et al., 1964)		← Fiber Optics → (Danel and Delhayer, 1971)
	Turn Diode (Perez-Blanco, 1980)		
	← Laser		→
	x-ray (Rowe, et al., 1962)		Laser Holography (Bankoff, et al., 1979) Multiple x-ray
Nuclear			3-Beam Gamma Densitometer (Wesley, 1977)
			Neutron Scattering (Smith, 1975) (Roussean and Riegel, 1978)
			Pulsed Neutron Activation (Rochau, 1979)
Ultrasonics			Ultrasonic (Arave, 1979)
Electrical	Capacitance (Daniel and Brachett, 1951) (Van Zoonen, 1962)		Concentric Ring Electrodes Electrical Impedance (Olsen, 1967)

Table 4 Determination of Phase Configuration

Methods of Approach	Gas-solid Flow	Liquid-solid Flow	Gas-Liquid Flow: Droplets	Gas-Liquid Flow: Bubbles
Mechanical	← Sample Collection — Microscopy	→	(Freezing) →	
Optical			Light Extinction (Walters, 1969)	Light Extinction (Calderbank, 1958)
	Forward Scattering (Rosenweig, et al., 1961)			
	Photometric Analysis (Knudson and White, 1945)		Multidetector Diffraction (Azzopardi, et al., 1978)	Light Scattering (Keller, 1974)
	Royco Counter (Lieberman, 1976)	Coulter Counter (Ullrich, 1960)		
	← Laser →		← Holography (Bankoff, et al., 1979)	→

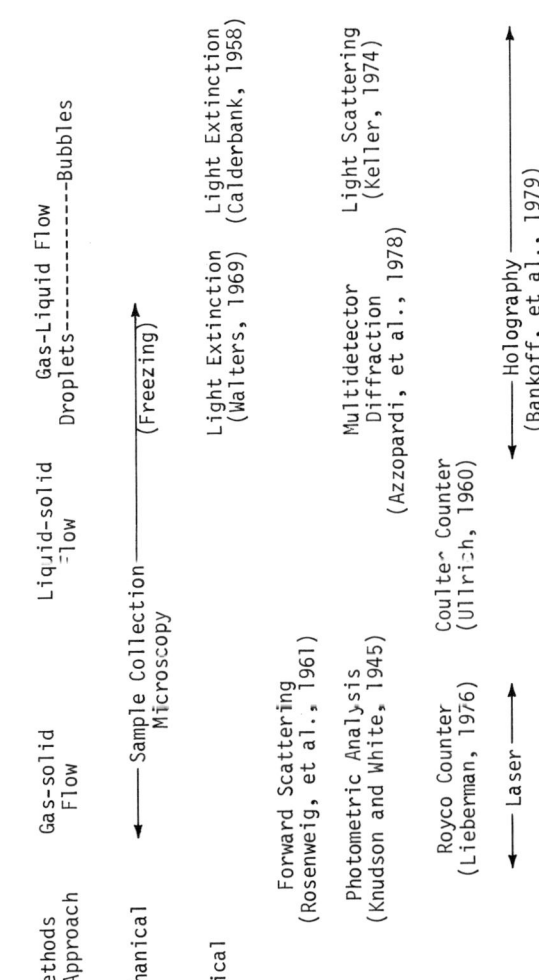

particles can be measured under static condition, but sizing of fine particles remains a continuing challenge. Small bubbles and droplets belong to the same category. Laser holography appears to be the ultimate of measuring phase configurations of various flow regimes. The challenge is in data processing and utilization.

It appears that a general, fool-proof and direct indicating type of instrumentation is not to become available in the immediate future.

REFERENCES

Alia, P., L. Cravarolo, A. Hassid, and E. Pedrocchi, 1968, "Phase and Velocity Distribution in Two-Phase Adiabatic Annular Dispersed Flow," Euratom Report, EUR 3759E.

Anderson, J. L., "Drag Devices for Two-Phase Mass Flow Measurements," International Colloquium on Two-Phase Flow Instrumentation, June 11-14, 1979, EG&G Idaho, Inc.

Arave, A. E., "Ultransonic Densitometer Development," NUREG/CP-0007, October 1979.

Azzopardi, B. J., G. Freeman, and P. B. Whalley, 1978, "Drop Sizes in Annular Two-Phase Flow," Report, AERE-R9074, UKAEA, Harwell.

Banerjee, S., T. R. Heidrick, J. R. Saltvold, and R. S. Flemons, 1978b, "Measurement of Void Fraction and Mass Velocity in Transient Two-Phase Flow," Transient Two-phase Flow, eds. S. Banerjee and K. R. Weaver, pp. 789-834, Atomic Energy Canada Ltd., Toronto.

Bankoff, S. G., M. C. Yuen, and R. S. Tankin, "Pilot Tube, Laser Doppler Anemomentry and Holography," NUREG/CP-0007, October 1979.

Calderbank, B. H., 1958, "Physical Rate Processes in Industrial Fermentation I: The Interfacial Area in Gas-Liquid Contracting with Mechanical Agitation." Trans. Inst. Chem. Engr., 36:443-463.

Chao, B. T., S. T. Leung, H. Perez-Blanco, J. H. Saunders, and S. L. Soo, 1979, "Modeling of Particle Suspension Flow System and Measurements," Proc. 1st Conf. in Pneumatic Conveying, Int. Powder Inst., Jan. 18, 1979, London, England.

Cheng, L., S. K. Tung, and S. L. Soo, 1970, "Electrical Measurements of Flow Rate of Pulverized Coal Suspensions," Trans. ASME, J. Engr. for Power, 92, (A-2), 135-149.

Chisholm, D., 1972, "The Compressible Flow of Two-Phase Mixtures through Orifices, Nozzles and Venturi Meters," NEL Report 549, pp. 66-79, National Engineering Lab., East Killbride, Scotland.

Class, G., K. Hain, et al., 1979, "Mass Flow Meter, Entwicking and Einsatz eines Massenstrom-Messgerates fur instaionare Zweiphasenstromungen," KfK-Bericht 2790.

Coney, M. W. E., and S. A. Fisher, 1976, "Instrumentation for Two-Phase Flow in use or under Development at the Central Electricity Research Laborotories," European Two-Phase Flow Group Meeting, Erlangen, Paper B2.

Cousins, L. B., and G. F. Hewitt, 1968a, "Liquid Phase Transfer in Annular Two-Phase Flow: Radial Liquid Mixing," Report AERE-R5693, UKAEA, Harwell.

Cousins, L. B., and G. F. Hewitt, 1968b, "Liquid Phase Transfer in Annular Two-Phase Flow: Droplet Deposition and Liquid Entrainment," Report AERE-R5657, UKAEA, Harwell.

Cousins, L. B., W. H. Denton, and G. F. Hewitt, 1965, "Liquid Mass Transfer in Annular Two-Phase Flow," Symp., Two-Phase Flow, Exeter, Paper C4; Report AERE-R4926, UKAEA, Harwell.

Danel, F., and J. M. Delhaye, 1971, "Sonde Optique pour Mesure du Taux de Presence Local en Ecoulement Diphasique," Mes. Regulation Autom., pp. 99-101.

Daniel, J. H., and F. S. Brackett, 1951, "Statistical Method to Investigating the Nature and Behavior of Small, Airborne, Charged Particles," J. Appl. Phys., 22, 542-554.

Delhaye, J. M., 1969, "Hot-Film Aemometry, in Two-Phase Flow Instrumentation," eds. B. W. LeTourneau and E. B. Bergles, pp. 58-69, ASME, New York.

Hewitt, G. F., 1978, "Liquid Mass Transport in Annular Two Phase Flow," invited lecture at the 1978 International Seminar of the International Centre for Heat and Mass Transfer, Dubrovnik.

Hewitt, G. F., 1982, "Overall Measurements," Handbook of Multiphase Systems, (G. Hestroni, ed.) Hemisphere Pub. Corp., Washington, D.C., October 13-18.

Hewitt, G. F., R. D. King, and P. C. Lovegrove, 1962, "Techniques for Liquid Film Pressure Drop Studies in Annular Two-Phase Flow," Report AERE-R3921, UKAEA, Harwell.

Hsu, Y. Y., and A. L. M. Hon, 1980, "Summary of the USNRL Sponsored Two-Phase Flow Instrumentation for Reactor Safety Research," Presented at the 26th Annual Meeting of the American Nuclear Society, June 8-12, 1980, Las Vegas, Nev.

Johnson, H. A., and A. H. Abou-Sabe, 1952, "Heat Transfer and Pressure Drop for Turbulent Flow of Air-Water Mixtures in a Horizontal Tube," Trans. ASME 74:977-987.

Keller, A., 1972, "The Influence of the Cavitation Nucleous Spectrum on Cabitation Inception Investigated with a Scattered-Light Counting Method," J. Basic Engr., 94:917-925.

Lieberman, A., 1982, "Aerosol Measurement and Analysis," Handbook of Multiphase Systems, (G. Hetstroni, ed.) Hemisphere Publ. Corp., Washington, D.C., 10-119 to 10-165.

Lin, J. S., 1981, "Particle-Tracking Studies for Solid Motion in a Gas Fluidized Bed," Ph.D. Thesis, University of Illinois at Urbana-Champaign, Urbana, IL 61801.

Newitt, D. M., J. F. Richardson, and C. A. Shook, 1962, "Hydraulic Conveying of Solids in Horizontal Pipes," Proc. Symp. of Interaction between Fluids and Particles, Int. Chem. Engrs., London, 87-91.

Olsen, H. O., 1967, "Theoretical and Experimental Investigation of Impedence Void Meters," Kjeller, Norway, Report 118.

Perez-Blanco, H., 1979, "Modeling of Dilute, Swirling Gas-Solid Suspensions," Ph.D. Thesis, University of Illinois at Urbana-Champaign, Urbana, IL 61801.

Rochau, G. E., "Development of a Pulsed Neutron Generator for Two-Phase Flow Measurement," NUREG/CP-0007, October 1979.

Rousseau, J. C., and B. Riegel, 1978, "Super-CANON Experiments," 2d OECD/NEA/CNSI Specialists Meet. Transient Two-Phase Flow, Paris.

Rowe, P. N., and B. A. Partridge, 1962, "Particle Motion caused by Bubbles in a Fluidized Bed," Proc. Symp. on Interaction between Fluids and Particles, Int. Chem. Engrs., London, 135-142.

Ryley, D. J., and G. A. Kirkman, 1967, "The Concurrent Measurement of Momentum and Stagnation Enthalpy in a High-Quality Wet Steam Flow," Proc. Inst. Mech. Engr., 182:250-257.

Saunders, J. H., Jr., 1980, "Scaling and Performance of a Device for Fluidized Bed Entrainment Reduction," Ph.D. Thesis, University of Illinois at Urbana-Champaign, Urbana, IL 61801.

Soo, S. L., 1967, Fluid Dynamics of Multiphase Systems, Marisdel Pub. Co., Waltham, Mass.

Soo, S. L., 1981, "Effects of Configuration of Phases on Dynamic Relations," AIChE Symposium Series 208, Vol. 77, Heat Transfer, (R. P. Stern, ed.), 152-160.

Soo, S. L., H. K. Ihrig, Jr., and A. F. El Kouh, 1960, "Experimental Determination of Statistical Properties of Two-Phase Turbulent Motion," Trans. ASME, J. Basic Engr., 82, (3), 609-621.

Soo, S. L., G. J. Trezek, R. C. Diminck and G. F. Hohnstreiter, 1964, "Concentration and Mass Flow Distribution in a Gas-Solid Suspension," Ind. Engr. Chem. Fund., 3, 98-106.

Turnage, K. G., and C. E. Davis, "Two-Phase Flow Measurements with Advanced Instrumented Spool Pieces and Local Conductivity Probes," NUREG/CP-0007, October 1979.

Vanoni, V. A., 1950, "Dynamics of Particulate Matter in Fluid Suspension," Report No. N-71.1, Hydrodynamics Laboratory, Cal. Tech., Pasadena, Cal.

Van Zoonen, D., 1962, "Measurements of Differential Phenomenon and Velocity Profiles in a Vertical Riser," Proc., Symp. on Interaction between Fluids and Particles, Int. Chem. Engrs., London, 64-82.

Walters, P. T., 1969, "Optical Methods for Measuring Water Droplets in Wet-Steam Flows," CEGB Report CERL RD/L/N-107/69.

Wesley, R. D., 1977, "Performance of Drag-Disc Turbine and Gamma Densitometer in LOFT," Proc. Meet. Review Group Two-Phase Flow Instrumentation, Report, NUREG--375, U. S. Nuclear Regulatory Commission.

Wicks, M., and A. E. Dukler, 1966, "In-situ Measurements of Dropsize Distribution in Two-Phase Flow--A New Method for Electrically Conducting Liquids," AIChE Symp., Two-Phase Flow, Chicago.

TOMOGRAPHIC RECONSTRUCTION OF THE TIME-AVERAGED DENSITY DISTRIBUTION IN TWO-PHASE FLOW

J. R. Fincke
Idaho National Engineering Laboratory
EG&G Idaho, Inc.
Idaho Falls, Idaho 83415

Abstract

The technique of reconstructive tomography has been applied to the measurement of time-average density and density distribution in a two-phase flow field. The technique of reconstructive tomography provides a model-independent method of obtaining flow-field density information.

A tomographic densitometer system for the measurement of two-phase flow has two unique problems: a limited number of data values and a correspondingly coarse reconstruction grid. These problems were studied both experimentally through the use of prototype hardware on a 3-in. pipe and analytically through computer generation of simulated data. The prototype data were taken on phantoms constructed of all Plexiglas and Plexiglas laminated with wood and polyurethane foam.

Reconstructions obtained from prototype data are compared with reconstructions from the simulated data. Also presented are some representative results in a horizontal air/water flow.

INTRODUCTION

In order to fully characterize an inhomogeneous two-phase flow field, information about local and global density and velocity are required. This paper addresses the problem of obtaining density information on a steady-state, two-phase flow.

Historically, local density has been measured through the use of very small intrusive probes, such as hot-film anemometers, and optical or conductivity probes. While these instruments have proven effective for some two-phase flow regimes, they are not foolproof, and the results depend heavily on the way in which the data are processed. Furthermore, measurements of this type are intrusive and suffer a problem of survivability when installed in a typical pressurized water reactor (PWR) environment of 560 K and 15.5 MPa.

The gamma densitometer, utilized either in a fixed-ray mode or as a single-beam scanning device, has also been used extensively for the characterization of two-phase flow. The gamma densitometer offers a distinct advantage over the local probe in that it is nonintrusive, thus avoiding flow disruption and transducer survivability problems. While the gamma densitometer has proven to be very successful for the measurement of global densities, information about local density can be obtained only by assuming some type of distribution model.[1]

The application of reconstruction tomography (RT), also known as computed tomography (CT) or computer assisted tomography (CAT), is the next logical step in the interpretation of densitometer data. The technique was developed in the medical radiology field to provide a clear image of a body section or layer without interference from other regions.[2] The main advantage of RT for the measurement of a two-phase density distribution is that the technique is model independent. If a sufficient number of views (or projections) are taken, the distribution of attenuation coefficients, which are proportional to density for a steam/water system, may be determined. The reconstruction of the density distribution image from its projections is a complex mathematical process generally performed on a computer.

Although many aspects of RT have been extensively studied in recent years, especially for medical applications, a practical system for the density measurement of two-phase flow has two unique problems that have only recently been investigated: a rather limited number of data values (or rays) per view, and the correspondingly coarse reconstruction grid. The number of grid elements (or pixels) on a side of the reconstruction grid are optimally determined by the number of data values, and the data values per view are ultimately limited by the system cost.

At the Idaho National Engineering Laboratory, we have studied the limited-data and coarse-reconstruction-grid problems both experimentally with a nine-beam prototype device on a 3-in. diameter pipe, and analytically through computer generation of simulated data.[3] The results obtained indicate the feasibility of using RT for the measurement of both average and local density. The beam geometry and representative pixel grid for this device appears in Fig. 1, while the actual hardware is shown in Fig. 2.

Based on our experience from the development and testing of the prototype for the 3-in. diameter pipe, we made a device for a 14-in. diameter pipe for operational use on a steady-state, two-phase test loop to provide time-averaged density profile information. This device (Fig. 3) is currently in use on a high-temperature, high-pressure, two-phase test facility at the Idaho National Engineering Laboratory.

A typical scan takes from 30 to 60 s. This scan period is determined by the number of views to be recorded, the scanning angle, and the sample period at each view location. The sample period is determined by the periodicity of the flow. For instance in a slug flow, one would want to average over several slugs at each angular location in the scan. The device is typically scanned through 180 deg using 9 deg increments. Further details appear in Reference 4.

We applied the two most widely used algorithms to the coarse reconstruction grid problem:[3] namely, DCON, a convolution (or filtered back-projection) algorithm similar to those used in all new CT scanners, and an algebraic reconstruction technique (ART), an iterative algorithm similar to the ones used on early CT scanners. These and other algorithms are reviewed in Reference 2.

Convolution algorithms are considered superior to iterative algorithms in the medical field because the algorithm generally executes faster and produces reconstructions that are often better than, or at least as good as, the interactive techniques, provided the data are plentiful, complete, accurate, and not of high contrast.[5] In addition, the speed of the convolution algorithm can be greatly increased by taking advantage of the ability of the algorithm to operate on the data from each view independently before back-projection and by using special purpose hardware, such as an array processor.

However, we found that the convolution algorithm produced artifacts when an incomplete set of projections were used. It normally requires data from views of regular angular increments for a total of 360 deg (or at least 180 deg) plus the angle of divergence of the gamma beam. Furthermore, the accuracy of the convolution method is degraded considerably when the data points are sparse. For these reasons, we subsequently used the ART algorithm for all reconstruction involving sparse data and coarse grids.

In this paper, I describe the ART iterative reconstruction algorithm and the prototype nine-beam densitometer. I first review simulated results and results from actual data obtained from phantoms made of all Plexiglas and Plexiglas laminated with wood and polyurethane foam. I then compare reconstructions from the experimental data with reconstructions from the simulated data. I then present some representative results obtained in steady-state air/water flow in a 3-inch horizontal pipe.

ART ALGORITHM

The ART is an iterative algorithm which, starting from an initial estimate of the reconstruction grid, applies corrections on a ray-by-ray basis to make projections (called pseudo-ray sums), calculated from the reconstruction grid, correspond more closely to the real or measured ray sums. The set of ray sums can be represented by

$$P_j = \sum_{i=1}^{N} W_{ij}\, f_i, \tag{1}$$

where W_{ij} is the weighting factor that represents the contribution of the ith pixel to the jth ray sum, P_j; and f_i is the pixel density. Note that most weighting factors are zero, since only a small number of pixels contribute to a given ray. The basic strategy of the iterative method is to apply corrections to arbitrary pixel densities in an attempt to match the measured rays or projections. The procedure is repeated until the calculated and measured projections agree within the desired accuracy. The accuracy is ideally limited only by the number of iterations.

The correction process during the kth iteration is described by the equation

$$f_i^k = f_i^{k-1} + \sum_{j=1}^{N} \Delta f_{ij}, \tag{2}$$

where f_i^{k-1} is the ith density before the kth iteration, f_i^k is the density after the kth iteration, and Δf_{ij} is the correction applied to the ith pixel from the jth ray. The version of ART used here is additive ART; that is, each pixel receives a correction in proportion to its weight, W_{ij}:

$$\Delta f_{ij} = \frac{R\, W_{ij}\, \Delta P_j}{\sum_{i=1}^{N} W_{ij}^2} \tag{3}$$

where R is a relaxation factor, and ΔP_j is the difference between the measured and calculated ray sum. The denominator is a normalizing factor to ensure that the total change in the ray sum, if all pixels are corrected, is ΔP_j.

The factors that can significantly affect convergence of the ART algorithm[2] are: the order of choosing rays and applying corrections, the relaxation coefficient R, the application of constraints, and the number of iterations. These factors are discussed in Reference 3.

DESCRIPTION OF HARDWARE

The tomographic densitometer is similar in theory and operation to standard densitometers. There are, however, several unique features that make this density measuring system unique. The nine-beam system has the ability to be rotated about the pipe center so that any of an infinite number of views may be obtained. The radiation detector is operated in the pulse mode, and through the use of special electronics, the system may be operated at very high count rates (in excess of 1×10^6 counts/s) to obtain low statistical error. The system is electronically stabilized against drift by a photopeak locking technique.

The detector geometry is detailed in Fig. 1; the on-pipe hardware is shown in Fig. 2. The radiation source is 0.5-Ci americium with a 60-keV gamma ray as the principal photo peak. The source active area seen by the detectors is 0.635 by 1.54 cm and is collimated into a fan beam whose total angle is 32 deg. The detectors have individual collimators. The entire source, collimator, and detector array rotates about the pipe center on two ball-bearing assemblies.

The radiation detector is a commercial sodium-iodide-crystal photomultiplier tube operated in the pulse mode. The detector, a Harshaw 4S4 with a 2.54- by 2.54-cm crystal, is coupled to an electronic package designed and built by EG&G Idaho, Inc. The electronics has been termed the megacount stabilized densitometer (MSD) (Fig. 4).

Gamma rays are randomly (Poisson distribution) emitted from the source and are attenuated in number proportional to the density of the fluid in the pipe. Each gamma ray entering the crystal produces a light pulse that is sensed and amplified by the photomultiplier tube. The resulting current pulse is converted to a voltage pulse and amplified by the preamplifier. The resulting pulse has a rise time of approximately 100 ns.

Each pulse has a long trailing edge that can overlap the leading edge of the next randomly produced pulse. If this overlap occurs, the following pulse (or pulses) is piled on top of the trailing tail of the first pulse. This pileup prevents counting of each subsequent pulse. A baseline restorer circuit delays the pulse and subtracts the delayed pulse from the undelayed pulse. This process eliminates the long tail of the pulse and returns piled-up pulses to a zero baseline. The restored pulse train is fed to the single channel analyzer (SCA), peak detector, and stabilizer circuitry as shown in Fig. 4.

The SCA produces an output logic pulse whenever the input pulse amplitude is between an upper and a lower voltage setting. This region is the SCA window and is established by two high-speed comparators, which are enabled by a control pulse from the peak detector circuit. The comparator output signal and the peak detector control pulse are ANDed to produce an output pulse for every input pulse that is within the preset amplitude window.

The peak detector circuit receives the restored pulse from the baseline restorer circuitry. The pulse is delayed, offset biased, and compared with the undelayed pulse to establish a pulse that controls the SCA comparators. This control pulse causes the SCA comparators to hold the logic state determined when the pulse is at its peak.

The stabilizer is a feedback circuit that compensates for the drift of the detector due to count rate and temperature changes. The stabilizer has two SCAs with narrow windows: one window set on the low side of the photopeak distribution and the other set on the upper side. Since the photopeak is approximately symmetrical, each SCA will have the same output count rate. If drift occurs, the photopeak will shift with respect to the SCA window settings, and the SCA output count rates will become unequal. The output count rates are divided and compared by an up/down counter. The resulting difference count is converted to an analog error signal by a digital-to-analog (D/A) converter. This error signal controls the high-voltage supply, which changes the gain of the photomultiplier tube and shifts the photopeak until the count rates are once again equal.

RECONSTRUCTIONS FROM SIMULATED AND ACTUAL DATA

Using both simulated and actual data, we evaluated the effects of using a limited number of data values (limited by both number of rays per view and number of views) and the correspondingly coarse reconstruction grids. Simulated data, which assumes a line beam and no error due to single- or multiple-photon scattering, is useful for optimizing beam geometry, data values, and reconstruction grid size and establishes a baseline against which to compare reconstructions from actual data.[3] The reconstructions from actual data include finite beam size and any counting errors due to single- or multiple-scattered events. Reconstructions were performed on both simulated and actual data for eight test phantoms; several of these phantoms are shown in Fig. 5. The phantons were fabricated of all Plexiglas and Plexiglas laminated with wood and polyurethane foam. By constraining the final foam volume and varying the amount of polyurethane used, different foam densities were obtained.

The quality of the reconstruction is judged both qualitatively by visual examination, and quantitatively by the pixel-by-pixel differences between the simulated-data reconstruction and the digitized phantom, and by comparison of the cross-sectional average densities between the phantom and reconstructions. An additional means of comparison, which has been used by several investigators to judge noisy reconstructions,[5] is the sum of the squares of the pixel-by-pixel differences. Since the iterative algorithms can be shown to converge in a least squares manner (with ideal, noiseless data),[5] this latter measure has a theoretical basis:

$$\mathrm{DIST} = \frac{1}{N^2} \frac{1}{(\mathrm{VARP})^{1/2}} \left[\sum_{i=1}^{N} \sum_{j=1}^{N} (r_{ij} - P_{ij})^2 \right]^{1/2} , \tag{4}$$

where

N = number of pixels

r_{ij} = reconstructed density of pixel ij

P_{ij} = density of phantom pixel ij

$$\mathrm{VARP} = \frac{1}{N^2} \sum_{i=1}^{N} \sum_{j=1}^{N} (P_{ij} - \bar{p})^2$$

$$\bar{p} = \frac{1}{N^2} \sum_{i=1}^{N} \sum_{j=1}^{N} P_{ij} .$$

We chose the eccentric-bubble phantom shown in Fig. 6 to study the effects of using a limited number of data values. This configuration was chosen because it contains two large, distinct regions where the effects of noise can be clearly seen. The number of detectors was fixed by hardware constraints, and the number of data values were varied by changing the number of views taken. Table 1 summarizes the results from the simulated-data study. In all cases, the value of average density obtained is reasonable. The noise of the reconstruction characterized by the quantity DIST is minimal for 30 views, acceptable for 15 views, slightly high for 10 views, and generally unacceptable for 5 views. The general conclusion is that, for the nine-detector prototype system, at least 10 views are required and 15 or more are desirable for a 12 by 12 pixel array. Ultimately the quality of the reconstruction is limited by the number of detectors. A further increase in the number of views is of limited usefulness so far as reconstruction quality is concerned.

Figure 6 shows a photograph of the eccentric-bubble Plexiglas phantom, the digitization of the phantom, the reconstruction from simulated data using 21 views over 180 deg, a plot of the pixel-by-pixel differences between the simulated-data reconstruction and the digitized phantom, and the reconstruction from actual data.

The phantom's average density is computed by summing over all pixels and dividing by the total number of pixels. For the eight phantoms tested, the root mean square (rms) error in average density for the simulated-data reconstructions was 0.0066 g/cm^3, while the rms error for the reconstructions from actual data was 0.0185 g/cm^3.

Table 1. Summary of Simulated Data Study Results[a]

Number of Views	Average Density (g/cm^3)	Distribution
30	0.5875	0.3689
15	0.5838	0.4249
10	0.5848	0.4778
5	0.5738	0.5785

a. Summarization of effects of limited number of data values on quality of reconstruction for 12- by 12-pixel array, 0.555-cm pixel size, and a phantom average density of 0.5886 g/cm^3 for five iterations. The phantom used is shown in Fig. 6.

RECONSTRUCTIONS OF AIR/WATER FLOW

In this section we examine briefly some representative results obtained in a horizontal air/water test loop. The facility consists of a centrifugal pump, separation tank, mixer, associated valves, pressure and temperature instruments, and reference turbines necessary to calculate the mass flow rates of air and water before mixing. Test loop operating parameters are 150 kPa and 300 K nominal. The maximum water flow rate is 4.5 L/s, and the maximum air flow rate is 25 L/s at 220 kPa. The test loop piping consists of 3-in. Schedule-160 piping, with a Plexiglas test section for visual observation. The straight run upstream of the test section is 38 diameters in length. Data acquisition is provided by a Hewlett-Packard 9825 programmable calculator and a NEFF-620 front end. Data are typically acquired at a rate of 20 samples/s for a period of several seconds. Temporal averages and standard deviations are then computed. The sample period was chosen such that the temporal averages and standard deviations did not change with an increase in sample period.

To demonstrate the utility of tomography in two-phase flow, we conducted a series of tests near various flow regime boundaries. Here, we examine only one of these transitions.

Figure 7 shows the details of the transition from wavy stratified to a thin-film, slumped annular flow. The superficial liquid velocity was held constant at 0.23 m/s, while the superficial air velocity was raised from 9.4 to 19.5 m/s. The wave structure of the stratified flow appears as a smearing of the liquid/gas interface. The annular flow consisted of a fast moving froth in the bottom of the pipe with a very thin (<0.5 mm) film covering the pipe wall. As the transition takes place, the liquid/gas mixture in the pipe bottom thins out and at the same time climbs the pipe walls resulting in a cupped interface. In most of our work conducted in horizontal 3-in. and larger piping, the annular regime is characterized as a slumped annular flow.

CONCLUSIONS

The results of performing tomographic reconstructions of phantoms from both simulated data and actual data indicate that, even with limited numbers of data values, the technique can provide useful two-phase-flow density-distribution data. The algebraic reconstruction algorithm proved to be superior to the convolution technique for the number of data values available. The ultimate quality of the density field reconstruction is limited by the number of detectors, not the number of views. For the prototype system described here, the errors due to multiple scattering and finite beam size are small with respect to the errors inherent in the reconstruction technique. Estimated global density accuracy is 2% of scale. The accuracy of the density distribution obtained depends on the flow regime. The technique is a significant advancement over current techniques in that it provides model-independent phase-distribution information.

REFERENCES

1. Lassahn, G. D., LOFT Three-Beam Densitometer Interpretation, TREE-NUREG-1111, October 1977.

2. Brooks, R. A. and DiChiro, G., "Principles of Computer Assisted Tomography (CAT) in Radiographic and Radioisotopic Imaging," Phys. in Med. and Biol., 21, 1976, pp. 689-732.

3. Fincke, J. R., Berggren, M. J., and Johnson, S. A., "The Application of Reconstructive Tomography to the Measurement of Density Distribution in Two-Phase Flow," Proceedings of the 26th International Instrumentation Symposium, Seattle, Washington, May 1980, pp. 235-243.

4. Fincke, J. R., Cheever, G. C., Fackrell, L. J., Scown, V. S., Thornton, B. V., and Ward, M. W., "The Development of Reconstructive Tomography for the Measurement of Density Distribution in Large Pipe Steady-State Multi-Phase Flows," NRC Review Group Conference on Advanced Instrumentation Research for Reactor Safety, NUREG/CP-0015, December 1980, pp. v.5-5 to v.5-23.

5. Herman, G. T. and Lenta, A., "Iterative Reconstruction Algorithms," Comput. Biol. Med., 6, 1976, pp. 273-294.

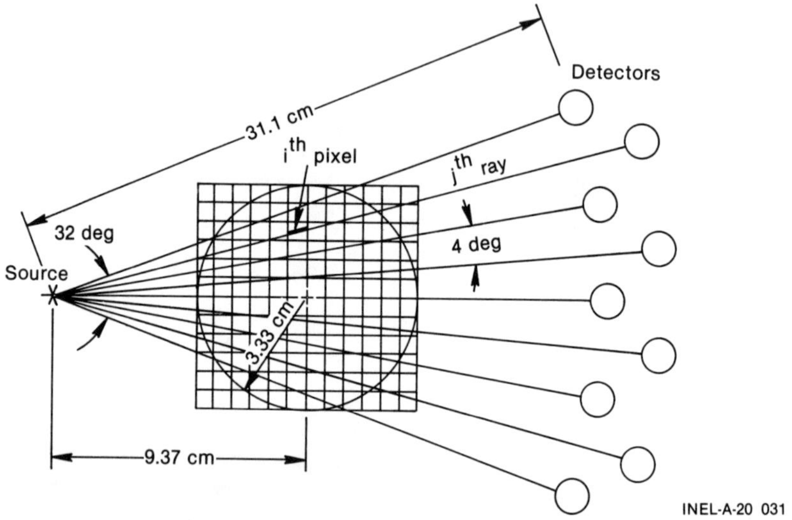

Fig. 1. Ray geometry and pixel array for prototype nine-beam tomographic densitometer used for iterative reconstruction. The density field is bounded by the circle, which contains N cells along a diameter.

Fig. 2. Prototype nine-beam tomographic densitometer mounted on 3-in. diameter pipe.

Fig. 3. Large operational tomographic densitometer in use on a two-phase test facility at the Idaho National Engineering Laboratory.

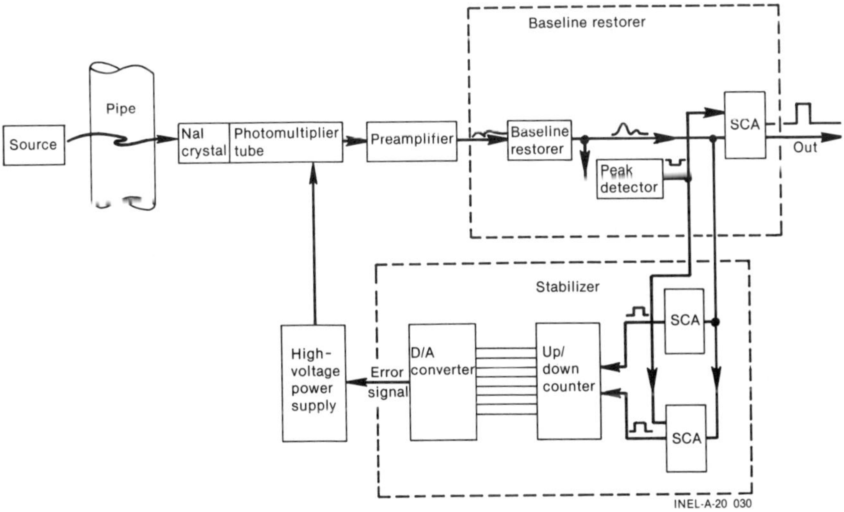

Fig. 4. Megacount stabilized densitometer electronics.

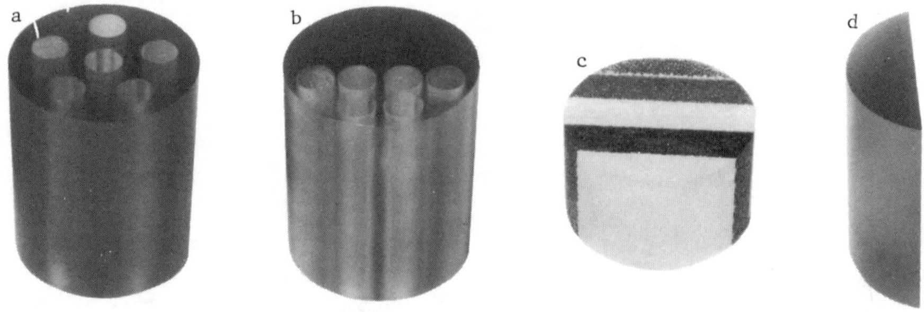

Fig. 5. Test-case phantoms: (a) six-hole symmetric, (b) six-hole clustered, (c) laminated Plexiglas, wood and polyurethane, and (d) stratified.

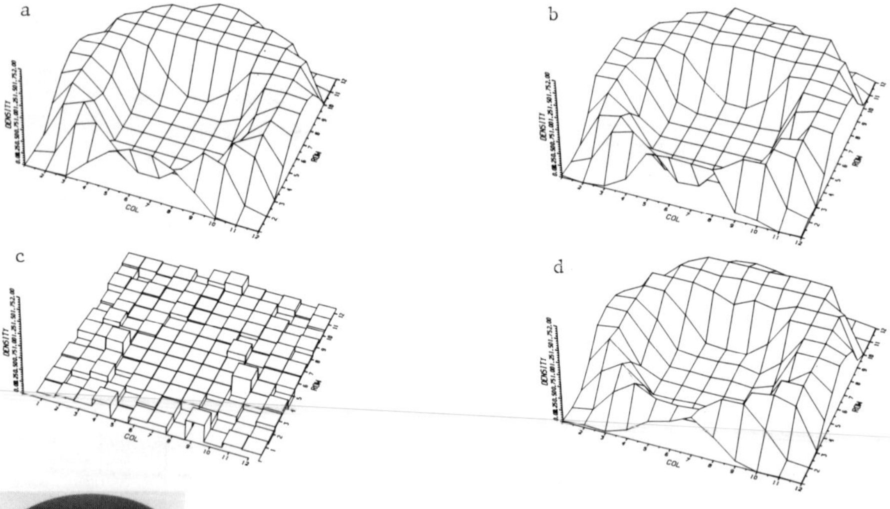

Plexiglas phantom

Fig. 6. Comparison of reconstructions from both simulated and actual data on eccentric-bubble Plexiglas phantom: (a) digitization of phantom, (b) reconstruction from simulated data, (c) pixel-by-pixel differences between simulated-data reconstruction and digitized phantom, and (d) reconstruction from actual data.

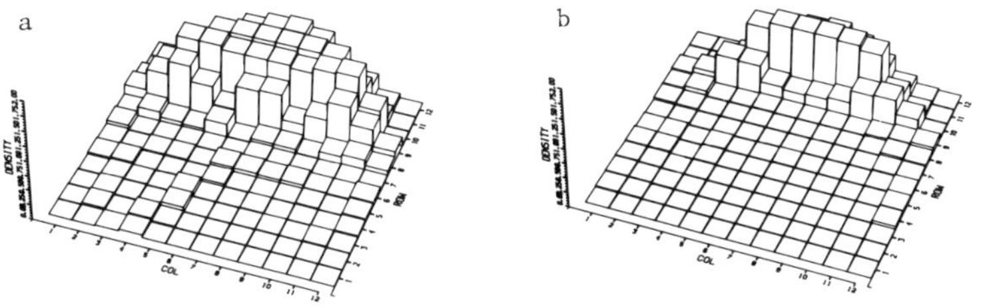

Fig. 7. Transition from wavy-stratified (a) to slumped-annular (b) flow.

NUMERICAL METHODS

SECOND ORDER ACCURATE FINITE DIFFERENCE
TECHNIQUES FOR DYNAMIC RESPONSE OF LOCKING MATERIALS

BY
S. HANAGUD* and S. CHANDRASHEKHARA**
Georgia Institute of Technology, Atlanta, GA. 30332

ABSTRACT

Locking materials have been proposed as candidate materials to attenuate peak shock loads and to protect structures. In this paper, a Lax's type of second order accurate finite difference scheme has been used to study the transient dynamic response of locking materials. Lax's method has been modified by using two step procedures due to McCormack, Richtmyer and Gottlieb. Results have been presented for specific cases of loading in aluminum locking materials.

INTRODUCTION

Locking materials[1,2] have been described by other terms, such as energy absorbing materials, foams, distended materials[3] or nonreactive porous solids[4]. A locking material is characterized by its special mechanical behavior when a uniform hydro-static pressure is applied. An idealized behavior is illustrated in Figure 1. For values of pressure p less than a certain value P_E, the material behaves like an isotropic, linear elastic solid. For values of pressure p greater than the value P_E, the pore spaces collapse and the material is said to lock at a density ρ_ℓ. The subsequent behavior of the locking material is that of an incompressible material. A more realistic locking behavior is as shown in Figure 2. After reaching the locking density ρ_ℓ, the mechanical behavior of the material approximately follows the pattern of the corresponding

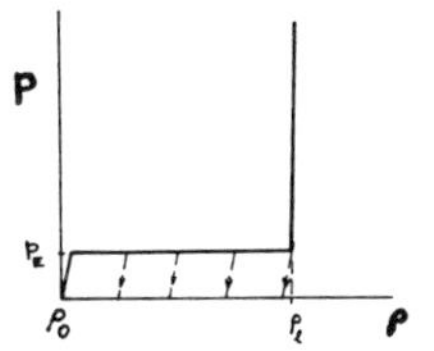

FIG. 1 an elastic locking solid

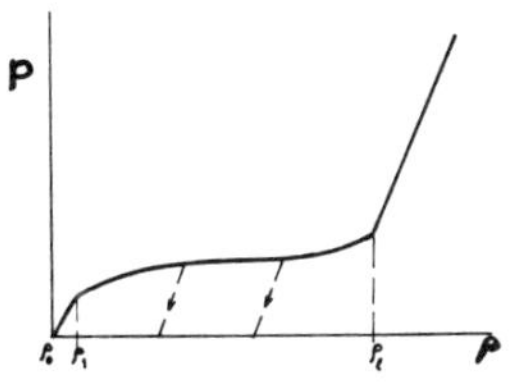

FIG. 2 experimental p-ρ diagram for locking materials

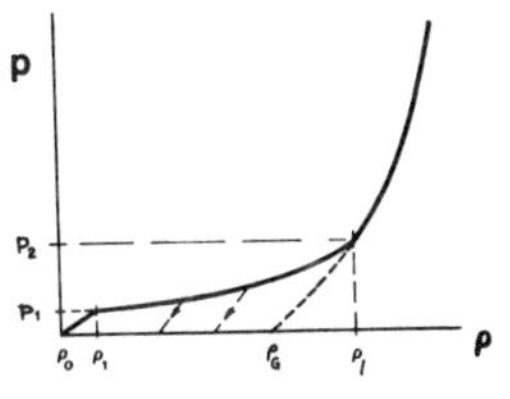

FIG. 3 pressure - density curve

*Professor, **Graduate student

parent solid from which the locking material has been produced. In most cases, the collapsed pore spaces can not be recovered. Thus, the unloading paths will be different from the loading paths.

Locking materials are effective peak stress attenuators. They can be used to protect important structures from impact loading. The superiority of the locking materials, as peak stress attenuators, is because of the large ratio of the unloading or rarefaction wave velocity to the loading wave or shock wave velocity. This phenomenon is discussed later in the paper. Also, as discussed in subsequent sections of this paper, the attenuation capabilities have been validated by appropriate experimental techniques.[3]

In the point of view of engineering design, an essential design tool is the computer simulation of the transient dynamic response of the locking material or locking material and structure combinations. Such a computer simulation can be used to check if the selected design parameters provide the desired margin of safety, to iterate the design parameters and to establish the reliability of the design. To date, such computer simulations have been attempted by the use of finite difference techniques developed by Von Neumann and Richtmyer[5,6,7]. Von Neumann's technique is a very powerful tool in numerical analysis. In fact, Cole[8] has described the method as a "new math" of mechanics. The technique provides stresses and particle velocities to a first order of accuracy. In order to achieve a desired degree of accuracy, small time steps and small spatial mesh sizes are necessary. Furthermore, as applied to locking materials, very small time steps are needed[4] to account for the phase transition from the initial density to the locked density. This process reduces the computational efficiency. Another difficulty associated with the numerical procedure for locking material is an accurate simulation of the precursor wave[4].

The difficulty associated with small time steps and hence computational inefficiency can be improved by using the developments during sixties and seventies in the field of numerical analysis of hyperbolic differential equations[9-14]. The second difficulty involving the accurate simulation of the location and steepness of stress waves can also be improved by using more recent improvements due to Lax[15] and Harten[16]. This paper is addressed to the first problem of improving the computational efficiency by the use of second order accurate difference techniques that were developed by Lax[9]. In computer programming of the Lax's procedure, two step formulations due to McCormack[14] and Richtmyer[6] and Gottlieb[12] have been used.

CANDIDATE MATERIALS

In principle, locking material can be produced from any parent solid material. The needed process of production requires the formation of a nonreactive porous solid of a density lower than that of the parent material. A commercial aluminum locking material MO-AK (Emerson and Cummings) has been available in the past. This is usually inhomogeneous. Homogeneous aluminum locking material can be produced by techniques[3] such as hot pressing of aluminum powders, cold pressing of aluminum powders followed by sintering or a repeated sequence hot pressing that is followed by sintering of aluminum powders and the use of silica microballons. Commercially, graphite locking materials are available from companies such as National Carbon Company. Of course, a very common locking material of comparatively low threshold pressure capability that is easily available is styro-foam. Depending on the environment and the peak stresses that need to be attenuated, different types of locking material can be chosen or designed for a particular application. For example, for high stress applications, a locking material made of steel may be an interesting possibility. In some cases, a combination of different locking materials or a sandwich construction may be more practical.

EXPERIMENTAL BACKGROUND

In the past, experimental investigations have been conducted[3] on several locking materials. In these investigations, metals, plastics, graphite and ceramics have been used as parent materials to produce locking materials. Experiments have been conducted by using light gas guns and flyer plates. In these experiments, attempts have been made to simulate conditions of one dimensional strain. In the point of view of the present study, a significant result from these studies concerns a comparison of the shock wave velocities in locking materials with the corresponding wave velocities in solid materials.

In most of these studies, a two wave pattern has been observed in both locking materials and solids. In locking materials, a forerunner wave carries stresses in the material before the phase transition from the initial density ρ_0 to the locked density ρ_ℓ. The second slow wave corresponds to the shock wave. This shock wave carried stresses that exceed the stresses necessary for phase transition. Solid materials also exhibit a two wave pattern. In this case, the stresses carried by the forerunner wave correspond to the yield limit and the elastic behavior of the solid before yield. This forerunner wave travels at a velocity corresponding to the irrotational wave velocity. The second wave carries stresses that exceed the elastic limit. The experiments confirm the fact that this second wave in the solid is much faster than the second shock wave in the locking material. For example, in aluminum locking materials, for initial density ρ_0 = 2.1 gms/c.c., the second shock wave velocity varies from 0.7 to 1.29 mm/μsec. The second wave velocity in the corresponding solid is in the range of 4.62 mm/μsec. Similarly, the first wave velocities in the locking materials are smaller than the first wave velocities in the corresponding solid. The first wave velocities, in aluminum locking materials of initial density ρ_0 = 2.1 gms/c.c. varies in the range 1.6 mm/μsec. to 2.0 mm/μsec. The corresponding first wave in the solid travelled with the velocity of 6.11 mm/μsec.

The theoretical foundation and the results of the experimental studies confirm the potential benefits that can be derived from locking materials when they are used as protective structures.

ANALYSIS UNDER CONDITIONS OF ONE DIMENSIONAL STRAIN

As explained in the introduction, the second order accurate Lax type of finite difference scheme has been used, in this paper, to solve the transient dynamic response of locking materials. In particular, this paper has restricted the application to one dimensional strain in x-direction. The equations of motion have been written in Lagrangian coordinates. The initial positions of the body have been selected as the Lagrangian coordinates. Large elastic-plastic deformations of the locking material have been considered. The resulting equations are as follows:

$$\frac{\partial}{\partial t} \{\underset{\sim}{u}\} = [A(\rho)] \frac{\partial}{\partial x} \{\underset{\sim}{u}\} \tag{1}$$

where

$$\{\underset{\sim}{u}\}^T = v(x,t),\ \rho(x,t),\ \sigma_x(x,t),\ \sigma_y(x,t) \tag{2}$$

and

$$[A(\rho)] = \begin{bmatrix} 0 & 0 & \rho^{-1} & 0 \\ -\rho & 0 & 0 & 0 \\ \rho f'(\rho) + \frac{4G}{3} & 0 & 0 & 0 \\ \rho f'(\rho) - \frac{2G}{3} & 0 & 0 & 0 \end{bmatrix} \tag{3}$$

In these equations, it has been assumed that the stress tensor can be separated into hydrostatic pressure and stress deviators. The hydrostatic pressure has been assumed to be related to the changes in density and follow a locking behavior as shown in Figure 3. The stress deviators are assumed to follow an elastic-ideal plastic behavior with Von Mises' yield condition.

All the assumptions of the preceding paragraph are approximations. However, these approximations have been used, in the past, to express the mechanical behavior of locking materials[3,4,17] and other solids[17] at very high stresses. On the basis of experimental results, the approximations have been found to be reasonable[17]. In the absence of any available constitutive relationships on the basis of second Piola-Kirchhoff stresses and Green-Lagrange strains, this paper has considered similar approximations. The constitutive relationships have been written in terms of Cauchy stresses, particle velocity gradients and density. Then,

$$\sigma_x = -p + S_x \tag{4}$$

$$\sigma_y = -p + S_y \tag{5}$$

and

$$\begin{aligned} p &= f_1(\rho) \qquad && \rho_o < \rho < \rho_1 \\ p &= f_2(\rho) && \rho_1 < \rho < \rho_\ell \\ p &= f_3(\rho) && \rho_\ell < \rho \end{aligned} \tag{6}$$

The quantities f_1, f_2 and f_3 are as follows (Figure 3):

$$f_1(\rho) = P_1 \frac{\rho_1}{\rho_o} \frac{\rho_o-\rho}{\rho_o-\rho_1} \left[1 + b \left(\frac{\rho_o}{\rho}\right) \left(\frac{\rho_1-\rho}{\rho_o-\rho_1}\right) \right]$$

$$f_2(\rho) = P_1 + \frac{K}{R} \left[\left(\frac{\rho}{\rho_1}\right)^R - 1\right]$$

$$f_3(\rho) = A\ (\rho/\rho_G - 1) + B\ (\rho/\rho_G - 1)^2 \tag{7}$$

Similarly,

$$\dot{S}_x = \frac{\dot{S}_y}{2} = \frac{\dot{S}_z}{2} = 4G \left(- \frac{\dot{\rho}}{\rho}\right) \tag{8}$$

with the yield condition

$$S_x^{\ 2} + S_y^{\ 2} - \frac{2}{3} Y^2 \leq 0 \tag{9}$$

NUMERICAL ANALYSIS

A numerical integration of the hyperbolic partial differential equation (1) by a procedure similar to that of Lax requires that the field variables $\{\underset{\sim}{u}\}$ at $t + \Delta t$ should be calculated from a knowledge of the field variables $\{\underset{\sim}{u}\}$ and its spatial derivatives at t. To a second order accuracy,

$$\{\underset{\sim}{u}\}^{t+\Delta t} = \{\underset{\sim}{u}\}^t + \left\{\frac{\partial \underset{\sim}{u}}{\partial t}\right\}^t \Delta t + \left\{\frac{\partial^2 \underset{\sim}{u}}{\partial t^2}\right\}^t \frac{\Delta t^2}{2} + 0(\Delta t)^3 \tag{10}$$

The differential equation (1) can be used to express the quantities $\partial \underset{\sim}{u}/\partial t$ and $\partial^2 \underset{\sim}{u}/\partial t^2$ in terms of the spatial derivatives of $\{\underset{\sim}{u}\}$. Then,

$$\{\underset{\sim}{u}\}^{t+\Delta t} = \{\underset{\sim}{u}\}^t + [A(\rho)] \left\{\frac{\partial \underset{\sim}{u}}{\partial x}\right\}^t + [A(\rho)]^2 \left\{\frac{\partial^2 \underset{\sim}{u}}{\partial x^2}\right\} \frac{\Delta t^2}{2}$$

$$+ [A(\rho)] \left[\frac{\partial A}{\partial x}\right] \left\{\frac{\partial \underset{\sim}{u}}{\partial x}\right\} \frac{\Delta t^2}{2} + \left[\frac{\partial A}{\partial x}\right] \left\{\frac{\partial \underset{\sim}{u}}{\partial x}\right\} \frac{\Delta t^2}{2} \tag{11}$$

The equation (11) is very complicated. It involves $[A(\rho)]^2$ and multiples of first and second partial derivatives of $\{\underset{\sim}{u}\}$ with respect to x. This equation can be considerably simplified by using a two step approximation due to McCormack, Richtmyer or Gottlieb. For example, McCormack two step formulation is as follows.

$$\{\underset{\sim}{u}\}^* = \{\underset{\sim}{u}\}^t + \left\{\frac{\partial \underset{\sim}{u}}{\partial t}\right\}^t \Delta t$$

$$\{\underset{\sim}{u}\}^{t+\Delta t} = \frac{1}{2} [\{\underset{\sim}{u}\} + \{\underset{\sim}{u}^*\}] + \frac{\Delta t}{2} \left\{\frac{\partial \underset{\sim}{u}^*}{\partial t}\right\} \tag{12}$$

It is easy to verify the accuracy of (12) by expansion. Now, the time derivatives on the right hand sides of (12) are replaced by spatial derivatives. Thus,

$$\{\underset{\sim}{u}\}^* = \{\underset{\sim}{u}\}^t + [A(\rho)] \left\{\frac{\partial \underset{\sim}{u}}{\partial x}\right\}^t \Delta t$$

$$\{\underset{\sim}{u}\}^{t+\Delta t} = \frac{1}{2} [\{\underset{\sim}{u}\} + \{\underset{\sim}{u}\}^*] + \frac{\Delta t}{2} [A^*] \left\{\frac{\partial \underset{\sim}{u}^*}{\partial x}\right\} \tag{13}$$

It is to be noted that $[A(\rho)]^*$ and $\{\partial \underset{\sim}{u}/\partial x\}^*$ are to be evaluated by using the star values of the field variables. Also, it can be seen that the formulation described in the equations (13) is much simpler than the formulation described in (11). To complete the finite difference formulation, the spatial derivatives of $\{\underset{\sim}{u}\}$ and $\{\underset{\sim}{u}\}^*$ are expressed in terms of their spatial finite difference approximations. A central, forward or backward difference is used depending on whether the point under consideration is an interior point, left boundary point or right boundary point.

STABILITY REQUIREMENTS

The equations (13) can be written in the form of finite difference operators. Then,

$$L^* = I + k[A] \Delta_x$$

$$L_1 = \frac{1}{2} I + k[A] \Delta_x \tag{14}$$

In these equations, Δ_x represents the spatial finite difference approximations. The quantity I is the identity operator and Δt has been replaced by k. Then,

$$\{\underset{\sim}{u}\}^{t+\Delta t} = (\frac{I}{2} + L_1 L^*) \{\underset{\sim}{u}\}^t \tag{15}$$

The operators I, L_1 and L^*, operating on $\{\underset{\sim}{u}\}^t$ in the manner shown in equation (15) change $\{\underset{\sim}{u}\}^t$ to $\{\underset{\sim}{u}\}^{t+\Delta t}$. This is an explicit scheme and is subject to the usual restrictions on $k = \Delta t$ to maintain the stability of the computational scheme. In particular, it can be shown that $\Delta t \leq \Delta x/c$ where c is the fastest of the local elastic wave velocities.

PLASTICITY EFFECTS AND YIELD CONDITIONS

The numerical method that has been used in this paper is an explicit method. The field variables at $t+\Delta t$ depend only on the field variables, at time t. The particle velocities and strains at $t+\Delta t$ can be computed from the current field variables at t. The stresses computed at $t+\Delta t$ may violate the yield condition (9). However, these stresses can be adjusted along appropriate normals to the yield surface back to the yield surface.[7] The already calculated velocities and strains that depend only on the field variables at t remain unchanged. The only regions where iterative calculations are needed are at the boundaries.

DISCUSSION OF RESULTS

In order to achieve the same accuracy as a first order accurate method, a time step equal to the square root of Δt that is used in the first order method is needed. The quantity Δt is usually less than one and hence the second order accurate method uses a larger time step. Hence, a larger Δt can be chosen to obtain the same order of accuracy as a first order method. This leads to a smaller number of the finite difference cells and increased computational efficiency. The developed computer program has been used to examine the efficiency of locking materials in attenuating stresses and protecting structures. In this analysis, specified items include the peak impact stress and the unloading pattern of the stress wave impinging on the locking materials. The locking density ρ_ℓ is usually fixed for a given locking material. The results of the numerical analysis can be used to explain the peak stress and the stress distribution for various times after impact and unloading at the boundary.

In the numerical example, aluminum locking materials have been selected. Some of the properties of aluminum locking materials have been extensively investigated[3]. A pressure density relationship as shown in Figure 3 has been assumed. The locking density ρ_ℓ is equal to the 2.72 gms/c.c. Various initial densities have been considered. Results have been presented for ρ_o = 1.39 and 2.1 gms/c.c. here. There are three distinct branches of the pressure-density relationship. Equations for these branches are the same as in equations (6) and (7).

The quantity ρ_G is the intersection of the p-ρ curve with ρ axis as shown in Figure 3. The appropriate constants are selected from reference 3. For all $\rho > \rho_\ell$, the unloading is assumed to follow the slope of the solid p-ρ curve as shown in Figure 3.

Two types of loading have been considered. The first type of loading consists of a step loading followed by step unloading. It has been assumed that a peak compressive stress of σ_1 is applied at time t=0 to the left boundary of the slab of a locking material. For purposes of illustration, a value of $-\sigma_1$ = 100,000 psi has been assumed. The applied stress is reduced to zero by step unloading at time t = 0.2 μsec. This loading pattern is illustrated in Figure 4 and will be called loading pattern 'a'. The second

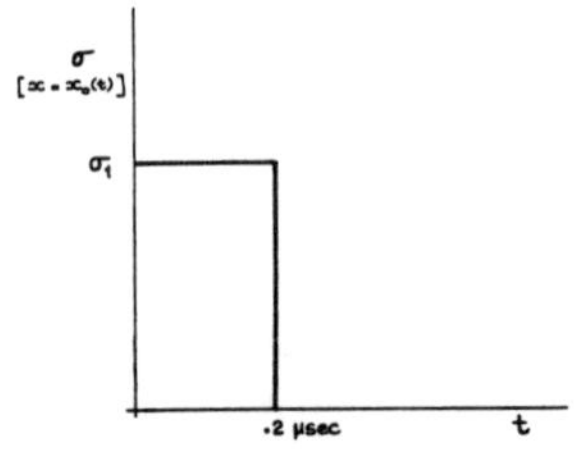

FIG. 4 step loading, pattern "a"

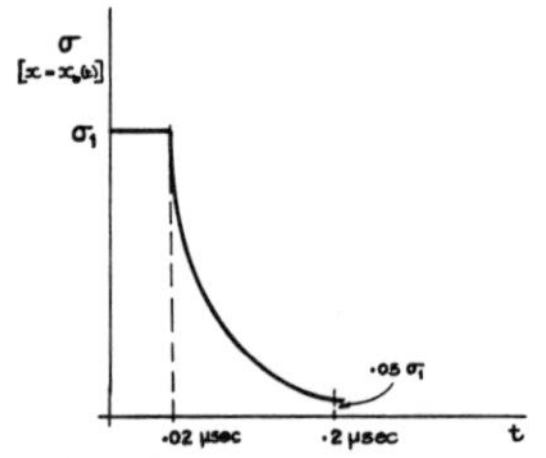

FIG. 5 exponential decrease of load, pattern "b"

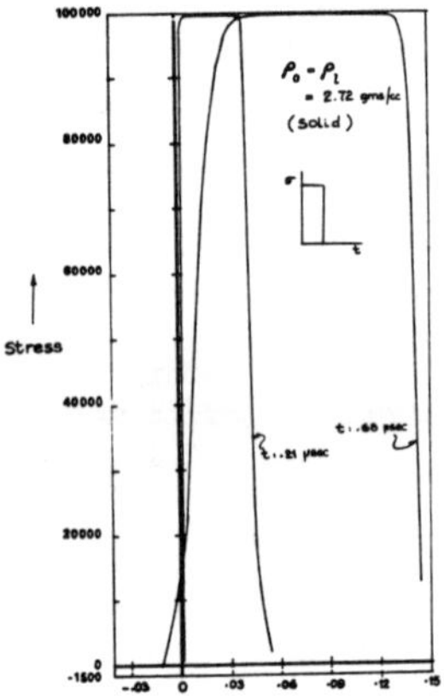

FIG. 6 stress distribution at various times

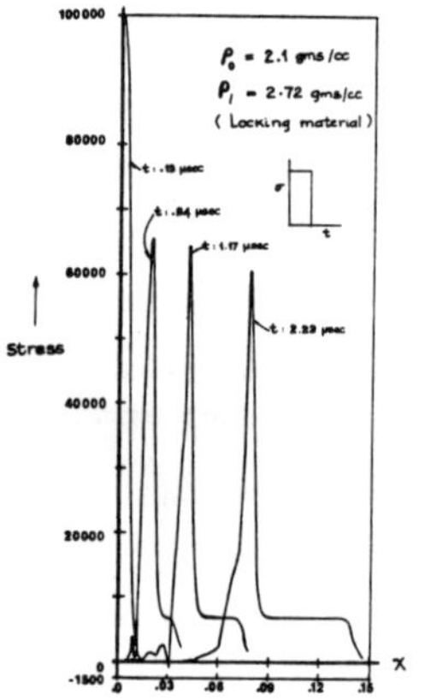

FIG. 7 stress distribution at various times

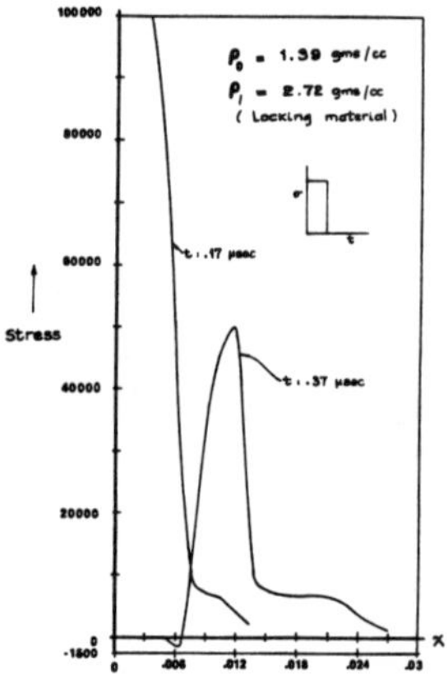

FIG. 8 stress distribution at various times

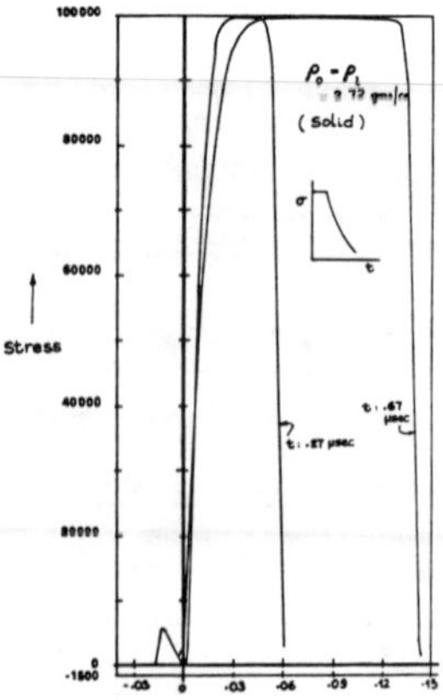

FIG. 9 stress distribution at various times

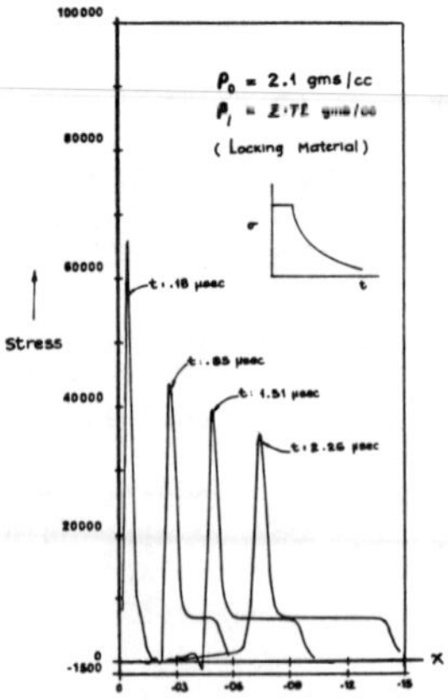

FIG. 10 stress distribution at various times

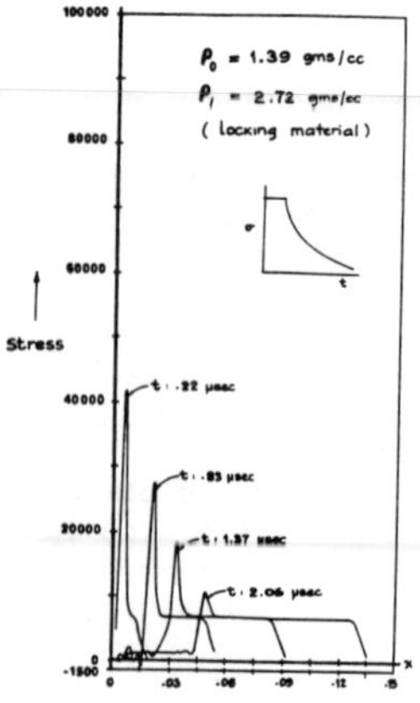

FIG. 11 stress distribution at various times

type of loading consists of a step loading of a compressive stress of magnitude $-\sigma_1$ at t=0 followed by an exponential unloading at the left boundary. For purposes of illustration, $-\sigma_1$ has been assumed to be equal to 100,000 psi. The applied stress is assumed to be maintained at $\sigma = -\sigma_1$ for a duration of 0.02 μsec. Then the exponential decay of the applied load at the left boundary decreases the magnitude of σ_1 to .05 σ_1 at 0.2 μsecs. This unloading pattern is illustrated in Figure 5 and will be called loading 'b'.

First, a solid material slab of initial thickness 0.5 inches has been considered. The thickness has been divided into 100 cells. In this case the solid density is ρ_ℓ = 2.72 gms/c.c. The transient response of this slab to the loading pattern 'a' has been studied by using the developed computer program and the results are illustrated in Figure 6. This figure is a plot of the stress σ_x as a function of the thickness at two different instants of time t=0.21 μsecs and t=0.68 μsecs. As seen in the figure, the stress at a distance of 0.14 inch from the initial left boundary is still approximately 100,000 psi. This is equal to the applied peak stress. No significant attenuation has taken place during the travel of the stress wave through the thickness equal to 0.14 inch.

Next, a locking material of initial density ρ_o = 2.1 gms/c.c. and locking density ρ_ℓ = 2.72 gms.c.c. has been considered. The thickness and division into cells are identical to those for the solid. The transient response to the loading pattern 'a' has been computed and illustrated in Figure 7. The stress distribution as a function of the distance from the left boundary has been illustrated for t=0.13 μsec, 0.54 μsec, 1.15 μsec, and 2.22 μsec. It can be seen that the peak stress has reduced by approximately 35% over a distance of 0.09 inch from the left boundary. In this figure, the elastic forerunner wave can also be seen. Similar stress distributions are observed for locking materials of initial density ρ_o = 1.818 gms/c.c., 1.604 gms/c.c. and 1.39 gms/c.c. If these locking materials are considered as countermeasures in front of a given structure, the peak impact stress has been reduced by an amount as much as 50% when the shock wave has propagated a distance of 0.013 inch (Figure 8). It can also be observed that the locking material of lower initial density has the potential of attenuating the impact peak stress by a larger percentage when compared with the locking material of higher initial density.

As a next step, the loading pattern 'b' has been considered. The results of the study of transient response through a solid has been illustrated in Figure 9. The results are similar to those for the loading pattern 'a'. The peak impact stress has not attenuated during the passage of the stress wave over a distance of 0.15 inch. Similar stress distribution for loading pattern 'b' has been illustrated in Figures 10 and 11 for different locking materials of initial density ρ_o = 2.1 gms/c.c. and 1.39 gms/c.c. The Figure 10 is for a locking material with initial density 2.1 gms/c.c. Over a shock wave traverse of .045 inch the peak stress has been reduced by 60%. However, the locking material with initial density of 1.39 gms/c.c. can attenuate the peak stress by almost 90%, during the shock traverse of 0.05". This is illustrated in Figure 11. The unloading waves and the elastic forerunners can be identified in all these figures.

CONCLUSIONS

This paper has reported the development of a second order accurate numerical technique that can be used to study the transient dynamic response of locking materials subjected to shock loads. The efficiency of the computational technique and the efficiency of the locking materials, in attenuating the peak impact loads have been investigated. A similar computer program in more than one space dimensions will be a very practical tool in the design of protective structures.

REFERENCES

1. Hanagud, S., Sudaer Report No. 152, Stanford University, California, 1963.
2. Hanagud, S., Proc. 5th U.S. National Congress, 1966. (Invited Sectional Lecture).
3. Linde, R.K., and D.N. Schmidt, AFWL-TR-66-13, 1966.
4. Hanagud, S., G.S. Sidhu and B. Ross, Israel Journal of Technology, 1969.
5. Von Neumann, J., and R.D. Richtmyer, J. Appl. Phys. Vol. 21, 1950, pp. 232-237.
6. Richtmyer, R.D., and K.W. Morton, Interscience - Wiley, New York, 1967.
7. Wilkins, M.L., UCRL-7322, Physics L. L. Laboratories, California, 1962.
8. Cole, J.D., Proc. U.S. Nat'l. Congr. Appl. Mech. 1970, pp. 3-10.
9. Lax, P., and B. Wendroff, Comm. Pure. Apply. Math, 17, 1964, pp. 381-390.
10. Strang, G., Numer. Math. 1964, p. 37.
11. Strang, G., SIAM. J. Numer. Anal., 5, 1968, p. 506.
12. Gottlieb, D., Anal., 9, 1972, pp. 650-661.
13. Gottlieb, D., and Turkel, E., J. Comp. Phys., Vol. 26, 1978.
14. Lerat, A., and Peyret, R., Comp. and Fluids, Vol. 2, 1974, p. 35.
15. Lax, P.D., and Harten, A.D., SIAM J. of Num. Amalysis, 1981.
16. Harten, A., AEC Report, CR 11-1-3077, 1979.
17. Varley, E. (ed.), AMD Vol. 17, ASME, New York, 1976.

ACKNOWLEDGEMENT

Authors acknowledge the support of this work by AFOSR and AMMRC.

SURFACE CONTROL TEMPERATURES FOR THE
BRIDGMAN-STOCKBARGER TECHNIQUE

Larry M. Foster, Ph.D.†
Science Applications, Inc.
Huntsville, Alabama

ABSTRACT

In Bridgman-Stockbarger crystal growth techniques, the working material's surface temperature cannot be arbitrarily fixed if a flat solid-melt interface is to be achieved. A method for computing the heater zone surface temperatures required for a flat solid-melt interface is investigated. An adiabatic zone containing the solid-melt interface is included. A test numerical case is given followed by a discussion of similar problems arising from float zone techniques.

1. INTRODUCTION

In the classical model for Bridgman-Stockbarger crystal growth, a semi-infinite cylinder ($x \leq 0$ and $0 \leq r \leq 1$) of molten material is adjoined to another semi-infinite cylinder ($x \geq 0$ and $0 \leq r \leq 1$) of solid material at the cylinder end caps to form a solid melt interface ($x=0$). In most studies, the surface ($r=1$) temperatures are assumed constant on each of the two (solid or liquid) regions. If the interior ($r<1$) temperature is assumed continuous across the solid-melt interface and if energy is conserved across the solid-melt interface, the resulting problem is mathematically well posed. If, however, the temperature at the solid-melt interface is also required to be the material melting temperature (a natural boundary condition), the problem is over posed and generally not solvable. In this paper we replace the constant molten region surface temperature by a nonconstant surface temperature (a control) such that the resulting boundary conditions are again compatible and the model is well posed. The principle thrust of this investigation is to develop a method for approximating the desired nonconstant control temperature.

---

† Work completed while a NASA/ASEE Summer Faculty Fellow, 1980-81, Space Sciences Laboratory, NASA, Marshall Space Flight Center.

More precisely, the scaled partial differential equations defining crystal growth by the Bridgman-Stockbarger technique [1] are, for $0 < r < 1$,

$$(1) \quad \nabla^2 T = P_\ell \frac{\partial T}{\partial x}, \; x < 0, \text{ and } \nabla^2 T = P_s \frac{\partial T}{\partial x}, \; x > 0$$

where P_ℓ and P_s are the liquid and solid Péclet numbers respectively, T is the working material temperature and x is the axial distance from an assumed flat solid-melt interface. To conserve energy at the interface,

$$(2) \quad -K_\ell \frac{\partial T}{\partial x}\Big)_{x=0^-} + K_s \frac{\partial T}{\partial x}\Big)_{x=0^+} + \mathscr{L} = 0$$

where K_ℓ and K_s are the respective liquid and solid conductivities and $\mathscr{L}$ is the product of the crystal growth rate, the melt density, and the latent heat of solidification. The boundary heating and cooling mechanisms in this idealized case are, at $r = 1$,

$$(3) \quad -K_\ell \frac{\partial T}{\partial r} = h_\ell (T - T_\ell), \; x < 0, \text{ and } -K_s \frac{\partial T}{\partial r} = h_s T, \; x > 0$$

where the ambient temperature for the solid region, $x > 0$, has been scaled to zero and h_ℓ and h_s are the liquid and solid heat transfer coefficients at crucible surface, r=1. Assuming T is continuous at x=0, equations (1)-(3) are mathematically properly posed and may be solved by the separation of variables method. However, if M is the working material's melting temperature and we adjoin to equations (1)-(3)

$$(4) \quad T\Big)_{x=0^-} = M = T\Big)_{x=0^+}, \; 0 < r < 1,$$

then the system (1)-(4) is generally overposed. If the system (1)-(4) has a solution, then the process parameters P_ℓ, P_s, K_ℓ, K_s, $\mathscr{L}$, h_ℓ, h_s, T_ℓ and M are necessarily functionally dependent [2]. In general, however, this necessary functional dependence is not sufficient to guarantee (1)-(4) has a solution. The problem is further complicated by the addition of a more realistic adiabatic zone from x=-Q to x=L containing the solid-melt interface. The resulting system, with labeled equations, is illustrated in Figure 1.

From Figure 1, let System I include equations (5)-(8), and (11)-(16). Then System I is properly posed if we include the realistic conditions

$$(19) \quad \frac{\partial T_i}{\partial r}\Big)_{r=0^+} = 0, \; i = 1, 2, \text{ and } 3$$

and

$$(20) \quad \max_{0<r<1} \lim_{x \to \infty} |T_1(x,r)| < \infty .$$

$$T_1 = T_2 \quad (5)$$

$$\frac{\partial T_1}{\partial x} = \frac{\partial T_2}{\partial x} \quad (6)$$

$$T_2 = M = T_3 \quad (7)$$

$$-K_\ell \frac{\partial T_3}{\partial x} + K_s \frac{\partial T_2}{\partial x} + \mathscr{L} = 0 \quad (8)$$

$$T_3 = T \quad (9)$$

$$\frac{\partial T}{\partial x} = \frac{\partial T_3}{\partial x} \quad (10)$$

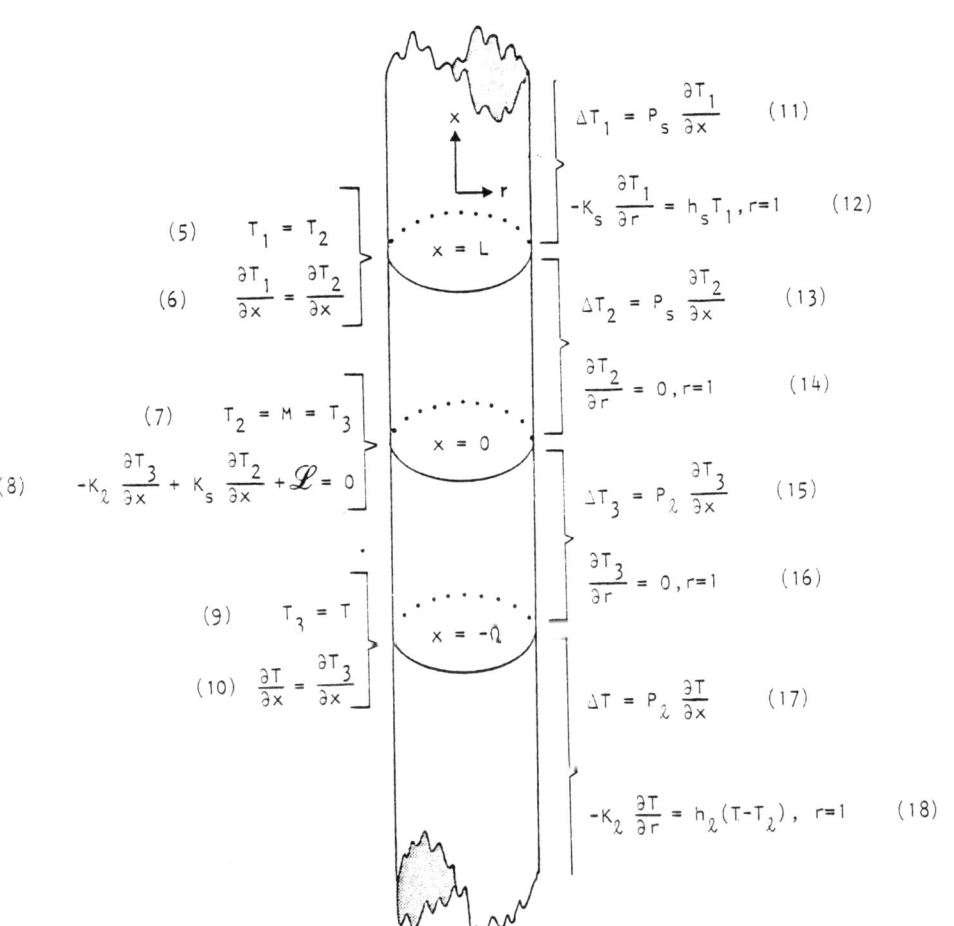

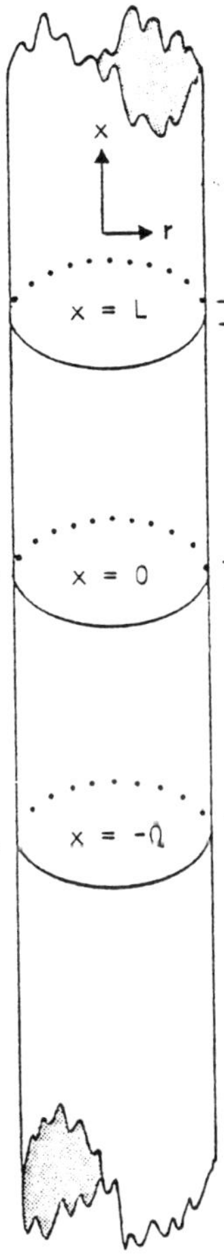

$$\Delta T_1 = P_s \frac{\partial T_1}{\partial x} \quad (11)$$

$$-K_s \frac{\partial T_1}{\partial r} = h_s T_1, r=1 \quad (12)$$

$$\Delta T_2 = P_s \frac{\partial T_2}{\partial x} \quad (13)$$

$$\frac{\partial T_2}{\partial r} = 0, r=1 \quad (14)$$

$$\Delta T_3 = P_\ell \frac{\partial T_3}{\partial x} \quad (15)$$

$$\frac{\partial T_3}{\partial r} = 0, r=1 \quad (16)$$

$$\Delta T = P_\ell \frac{\partial T}{\partial x} \quad (17)$$

$$-K_\ell \frac{\partial T}{\partial r} = h_\ell (T - T_\ell), \quad r=1 \quad (18)$$

Figure 1

However, if (9), (10), (17) and (18) are adjoined to System I, the resulting system will be overposed and generally insolvable. To circumvent this dilemma, we propose the following. First solve System I and define

$$A(r) = T_3(x,r) \quad \text{and} \quad B(r) = \frac{\partial T_3}{\partial x}(x,r) \quad \text{for } x = -Q \text{ and } 0 < r < 1.$$

Then replace condition (18) by $T(x,1)=f(x)$, where $f(-Q)=A(1)$ and $f'(-Q)=B(1)$, but $f(x)$ otherwise is unknown. The control $f(x)$ is constructed such that the solution of the well posed System II:

(21) $\nabla^2 T = P_\ell \dfrac{\partial T}{\partial x}$, $x < -Q$, $0 < r < 1$

(22) $T(x,r) = A(r)$, $x = -Q$, $0 < r < 1$

(23) $T(x,r) = f(x)$, $x < -Q$, $r = 1$

also satisfies

(24) $\dfrac{\partial T}{\partial x}(x,r) = B(r)$, $x = -Q$, $0 < r < 1$.

In other words, we wish to construct a heating control $f(x)$ such that if (18) is replaced by (23), the system has a solution. If, in addition, a Newton heating control $g(x)$ is desired, i.e.,

$$-K_\ell \frac{\partial T}{\partial r} = h_\ell \left(T - g(x)\right), \quad x < -Q, \quad r = 1$$

instead of an infinite sink heating control $f(x)$, then we need only solve

$$-K_\ell \frac{\partial T}{\partial r} = h_\ell \left(f(x) - g(x)\right), \quad x < -Q, \quad r = 1$$

for $g(x)$ since $f(x)$ and T $\left(\text{and hence } \dfrac{\partial T}{\partial r}\right)$ have already been found.

In Section 2 of this paper, the control $f(x)$ is constructed such that the system (21)-(24) has a solution. The solution of System I is constructed in Section 3 and coupled with the results of Section 2. A numerical case study is given in Section 4 followed by various remarks and conclusions in Section 5.

2. CONSTRUCTION OF THE CONTROL SURFACE

If $f(x)$ is the desired control such that (21)-(24) has a solution, then define $\Theta(x,r)$ by $T(x,r) = \Theta(x,r) + f(x)$, $x < -Q$ and $0 < r < 1$. If we define $G(x) = P_\ell f' - f''$, $\mathfrak{a}(r) = A(r) - A(1)$ and $\mathfrak{b}(r) = B(r) - B(1)$, then

However, if (9), (10), (17) and (18) are adjoined to System I, the resulting system will be overposed and generally insolvable. To circumvent this dilemma, we propose the following. First solve System I and define

$$A(r) = T_3(x,r) \quad \text{and} \quad B(r) = \frac{\partial T_3}{\partial x}(x,r) \quad \text{for } x = -Q \text{ and } 0 < r < 1.$$

Then replace condition (18) by $T(x,1)=f(x)$, where $f(-Q)=A(1)$ and $f'(-Q)=B(1)$, but $f(x)$ otherwise is unknown. The control $f(x)$ is constructed such that the solution of the well posed System II:

$$(21) \quad \nabla^2 T = P_\ell \frac{\partial T}{\partial x}, \; x < -Q, \; 0 < r < 1$$

$$(22) \quad T(x,r) = A(r), \; x = -Q, \; 0 < r < 1$$

$$(23) \quad T(x,r) = f(x), \; x < -Q, \; r = 1$$

also satisfies

$$(24) \quad \frac{\partial T}{\partial x}(x,r) = B(r), \; x = -Q, \; 0 < r < 1 \; .$$

In other words, we wish to construct a heating control $f(x)$ such that if (18) is replaced by (23), the system has a solution. If, in addition, a Newton heating control $g(x)$ is desired, i.e.,

$$-K_\ell \frac{\partial T}{\partial r} = h_\ell \big(T - g(x)\big), \; x < -Q, \; r = 1$$

instead of an infinite sink heating control $f(x)$, then we need only solve

$$-K_\ell \frac{\partial T}{\partial r} = h_\ell \big(f(x) - g(x)\big), \; x < -Q, \; r = 1$$

for $g(x)$ since $f(x)$ and T $\left(\text{and hence } \frac{\partial T}{\partial r}\right)$ have already been found.

In Section 2 of this paper, the control $f(x)$ is constructed such that the system (21)-(24) has a solution. The solution of System I is constructed in Section 3 and coupled with the results of Section 2. A numerical case study is given in Section 4 followed by various remarks and conclusions in Section 5.

2. CONSTRUCTION OF THE CONTROL SURFACE

If $f(x)$ is the desired control such that (21)-(24) has a solution, then define $\Theta(x,r)$ by $T(x,r) = \Theta(x,r) + f(x)$, $x < -Q$ and $0 < r < 1$. If we define $G(x) = P_\ell f' - f''$, $\mathfrak{a}(r) = A(r) - A(1)$ and $\mathfrak{b}(r) = B(r) - B(1)$, then

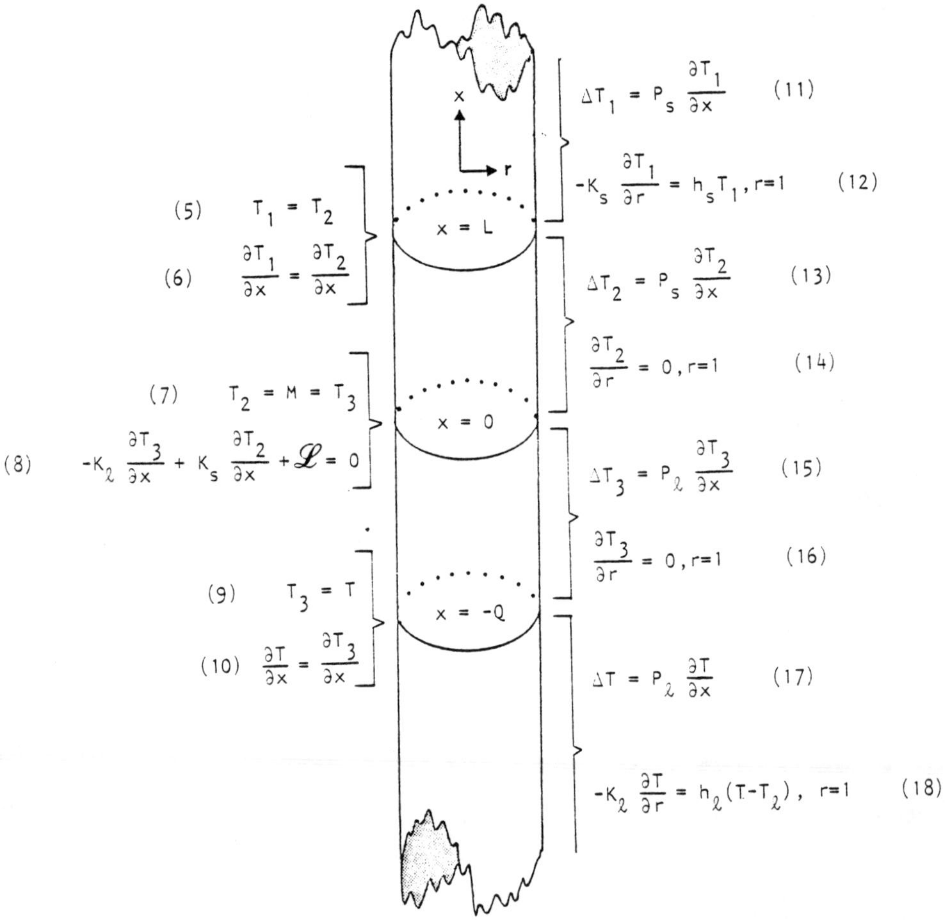
x
r
x = L
x = 0
x = -Q
(5) $T_1 = T_2$
(6) $\frac{\partial T_1}{\partial x} = \frac{\partial T_2}{\partial x}$
(7) $T_2 = M = T_3$
(8) $-K_\ell \frac{\partial T_3}{\partial x} + K_s \frac{\partial T_2}{\partial x} + \mathscr{L} = 0$
(9) $T_3 = T$
(10) $\frac{\partial T}{\partial x} = \frac{\partial T_3}{\partial x}$
$\Delta T_1 = P_s \frac{\partial T_1}{\partial x}$ (11)
$-K_s \frac{\partial T_1}{\partial r} = h_s T_1, r=1$ (12)
$\Delta T_2 = P_s \frac{\partial T_2}{\partial x}$ (13)
$\frac{\partial T_2}{\partial r} = 0, r=1$ (14)
$\Delta T_3 = P_\ell \frac{\partial T_3}{\partial x}$ (15)
$\frac{\partial T_3}{\partial r} = 0, r=1$ (16)
$\Delta T = P_\ell \frac{\partial T}{\partial x}$ (17)
$-K_\ell \frac{\partial T}{\partial r} = h_\ell (T - T_\ell), \; r=1$ (18)

Figure 1

(25) $\Theta_{xx} + \Theta_{rr} + \frac{1}{r}\Theta_r = P_\ell \Theta_x + G(x)$, $x < -Q$, $0 < r < 1$

(26) $\Theta(x,r) = a(r)$, and $\Theta_x(x,r) = b(r)$, $x = -Q$, $0 < r < 1$

and

(27) $\Theta(x,r) = 0$, $x < -Q$, $r = 1$.

Although the internal heat generation term $G(x)$ in (25) complicates the partial differential equation, the reduction of the nonhomogenous boundary condition (23) to that of a homogenous one (27) greatly simplifies the solution process. To solve (25), we first solve the Helmholtz eigenvalue problem:

$$(28)\quad \begin{cases} \nabla^2\psi(r) + \lambda^2\psi(r) = 0\ ,\ 0 < r < 1 \\ \psi(1) = 0\ ,\ \text{and}\ \psi'(0) = 0\ . \end{cases}$$

The solutions of (28) are the Bessel function $\psi_n(r)=J_o(\lambda_n r)$, $n=1,2,\ldots$, where the eigenvalues of (28) are the real positive roots of $J_o(\lambda_n)=0$. Moreover,

$$(29)\quad \int_0^1 \psi_N(r)\psi_M(r)\,r\,dr = \begin{cases} 0 \text{ if } N \neq M \\ \frac{1}{2} J_1^{\,2}(\lambda_M) \text{ if } M = N \end{cases}$$

Hence, if we let

$$(30)\quad \Theta(x,r) = \sum_{n=1}^{\infty} C_n(x)\psi_n(r)$$

then

$$(31)\quad \int_0^1 \Theta(x,r)\psi_m(r)\,r\,dr = C_m(x)\,\frac{J_1^{\,2}(\lambda_m)}{2}$$

Equations (30) and (31) form the dual integral transform pair of $\Theta(x,r)$:

$$\Theta(x,r) = \sum_{m=1}^{\infty} \frac{2\psi_m(r)\overline{\Theta}_m(x)}{J_1^{\,2}(\lambda_m)} \quad \text{and} \quad \overline{\Theta}_m(x) = \int_0^1 \Theta(x,r)\psi_m(r)\,r\,dr\ .$$

If we define

$$\overline{G}_m(x) = \int_0^1 G(x)\psi_m(r)\,r\,dr,$$

multiply (25) by $\psi_m(r)rdr$, integrate each side of the resulting equation from $r = 0$ to $r = 1$ and apply Green's Theorem, then

(32) $$-\lambda_m^2 \bar{\Theta}_m + \bar{\Theta}_m'' = P_\ell \bar{\Theta}_m' + \bar{G}_m \ .$$

Since

(33) $$\bar{\Theta}_m(-Q) = \int_0^1 a(r)\psi_m(r)\,r\,dr \quad \text{and} \quad \bar{\Theta}_m'(-Q) = \int_0^1 b(r)\psi_m(r)\,r\,dr \ ,$$

we need only solve the initial value problem (32) and (33) to compute $\bar{\Theta}_m(x)$. Because $a(1) = b(1) = 0$, we may expand $a(r)$ and $b(r)$ as

(34) $$a(r) = \sum_{n=1}^{\infty} a_n J_o(\lambda_n r) \quad \text{and} \quad b(r) = \sum_{n=1}^{\infty} b_n J_o(\lambda_n r)$$

where $a_n = \frac{2}{J_1^{\,2}(\lambda_n)} \int_0^1 J_o(\lambda_n r) a(r)\,r\,dr$ and $b_n = \frac{2}{J_1^{\,2}(\lambda_n)} \int_0^1 J_o(\lambda_n r) b(r)\,r\,dr$.

Combining (29), (33) and (34),

(35) $$\bar{\Theta}_m(-Q) = a_m \frac{J_1^{\,2}(\lambda_m)}{2} \quad \text{and} \quad \bar{\Theta}_m'(-Q) = b_m \frac{J_1^{\,2}(\lambda_m)}{2} \ .$$

For notational convenience, let $S_n = \sqrt{P_\ell^2 + 4\lambda_n^2}$. Then the solution of (32) is

(36) $$\bar{\Theta}_m(x) = \frac{J_1^2(\lambda_m)}{2S_m}\left(b_m - \frac{P_\ell - S_m}{2} a_m\right) \operatorname{Exp}\left(\frac{P_\ell + S_m}{2}(x + Q)\right)$$
$$+ e^{\frac{P_\ell + S_m}{2}x} \int_{-Q}^{x} \frac{\bar{G}_m(t)}{S_m} e^{-\frac{P_\ell + S_m}{2}t}\,dt$$
$$+ \frac{J_1^2(\lambda_m)}{2\,S_m}\left(\frac{P_\ell + S_m}{2} a_m - b_m\right) e^{\frac{P_\ell - S_m}{2}(x + Q)}$$
$$- \operatorname{Exp}\left(\frac{P_\ell - S_m}{2}x\right) \int_{-Q}^{x} \frac{\bar{G}_m(t)}{S_m} e^{-\frac{P_\ell - S_m}{2}t}\,dt \ .$$

Since we expect $\lim_{x \to -\infty} T(x,r) - f(x) = 0$, we require $\lim_{x \to -\infty} \bar{\Theta}_m(x) = 0$. Hence, if we assume $G(x) = P_\ell f' - f'' \to 0$ as $x \to -\infty$ (a realistic assumption), then (36) implies we must also require

$$(37)\qquad \lim_{x\to-\infty}\left[\frac{J_1^2(\lambda_m)}{2S_m}\left(\frac{P_\ell+S_m}{2}a_m-b_m\right)\operatorname{Exp}\left(\frac{P_\ell-S_m}{2}Q\right)-\int_{-Q}^{x}\frac{\overline{G}_m(t)}{S_m}e^{-\frac{P_\ell-S_m}{2}t}\,dt\right]e^{\frac{P_\ell-S_m}{2}x}=0\ .$$

For notational convenience, let $G(x-Q) = E(-x)$, $x < 0$. Then (37) implies

$$(38)\qquad \frac{\lambda_m J_1(\lambda_m)}{2}\left(b_m-\frac{P_\ell+S_m}{2}a_m\right)=\int_0^\infty E(t)\,e^{-\frac{S_m-P_\ell}{2}t}\,dt\ .$$

Because we assume $G(x) \to 0$ as $x \to \infty$, we expand $E(t) = \sum_{k=1}^{\infty} a_k e^{-kt}$.

However, for computational purposes, we replace $E(t)$ in (38) by $\sum_{k=1}^{N} a_k e^{-kt}$ and then solve for $a_1, a_2, \ldots, a_N$. To approximate $f(x)$, we then solve

$$\begin{cases}\sum_{k=1}^{N} a_k e^{-kt} = -P_\ell g'(t) - g''(t),\ t > 0\\ g(0) = A(1)\ ,\ \text{and}\ g'(0) = -B(1)\end{cases}$$

for $g(t)$ and set $f(t) = g(-Q-t)$, $t < -Q$.

3. SOLUTION OF SYSTEM I

To complete the method described in Section 2, the functions $A(r)$ and $B(r)$ must be known. Because

$$T_3(x,r) = A(r)\quad\text{and}\quad \frac{\partial T_3}{\partial x}(x,r) = B(r)\ \text{at}\ x = -Q\ ,$$

we need only solve System I (equations (5)-(8), (11)-(16)), combined with (19) and (20). Assume $h_s \neq 0$ and let β_n and γ_n be the real roots of

$$K_s\beta_n J_1(\beta_n) - h_s J_0(\beta_n) = 0\quad\text{and}\quad J_1(\gamma_n) = 0\ \text{with}\ \gamma_1 = 0\ .$$

Table 1

n	B_n	C_n	D_n	E_n
1	.6146+02	-.5146+02	.3073+02	-.2073+02
2	-.1017-01	.1017-01	-.1016-01	.1016-01
3	.1928-03	-.1928-03	.1927-03	-.1927-03
4	-.4959-05	.4959-05	-.4960-05	.4959-05
5	.1463-06	-.1463-06	.1463-06	-.1463-06
6	-.4670-08	.4670-08	-.4700-08	.4670-08
7	.1570-09	-.1570-09	.1570-09	-.1570-09

To avoid numerical instabilities, we approximate the coefficients $a_1,\ldots,a_5$, b_1, .., and b_5 in (34) by solving the overposed systems of linear equations

$$\sum_{n=1}^{5} a_n J_0(\lambda_n r_k) = \tilde{a}(r_k) \quad \text{and} \quad \sum_{n=1}^{5} b_n J_0(\lambda_n r_k) = \tilde{b}(r_k),\ k = 1,\ \ldots,\ 10$$

for $a_1,\ldots,a_5,b_1,\ldots,b_5$ by least squares methods where $\tilde{a}(r) = \tilde{A}(r) - \tilde{A}(1)$, $\tilde{b}(r) = \tilde{B}(r) - \tilde{B}(1)$ (see (42)) and $r_k = k(k-1)/10$.

For N = 3, the control f(x) generated by the method described in Section 2 is tabulated in Table 2.

Table 2

x	f(x)	x	f(x)
-1.0	14.055	- 6.5	107.70
-1.5	14.967	- 7.0	118.57
-2.0	13.405	- 7.5	128.55
-2.5	14.701	- 8.0	137.65
-3.0	20.701	- 8.5	145.94
-3.5	30.600	- 9.0	153.47
-4.0	42.918	- 9.5	160.31
-4.5	56.359	-10.0	166.50
-5.0	70.012	-10.5	172.11
-5.5	83.307	-11.0	177.19
-6.0	95.922		

Separating variables, the approximate solutions of (11)-(16), (19) and (20) are of the form

$$(39)\quad T_1(x,r) = \sum_{n=1}^{K} A_n J_0(\beta_n r)\,\mathrm{Exp}\left(\frac{P_s - \sqrt{P_s^2 + 4\beta_n^2}}{2}\,x\right)$$

$$(40)\quad T_2(x,r) = \sum_{n=1}^{K} J_0(\gamma_n r)\left[B_n\, e^{\frac{P_s - \sqrt{P_s^2 + 4\gamma_n^2}}{2}x} + C_n\, e^{\frac{P_s + \sqrt{P_s^2 + 4\gamma_n^2}}{2}x}\right]$$

and

$$(41)\quad T_3(x,r) = \sum_{n=1}^{K} J_0(\gamma_n r)\left[D_n\, e^{\frac{P_\ell - \sqrt{P_\ell^2 + 4\gamma_n^2}}{2}x} + E_n\, e^{\frac{P_\ell + \sqrt{P_\ell^2 + 4\gamma_n^2}}{2}x}\right]$$

Substituting (39)-(41) in (5)-(8) and using the orthogonality of J_0, the coefficients A_n, B_n, C_n, D_n and E_n in (39)-(41) are easily computed as the solution of a system of linear equations.

The two functions, A(r) and B(r), required in Section 2, are then computationally approximated by

$$(42)\quad \begin{cases} \tilde{A}(r) = \displaystyle\sum_{n=1}^{K} J_0(\gamma_n r)\left[D_n\, e^{-\frac{P_\ell - \sqrt{P_\ell^2 + 4\gamma_n^2}}{2}Q} + E_n\, e^{-\frac{P_\ell + \sqrt{P_\ell^2 + 4\gamma_n^2}}{2}Q}\right] \\ \tilde{B}(r) = \displaystyle\sum_{n=1}^{K} J_0(\gamma_n r)\left[D_n \frac{P_\ell - \sqrt{P_\ell^2 + 4\gamma_n^2}}{2}\,\mathrm{Exp}\left(-\frac{P_\ell - \sqrt{P_\ell^2 + 4\gamma_n^2}}{2}Q\right)\right. \\ \qquad\left. + E_n \frac{P_\ell + \sqrt{P_\ell^2 + 4\gamma_n^2}}{2}\,\mathrm{Exp}\left(-\frac{P_\ell + \sqrt{P_\ell^2 + 4\gamma_n^2}}{2}Q\right)\right] \end{cases}$$

4. NUMERICAL CASE STUDY

In this section we present a numerical case study of the methods described in Sections 2 and 3. Let $P_\ell = .1$, $P_s = .2$, $h_s = K_s = K_\ell = h_\ell = L = Q = \mathscr{L} = 1.0$, and $M = 10.0$. For $K = 7$, the various B_n, C_n, D_n and E_n are given in Table 1.

5. CONCLUSIONS

In this paper we have presented a method for constructing a control surface temperature such that the partial differential equations and boundary conditions describing Bridgman-Stockbarger crystal growth are not overposed if we require the solid-melt interface to be flat at x=0. This method can also be applied to symmetric float zone methods described by:

$$(43)\quad \nabla^2 T_s = P_s \frac{\partial T_s}{\partial x}, \quad 0 < r < 1, \quad x < -Q \text{ or } x > Q$$

$$(44)\quad \nabla^2 T_\ell = P_\ell \frac{\partial T_\ell}{\partial x}, \quad -Q < x < Q, \quad 0 < r < 1$$

$$(45)\quad -K_\ell \frac{\partial T_\ell}{\partial r} = h_\ell\big(T_\ell - C(x)\big), \quad -Q < x < Q$$

$$(46)\quad -K_\ell \frac{\partial T_\ell}{\partial x} + K_s \frac{\partial T_s}{\partial x} + \mathscr{L}_{-Q} = 0, \quad x = -Q, \ 0 < r < 1$$

$$(47)\quad -K_\ell \frac{\partial T_\ell}{\partial x} + K_s \frac{\partial T_s}{\partial x} + \mathscr{L}_{Q} = 0, \quad x = Q, \ 0 < r < 1$$

and

$$(48)\quad T_\ell = M = T_s \quad \text{at} \quad x = \pm Q .$$

where C(x) is a priori known. Equations (44), (45) and (48) can be solved by the dual integral transform methods.

Setting $B(r) = \frac{\partial T_s}{\partial x}$, (see (46)) and A(r) = M, we may use the method of Section 2 to construct the cooler control f(x), i.e., $T_s = f(x)$, r = 1, x < -Q.

As a numerical aside, the reader is warned not to use large N in the approximation of E(t) because the resulting linear system (see (38)) is similar to the Hilbert matrix system for such N. This, of course, is due to our method of expanding E(t) in terms of e^{-kt}, k = 1,2,···.

6. REFERENCES

1. Chang, C. E., and Wilson, W.R., "Control of Interface Shape in the Vertical Bridgman-Stockbarger Technique," Journal of Crystal Growth, Vol. 21, pp. 135-140 (1974).

2. Foster, L. M., "Material and Process Constraints for a Flat Interface in the Bridgman-Stockbarger Technique," NASA CR-161511, Sec. VIII, (October 1980).

THREE DIMENSIONAL FINITE ELEMENT ANALYSIS OF DEFLECTION AND BENDING MOMENT DISTRIBUTION IN A DOUBLE CIRCULAR ARC GEAR TOOTH

Dr.M.A.K. FAHMY

INSPECTORATE
International Inspection and Consulting Gmbh and Co KG
Grupellostr. 19
D-4000 DÜSSELDORF 1
W. Germany

ABSTRACT :

The previous results of stress distribution analysis in involute gear profiles, both symmetric and assymmetric, using 2-D finite elements are reviewed.

The necessity for three-dimensional study of the deflection and bending moment distribution in gears, particularly the double circular arc gears, is explained and the suitability of the 3-D finite element method for such a special problem is discussed and executed using SAP-V computer program. Results show that the values obtained by this method are very close to those obtained by holographic interferometry.

A comparision with prvious results obtained on conventional helical involute gears and single-or double circular arc tooth profile gears, shows that the latter is able to carry 1.5 to 2.0 times the load carried by the single circular arc tooth profile gears.

A form factor for the double circular arc tooth profile gears is experimentally and theoretically derived and introduced into the load capacity formula.

THREE DIMENSIONAL FINITE ELEMENT ANALYSIS OF DEFLECTION AND BENDING MOMENT DISTRIBUTION IN A DOUBLE CIRCULAR ARC GEAR TOOTH

Dr. M.A.K. FAHMY

SUMMARY :

The previous results of stress distribution analysis in involute gear profiles, both symmetric and assymmetric, using 2-D finite elements are reviewed.

The necessity for three-dimensional study of the deflection and bending moment distribution in gears, particularly the double circular arc gears, is explained and the suitability of the 3-D finite element method for such a special problem is discussed and executed using SAP-V computer program. Results show that the values obtained by this method are very close to those obtained by holographic interferometry.

A comparison with previous results obtained on conventional helical involute gears and single- or double circular arc tooth profile gears, shows that the latter is able to carry 1.5 to 2.0 times the load carried by the single circular arc tooth profile gears.

A form factor for the double circular arc tooth profile gears is experimentally and theoretically derived and introduced into the load capacity formula.

INTRODUCTION :

In the solution of stress problems, the finite element method is considered to be a versatile and efficient solution technique. The static problem is formulated in variational form :

$$\delta \pi = \delta U - \delta W$$

the potential energy being :

$$U = 1/2 \int \bar{\bar{\varepsilon}}\, \bar{\bar{\sigma}}\ dv$$

and

$$W = \int_v [u] \{F\}\ dv$$

This variational problem must be reduced to a set of linear equations by the finite element displacement method. The displacement in an element u is expressed in terms of nodal displacements $[\delta_i]$

$$[u] = [N] [\delta_i]$$

This leads then to an approximated problem

$$[K] [\delta_i] = \{F\}$$

This is the equilibrium equation of the structure and is to be solved by appropriate methods. The structure is partitioned in a number of finite elements depending on the type of structure.

FINITE ELEMENT METHOD APPLIED TO GEAR PROBLEMS :

Although the finite element method has been used successfully in several stress analysis methods, it has not found a wide application in gear problems. The only knownapplication of the finite element method to gear design is the work of Wilcox (5), Wilcox and Coleman (6) and Wallace and Seireg (7), where this technique has been used to study the fillet stress and stress distribution in involute gear profiles, treating them as a two dimensional problem, because an involute gear has gear contact characteristics independent of the gear width.

Numerical results obtained for the problems considered show good agreement with experimental values, and were the basis for improved analytical formulas.

A type of gear discretization, as used in (6,7), is shown in fig.1.2

FINITE ELEMENT STUDY OF THE WILDHABER-NOVIKOV CIRCULAR ARC GEARS :

Wildhaber-Novikov gears are helical, with circular arc tooth profiles. They have proved to be at least 3 to 5 times stronger than involute gears (11, 12, 13) and are actually replacing involute gears in many aircraft, marine and industrial applications calling for high load-carrying capacity. The complexity of the bending stress distribution in double circular arc tooth profile gears and the stubbiness of the tooth, the effect of gear deformation, the large face width, the convex-concave profile of the tooth, the semi-ellipsoidal contact load distribution and the varying tooth stiffness due to the change in the helix angle make the problem rather complicated and no theoretical analysis is known to take all these factors into consideration. Thus it was necessary to make a full three-dimensional finite element analysis of the problem, in order to evaluate maximum deflection and stress and also the bending moment distribution in the tooth flanks.

In this research the analysis has been carried out for static conditions only and the effect of tangential frictional forces at the contact surface, though small in magnitude for Wildhaber-Novikov gears, was neglected.

A 3-dimensional solid element, type PR - 20, fig. 3, was chosen in an earlier study of the single circular arc gear, the gear tooth being discretised in 40 type PR - 20 elements. For a better representation of the varying elliptical loads on the tooth surface, the authors prefer to use 80 PR - 20 elements, as shown on fig. 4.

The problem was also studied with 120 PR - 20 elements but results were not significantly different from those obtained with 80 PR - 20 elements while requiring extended core memory of the computer.
A central memory of 220 000 words was necessary and calculation time amounted to 2000 CP seconds. The program used was SAP- , (4).

RESULTS OBTAINED FOR DIFFERENT LOADING CONDITIONS :

The results obtained from this 3-dimensional finite element study are reasonably close to photoelastic results and to those obtained by 3-D-holographic interferometry (14), fig.5, show the differences in results obtained.
The application of this method seems justified by the reasonable number of elements used for the analysis and the computer time required.

Fig.6a, 6b, 6c, 6d, show the deflection of the tooth model at starting of contact under a load of 1 newton applied in addendum and dedendum points of contact for four loading cases starting from 0.25 times full load to full load for helix angles of 40, 30, 20, 10 degrees. It is clear from the curves that the deflection is maximum at the start of tooth contact due to the low tooth stiffness, while the tilting effect is reduced by decreasing the helix angle due to the fact that the load is uniformly distributed over a wider area of contact at lower values of the helix angle.

Fig. 7a, 7b, 7c, 7d, show deflection of the tooth in the central zone of contact under a load of 1 newton applied in addendum points of contact for load cases starting from 0.25 times full load to full load at helix angles of 10, 20, 30, 40 degrees.

The curves show that the deflection is much less than that for start or end of contact and this is due to the larger stiffness in the central area of contact and also the deflection is increased with the increase of the helix angle.

The distribution of deflection along the path of contact when the load moves from start to end of contact under 1 newton load at addendum and dedendum points of contact and 30 degrees helix angle is represented on fig. 8.

The curves show clearly the effect of stiffness variation, which is less at start, giving higher values of deflection, while being higher in the middle, giving lower values of deflection in the central zone of contact.

The three-dimensional tooth deformation for load applied at start and central zones of tooth contact is dramatically demonstrated on fig. 9a and 9b.

RELATION BETWEEN DEFLECTION AND APPLIED LOAD :

The deflection under the same conditions of pressure angle, helix angle and for the same material are considered to be proportional to the applied load and inversely proportionally to the normal module.

This proportionality has been studied for more than 80 different load cases and more than 45 000 values were obtained for the tooth deflection. From these results a simple mathematical model for tooth deflection was obtained :

$$W = S \cdot \eta \cdot P \cdot m_n^{-1} \quad \mu m$$

With S material factor, η deflection coefficient m_n normal module, P the combined load in addendum and dedendum points of contact and W the deflection.

Comparison with previous experimental values obtained from three-dimensional holographic interferometry (14) shows an estimated difference of 10 percent. Comparison was also made with photoelastic and strain gauge measurements and here also reasonable agreement was found.

RESULTS OF THE STRESS CALCULATIONS ON THE DOUBLE CIRCULAR ARC TOOTH PROFILE :

Contact and bending stresses have been studied for the various load cases and various helix angles and fig. 10 shows the results of the stress calculations for a double circular arc tooth profile gear.

Line (ß) shows the bending stress for contact at the centre of face width.

Line (1) is the result of simplified calculations corresponding to the line (ß) and is obtained assuming that the force is concentrated at the centre of addendum and dedendum areas of contact.

Line (β_{WE}) is obtained from Wellauer's method mentioned in (15).

Curve (β_T) which is calculated assuming that the load is applied along the width of the addendum elliptical contact area shows better results than Wallauers method.

From these results it follows that for small helix angle the bending moment at the tooth root must be calculated considering the area where the tooth load on addendum and dedendum is applied.

COMPARISON WITH STRESSES OF INVOLUTE AND CIRCULAR ARC GEARS :

Fig. 11 a, 11 b, show bending and contact stresses of single and double circular arc and involute profile gears, for various helix angles.

There are many methods of stress calculation for involute helical gears, but in this research comparison was made with the stresses calculated from accepted formulas used in gear design and the values obtained from 3-dimensional finite element analysis of stresses in the root of the tooth model. For the involute gears used for comparison purposes, helix angle and normal module are chosen to be equal to the corresponding values of the single and double circular arc tooth profile gears, and to be of standard tooth dimensions and 25 degrees normal pressure angle.

The contact stress of single and double circular arc gears is larger than that of the involute gears for large helix angle. But in the case of smaller helix angle, the contact stress becomes smaller than that of the involute gears.

From the above mentioned stress calculations on double circular arc tooth profile model and for lower helix angles the stresses calculated show that this type has a load carrying capacity 1.5-2.0 times that of single circular arc tooth profile gears.

LOAD CARRYING CAPACITY :

The double circular arc tooth gear load capacity was studied in this research under variable static loading conditions and the contact and bending stresses are calculated by SAP-5, in addition to the deflection as determined from the 3-D finite element model.

The study was extended to show the effect of a change of helix angle, normal module, face width and profile addendum and dedendum radius.

All the values obtained from the measurements of deflection and contact-bending stresses were compared to the theoretical and experimental values obtained from 3-dimensional holographic interferometry and strain gauge measurements and 3-dimensional photoelastic measurements.
The results found allow to set up a load capacity formula indicating that the main factors affecting the design of double circular arc gears are the face width helix angle, pitch circle diameter and number of teeth, normal module and finally a form factor depending on the addendum and dedendum profile radius, which was determined during measurements and found to be Y = 0.752 for a helix angle of 28.3 degree, a normal module of 4 mm, and 27 degree pressure angle in the normal plane.

The maximum load capacity of double circular are profile teeth is given by:

$$W_{t_{max}} = 0.465 \cdot S_{b_{max}} \cdot Y \cdot \frac{F^2}{N} \cdot \frac{D_p}{m_N} \cdot \cos^2 \beta$$

where $S_{b_{max}}$ = Maximum bending stress.

CONCLUSION :

The main results obtained during this investigation are summarized as follows :

1. The three dimensional finite element analysis for this complex geartooth problem gives results comparable to the experimental values obtained so far, but produces more information on the actual tooth deformation when using a reasonable number of finite elements.

2. The value of bending stress calculated from E.G. Wallauer's method is always greater than that obtained from experimental methods, when the helix angle is small, the difference becomses very large.

3. Calculated bending stress of single and double circular arc tooth profile gears is always greater than that of the involute gears in the range of helix angles 10 to 40 degrees.

4. Contact stress becomes smaller than that of involute gears for a helix angle smaller than 30 degrees.

5. Decreasing the helix angle reduces the tilting effect, leading to better contact accuracy.

6. A maximum load capacity formula for the double circular arc geartooth was experimentally and theoretically derived and a form factor determined.

7. Comparative study between conventional involute gears and single or double circular arc tooth profile gears shows that the latter can carry 1.5 - 2.0 times the load which single circular arc gears of the same size and material can carry.

REFERENCES :

1. Zienkiewicz. O.C. "The Finite-Element Method" McGraw-Hill Book Co., Inc., New York, 1977.
2. Huebner, K.H., "The Finite-Element Method for Engineers" John Wiley and Sons, N.Y., 500 pages, 1975.
3. Bathe, K.S., and Wilson, W.L., "Numerical Methods in Finite-Element Analysis "Prentice Hall, New York, 1976.
4. Bathe, K.J., Wilson, E.L., Peterson, E.F., "SAP-IV a structural analysis program for static and dynamic response of linear systems "Report No. 73-11, College of engineering Berkely, USA, 1973.
5. Wilcox, L.E., "Generation of simplified stress formula using the Method of Finite-Elements " Gleason Work's Engineering Research Report, December 1970.
6. Wilcox, L.E., and Coleman, W., "Application of Finite-Elements to the analysis of gear tooth stresses". Transactions of ASME, Journal of Engineering for industry Vol 95, series B, No. 4, November 1973, P1139.
7. Wallace, D.B. and Seireg, A., "Computer Simulation of dynamic stress, deformation and fracture of gear teeth" Transactions of the ASME, Journal of Engineering for Industry, Vol 95, series B, no. 4. November 1973, P 1108
8. Othe, S., "Finite-Element Analysis of elastic Contact Problems" Bulletin of ASME, Vol 16, No. 95, May 1973, P 979.
9. Cangal, M.D. "Direct Finite - Element Analysis of elastic contact problems" International Journal of numerical methods in engineering, Vol. 5, No. 1, P 147
10 Wilson, E.A., and Parson, B., "Finite - Element Analysis of Elastic Contact Problems Using Differential Displacements" ibid, Vol 2, 1970, P 387.
11 C.J. Klein "The Wildhaber-Novikov system of Gearing" DME/NAE Quarterly Bulletin, National Research Council of Canada, Ottawa, April 1965.
12 T. Allan, "Some Aspects of the design and performance of Wildhaber-Novikov Gearing. Proceedings of the Institution of Mechanical Engineers, Vol. 179 P I 1, 1964 - 65
13 NP Chironis "Design of Novikov Gears" Product engineering, September 17, 1962, P 91
14 Fahmy, M.A.K., Jonckheere, R.E., "Deflection and Bending moment distribution of gear teeth of double circular arc profile using three-dimensional holographic interferometry" ASME, Transmissions and Gearing Conference, San Francisco, California, August 18-21, 1980
15 Wellauer, E.G., and Seireg. A, "Bending strength of gear teeth by Cantilever plate theory" Trans. ASME series B, Vol. No. 82, No. 3, 1960.

NOMENCLATURE.

D_p : pitch circle diameter
F : face width
m_N : normal module, mm
N : number of teeth
P : total load of addendum and dedendum
S_{bmax} : maximum bending stress
U : strain energy
w : deflection, µm
W : energy of loading
W_{tmax} : maximum tangential load, N
Y : form factor
$[u]$: displacement vector
$[\delta_i]$: nodal displacement vector
$[N]$: matrix of shape functions
$[K]$: stiffness matrix
$\{F\}$: force vector
π : total potential energy
$\bar{\bar{\varepsilon}}$: strain tensor
$\bar{\bar{\sigma}}$: stress tensor
β : helix angle
η : deflection coefficient, $\mu m\ N^{-1}\ m_N$

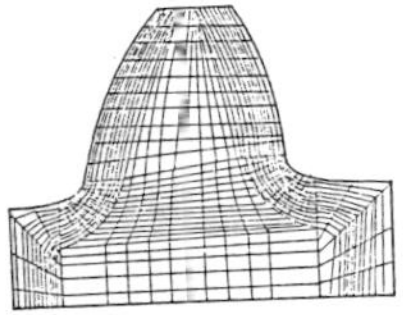

NO. OF ELEMENTS USED : 840
(WILCOX AND COLEMAN)
Fig. 1

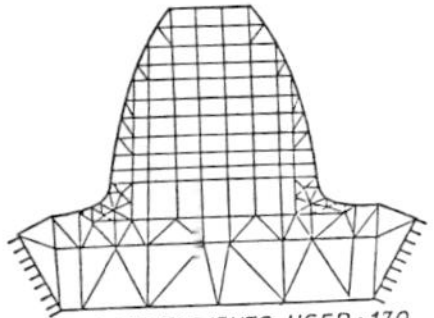

NO. OF ELEMENTS USED : 170
NO. OF NODES : 147
(WALLACE AND SEIREG)
Fig 2

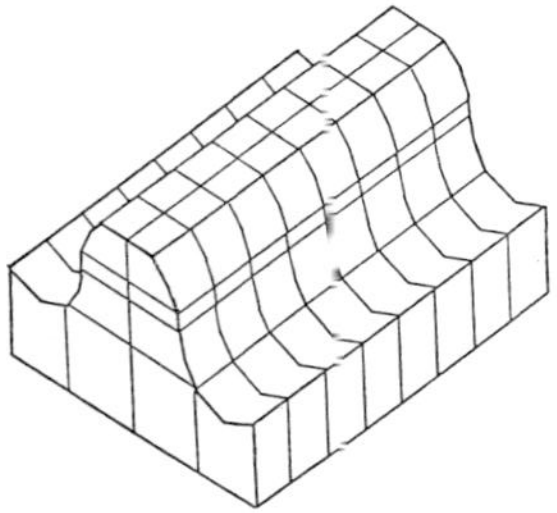

Fig.4 80 P R - 20 Elements model before loading

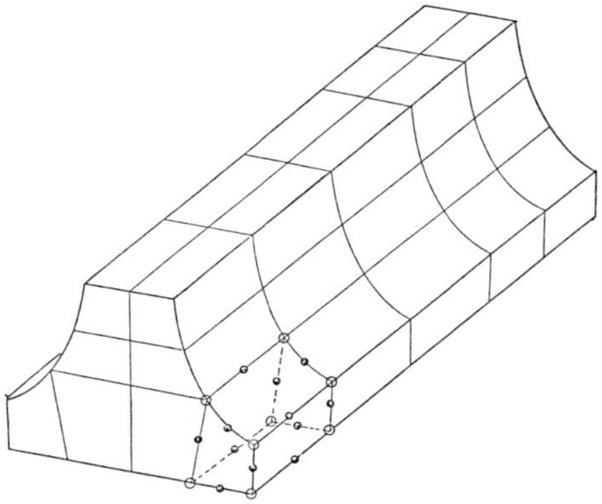

Fig.3 Simple finite element model of a conformal tooth showing one 20 node hexahedral element as used for the finite element stress analysis

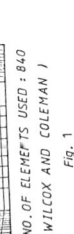

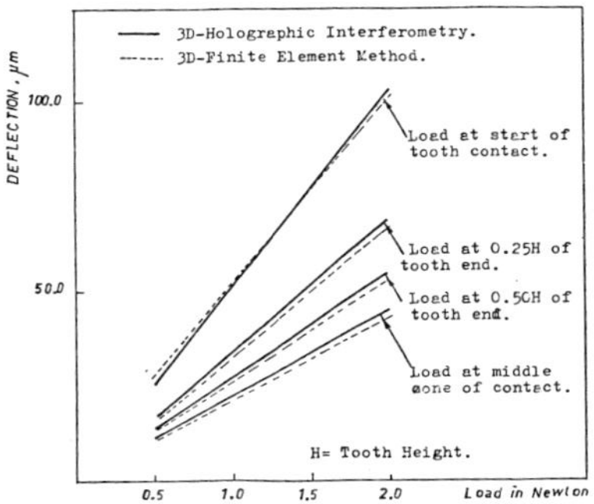

Fig.5 Shows the differences in measurements of deflection with load at addendum and dedendum pt. of contact at different positions of load when using 3D-holographic interferometry and 3D-finite element method.

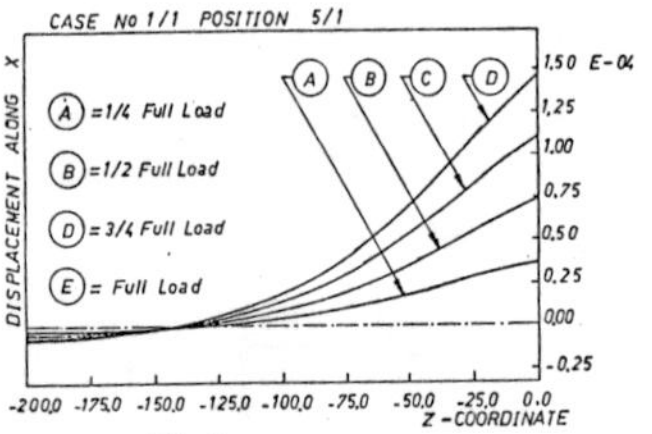

Fig.6a

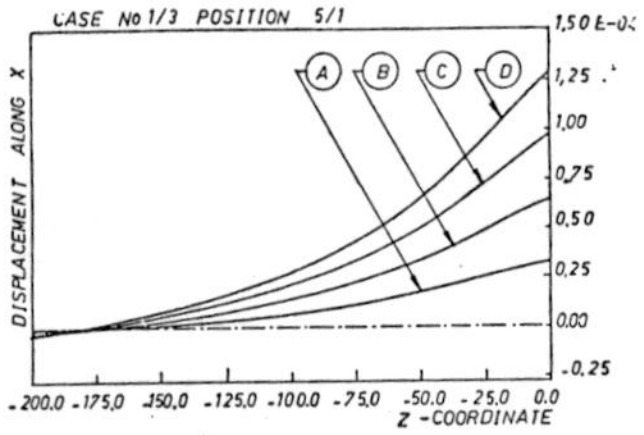

Fig.6c

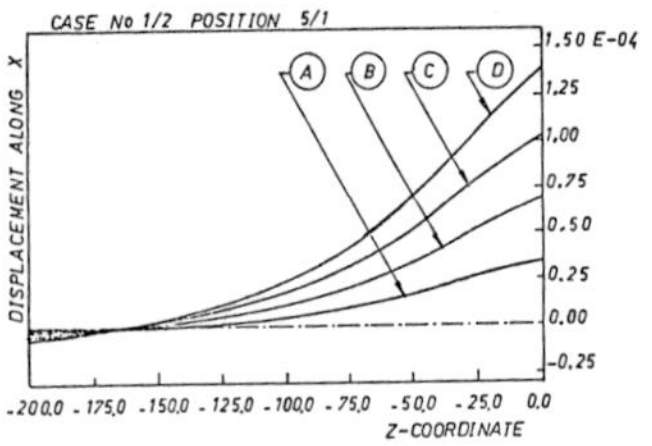

Fig.6b

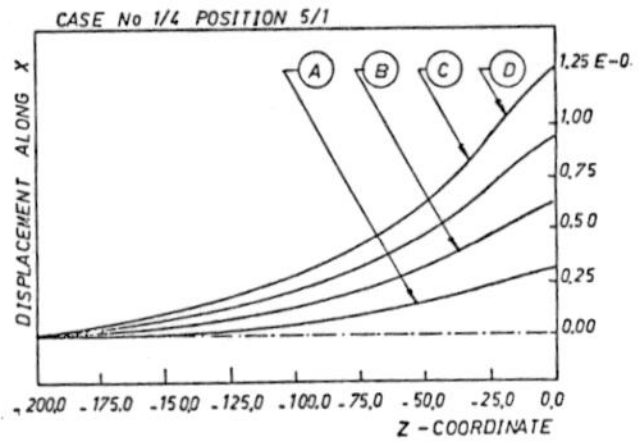

Fig.6d

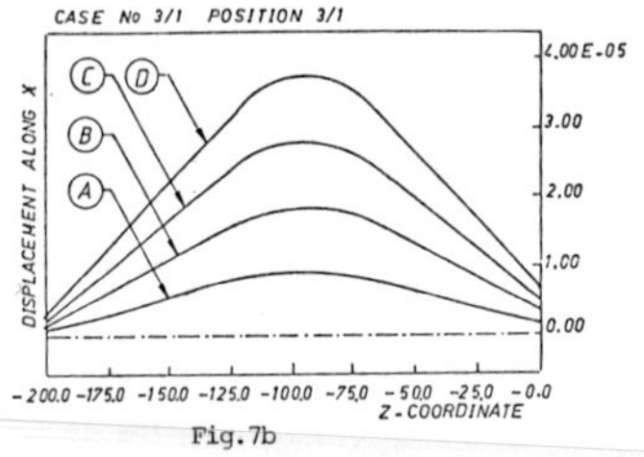

Fig.7b

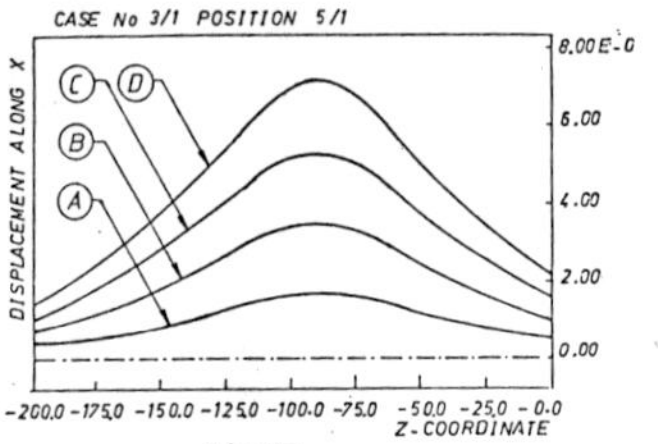

Fig.7d

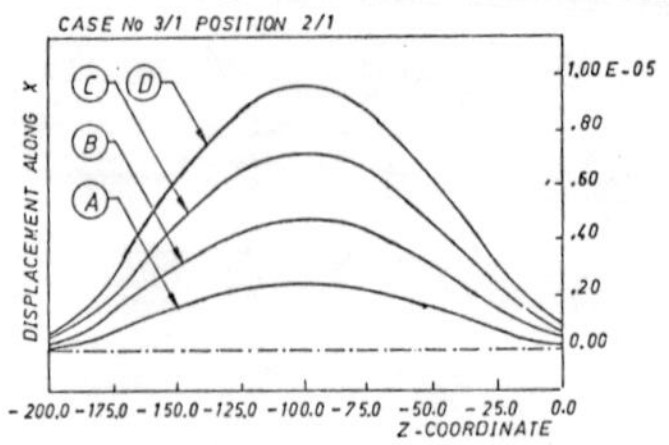

Fig.7a

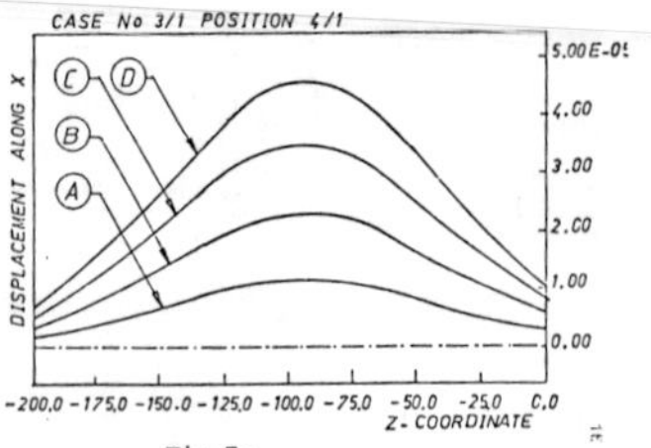

Fig.7c

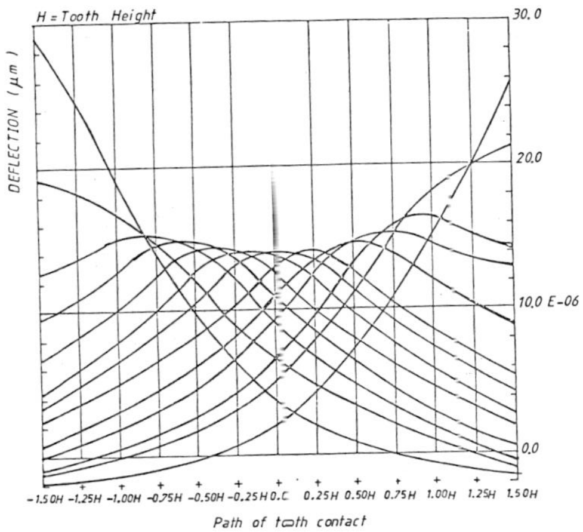

Fig. 8 Distribution of deflection along the path of tooth contact

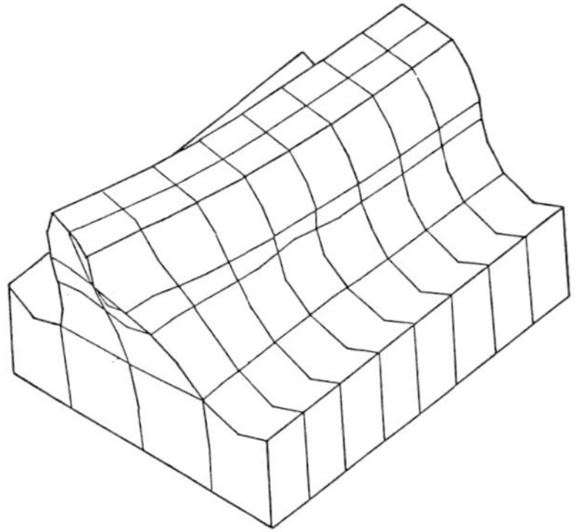

Fig. 9a Three-dimensional finite element model deformed under load applied at start of tooth contact

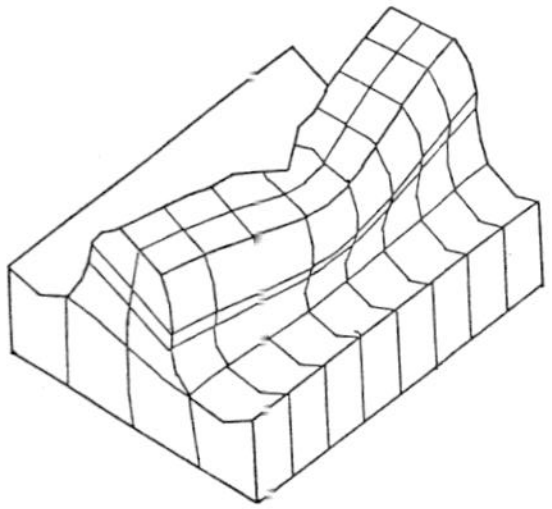

Fig. 9b Three-dimensional finite element model deformed under load applied at middle zone of contact

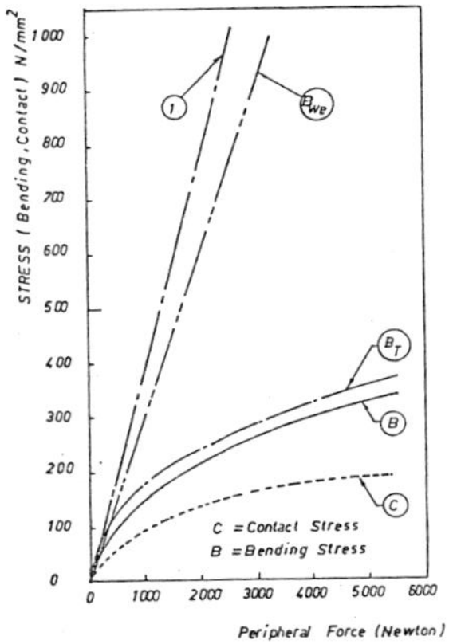

Fig. 10 Results of stress calculations on double circular - arc tooth - profile steel gears.

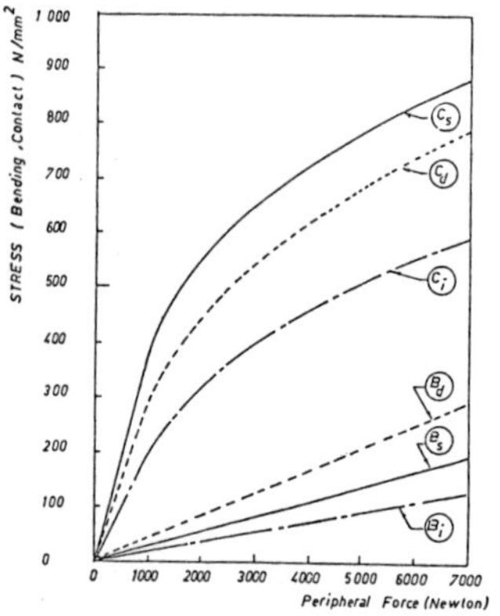

Fig.11a Bending and contact stresses of both single -double circular - arc tooth - and involute profile gears in case of large helix angle .

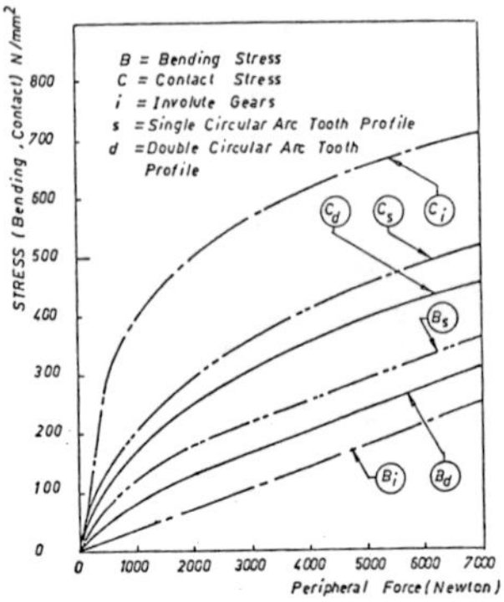

Fig.11b Bending and contact stresses of both single -double circular - arc tooth - and involute profile gears in case of smaller helix angle .

INCREMENTAL STIFFNESS MATRICES FOR NONLINEAR STRUCTURAL ANALYSIS

by Robert K. Wen
Department of Civil Engineering
Michigan State University

ABSTRACT

It is well known that when the strain energy of a nonlinear elastic structure is written as a quartic function of the generalized coordinates, the strain energy, equilibrium equation, and incremental equilibrium equation can all be related in terms of (in addition to the linear stiffness matrix) two incremental stiffness matrices [N1] and [N2] which contain, respectively, linear and quadratic terms of the generalized coordinates. Although the relations seem plausible and have been used by researchers, their validity is not obvious and no proof seems to exist in the literature. A formal proof is presented herein for the general case in which the strain energy function may be a function of the generalized coordinates of higher order than four. Numerical results illustrating the use of these matrices for certain new nonlinear finite elements in structural analysis are also presented. They include the equilibrium paths of an arch space frame and a curved beam, and the vertical "buckling load" of an arch as affected by a constant horizontal load.

INTRODUCTION

In discussing several approaches that may be taken for the analysis of nonlinear elastic structures subjected to conservative loads using Lagrangian displacement coordinates, Mallet and Marcal [1] introduced the notion [N1] and [N2] to represent the "incremental stiffness matrices." Both matrices have the property of "repeatability" in the following sense.

Potential Energy Expression:

$$\pi = \{q\}^T \left(\frac{1}{2}[K] + \frac{1}{6}[N1] + \frac{1}{12}[N2]\right)\{q\} - \{q\}^T\{P\} \tag{1}$$

Equilibrium Equation:

$$\left([K] + \frac{1}{2}[N1] + \frac{1}{3}[N2]\right)\{q\} - \{P\} = \{0\} \tag{2}$$

Linear Incremental Equilibrium Equation:

$$([K]+[N1]+[N2])\{\Delta q\} - \{\Delta P\} = \{0\} \tag{3}$$

in which {q} is the displacement vector, {P} is the load vector, [K] is the linear stiffness matrix, and [N1] and [N2] contain, respectively, linear and quadratic terms of the displacement variables.

The notation has provided a tie among the several formulations of the nonlinear elastic structural problem and can also aid in the setting up of the appropriate equations for computations. It has received considerable attention and has been adopted by a number of researchers [2,3,4,5,6]. However, no method was given in Ref. 1 for the construction of these matrices. As was pointed out in Ref. 7, for a given expression of the strain energy, there is more than one way in which [N1] and [N2] may be written for Eq. 1, and among them there is but one way that Eqs. 2 and 3 are also valid simultaneously. In Ref. 7 is shown a method to construct the correct [N1] and [N2]. It had also been mentioned in Ref. 3 that

$$[N1] = [n_{1,ij}] \; ; \; n_{1,ij} = \partial^2 U_3/\partial q_i \partial q_j \tag{4a}$$

$$[N2] = [n_{2,ij}] \; ; \; n_{2,ij} = \partial^2 U_4/\partial q_i \partial q_j \tag{4b}$$

in which U_3 and U_4 are parts of the strain energy defined in Eq. 5. By considering the second variation of the potential energy function, the preceding definitions lead directly to Eq. 3, but it does not seem obvious that Eqs. 1 and 2 would necessarily follow. (In this connection it may be mentioned that similar expressions to Eqs. 4, given in Ref. 8 and also quoted in Ref. 7, contained typographical errors of factors of 1/2 and 1/3.) As far as is known to the author, no proof of the relation can be found in the existing literature.

The purpose of this paper is to present a formal proof of the repeatability property of the incremental stiffness matrices for the general case in which the strain energy may contain terms higher than fourth order, and also to present numerical results illustrating the use of these matrices connected with certain new finite elements for structural analysis. The results include the equilibrium paths of an arch space frame and a curved beam, and the vertical "buckling load" of an arch as affected by a constant horizontal load.

STATEMENT OF PROBLEM

Let the strain energy of a nonlinear elastic structure with n degrees of freedom be denoted by U, which contains nonlinear terms to order p.

$$U = U_2 + U_3 + U_4 + \dots + U_m + \dots + U_p \tag{5}$$

in which U_i, i=2, ..., p denotes the sum of those terms in U for each of which the sum of the exponents of the variables is equal to i. It is desired to show that:

$$\text{given} \quad [N_{m-2}] = [n_{m-2,jk}]$$

$$n_{m-2,jk} = \frac{\partial^2 U_m}{\partial q_j \partial q_k}, \quad 2 \leq m \leq p \tag{6}$$

then the strain energy, the equilibrium equations, and the linear incremental equations are given, respectively, by:

$$U = \{q\}^T \left(\sum_{i=0}^{p-2} \frac{[N_i]}{(i+2)(i+1)} \right) \{q\} \tag{7}$$

$$\left(\sum_{i=0}^{p-2} \frac{1}{i+1} [N_i] \right) \{q\} = \{P\} \tag{8}$$

$$\left(\sum_{i=0}^{p-2} [N_i] \right) \{\Delta q\} = \{\Delta P\} \tag{9}$$

For uniformity of notation the linear stiffness matrix has been represented by N_0 instead of the usual symbol K.

Eq. 9 follows immediately from the second variation of the potential energy and the definition given by Eq. 6. In order to show that Eqs. 7 and 8 follow from Eq. 6, it suffices to show that, for a representative term in Eq. 5, say U_m, the following holds:

$$\text{given} \quad [N_{m-2}] = \left[\frac{\partial^2 U_m}{\partial q_i \partial q_j}\right] \tag{10}$$

$$\text{then} \quad \frac{1}{m(m-1)} \{q\}^T [N_{m-2}] \{q\} = U_m \tag{11}$$

$$\left(\frac{1}{m-1}\right) [N_{m-2}] \{q\} = \left\{\frac{\partial U_m}{\partial q}\right\} \tag{12}$$

PROOF

The expression U_m may be written, by definition,

$$U_m = \sum_{r_1=0}^{m} \sum_{r_2=0}^{m} \cdots \sum_{r_n=0}^{m} a_{r_1 r_2 \cdots r_n} q_1^{r_1} q_2^{r_2} \cdots q_n^{r_n} \tag{13}$$

in which $\sum_{j=1}^{m} r_j = m$. It can also be written as, using the summation convention of repeated indices:

$$U_m = c_{i_1 i_2 \cdots i_m} q_{i_1} q_{i_2} \cdots q_{i_m} \tag{14}$$

in which $c_{i_1 i_2 \cdots i_m}$ is a scaler coefficient; and for a specific set of values $i_1, i_2, \cdots, i_m$, each of the i's may independently assume any integer value from 1 to n; thus, e.g., for n=3, m=6, c_{112133} corresponds to $q_1 q_1 q_2 q_1 q_3 q_3$, and c_{213131} corresponds to $q_2 q_1 q_3 q_1 q_3 q_1$. Both latter terms form part of the term representing $q_1^3 q_2 q_3^2$ in Eq. 13, with $r_1=3$, $r_2=1$, and $r_3=2$.

Taking the derivative of Eq. 14 with respect to q_s

$$\begin{aligned} \frac{\partial U_m}{\partial q_s} &= c_{i_1 i_2 \cdots i_m} [q_{i_1} \frac{\partial}{\partial q_s} (q_{i_2} q_{i_3} \cdots q_{i_m}) \\ &\qquad + (q_{i_2} q_{i_3} \cdots q_{i_m}) \frac{\partial q_{i_1}}{\partial q_s}] \\ &= c_{i_1 i_2 \cdots i_m} [q_{i_1} q_{i_2} \frac{\partial}{\partial q_s} (q_{i_3} q_{i_4} \cdots q_{i_m}) \\ &\qquad + q_{i_1} \frac{\partial q_{i_2}}{\partial q_s} (q_{i_3} q_{i_4} \cdots q_{i_m}) + \frac{\partial q_{i_1}}{\partial q_s} (q_{i_2} q_{i_3} \cdots q_{i_m})] \\ &= c_{i_1 i_2 \cdots i_m} [\frac{\partial q_{i_1}}{\partial q_s} q_{i_2} q_{i_3} \cdots q_{i_m} + \frac{\partial q_{i_2}}{\partial q_s} q_{i_1} q_{i_3} \cdots q_{i_m} \\ &\qquad + \cdots + \frac{\partial q_{i_m}}{\partial q_s} q_{i_1} q_{i_2} \cdots q_{i_{m-1}}] \end{aligned} \tag{15}$$

Noting that $\frac{\partial q_i}{\partial q_s} = \delta_{i_1 s}$ (the Kronecker delta),

$$\begin{aligned} \frac{\partial U_m}{\partial q_s} &= \cdots c_{s\, i_2 \cdots i_m} q_{i_2} q_{i_3} \cdots q_{i_m} + c_{i_1 s \cdots i_m} q_{i_1} q_{i_3} \cdots q_{i_m} \\ &\quad + \cdots + c_{i_1 i_2 \cdots i_{m-1} s}\, q_{i_1} q_{i_2} \cdots q_{i_{m-1}} \end{aligned} \tag{16}$$

In each of the preceding terms there are (m-1) sums and (m-1) factors involving the q's. Except for s, all the indices i_1, i_2, etc., are "dummy" ones, and they may be rearranged. Thus,

$$\frac{\partial U_m}{\partial q_s} = c_{s\, i_1 \cdots i_{m-1}}\, q_{i_1} q_{i_2} \cdots q_{i_{m-1}} + c_{i_1 s\, i_2 \cdots i_{m-1}}\, q_{i_1} q_{i_2} \cdots q_{i_{m-1}}$$

$$+ \cdots + c_{i_1 i_2 \cdots i_{m-1} s}\, q_{i_1} q_{i_2} \cdots q_{i_{m-1}}$$

$$= (c_{s\, i_1 \cdots i_{m-1}} + c_{i_1 s\, i_2 \cdots i_{m-1}} + \cdots + c_{i_1 i_2 \cdots i_{m-1} s}) \cdot$$

$$(q_{i_1} q_{i_2} \cdots q_{i_{m-1}}) \tag{17}$$

Without loss of generality, we can set (see Appendix I)

$$c_{s\, i_1 \cdots i_{m-1}} = c_{i_1 s\, i_2 \cdots i_{m-1}} = c_{i_1 i_2 \cdots i_{m-1} s} \quad \text{etc.} \tag{18}$$

$$\frac{\partial U_m}{\partial q_s} = m\, c_{s\, i_1 \cdots i_{m-1}}\, q_{i_1} q_{i_2} \cdots q_{i_{m-1}} \equiv m\, U_{m-1,s} \tag{19}$$

in which it is seen that $U_{m-1,s}$ (the subscript s does not denote derivative) is similar in structure to U_m, excepting that it is one order lower and has a free index s in the c's. Consider

$$\frac{\partial U_{m-1,s}}{\partial q_t} = (m-1)\, c_{t\, s\, i_1 i_2 \cdots i_{m-2}}\, q_{i_1} q_{i_2} \cdots q_{i_{m-2}}$$

$$\frac{\partial^2 U_m}{\partial q_t \partial q_s} = m(m-1)\, c_{t\, s\, i_1 i_2 \cdots i_{m-2}}\, q_{i_1} q_{i_2} \cdots q_{i_{m-2}}$$

$$= (m)(m-1)\, A_{ts} = N_{m-2,ts} \tag{20}$$

Substituting the above into the left-hand side of Eq. 11, one has

$$\frac{1}{m(m-1)} \{q\}^T [N_{m-2}] \{q\} = \sum_t \sum_s A_{ts}\, q_t q_s$$

$$= c_{tsi_1 i_2 \cdots i_{m-2}}\, q_{i_1} q_{i_2} \cdots q_{m-2} q_t q_s = U_m \tag{21}$$

The last step follows from Eq. 14. Thus Eq. 11 is proved.

To prove Eq. 12, upon substitution of Eq. 20 its left-hand side is

$$\left(\frac{1}{m-1}\right)(m)(m-1)\ A_{ts}\ \{q\}$$

Its sth row is

$$m\,A_{ts}q_t = m\,c_{tsi_1\ldots i_{m-2}}\,q_t q_{i_1}\ldots q_{i_{m-2}} = \frac{\partial U_m}{\partial q_s} \tag{22}$$

The last step follows from Eq. 19, noting that all indices except s are dummy ones. Thus, Eq. 12 is proved.

NUMERICAL EXAMPLES

Beam-Column Elements in Three-Dimensional Space. -- The element is assumed to be composed of linearly elastic material, prismatic and have doubly symmetric cross-sections. The strain energy is:

$$U = \int_o^L \frac{1}{2} AE\,\varepsilon^2\,dx + \int_o^L \frac{1}{2} GJ\left(\frac{d\phi}{dx}\right)^2 dx \tag{23}$$

in which E and G are the Young's and shear moduli; A, J, and L are the cross-sectional area, torsional constant and length; x is the longitudinal coordinate and ϕ is the angle of twist. The normal strain ε is taken to be:

$$\varepsilon = \frac{du}{dx} + \frac{1}{L}\int_o^L \frac{1}{2}\left[\left(\frac{dv}{dx}\right)^2 + \left(\frac{dw}{dx}\right)^2\right]dx + \eta\frac{d^2v}{dx^2} + \zeta\frac{d^2w}{dx^2} \tag{24}$$

in which u, v and w are the longitudinal and transverse displacements, and η and ζ are distances measured from the neutral axes in the two principal planes of bending. Note that in this model [9], as represented by Eq. (24), the nonlinear part of the normal strain due to bending, i.e.,

$$\frac{1}{2}\left[\left(\frac{dv}{dx}\right)^2 + \left(\frac{dw}{dx}\right)^2\right],$$

is averaged over the length. The interpolation functions used are polynomials, cubic for v and w, and linear for u and ϕ. They are successively substituted into Eq. (24) and Eq. (23). The incremental stiffness matrices [N1] and [N2] are then obtained in accordance with Eqs. (4a) and (4b).

In Fig. 1 is shown a space arch frame subjected to four vertical loads, each denoted by P, and two lateral loads, 0.001 P each. The major principal axis of inertia x is normal to the planes of the arch ribs. Each member is represented by a single beam-column element. Following the usual stiffness method of structural analysis, the stiffnesses of the members were assembled, and the load-displacement curve was obtained by use of Eq. (3), using a solution procedure of incremental loading ($\Delta P = 5$ units). For each load increment, the Newton-Raphson algorithm was employed. A fixed coordinate system was used. As the distortions are not large, the results are not significantly different from those based on a moving coordinate system. It is seen that the structure softens with increasing load and an approximate value of the bifurcation load may be taken to be

$$P \simeq 2\, E\, (I_x)_1 / L^2 .$$

The load deflection curves of a cantilever beam curves in the horizontal plane and subjected to a vertical load are shown in Fig. 2. For this problem the coordinate system is updated after every load increment and iteration to convergence was also carried out by the Newton-Raphson algorithm. Here the structure stiffens under increasing loads. The curved member is represented by four straight beam elements. Curved beam elements are used in the following example.

Curved Beam Elements in Three Dimensional Space. -- The element is presumed to lie in a plane but is deformable in the three dimensional space [10]. Its geometry is represented by a fourth order polynomial. Thus a curved structural member can be represented by these elements with continuous curvature as well as slopes at the nodes. The details of the nonlinear strain-displacement relations and the interpolation functions may be found in Ref. [6]. By assuming that the load displacement relation is linear and setting $\{\Delta P\} = \{0\}$ in Eq. (3), a quadratic eigenvalue problem can be formulated as follows:

$$([K] + p_{cr}\, [N1] + p_{cr}^2\, [N2])_{(\{q_o\})} \{\Delta q\} = \{0\} \tag{25}$$

In the preceding equation [N1] and [N2] are to be evaluated at $\{q_o\}$ which denotes the linear displacement vector corresponding to some reference load vector $\{P_o\}$. The "buckling load" vector is taken to be the product of the smallest eigenvalue p_{cr} and $\{P_o\}$. It is the smallest load corresponding to a state of neutral equilibrium.

The effect of a prescribed constant horizontal load {H} on the vertical buckling load of a parabolic arch is shown in Fig. 3. At each end the arch is restrained but free to rotate about the axis normal to its plane. In consistent units, the Young's modulus and Poisson's ratio are 4,176,000 and 0.3; the cross-sectional area, in-plane and out-of-plane moments of inertia and the torsional constant are 1.67, 13.16, 1.88 and 6.34. In this case, the linear stiffness matrix [K] in Eq. (25) should be replaced by:

$$[K_m] = [K] + ([N1] + [N2])_{(\{q_H\})} \tag{26}$$

in which $\{q_H\}$ is the linear displacement vector corresponding to {H} [10]. As expected, the vertical buckling load decreases rapidly as the magnitude of {H} increases. The buckling mode is out-of-plane and symmetric. For the case of zero horizontal load, the finite element solution is very close to the correct bifurcation load. For finite horizontal loads, probably no bifurcation load exists. The vertical "buckling load" as obtained above from the eigensolution could be taken as an approximation to the limit load. However, its accuracy needs to be studied.

ACKNOWLEDGEMENT

This work is supported in part by the National Science Foundation under Grant No. ENG-7822478.

APPENDIX I -- PROOF OF EQUATION 18

Corresponding to a specific set of the indices $(i_1, i_2, \ldots, i_{m-1}, s)$, there are a number of distinct coefficients $c_{j_1 j_2 \ldots j_m}$ in which the j's represent different arrangements of the fixed set of indices. Each term $c_{j_1 j_2 \ldots j_m} q_{j_1} q_{j_2} \ldots q_{j_m}$ is a part of the term $a_{r_1 r_2 \ldots r_n} q_1^{r_1} q_2^{r_2} \ldots q_n^{r_n}$ in which r_1 is equal to the number of 1's, r_2 the number of 2's, etc., in the set of the indices $(i_1, i_2, \ldots, i_{m-1}, s)$. Thus, Eq. 18 is valid as each of the c's may be set equal to $a_{r_1 r_2 \ldots r_n} \cdot (r_1 ! \, r_2 ! \ldots r_n ! \, / \, m !)$.

APPENDIX II -- REFERENCES

1. Mallett, R. H., and Marcal, P. V., "Finite Element Analysis of Nonlinear Structures," Journal of the Structural Division, ASCE, Vol. 94, Sept. 1968, pp. 2081-2105.

2. Stricklin, J. A., Von Riesemann, W. A., Tillerson, J. R., and Haisler, W. E., "Static Geometric and Material Nonlinear Analysis," Advances in Computational Methods in Structural Mechanics and Design, University of Alabama, Huntsville, 1972, pp. 301-324.

3. Yang, T. Y., "Finite-Displacement Plate Flexure by the Use of Matrix Incremental Approach," International Journal for Numerical Methods in Engineering, Vol. 4, No. 3, 1972, pp. 415-432.

4. Gallagher, R. H., "The Finite Element Method in Shell Stability Analysis," Computers and Structures, Vol. 3, 1973, pp. 543-557.

5. Jagannathan, D. S., Epstein, H. I., and Christiano, P., "Nonlinear Analysis of Reticulated Space Trusses," Journal of the Structural Division, ASCE, Dec. 1975, pp. 2641-2658.

6. Wen, R. K., and Lange, J., "A Curved Beam Element for Arch Buckling Analysis," to be published in the Journal of the Structural Division, ASCE, Nov. 1981.

7. Rajasekaran, S., and Murray, D. W., "Incremental Finite Element Matrices," Journal of the Structural Division, ASCE, Dec. 1973, pp. 2423-2438.

8. Yang, H. T. Y., "Flexible Plate Finite Element on Elastic Foundation," Journal of the Structural Division, ASCE, Vol. 96, Oct. 1970, pp. 2083-2101.

9. Rahimzadeh, H. J., "Nonlinear Elastic Frame Analysis by Finite Element," Ph.D. Thesis, Department of Civil Engineering, Michigan State University, 1981.

10. Lange, J., "Elastic Buckling of Arches by Finite Element Method," Ph.D. Thesis, Department of Civil Engineering, Michigan State University, 1980.

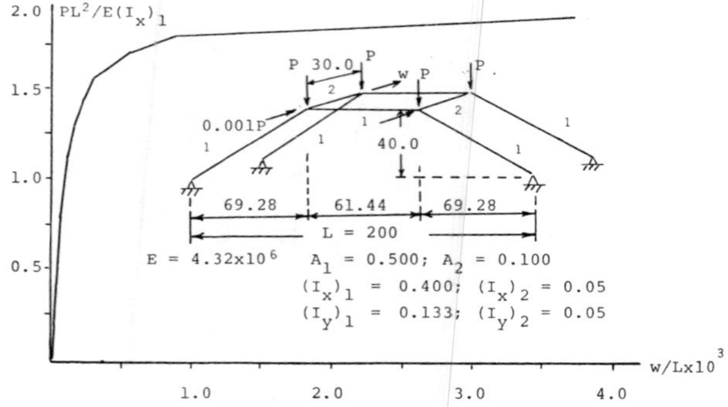

FIGURE 1. Space Arch Frame

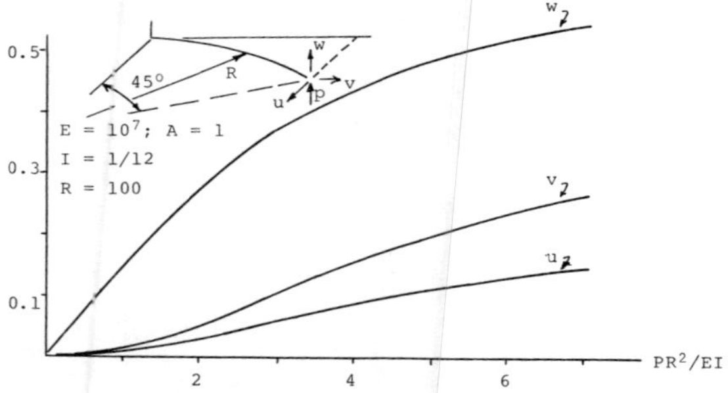

FIGURE 2. Curved Cantilever Beam

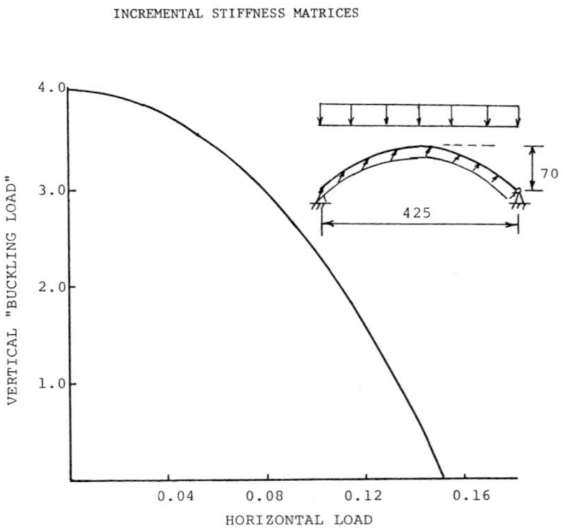

FIGURE 3. Effect of Horizontal Load on Vertical "Buckling Load"

AUTHOR INDEX